# p-ADIC MATHEMATICAL PHYSICS

To learn more about the AIP Conference Proceedings, including the
Conference Proceedings Series, please visit the webpage
**http://proceedings.aip.org/proceedings**

# p-ADIC MATHEMATICAL PHYSICS

2nd International Conference

Belgrade, Serbia and Montenegro  15 – 21 September 2005

*EDITORS*

Andrei Yu. Khrennikov
*Växjö University, Sweden*

Zoran Rakić
*University of Belgrade, Serbia and Montenegro*

Igor V. Volovich
*Steklov Mathematical Institute, Moscow, Russia*

SPONSORING   ORGANIZATIONS
UNESCO, Italy
Regional Bureau for Science in Europe - (ROSTE)
Ministry of Science and Environmental Protection, Republic of Serbia
The Abdus Salam International Centre for Theoretical Physics, Italy
International Center for Mathematical Modeling in Physics,
   Engineering, and Cognitive Science, Sweden

Melville, New York, 2006
AIP CONFERENCE PROCEEDINGS ■ VOLUME 826

**Editors:**

Andrei Yu. Khrennikov
International Center for Mathematical Modeling in Physics,
Engineering, and Cognitive Science MSI
Växjö University
Vejdes plats 7
SE-35195 Växjö
Sweden
E-mail: andrei.khrennikov@msi.vxu.se

Zoran Rakić
Faculty of Mathematics
University of Belgrade
Studentski trg 16, P.O. Box 550
11 001 Belgrade
Serbia and Montenegro
E-mail: zrakic@matf.bg.ac.yu

Igor V. Volovich
Steklov Mathematical Institute
Russian Academy of Sciences
Department of Mathematical Physics
Gubkin st. 8
Moscow, 119991
Russia
E-mail: volovich@mi.ras.ru

Authorization to photocopy items for internal or personal use, beyond the free copying permitted under the 1978 U.S. Copyright Law (see statement below), is granted by the American Institute of Physics for users registered with the Copyright Clearance Center (CCC) Transactional Reporting Service, provided that the base fee of $23.00 per copy is paid directly to CCC, 222 Rosewood Drive, Danvers, MA 01923, USA. For those organizations that have been granted a photocopy license by CCC, a separate system of payment has been arranged. The fee code for users of the Transactional Reporting Services is: 0-7354-0318-X/06/$23.00

© 2006 American Institute of Physics

Permission is granted to quote from the AIP Conference Proceedings with the customary acknowledgment of the source. Republication of an article or portions thereof (e.g., extensive excerpts, figures, tables, etc.) in original form or in translation, as well as other types of reuse (e.g., in course packs) require formal permission from AIP and may be subject to fees. As a courtesy, the author of the original proceedings article should be informed of any request for republication/reuse. Permission may be obtained online using Rightslink. Locate the article online at http://proceedings.aip.org, then simply click on the Rightslink icon/"Permission for Reuse" link found in the article abstract. You may also address requests to: AIP Office of Rights and Permissions, Suite 1NO1, 2 Huntington Quadrangle, Melville, NY 11747-4502, USA; Fax: 516-576-2450; Tel.: 516-576-2268; E-mail: rights@aip.org.

L.C. Catalog Card No. 2006922354
ISBN 0-7354-0318-X
ISSN 0094-243X

Printed in the United States of America

# CONTENTS

Dedication ............................................................. vii
Preface ................................................................. ix
Sponsors ............................................................... xi
Committees ........................................................... xiii
Conference Photograph ............................................... xv

## p-ADIC AND ULTRAMETRIC MODELS

Ergodic Transformations in the Space of $p$-Adic Integers ..................... 3
    V. Anashin
$p$-Adic and Adelic Cosmology: $p$-Adic Origin of Dark Energy and
Dark Matter ............................................................. 25
    B. Dragovich
On the Chaotic Properties of Quadratic Maps over Non-Archimedean
Fields .................................................................. 43
    V. Dremov, G. Shabat, and P. Vytnova
On Ultrametricity and a Symmetry between Bose-Einstein and
Fermi-Dirac Systems ..................................................... 55
    A. A. Ezhov and A. Yu. Khrennikov
$p$-Adic Strings and Their Applications ..................................... 65
    P. G. O. Freund
Proof of the Kurlberg-Rudnick Rate Conjecture ............................. 74
    S. Gurevich and R. Hadani
$p$-Adic Description of Hierarchical Systems Dynamics ...................... 81
    K. Lukierska-Walasek and K. Topolski
Capacities and Function Spaces on the Local Field ......................... 91
    H. Kaneko
$p$-Adic Probability Theory and Its Generalizations ........................ 105
    A. Yu. Khrennikov
Ultrametric Analysis and Interbasin Kinetics ............................. 121
    S. V. Kozyrev
Critical Exponents in $p$-Adic $\varphi^4$-Model ........................... 129
    M. D. Missarov and R. G. Stepanov
On Phase Transitions for $p$-Adic Potts Model with Competing
Interactions on a Cayley Tree ........................................... 140
    F. M. Mukhamedov, U. A. Rozikov, and J. F. F. Mendes
From Data to the Physics Using Ultrametrics: New Results in High
Dimensional Data Analysis .............................................. 151
    F. Murtagh
The Arithmetic of Discretized Rotations ................................. 162
    F. Vivaldi
$p$-Adic Models of Turbulence ........................................... 174
    S. Fischenko and E. Zelenov

## p-ADIC ANALYSIS

Pseudo-Differential Operators in the *p*-Adic Lizorkin Space . . . . . . . . . . . . . . . . . 195
    S. Albeverio, A. Yu. Khrennikov, and V. M. Shelkovich
Sequence-Spaces and Applications . . . . . . . . . . . . . . . . . . . . . . . . . . . . . . . . . . . . . . 206
    N. De Grande-De Kimpe
Ultrametric Gelfand Transforms . . . . . . . . . . . . . . . . . . . . . . . . . . . . . . . . . . . . . . . 214
    A. Escassut and N. Maïnetti
*p*-Adic Multiple Zeta Values: A Précis . . . . . . . . . . . . . . . . . . . . . . . . . . . . . . . . . . 222
    H. Furusho
Aspects of *p*-Adic Non-Linear Functional Analysis . . . . . . . . . . . . . . . . . . . . . . . . 237
    H. Glöckner
Umbral Calculus and Holonomic Modules in Positive Characteristic . . . . . . . . . 254
    A. N. Kochubei
Point on Curves Whose Coordinates are *p*-Adic *U*-Numbers . . . . . . . . . . . . . . . . 267
    H. Menken and K. R. Mamedov
Infinitesimals in Nonstandard Analysis versus Infinitesimals in *p*-Adic
Fields . . . . . . . . . . . . . . . . . . . . . . . . . . . . . . . . . . . . . . . . . . . . . . . . . . . . . . . . . . . . . . 274
    Ž. Mijajlović, M. Milošević, and A. Perović
Barrelledness of *p*-Adic $C^1$-Function Spaces . . . . . . . . . . . . . . . . . . . . . . . . . . . . . 280
    W. H. Schikhof
Local *p*-Adic Differential Equations . . . . . . . . . . . . . . . . . . . . . . . . . . . . . . . . . . . . 291
    M. van der Put and L. Taelman

## RELATED TOPICS

Nonlocal String Tachyon as a Model for Cosmological Dark Energy . . . . . . . . . . 301
    I. Ya Aref'eva
A Note on Weil's Explicit Formula . . . . . . . . . . . . . . . . . . . . . . . . . . . . . . . . . . . . . 312
    M. Avdispahić and L. Smajlović
The Use of Path Integral Ideals: Deriving the Euler Summation
Formula for Path Integrals . . . . . . . . . . . . . . . . . . . . . . . . . . . . . . . . . . . . . . . . . . . . 320
    A. Bogojević, A. Balaž, and A. Belić
Analysis of Business Connections Utilizing Theory of Topology of
Random Graphs . . . . . . . . . . . . . . . . . . . . . . . . . . . . . . . . . . . . . . . . . . . . . . . . . . . . 330
    J. Q. Trelewicz and I. V. Volovich
On Quantum Cryptography and Number Theory . . . . . . . . . . . . . . . . . . . . . . . . . 345
    A. S. Trushechkin and I. V. Volovich

## APPENDIX

Open Problems . . . . . . . . . . . . . . . . . . . . . . . . . . . . . . . . . . . . . . . . . . . . . . . . . . . . . 357
List of Participants . . . . . . . . . . . . . . . . . . . . . . . . . . . . . . . . . . . . . . . . . . . . . . . . . . 365
Author Index . . . . . . . . . . . . . . . . . . . . . . . . . . . . . . . . . . . . . . . . . . . . . . . . . . . . . . 367

In 2005 Professor **Branko Dragovich** had his 60th birthday. He is one of the founders of p-adic mathematical physics. His contributions to p-adic series, p-adic and adelic cosmology and string theory are really remarkable, and his series of papers on p-adic Feynman integration played a fundamental role in the foundations of p-adic quantum theory and adelic quantum mechanics. On this occasion the Editors as well as other participants of the Conference warmly congratulate our friend Branko Dragovich on his anniversary and wish him new outstanding achievements! The Proceedings of the Conference on p-adic mathematical physics are dedicated to Prof. Branko Dragovich.

*Editors*
*Andrei Yu. Khrennikov*
*Zoran Rakić*
*Igor V. Volovich*

# Preface

This volume contains the proceedings of the "2nd International Conference on p-Adic Mathematical Physics" held at the Institute of Physics, Belgrade, Serbia and Montenegro, 15-21 September 2005 (see also website: http://www.p-adic-mathphys2005.phy.bg.ac.yu/). The number of participants was 54 from 18 countries.

The "1st International Conference on p-Adic Mathematical Physics" was held at the Steklov Institute of Mathematics, Russian Academy of Sciences, Moscow, 1–4 October 2003. Papers from this conference are published in the Proceedings of V.A. Steklov Institute of Mathematics, Vol. **245** (2004). There is a plan to continue with this series of conferences on p-adic mathematical physics and hold meetings every two years in the future.

p-Adic mathematical physics was born in 1987 as a result of attempts to find a non-Archimedean approach to space-time at the Planck scale as well as to strings. For this period of time many directions of application of p-adic numbers and adeles in physics and some other sciences have appeared. It is worth mentioning some of them: p-adic and adelic string theory; p-adic and adelic quantum mechanics and quantum field theory; ultrametricity of spin glasses, biological and hierarchical systems; p-adic dynamical systems; p-adic probability theory; p-adic models of cognitive processes and cryptography; and p-adic and adelic cosmology. These applications have stimulated various developments in p-adic analysis and other related mathematical fields. Presently p-adic mathematical physics is an active, rapidly developing and promising branch of modern mathematical physics.

The subject of the conference were recent developments in p-adic mathematical physics, and also in related mathematical and physical areas. There were interesting and fruitful evening discussions, and one of them was devoted to the present status and future prospects of p-adic mathematical physics.

In addition to articles we included also in these proceedings some open problems presented by participants whose solutions are important for further progress in p-adic mathematical physics.

We hope that all participants enjoyed their stay at the Conference as well as in Belgrade. The Editors are grateful to all speakers for the efforts to prepare their contributions for the Proceedings. Hopefully, the presented collection of papers will be useful for all researchers interested in applications of p-adic numbers, adeles, non-Archimedean analysis and ultrametric methods in physics and sciences. Also, results presented at the Conference and contained in the Proceedings should stimulate further progress in p-adic mathematical physics and related topics.

*Editors*
*Andrei Yu. Khrennikov*
*Zoran Rakić*
*Igor V. Volovich*

# Sponsors:

- UNESCO Office in Venice - Regional Bureau for Science in Europe (ROSTE), Venice, Italy
- Ministry of Science and Environmental Protection, Republic of Serbia
- The Abdus Salam International Centre for Theoretical Physics, Trieste, Italy
- International Center for Mathematical Modeling in Physics, Engineering, and Cognitive Science, Växjö University, Sweden

International Advisory Committee:

Alain Escassut (France)
Shai Haran (Israel)
Yuri I. Manin (Germany)
Giorgio Parisi (Italy)
Franco Vivaldi (UK)
Vasilii S. Vladimirov (Russia)

International Organizing Committee:

Branko Dragovich, *Co-Chairman* (SCG)
Peter G. O. Freund (USA)
Hiroshi Kaneko (Japan)
Andrei Khrennikov (Sweden)
Anatoly Kochubei (Ukraine)
Sergei Kozyrev (Russia)
Wim Schikhof (Netherlands)
Veeravalli S. Varadarajan (USA)
Igor V. Volovich, *Co-Chairman* (Russia)

Local Organizing Committee:

Vladica Andrejić
Sanja Ćirković
Branko Dragovich, *Chairman*
Dejan Joković
Zoran Rakić

# p-ADIC AND ULTRAMETRIC MODELS

# Ergodic Transformations in the Space of $p$-Adic Integers

## Vladimir Anashin

*Faculty of Information Security, Russian State University for the Humanities, Kirovogradskaya Str., 25/2, Moscow 113534, RUSSIA*
*emails:* anashin@rsuh.ru, vs-anashin@yandex.ru

**Abstract.** Let $\mathscr{L}_1$ be the set of all mappings $f\colon \mathbb{Z}_p \longrightarrow \mathbb{Z}_p$ of the space of all $p$-adic integers $\mathbb{Z}_p$ into itself that satisfy Lipschitz condition with a constant 1. We prove that the mapping $f \in \mathscr{L}_1$ is ergodic with respect to the normalized Haar measure on $\mathbb{Z}_p$ if and only if $f$ induces a single cycle permutation on each residue ring $\mathbb{Z}/p^k\mathbb{Z}$ modulo $p^k$, for all $k = 1,2,3,\ldots$. The multivariate case, as well as measure-preserving mappings, are considered also.
Results of the paper in a combination with earlier results of the author give explicit description of ergodic mappings from $\mathscr{L}_1$. This characterization is complete for $p = 2$.
As an application we obtain a characterization of polynomials (and certain locally analytic functions) that induce ergodic transformations of $p$-adic spheres. The latter result implies a solution of a problem due to A. Khrennikov about the ergodicity of a perturbed monomial mapping on a sphere.

**Keywords:** Ergodic transformation, measure-preserving transformation, $p$-adic integers.
**PACS:** 05.45.-a, 05.90.+m, 02.30.-f, 02.30.Cj, 02.30.Sa.

## 1. INTRODUCTION

Let $\mathscr{L}_1$ be the set of all functions $f\colon \mathbb{Z}_p \longrightarrow \mathbb{Z}_p$ defined on (and valuated in) the space $\mathbb{Z}_p$ of all $p$-adic[1] integers that satisfy Lipschitz condition with a constant 1 with respect to the $p$-adic metric $\|\cdot\|_p$: $\|f(x) - f(y)\|_p \le \|x-y\|_p$ for all $x,y \in \mathbb{Z}_p$. For $p=2$ this class is of particular practical importance for computer science since it includes all mappings combined of standard microprocessor instructions, such as arithmetic ones (integer addition, multiplication, etc.) and bitwise logical ones (such as AND, bitwise logical 'and'; OR, bitwise logical 'or', etc.); see [5] and [4] for details.

Any mapping $f \in \mathscr{L}_1$ naturally induces a well-defined mapping $\bar{f}_k = f \bmod p^k\colon \mathbb{Z}/p^k\mathbb{Z} \longrightarrow \mathbb{Z}/p^k\mathbb{Z}$ of the residue ring $\mathbb{Z}/p^k\mathbb{Z}$ into itself by letting $\bar{f}_k(z) = f(z) \bmod p^k$, the least non-negative residue of $f(z)$ modulo $p^k$. That is, $\bar{f}_k(z)$ is the smallest non-negative rational integer $v$ such that $\|v - f(z)\|_p \le p^{-k}$ or, in other words,

$$\bar{f}_k(z) = v_0 + v_1 \cdot p + v_2 \cdot p^2 + \cdots + v_{k-1} \cdot p^{k-1},$$

---

[1] Throughout the paper $p$ is a prime.

whenever $f(z) = v_0 + v_1 \cdot p + v_2 \cdot p^2 + \cdots + v_{k-1} \cdot p^{k-1} + \cdots$ is a canonic $p$-adic representation of $f(z)$; $v_i = \delta_i(f(z)) \in \{0, 1, \ldots, p-1\}$, $i = 0, 1, 2, \ldots$. In view of what has been just said, $x \equiv y \pmod{p^k}$ for $x, y \in \mathbb{Z}_p$ means that $\|x - y\|_p \leq p^{-k}$. We use the same notation in the multivariate case also, i.e. for functions $F \colon \mathbb{Z}_p^n \longrightarrow \mathbb{Z}_p^m$, $(m \leq n)$ that satisfy Lipschitz condition with a constant 1.

Note that under this notation, the function $f \colon \mathbb{Z}_p \longrightarrow \mathbb{Z}_p$ satisfy Lipschitz condition with a constant 1 if and only if $f(x) \equiv f(y) \pmod{p^k}$ whenever $x \equiv y \pmod{p^k}$. Thus, functions that satisfy Lipschitz conditions with a constant 1 are exactly those ones that preserve all congruences of the ring $\mathbb{Z}_p$; i.e., they map cosets into cosets: $f(a + p^k \mathbb{Z}_p) \subset f(a) + p^k \mathbb{Z}_p$ for any $a \in \mathbb{Z}_p$ and any $k = 1, 2, \ldots$.

In algebra, functions which preserve all congruences of an algebraic system are called *compatible*; so throughout the paper we use for short the term 'compatible' instead of 'satisfying Lipschitz condition with a constant 1'. Note that a coset $a + p^k \mathbb{Z}_p$ of the ring $\mathbb{Z}_p$ with respect to the ideal $p^k \mathbb{Z}_p$ is a ball of radius $p^{-k}$ in the space $\mathbb{Z}_p$. Hence, in our case compatible mappings are exactly ones that map balls into balls. This is an exercise to prove that an analytic function which is defined by a power series $\sum_{i=0}^{\infty} a_i x^i$ (with $a_i \in \mathbb{Z}_p$ for all $i = 0, 1, 2, \ldots$) that converges everywhere on $\mathbb{Z}_p$, is compatible. We denote this class of analytic functions via $\mathscr{C}$. Natural examples of these functions are polynomials over $\mathbb{Z}_p$, certain $p$-adic logarithms (e.g., $\ln_p(1 + px) = \sum_{i=1}^{\infty}(-1)^{i+1}\frac{p^i x^i}{i}$), some rational functions (e.g., $\frac{1}{1+px} = \sum_{i=0}^{\infty}(-1)^i p^i x^i$), etc.

**1.1 Definition.** We say that $f \colon \mathbb{Z}_p \longrightarrow \mathbb{Z}_p$ is *bijective modulo* $p^k$ whenever $f \bmod p^k$ is a permutation of elements of the ring $\mathbb{Z}/p^k\mathbb{Z}$; and we say that $f$ is *transitive modulo* $p^k$ whenever $f \bmod p^k$ is a permutation with a single cycle.

We say that the multivariate function $F \colon \mathbb{Z}_p^n \longrightarrow \mathbb{Z}_p^m$ ($m \leq n$) is *balanced modulo* $p^k$ whenever the induced mapping $\bar{F}_k = F \bmod p^k \colon (\mathbb{Z}/p^k\mathbb{Z})^n \longrightarrow (\mathbb{Z}/p^k\mathbb{Z})^m$ of the corresponding Cartesian powers of the residue ring modulo $p^k$ satisfy the following condition: For each $v \in (\mathbb{Z}/p^k\mathbb{Z})^m$ the cardinality $\#\bar{F}_k^{-1}(v)$ of the full preimage $\bar{F}_k^{-1}(v) = \{w \in (\mathbb{Z}/p^k\mathbb{Z})^n \colon \bar{F}_k(w) = v\}$ of $v$ does not depend on $v$; that is $\#\bar{F}_k^{-1}(v) = \#\bar{F}_k^{-1}(w)$ for any two $v, w \in (\mathbb{Z}/p^k\mathbb{Z})^m$.[2]

Further in the paper we say that the function $f \colon \mathbb{Z}_p \longrightarrow \mathbb{Z}_p$ (or $F \colon \mathbb{Z}_p^n \longrightarrow \mathbb{Z}_p^m$) is *measure-preserving* whenever it preserves the unique Haar measure $\mu_p$, which is normalized so that the measure of the whole space is 1. Accordingly, we say that $f$ is *ergodic* whenever $f$ is ergodic with respect to $\mu_p$.

The paper study measure-preserving (in particular, ergodic) transformations of the space of $p$-adic integers; within this context the paper is a contribution to the theory of $p$-adic dynamical systems. The latter are of growing interest now because of their possible applications in different areas: For instance, applications of the $p$-adic dynamics to physics, cognitive sciences, and neural networks are discussed in [17]. Recently ergodic trans-

---

[2] We used the term *equiprobable* instead of balanced in [5]; however, the latter is more common in cryptographic literature.

formations of the space of 2-adic integers were successfully applied to pseudorandom number generation for computer simulations and especially for cryptography (stream cipher design), see [2], [10] as well as [7, 8, 6]. The following theorem was announced in [5]:

**1.2 Theorem.** *For $m = n = 1$, a compatible function $F\colon \mathbb{Z}_p^n \longrightarrow \mathbb{Z}_p^m$ is measure-preserving (or, accordingly, ergodic) if and only if it is bijective (accordingly, transitive) modulo $p^k$ for all $k = 1, 2, 3, \ldots$.*

*For $n \geq m$, the function $F$ is measure-preserving if and only if it is balanced modulo $p^k$, for all $k = 1, 2, 3, \ldots$.*

In the paper we prove this theorem, see Sections 2, 3 and 4. It worth notice here that from further considerations it follows that a compatible measure-preserving function $F\colon \mathbb{Z}_p^n \longrightarrow \mathbb{Z}_p^n$ is an isometry, see Note 2.5. Theorem 1.2 in a combination with earlier results of the author on transitivity modulo $p^k$ (see [3], [5] and [9]) is used further to obtain the following characterization of ergodic transformations of spheres:

**1.3 Theorem.** *Let $f$ be a $\mathscr{C}$-function (e.g., a polynomial over the ring $\mathbb{Z}_p$). In case $p$ odd, the mapping $z \mapsto f(z)$ is an ergodic [3] transformation of each sufficiently small sphere with a center at $y \in \mathbb{Z}_p$ if and only if the following two conditions hold simultaneously:*

- *$f(y) = y$, and*
- *the derivative $f'(y)$ of the function $f$ at the point $y \in \mathbb{Z}_p$ generates modulo $p^2$ the whole group of units $(\mathbb{Z}/p^2\mathbb{Z})^*$ of the residue ring $\mathbb{Z}/p^2\mathbb{Z}$.[4]*

*In case $p = 2$ no $\mathscr{C}$-function exists such that the mapping $z \mapsto f(z)$ is ergodic on all spheres around $y \in \mathbb{Z}_2$ of radii less than $\varepsilon$, whatever $\varepsilon > 0$ is taken.*

As a matter of fact, Theorem 1.3 remains true for a class $\mathscr{B}$ of functions that is wider than $\mathscr{C}$, and even for a class $\mathscr{A}$ that is bigger than $\mathscr{B}$. Both these classes $\mathscr{A}$ and $\mathscr{B}$ contain functions that are not necessarily analytic $\mathbb{Z}_p$, yet only locally analytic of order 1. Moreover, Theorem 1.3 is an immediate consequence of a more general Theorem 5.7 dealing with the ergodicity on a single sphere around $y \in \mathbb{Z}_p$ rather than on all sufficiently small spheres around $y \in \mathbb{Z}_p$, see Section 5 for details.

Earlier in [15] and [14] ergodicity of monomial mappings $z \mapsto z^\ell$ on spheres $S_{p^{-r}}(1)$ of a radius $p^{-r}$ with a center at 1 was studied: It was shown that for odd $p$ and $r > 1$ the mapping is ergodic iff $\ell$ is a generator of the group $(\mathbb{Z}/p^2\mathbb{Z})^*$. Mentioned Theorem 5.7 is a generalization of that result. Moreover, with the use of this theorem we are able to solve a problem that was put at the 2nd Int'l Conference on $p$-adic Mathematical Physics by Professor Andrei Khrennikov (see also [15], [14], and [16]):

> We know for which $\ell$ and $p$ the dynamical system $f(x) = x^\ell$ is ergodic on the sphere $S_{p^{-r}}(1)$. Let us consider the ergodicity of a perturbed system

---

[3] With respect to the induced measure.
[4] In this case they also say that $f'(y)$ is *primitive modulo $p^2$*, or $f'(y)$ is a *generator of the multiplicative group* $(\mathbb{Z}/p^2\mathbb{Z})^*$ of the residue ring $\mathbb{Z}/p^2\mathbb{Z}$.

$f(x) = x^\ell + q(x)$ for some polynomial $q(x) \in \mathbb{Z}_p[x]$ such that all coefficients of $q(x)$ are $p$-adically smaller than $p^{-r}$. This condition is necessary in order to guarantee that $S_{p^{-r}}(1)$ is invariant. For such a system to be ergodic it is necessary that $\ell$ is a generator of $(\mathbb{Z}/p^2\mathbb{Z})^*$. Is this sufficient?

We prove that *the answer is affirmative* if the radius $p^{-r}$ is sufficiently small (actually, if $r > 1$), see Proposition 5.10. Note that in view of Theorem 1.3 the mentioned perturbed mapping is ergodic on *all* spheres around 1 of radii less than $p^{-r}$ if and only if one more condition holds: 1 is a root of the polynomial $q(x)$.

It worth notice also that with the use of Theorem 5.7 it is possible to prove the ergodicity of the 'perturbed' analogs of mappings considered in [11] and [12] on all sufficiently small spheres, namely, of mappings $z \mapsto az^\ell + q(z)$ and $z \mapsto az + b + q(z)$, where $q$ is a '$p$-adically small' perturbation. See Section 5 for details.

## 2. MEASURE-PRESERVING ISOMETRIES

In this section we prove that *a compatible function* $F: \mathbb{Z}_p^n \longrightarrow \mathbb{Z}_p^n$ *preserves measure if and only if it is bijective modulo* $p^k$, *for all* $k = 1, 2, \ldots$. We consider a case $n = 1$ just to simplify notation; all statements of this section hold for a general case, their proofs are quite similar to ones of the case $n = 1$. It worth notice here that the main result of this section could be deduced also from a more general result of Section 3. However, we present a separate proof for the considered case since the proof gives us some extra information about the functions of considered type.

**2.1 Proposition.** *A compatible and measure-preserving function* $f: \mathbb{Z}_p \longrightarrow \mathbb{Z}_p$ *is a bijection of* $\mathbb{Z}_p$ *onto itself.*

*Proof.* We prove that $f$ is both injective and surjective.

Claim 1: *Under conditions of Proposition 2.1 the function $f$ is injective.*
Indeed, if there exist $a, b \in \mathbb{Z}_p$ ($a \neq b$) such that $f(a) = f(b) = z$ then for some $k$ the balls $a + p^k\mathbb{Z}_p$ and $b + p^k\mathbb{Z}_p$ are disjoint, whereas $f(a + p^k\mathbb{Z}_p), f(a + p^k\mathbb{Z}_p) \subset z + p^k\mathbb{Z}_p$. Hence $\mu_p(f^{-1}(z + p^k\mathbb{Z}_p)) \geq 2 \cdot p^{-k}$ since $f^{-1}(z + p^k\mathbb{Z}_p)) \supset f^{-1}(a + p^k\mathbb{Z}_p)), f^{-1}(b + p^k\mathbb{Z}_p))$; so $f$ does not preserve $\mu_p$.

Claim 2: *Under conditions of Proposition 2.1 the function $f$ is bijective modulo $p^k$ for all $k = 1, 2, \ldots$.*
Otherwise for suitable $a, b \in \mathbb{Z}_p$ ($a \neq b$), and $k$ the balls $a + p^k\mathbb{Z}_p$ and $b + p^k\mathbb{Z}_p$ are disjoint, whereas $f(a + p^k\mathbb{Z}_p), f(a + p^k\mathbb{Z}_p) \subset z + p^k\mathbb{Z}_p$. Yet this leads to a contradiction, see Claim 1.

Claim 3: *Under conditions of Proposition 2.1 the function $f$ is surjective.*
Take arbitrary $z \in \mathbb{Z}_p$. Then in view of Claim 2 there exists exactly one $x_1 \in \mathbb{Z}/p\mathbb{Z}$ such that $f(x_1) \equiv z \pmod{p}$ (here and further we identify elements of the residue ring $\mathbb{Z}/p^k\mathbb{Z}$ with non-negative rational integers $0, 1, \ldots, p^k - 1$ in an obvious way). Similarly,

there exists exactly one $x_2 \in \mathbb{Z}/p^2\mathbb{Z}$ such that $f(x_2) \equiv z \pmod{p^2}$; whence necessarily $x_2 \equiv x_1 \pmod{p}$, etc.

So we obtain a sequence $x_2, x_2, \ldots$ such that $\|f(x_i) - z\|_p \leq p^{-i}$ and $\|x_{i+1} - x_i\|_p \leq p^{-i}$ for $i = 1, 2, \ldots$. It is an exercise to show now that the sequence $x_2, x_2, \ldots$ is a Cauchy sequence (which hence converges to some $x \in \mathbb{Z}_p$), and that $f(x) = z$. □

**2.2 Note.** As a bonus we have that whenever a compatible function $g: \mathbb{Z}_p \longrightarrow \mathbb{Z}_p$ is bijective modulo $p^k$ for all $k = 1, 2, \ldots$, it is a bijection of $\mathbb{Z}_p$ onto $\mathbb{Z}_p$, see proofs of Claims 2 and 3 of the proof of Proposition 2.1.

**2.3 Proposition.** *Let a compatible function* $g: \mathbb{Z}_p \longrightarrow \mathbb{Z}_p$ *be bijective modulo $p^k$ for all $k = 1, 2, \ldots$. Then $g$ preserves measure.*

*Proof.* In view of Note 2.2 the function $g$ is a bijection of $\mathbb{Z}_p$ onto $\mathbb{Z}_p$; whence, there exist an inverse function $f = g^{-1}$, which is also a bijection of $\mathbb{Z}_p$ onto $\mathbb{Z}_p$. Moreover, $f$ is continuous since $g$ is continuous.

Claim 1: *$f$ is compatible.*
If there are $a, b \in \mathbb{Z}_p$ such that $a \equiv b \pmod{p^k}$ and $f(a) \not\equiv f(b) \pmod{p^k}$ then assuming $a = g(u)$, $b = g(v)$ for uniquely defined $u, v \in \mathbb{Z}_p$ we have $g(u) \equiv g(v) \pmod{p^k}$ and $f(g(u)) \not\equiv f(g(v)) \pmod{p^k}$; that is, $g(u) \equiv g(v) \pmod{p^k}$ and $u \not\equiv v \pmod{p^k}$. The latter contradicts conditions of Proposition 2.3.

Claim 2: $f(a + p^k\mathbb{Z}_p) = f(a) + p^k\mathbb{Z}_p$ *for every $a \in \mathbb{Z}_p$ and every $k = 1, 2, \ldots$.*
In view of Claim 1, $f(a + p^k\mathbb{Z}_p) \subset f(a) + p^k\mathbb{Z}_p$. To prove the inverse inclusion, denote $f(a) = b$; then $g(b) = a$. Since $g$ is compatible, $g(b + p^k\mathbb{Z}_p) \subset g(b) + p^k\mathbb{Z}_p$. Applying a bijection $f$ to the both sides of this inclusion, one obtains $b + p^k\mathbb{Z}_p \subset f(g(b) + p^k\mathbb{Z}_p)$, since $f$ is compatible (see Claim 1); that is, $f(a) + p^k\mathbb{Z}_p \subset f(a + p^k\mathbb{Z}_p)$, the needed inverse inclusion.

Claim 3: *$f$ is bijective modulo $p^k$ for all $k = 1, 2, \ldots$.*
Assuming there exist $u, v \in \mathbb{Z}_p$ and $k \in \{1, 2, \ldots\}$ such that $u \equiv v \pmod{p^k}$ and $f(u) \not\equiv f(v) \pmod{p^k}$ one obtains that

$$u + p^k\mathbb{Z}_p = v + p^k\mathbb{Z}_p, \quad \text{yet} \quad f(u) + p^k\mathbb{Z}_p \neq f(v) + p^k\mathbb{Z}_p,$$

a contradiction in view of Claim 2.

Claim 4: *$f$ satisfies conditions of Proposition 2.3.*
See Claims 1 and 3.

Claim 5: $g(a + p^k\mathbb{Z}_p) = g(a) + p^k\mathbb{Z}_p$ *for every $a \in \mathbb{Z}_p$ and every $k = 1, 2, \ldots$.*
See Claim 4.

Claim 6: $\mu_p(g(M)) = \mu_p(M)$, *for every measurable $M \subset \mathbb{Z}_p$.*
Since $M$ is measurable, then

$$\mu_p(M) = \inf\{\mu_p(V) : V \supset M, V \text{ is open in } \mathbb{Z}_p\}.$$

Since $V$ is open, it is a disjoint union of a countable number of balls $V_j$ of non-zero radius each: $V = \bigcup_{j \in J} V_j$. Then $g(V) = \bigcup_{j \in J} g(V_j)$, since $g$ is a bijection. Note that in view of Claim 5, each $g(V_j)$ is a ball of a radius that is equal to the one of the ball $V_j$; that is, $\mu_p(g(V_j)) = \mu_p(V_j)$, for all $j \in J$. Moreover, the balls are disjoint: $g(V_i) \cap g(V_j) = \emptyset$ whenever $i \neq j$ (since $f(g(V_i) \cap g(V_j)) = V_i \cap V_j$ in view of Claim 2). This implies that $\mu_p(g(V)) = \mu_p(V)$. Note that $g(V)$ is open since $g$ is a continuous bijection. Hence,

$$\mu_p(g(M)) \leq \inf\{\mu_p(g(V)) \colon V \supset M, V \text{ is open in } \mathbb{Z}_p\} = \mu_p(M).$$

In view of Claim 4, one has then $\mu_p(f(R)) \leq \mu_p(R)$, for every measurable $R \subset \mathbb{Z}_p$. Now we take $R = g(M)$ (whence $f(R) = M$) and obtain $\mu_p(M) \leq \mu_p(g(M))$, thus proving the Proposition. $\square$

**2.4 Corollary.** *A compatible function $f \colon \mathbb{Z}_p \longrightarrow \mathbb{Z}_p$ preserves measure if and only if it is bijective modulo $p^k$ for all $k = 1, 2, \ldots$.*

*Proof.* Necessity of the conditions is proved by Claim 2 of Proposition 2.1, whereas their sufficiency is proved by Proposition 2.3. $\square$

2.5 Note. As a bonus we have that *every compatible measure-preserving function $f \colon \mathbb{Z}_p \longrightarrow \mathbb{Z}_p$ is an isometry*: A distance between two points is just a radius of the smallest ball that contains them both; however, as it was shown, a measure-preserving compatible mapping is a bijection that merely permutes balls of the same radius.

## 3. MEASURE-PRESERVING FUNCTIONS

In this section we prove that *a compatible function $F \colon \mathbb{Z}_p^n \longrightarrow \mathbb{Z}_p^m$, $m \leq n$, preserves measure if and only if it is balanced modulo $p^k$, for all $k = 1, 2, \ldots$.*

**3.1 Lemma.** *Let a compatible function $F \colon \mathbb{Z}_p^n \longrightarrow \mathbb{Z}_p^m$, $m \leq n$, be balanced modulo $p^k$, for all $k = 1, 2, \ldots$. Then for every $b \in \mathbb{Z}_p^m$ a full preimage $F^{-1}(b + p^s \mathbb{Z}_p^m)$ is a union of $p^{s(n-m)}$ pairwise disjoint balls $a_j + p^s \mathbb{Z}_p^n$, $j = 1, 2, \ldots, p^{s(n-m)}$.*

*Proof.* We start with proving the lemma 'modulo $p^k$'.

Claim 1. For every $\bar{b}_k \in (\mathbb{Z}/p^k)^m$, a full preimage $\bar{F}_k^{-1}(\bar{b}_k + p^s(\mathbb{Z}/p^k\mathbb{Z})^m)$ of the coset $\bar{b}_k + p^s(\mathbb{Z}/p^k\mathbb{Z})^m \subset (\mathbb{Z}/p^k\mathbb{Z})^m$ (modulo the ideal $p^k(\mathbb{Z}/p^k\mathbb{Z})^m$ of the ring $(\mathbb{Z}/p^k\mathbb{Z})^m$) is a disjoint union of $p^{s(n-m)}$ suitable pairwise disjoint cosets (modulo the ideal $p^s(\mathbb{Z}/p^k\mathbb{Z})^n$ of the ring $(\mathbb{Z}/p^k\mathbb{Z})^n$):

$$\bar{F}_k^{-1}(\bar{b}_k + p^s(\mathbb{Z}/p^k\mathbb{Z})^m) = \bigcup_{j=1}^{p^{s(n-m)}} (\bar{a}_{k,j} + p^s(\mathbb{Z}/p^k\mathbb{Z})^n).$$

Here and further we assume that $s \leq k$. In this case $\#(\bar{b}_k + p^s(\mathbb{Z}/p^k\mathbb{Z})^m) = p^{m(k-s)}$, and since $F$ is balanced modulo $p^k$, then

$$\#F_k^{-1}(\bar{b}_k + p^s(\mathbb{Z}/p^k\mathbb{Z})^m) = p^{k(n-m)} \cdot p^{m(k-s)} = p^{kn-ms}. \tag{1}$$

Further, since $F$ is balanced modulo $p^s$, then $\#F_s^{-1}(\bar{b}_s) = p^{s(n-m)}$, for every $\bar{b}_s \in \{0, 1, \ldots, p^s - 1\}^m = (\mathbb{Z}/p^s\mathbb{Z})^m$. Take $\bar{b}_s \equiv \bar{b}_k \pmod{p^s}$ and let

$$F_s^{-1}(\bar{b}_s) = \{\bar{a}_{s,1}, \ldots, \bar{a}_{s,p^{s(n-m)}}\} \subset (\mathbb{Z}/p^s\mathbb{Z})^n = \{0, 1, \ldots, p^s - 1\}^n.$$

For $j = 1, 2, \ldots, p^{s(n-m)}$ choose (and fix) $\bar{a}_{k,j} \in (\mathbb{Z}/p^k\mathbb{Z})^n$ so that $\bar{a}_{k,j} \equiv \bar{a}_{s,j} \pmod{p^s}$. Note that the latter congruence, in accordance with what has been agreed in Section 1, just means that $\|\bar{a}_{k,j} - \bar{a}_{s,j}\|_p \leq p^{-s}$; that is $\bar{a}_{k,j}^{(i)} \equiv \bar{a}_{s,j}^{(i)} \pmod{p^s}$ for each $i^{\text{th}}$ component $\bar{a}_{k,j}^{(i)}$ of $\bar{a}_{k,j} \in (\mathbb{Z}/p^k\mathbb{Z})^n = \{0, 1, \ldots, p^k - 1\}^n$, $i = 1, 2, \ldots, n$.

Now for $j = 1, 2, \ldots, p^{s(n-m)}$ take $\hat{a}_{k,j} \in (\mathbb{Z}/p^k\mathbb{Z})^n$ so that $\hat{a}_{k,j} \equiv \bar{a}_{s,j} \pmod{p^s}$; that is, $\hat{a}_{k,j} \in \bar{a}_{k,j} + p^s(\mathbb{Z}/p^k\mathbb{Z})^n$, and vice versa. Since $F$ is compatible, $\bar{F}_k(\hat{a}_{k,j}) \equiv \bar{b}_s \pmod{p^s}$; thus, $\bar{F}_k(\hat{a}_{k,j}) \in \bar{b}_k + p^s(\mathbb{Z}/p^k\mathbb{Z})^m$ (recall that $\bar{b}_s \equiv \bar{b}_k \pmod{p^s}$ by our choice). So every $\hat{a}_{k,j}$ is an $\bar{F}_k$-preimage of a certain element of the coset $b_k + p^s(\mathbb{Z}/p^k\mathbb{Z})^m$, and there are exactly $p^{s(n-m)} \cdot p^{n(k-s)} = p^{nk-ms}$ these elements $\hat{a}_{k,j}$. Comparing this number with what is given by equation (1), we conclude that all these $\hat{a}_{k,j}$ constitute the full preimage $\bar{F}_k^{-1}(\bar{b}_k + p^s(\mathbb{Z}/p^k\mathbb{Z})^m)$, which is then just the union of cosets $\bar{a}_{k,j} + p^s(\mathbb{Z}/p^k\mathbb{Z})^n$ over $j \in \{1, \ldots, p^{s(n-m)}\}$. These cosets are disjoint since all $\bar{a}_{k,j}$ are different modulo $p^s$.

<u>Claim 2.</u> For $j = 1, 2, \ldots, p^{s(n-m)}$ fix $a_j \in \mathbb{Z}_p^n$ such that $a_j \equiv \bar{a}_{s,j} \pmod{p^s}$, where $\bar{a}_{s,j}$ are defined as above for $\bar{b}_k \equiv b \pmod{p^k}$. Then

$$F^{-1}(b + p^s \mathbb{Z}_p^m) = \bigcup_{j=1}^{p^{s(n-m)}} (a_j + p^s \mathbb{Z}_p^n).$$

First note that in this setting the definition of $\bar{a}_{s,j}$ (whence, of $a_j$) does not depend on $k$, only on $b$ and $s$, since for $\bar{b}_k \equiv b \pmod{p^k}$ the set $\{\bar{a}_{s,1}, \ldots, \bar{a}_{s,p^{s(n-m)}}\}$ is just a full $\bar{F}_s$-preimage of $(b \bmod p^s)$; here $(b \bmod p^s)$ is a unique non-negative rational integer that lays at the distance $p^{-s}$ from the point $b$; an approximation of $b$ by a non-negative rational integer with precision $p^{-s}$ with respect to a $p$-adic metric. In other words, given $b \in \mathbb{Z}_p^m$, we put $\bar{b}_s \equiv b \pmod{p^s}$, where $\bar{b}_s \in \{1, 2, \ldots, p^s - 1\}^m$, then take all solutions $\bar{a}_{s,j} \in \{1, 2, \ldots, p^s - 1\}^n$ of the congruence $\bar{F}_s(x) \equiv \bar{b}_s \pmod{p^s}$ in indeterminate $x$, and after that, for each of these $p^{s(n-m)}$ solutions $\bar{a}_{s,j}$, we choose an arbitrary $a_j \in \mathbb{Z}_p^n$ so that $a_j \equiv \bar{a}_{s,j} \pmod{p^s}$.

Form the definition of $\bar{a}_j$ it follows immediately that for every $h \in (\mathbb{Z}_p)^n$, $F(a_j + p^s \cdot h) \equiv b \pmod{p^s}$ since $F$ is compatible; whence $F^{-1}(b + p^s \mathbb{Z}_p^m) \supset \bigcup_{j=1}^{p^{s(n-m)}} (a_j + p^s \mathbb{Z}_p^n)$. Thus, we must prove the inverse inclusion only.

Given $c \in b + p^s \mathbb{Z}_p^m$, then for every $k \geq s$ from Claim 1 it follows immediately that $F^{-1}(c) \in \bar{F}_k^{-1}(c \bmod p^k) + p^k \mathbb{Z}_p^n$, where $\bar{F}_k^{-1}(c \bmod p^k)$ is a subset of a finite set $\bigcup_{j=1}^{p^{s(n-m)}} (\bar{a}_{k,j} + p^s \cdot \{0, 1, \ldots, p^{k-s} - 1\}^n)$.

Thus, applying Claim 1 we obtain:

$$F^{-1}(c) \in \bigcap_{k=s}^{\infty} (\bar{F}_k^{-1}(c \bmod p^k) + p^k \mathbb{Z}_p^n)$$

$$\subset \bigcap_{k=s}^{\infty} \bigcup_{j=1}^{p^{s(n-m)}} (\bar{a}_{k,j} + p^s \cdot \{0, 1, \ldots, p^{k-s} - 1\}^n + p^k \mathbb{Z}_p^n)$$

$$= \bigcup_{j=1}^{p^{s(n-m)}} \bigcap_{k=s}^{\infty} (\bar{a}_{k,j} + p^s \cdot \{0, 1, \ldots, p^{k-s} - 1\}^n + p^k \mathbb{Z}_p^n)$$

$$= \bigcup_{j=1}^{p^{s(n-m)}} \bigcap_{k=s}^{\infty} (\bar{a}_{s,j} + p^s \cdot \{0, 1, \ldots, p^{k-s} - 1\}^n + p^k \mathbb{Z}_p^n)$$

$$= \bigcup_{j=1}^{p^{s(n-m)}} (\bar{a}_{s,j} + p^s \mathbb{Z}_p^n) = \bigcup_{j=1}^{p^{s(n-m)}} (a_j + p^s \mathbb{Z}_p^n)$$

This finishes the proof of Lemma 3.1. □

**3.2 Corollary.** $\mu_p(F^{-1}(b + p^s \mathbb{Z}_p^m)) = \sum_{j=1}^{p^{s(n-m)}} \mu_p(a_j + p^s \mathbb{Z}_p^n) = p^{s(n-m)} \cdot p^{-sn} = p^{-sm} = \mu_p(b + p^s \mathbb{Z}_p^m)$.

**3.3 Proposition.** *Under conditions of Lemma 3.1, the function F preserves measure.*

*Proof.* Balls of form $b + p^s \mathbb{Z}_p^m$ constitute a base of a $\sigma$-ring of all measurable sets of the space $\mathbb{Z}_p^m$. In view of Corollary 3.2, $F$ is then a measurable mapping; that is, any preimage of a measurable set is measurable. Now let's find $\mu_p(F^{-1}(M))$ for a measurable $M \subset \mathbb{Z}_p^m$.

Any open measurable subset $A \subset \mathbb{Z}_p^m$ is a disjoint union of such balls; hence, $F^{-1}(A)$ is open measurable subset of $\mathbb{Z}_p^n$, and $\mu_p(F^{-1}(A)) = \mu_p(A)$ in view of Corollary 3.2. Further, for a measurable $M$ one has $\mu_p(M) = \inf\{\mu_p(V) : V \supset M, V \text{ is open in } \mathbb{Z}_p^m\}$; thus,

$$\mu_p(F^{-1}(M)) \leq \inf\{\mu_p(F^{-1}(V)) : V \supset M, V \text{ is open in } \mathbb{Z}_p^m\} = \mu_p(M).$$

On the other hand, $\mu_p(M) = \sup\{\mu_p(W) : W \subset M, W \text{ is closed in } \mathbb{Z}_p^m\}$. Since each ball $b + p^s \mathbb{Z}_p^m$ is closed in $\mathbb{Z}_p^m$, each closed subset $W \subset \mathbb{Z}_p^m$ is a countable union of such balls (and, maybe, points); hence, the union is disjoint, whence $\mu_p(F^{-1}(W))$ is a closed subset of $\mathbb{Z}_p^n$, and $\mu_p(F^{-1}(W)) = \mu_p(W)$ in view of Corollary 3.2. Thus,

$$\mu_p(F^{-1}(M)) \geq \sup\{\mu_p(F^{-1}(W)) : W \subset M, W \text{ is closed in } \mathbb{Z}_p^m\} = \mu_p(M).$$

Finally we get $\mu_p(F^{-1}(M)) = \mu_p(M)$, thus proving the Proposition. □

To finish considerations of this Section, we must now prove the inverse statement.

**3.4 Proposition.** *Any compatible measure-preserving function $F\colon \mathbb{Z}_p^n \longrightarrow \mathbb{Z}_p^m$ is balanced modulo $p^k$, for all $k = 1, 2, \ldots$.*

*Proof.* Let for some $k$ there exists $\bar{x}, \bar{y} \in (\mathbb{Z}/p^k)^m = \{0, 1, \ldots, p^k - 1\}^m$ such that $\#\bar{F}_k^{-1}(\bar{x}) \neq \#\bar{F}_k^{-1}(\bar{y})$; note that both $\bar{F}_k^{-1}(\bar{x})$ and $\bar{F}_k^{-1}(\bar{y})$ lie in a finite set $(\mathbb{Z}/p^k)^n = \{0, 1, \ldots, p^k - 1\}^n$. Consider two balls $\bar{x} + p^k \mathbb{Z}_p^m$ and $\bar{y} + p^k \mathbb{Z}_p^m$ in $\mathbb{Z}_p^m$. Then

$$F^{-1}(\bar{x} + p^k \mathbb{Z}_p^m) = \bigcup_{z \in \bar{F}_k^{-1}(\bar{x})} (z + p^k \mathbb{Z}_p^n), \qquad F^{-1}(\bar{y} + p^k \mathbb{Z}_p^m) = \bigcup_{z \in \bar{F}_k^{-1}(\bar{y})} (z + p^k \mathbb{Z}_p^n).$$

Thus, $\mu_p(F^{-1}(\bar{x} + p^k \mathbb{Z}_p^m)) \neq \mu_p(F^{-1}(\bar{y} + p^k \mathbb{Z}_p^m))$; a contradiction. □

## 4. ERGODIC FUNCTIONS

In dynamical systems theory an ergodic mapping is, by the definition, a metric endomorphism $T$ (i.e., a measure-preserving mapping of a measurable space $X$ into itself) that has no non-trivial (that is, of positive measure $< 1$) invariant sets (we assume as usual that the measure is normalized so that the measure of $X$ is 1). In this section we characterize ergodic functions among all compatible functions $F\colon \mathbb{Z}_p^n \longrightarrow \mathbb{Z}_p^n$.

**4.1 Proposition.** *A compatible function $F\colon \mathbb{Z}_p^n \longrightarrow \mathbb{Z}_p^n$ is ergodic if and only if $F$ is transitive modulo $p^k$, for all $k = 1, 2, \ldots$.*

*Proof.* We start with the 'if' part of the statement. By the definition, the function $F$ is ergodic whenever $F^{-1}(A) = A$ implies either $\mu_p(A) = 1$ or $\mu_p(A) = 0$, for any measurable $A \subset \mathbb{Z}_p^n$. Let $F$ be transitive modulo $p^k$ for every $k = 1, 2, \ldots$, yet let $F$ be not ergodic. That is, let there exist a measurable non-empty $A \subset \mathbb{Z}_p^n$ such that $0 < \mu_p(A) < 1$ and $F^{-1}(A) = A$ (whence $F(A) = A$, since $F$ is a bijection, see Section 2).

We claim that then there exists a closed $F$-invariant subset $\overline{C} \subset A$ (that is, $F^{-1}(\overline{C}) = \overline{C}$) such that $1 > \mu_p(\overline{C}) > 0$. Moreover, this closed subset $\overline{C}$ is a union of some finite number of balls of pairwise equal radii.

Indeed, as any open subset of $\mathbb{Z}_p^n$ is a countable union of balls, and since a complement of a ball of a positive radius $r$ is a union of a finite number of balls of this radius $r$ each, every closed subset of $\mathbb{Z}_p^n$ is a countable union of balls, some of which are, maybe, of zero radius (i.e., points). However,

$$\mu_p(A) = \sup\{\mu_p(S)\colon S \subset A, S \text{ is closed in } \mathbb{Z}_p^n\},$$

since $\mu_p$ is a regular measure. Thus, there exists a closed subset $B \subset A$ such that $\mu_p(B) > 0$ since $\mu_p(A) > 0$. Hence, there exists a subset $C \subset B$, which is a ball of a positive radius $r$; thus, $\mu_p(C) > 0$. Since in force of Section 2 the mapping $F$ is a compatible and measure-preserving bijection, both $F^{-1}(C)$ and $F(C)$ are balls of the same radius $r$. Thus, the set $\overline{C} = \bigcup_{s=-\infty}^{\infty} F^s(C)$ is an $F$-invariant subset of $A$: $F^{-1}(\overline{C}) = \overline{C}$, and $\overline{C} \subset A$. As the union $\bigcup_{s=-\infty}^{\infty} F^s(C)$ is a union of balls of the same radius $r$, then $\overline{C}$ is a union of a finite number of balls of radius $r$, since there are only finitely many balls of the radius $r$. Obviously, $\mu_p(\overline{C}) < 1$ since $\mu_p(A) < 1$ by our assumption. Also, $\mu_p(\overline{C}) \geq \mu_p(C) > 0$.

Now, to prove the 'if' part of the proposition we may additionally suggest that $A$ is either a ball (of radius, say, $1 > p^k > 0$), or $A$ is not a ball, yet a union of a finite number of balls of radius $r = p^k > 0$ each. In all cases the mapping $\bar{F}_k$ is not transitive since it has a proper invariant subset, which consists of all images modulo $p^k$ of these balls. Yet this contradicts our assumption that $F$ is transitive modulo $p^k$ for all $k = 1, 2, \ldots$.

Now we prove the 'only if' part of the proposition. Let $F$ be ergodic. Then $F$ preserves measure, so in view of Section 2 for each $k = 1, 2, \ldots$ the mapping $\bar{F}_k$ is a permutation of the elements of the ring $(\mathbb{Z}/p^k\mathbb{Z})^n$. In case for some $k$ the permutation $\bar{F}_k$ has more than one cycle, we have that there exists a proper subset $\bar{A} \subset (\mathbb{Z}/p^k\mathbb{Z})^n = \{0, 1, \ldots, p^k - 1\}^n$ such that $\bar{F}_k(\bar{A}) = \bar{A}$. This implies that $F(\bar{A} + p^k\mathbb{Z}_p^n) = \bar{A} + \mathbb{Z}_p^n$, i.e. $F^{-1}(\bar{A} + p^k\mathbb{Z}_p^n) = \bar{A} + p^k\mathbb{Z}_p^n$, since $F$ is a bijection, see Section 2. Yet $\mu_p(\bar{A} + p^k\mathbb{Z}_p^n) = (\#\bar{A}) \cdot p^{-kn}$, and $0 < (\#\bar{A}) \cdot p^{-kn} < 1$, since $\bar{A}$ is a proper subset in $\{0, 1, \ldots, p^k - 1\}^n$. This contradicts our assumption that $F$ is ergodic. □

## 5. THE ERGODICITY ON SPHERES

In this section we study compatible ergodic transformations of spheres centered at $y \in \mathbb{Z}_p$. Let $S_{p^{-r}}(y)$ be a sphere of radius $\frac{1}{p^r} < 1$ with a center at $y \in \mathbb{Z}_p$; that is

$$S_{p^{-r}}(y) = \left\{ z \in \mathbb{Z}_p : \|z - y\|_p = \frac{1}{p^r} \right\}.$$

Note that this sphere is a disjoint union of balls of radius $\frac{1}{p^{r+1}}$ each,

$$S_{p^{-r}}(y) = \bigcup_{s=1}^{p-1} (y + p^r s + p^{r+1}\mathbb{Z}_p), \qquad (2)$$

since $S_{p^{-r}}(y)$ is a set-theoretic complement of the ball $y + p^{r+1}\mathbb{Z}_p$ to the ball $y + p^r \mathbb{Z}_p$. So $S_{p^{-r}}(y)$ is a closed and simultaneously an open (whence, a measurable) subset of $\mathbb{Z}_p$. We consider a measure $\hat{\mu}_p$ induced on $S_{p^{-r}}(y)$ by the Haar measure $\mu_p$ on the whole space $\mathbb{Z}_p$; we assume that $\hat{\mu}_p$ is normalized so that $\hat{\mu}_p(S_{p^{-r}}(y)) = 1$. Now, if $f \in \mathscr{L}_1$ is a compatible mapping of $\mathbb{Z}_p$ into $\mathbb{Z}_p$ such that the sphere $S_{p^{-r}}(y)$ is invariant under the action of $f$ (that is, $f(S_{p^{-r}}(y)) \subset S_{p^{-r}}(y)$), we can consider a restriction of $f$ (which we denote by the same symbol $f$) on the sphere $S_{p^{-r}}(y)$ and study ergodicity of the restriction $f$ with respect to the measure $\hat{\mu}_p$. We say then that $f$ is *ergodic on the sphere* $S_{p^{-r}}(y)$ whenever $S_{p^{-r}}(y)$ is invariant under action of $f$, and the action is ergodic with respect to $\hat{\mu}_p$, in the above mentioned meaning.

The following easy proposition holds:

**5.1 Proposition.** *Whenever the set $S_{p^{-r}}(y)$ is invariant under action of $f \in \mathscr{L}_1$, $f(y) \equiv y \pmod{p^r}$.*

*Proof.* Since the set $S_{p^{-r}}(y)$ is invariant, and since $f$ maps balls into balls, $f(y + p^r s + p^{r+1} \mathbb{Z}_p) \subset y + p^r \hat{s} + p^{r+1} \mathbb{Z}_p$ for a suitable $\hat{s} \in \{1, 2, \ldots, p-1\}$ (see (2)). However, $f(y + p^r s) \equiv f(y) \pmod{p^r}$ since $f \in \mathscr{L}_1$, and the result follows. $\square$

From this Proposition we immediately get the following

**5.2 Corollary.** *Let all spheres around $y \in \mathbb{Z}_p$ of radii less than $\varepsilon > 0$ are invariant under action of $f \in \mathscr{L}_1$. Then $f(y) = y$.*

Further, as it follows from their proofs, all results of preceding sections hold not only for the whole space $\mathbb{Z}_p$, but (up to a proper re-statement) for any finite disjoint union of balls of pairwise equal radii as well[5]. This implies the following important note:

*5.3 Note.* A compatible mapping $f \colon \mathbb{Z}_p \longrightarrow \mathbb{Z}_p$ is ergodic on the sphere $S_{p^{-r}}(y)$ if and only if it induces on the residue ring $\mathbb{Z}/p^{k+1}\mathbb{Z}$ a mapping which acts on the subset

$$S_{p^{-r}}(y) \bmod p^{k+1} = \{y + p^r s + p^{r+1} \mathbb{Z} \colon s = 1, 2, \ldots, p-1\} \subset \mathbb{Z}/p^{k+1}\mathbb{Z}$$

as a permutation with a single cycle, for all $k = r, r+1, \ldots$.

It worth notice also that whenever a compatible mapping $f$ is ergodic on the sphere $S_{p^{-r}}(y)$, $f$ is a bijection of this sphere onto itself; moreover, it is an isometry of this sphere, see Notes 2.2 and 2.5. The same holds for balls.

From these notices we deduce the following lemma:

**5.4 Lemma.** *A compatible mapping $f \colon \mathbb{Z}_p \longrightarrow \mathbb{Z}_p$ is ergodic on the sphere $S_{p^{-r}}(y)$ if and only if the following two conditions hold simultaneously:*

*1) the mapping $z \mapsto f(z) \bmod p^{r+1}$ permutes cyclically elements of the set*

$$S_{p^{-r}}(y) \bmod p^{r+1} = \{y + p^r s \colon s = 1, 2, \ldots, p-1\} \subset \mathbb{Z}/p^{r+1}\mathbb{Z};$$

*2) the mapping $z \mapsto f^{p-1}(z) \bmod p^{r+t+1}$ permutes cyclically elements of the set*

$$B_{p^{-(r+1)}}(y + p^r s) \bmod p^{r+t+1} = \{y + p^r s + p^{r+1} S \colon S = 0, 1, 2, \ldots, p^t - 1\},$$

---

[5] Moreover, following the ideas of these proofs the corresponding results could be proved for arbitrary measurable subset of $\mathbb{Z}_p$ of a positive measure, instead of the whole space $\mathbb{Z}_p$.

for all $t = 1, 2, \ldots$ and some (equivalently, all) $s \in \{1, 2, \ldots, p-1\}$. Here $f^k$ stands for the $k^{th}$ iterate of $f$

$$f^k(a) = \underbrace{f(f\ldots(f(a))\ldots)}_{k \text{ times}}.$$

*Condition 2) holds if and only if $f^{p-1}$ is an ergodic transformation of the ball $B_{p^{-(r+1)}}(y + p^r s) = y + p^r s + p^{r+1} \mathbb{Z}_p$ of radius $\frac{1}{p^{r+1}}$ with center at the point $y + p^r s$, for some (equivalently, all) $s \in \{1, 2, \ldots, p-1\}$.*

*Proof.* As every compatible and ergodic transformation $f$ of the sphere is bijective on this sphere, and $f$ is an isometry on this sphere as well (see above notions), $f(a + p^k \mathbb{Z}_p) = f(a) + p^k \mathbb{Z}_p$, for all $a \in \mathbb{Z}_p$ and all $k = 1, 2, \ldots$. Thus, the mapping $z \mapsto f(z) \bmod p^{k+1}$ ($k > r$) permutes cyclically elements of the set

$$S_{p^{-r}}(y) \bmod p^{k+1} = \{y + p^r s + p^{r+1} S \colon s = 1, 2, \ldots, p-1; S = 0, 1, 2, \ldots, p^{k-r} - 1\}$$

if and only if conditions 1) and 2) hold simultaneously for $t = k - r$. This proves the first part of the statement of the lemma, in view of Note 5.3. The second part of the statement is just an analogue of Note 5.3 for balls instead of spheres. □

To state the central result of this section, which describes ergodic mappings of a sphere into itself in a rather wide class $\mathscr{B}$ of compatible mappings, we introduce this class first: Consider the following class $\mathscr{B}$ of mappings from $\mathbb{Z}_p$ into $\mathbb{Z}_p$

$$\mathscr{B} = \left\{ f(x) = \sum_{i=0}^{\infty} a_i \binom{x}{i} \colon \frac{a_i}{i!} \in \mathbb{Z}_p, \ i = 0, 1, 2, \ldots \right\}, \tag{3}$$

which was studied in detail in [5]. In view of the well-known criterion for the convergence of Mahler's series (see e.g. [20]), the series of the definition of $\mathscr{B}$ is convergent everywhere on $\mathbb{Z}_p$ and defines a uniformly continuous function on $\mathbb{Z}_p$. Note that, obviously, $\mathscr{B}$ is the class of all functions that could be represented by 'descending factorial' power series with $p$-adic integer coefficients, that is, $f \in \mathscr{B}$ if and only if $f(x) = \sum_{i=0}^{\infty} b_i x^{\underline{i}}$, ($b_i \in \mathbb{Z}_p$), where $x^{\underline{0}} = 1$, $x^{\underline{i}} = x(x-1)\cdots(x-i+1)$.
The class $\mathscr{B}$ is endowed with a non-Archimedean norm $\max_{z \in \mathbb{Z}_p} \|f(z)\|_p$, which defines a metric $D_p$ on $\mathscr{B}$. The following is proved in [5]:

- $\mathscr{B} \subset \mathscr{L}_1$, i.e., all functions of $\mathscr{B}$ are compatible;
- $\mathscr{B}$ is a completion (with respect to the metric $D_p$) of the class $\mathscr{P}$ of all polynomials over $\mathbb{Z}_p$;
- the class $\mathscr{C}$ of all analytic on $\mathbb{Z}_p$ functions that could be represented by convergent power series with coefficients of $\mathbb{Z}_p$, is a proper subclass of $\mathscr{B}$;
- $\mathscr{B}$ is closed with respect to addition, multiplication, compositions, and derivations of functions.

We stress that, in a contrast to the class $\mathscr{C}$, which consists of analytic functions, the class $\mathscr{B}$ is closed under compositions of functions. Further we intensively use this property without special remarks.

Despite among $\mathscr{B}$-functions there exist functions that are not analytic on $\mathbb{Z}_p$ (e.g., the function $\sum_{i=0}^{\infty} i! \binom{x}{i} = \sum_{i=0}^{\infty} x^i$), all $\mathscr{B}$-functions are analytic on all balls of radii less than 1; namely, the following theorem holds:

**5.5 Theorem** (Taylor theorem for $\mathscr{B}$-functions). *For every $f \in \mathscr{B}$, $a, h \in \mathbb{Z}_p$ and $k = 1, 2, 3, \ldots$ the following equality holds:*

$$f(a + p^k h) = f(a) + f'(a) \cdot p^k h + \frac{f''(a)}{2!} \cdot p^{2k} h^2 + \frac{f'''(a)}{3!} \cdot p^{3k} h^3 + \cdots, \qquad (4)$$

*where, as usual, $f^{(j)}(a)$ stands for the $j^{th}$ derivative of the function $f$ at the point $a \in \mathbb{Z}_p$. Moreover, all $\frac{f^{(j)}(a)}{j!}$ are p-adic integers, $j = 0, 1, 2, \ldots$.*

*Proof.* We prove the second claim of the theorem first:

**5.6 Lemma.** *Under conditions of Theorem 5.5, all $\frac{f^{(j)}(a)}{j!}$ are p-adic integers.*

*Proof of Lemma 5.6.* As we have demonstrated in [5], for every $f \in \mathscr{B}$ and every $x \in \mathbb{Z}_p$

$$f'(x) = \sum_{i=1}^{\infty} (-1)^{i+1} \frac{\Delta^i f(x)}{i}, {}^6 \qquad (5)$$

where $\Delta$ is a difference operator; $\Delta f(x) = f(x+1) - f(x)$. Thus, as $\Delta \binom{x}{i} = \binom{x}{i-1}$, from (3) we have $f'(x) = \sum_{k=0}^{\infty} \binom{x}{k} \sum_{i=1}^{\infty} (-1)^{i+1} \frac{a_{k+i}}{i}$ and further by induction,

$$f^{(n)}(x) = \sum_{k=0}^{\infty} \binom{x}{k} \sum_{i_1, i_2, \ldots, i_n \geq 1} \frac{a_{k+i_1+i_2+\cdots+i_n}}{i_1 \cdot i_2 \cdots i_n} (-1)^{n+i_1+i_2+\cdots+i_n}.$$

However,

$$\sum_{i_1, i_2, \ldots, i_n \geq 1} \frac{a_{k+i_1+i_2+\cdots+i_n}}{i_1 \cdot i_2 \cdots i_n} (-1)^{n+i_1+i_2+\cdots+i_n} = \sum_{s=n}^{\infty} \sum_{\substack{i_1, i_2, \ldots, i_n \geq 1 \\ i_1+i_2+\ldots+i_n=s}} \frac{a_{k+s}}{i_1 \cdot i_2 \cdots i_n} (-1)^{n+s}, \qquad (6)$$

and $\frac{a_{k+s}}{i_1 \cdot i_2 \cdots i_n} = \frac{a_{k+s}}{s!} \frac{s!}{i_1 \cdot i_2 \cdots i_n} \in \mathbb{Z}_p$ since both $\frac{(i_1+i_2+\cdots+i_n)!}{i_1 \cdot i_2 \cdots i_n} \in \mathbb{Z}$ and $\frac{a_{k+s}}{(k+s)!} \in \mathbb{Z}_p$, see the definition of a $\mathscr{B}$-function (3) for the latter. Thus, the sum

$$\sigma_s = \sum_{\substack{i_1, i_2, \ldots, i_n \geq 1 \\ i_1+i_2+\ldots+i_n=s}} \frac{a_{k+s}}{i_1 \cdot i_2 \cdots i_n} (-1)^{n+s}$$

in the right-hand side of (6) is a $p$-adic integer. Moreover, as $\frac{a_{k+s}}{i_1 \cdot i_2 \cdots i_n} = \frac{a_{k+s}}{j_1 \cdot j_2 \cdots j_n}$ whenever $j_1, j_2, \ldots, j_n$ is a permutation of $i_1, i_2, \ldots, i_n$, the sum $\sigma_s$ is a multiple of $n!$, i.e., $\frac{\sigma_s}{n!} \in \mathbb{Z}_p$. This proves the lemma. $\square$

---

[6] However, it is well known that whenever the left-hand side is convergent, it converges to $f'(x)$.

The rest of the proof of the theorem follows from a general result of Y. Amice, [1]: The result implies that any $\mathscr{B}$-function is analytic of order 1; this constitutes the first claim of Theorem 5.5.

Indeed, according to [1, Ch. III, Sec. 10, Th. 3, Cor. 1(c)] the function $f(x) = \sum_{i=0}^{\infty} a_i \cdot i! \binom{x}{i}$ ($a_i \in \mathbb{Q}_p$) is locally analytic of order $n$ on $\mathbb{Z}_p$ (that is, $f(a+p^n h) = \sum_{i=0}^{\infty} p^{in} h^i \frac{f^{(n)}(a)}{i!}$ for $h \in \mathbb{Z}_p$) if and only if

$$\lim_{i \to \infty} \left( \frac{i}{p-1} \cdot \left(1 - \frac{1}{p^n}\right) - \log_p \|a_i\|_p \right) = +\infty,$$

which obviously holds with $n = 1$ for any $\mathscr{B}$-function $f$ in force of the definition of the class $\mathscr{B}$, see (3). $\square$

Now we state the main result of the section.

**5.7 Theorem.** *Let the function $f$ lie in $\mathscr{B}$. The function $f$ is ergodic on the sphere $S_{p^{-r}}(y)$ of sufficiently small [7] radius $p^{-r}$ if and only if one of the following alternatives holds:*

1. *Whenever $p$ is odd, then simultaneously*
   - $f(y) \equiv y \pmod{p^{r+1}}$,
   - $f'(y)$ *generates the whole group of units modulo $p^2$.*
2. *Whenever $p = 2$, then simultaneously*
   - $f(y) \equiv y \pmod{2^{r+1}}$,
   - $f(y) \not\equiv y \pmod{2^{r+2}}$,
   - $f'(y) \equiv 1 \pmod{4}$.

*Proof.* As it immediately follows from Theorem 5.5, for every $g \in \mathscr{B}$ and all $k \in \mathbb{Z}_p$, $k = 1, 2, 3, \ldots$ the following equality holds

$$g(a + p^k h) = g(a) + g'(a) \cdot p^k h + p^{2k} h^2 \cdot \hat{g}(h), \tag{7}$$

for a suitable $\mathscr{C}$-function $\hat{g}$ of variable $h$. [8]

Since $f(y) = y + p^r z$ for a suitable $z \in \mathbb{Z}_p$ in view of Proposition 5.1, we deduce from (7) that the following equalities hold:

$$f(y + p^r s + p^{r+1} S) = f(y) + (p^r s + p^{r+1} S) \cdot f'(y) + p^{2r} \cdot (s + pS)^2 \cdot \hat{w}(s + pS) =$$
$$y + p^r z + p^r s \cdot f'(y) + p^{r+1} S \cdot f'(y) + p^{2r} \cdot v(s) + p^{2r+1} \cdot w(S), \tag{8}$$

where $v$, $\hat{w}$ and $w$ are $\mathscr{C}$-functions in the respective variables (note that we have used (7) twice; with $g = f$, $a = y$, $p^k h = p^r s + p^{r+1} S$, for the first time, and with $g = w$, $a = s$, $p^k h = p S$), for the second time. Note that $w$ depends also on $s$, yet this is of no importance in future argument.

---

[7] $p^{-r} < 1$ in case $p > 3$, and $p^{-r} < \frac{1}{p}$ in case $p \leq 3$.
[8] Of course, coefficients of series (3) that represents the function $p^{2k} \cdot g \in \mathscr{B}$ depend also on $a$ and $k$, but this is of no importance at the moment.

Iterating (8) we obtain

$$f^{p-1}(y+p^r s+p^{r+1}S) = y+p^r z \sum_{i=0}^{p-2}(f'(y))^i + p^r s \cdot (f'(y))^{p-1}$$
$$+ p^{r+1}S \cdot (f'(y))^{p-1} + p^{2r} \cdot \check{v}(s) + p^{2r+1} \cdot \check{w}(S), \quad (9)$$

for suitable $\check{v}$ and $\check{w}$, which are $\mathscr{B}$-functions now (since they are obtained with the use of compositions of $\mathscr{C}$-functions).

Now, to satisfy condition (2) of Lemma 5.4, the ball $y+p^r s+p^{r+1}\mathbb{Z}_p$ must be invariant under action of $f^{p-1}$, and $f^{p-1}$ must act ergodically on this ball. However, 9 implies that the ball is invariant if and only if

$$\sigma(z,s) = z\sum_{i=0}^{p-2}(f'(y))^i + s \cdot (f'(y))^{p-1} \equiv s \pmod{p}. \quad (10)$$

Assuming the ball is invariant, we have $\sigma(z,s) = s + p \cdot \gamma(z,s)$ for a suitable $p$-adic integer $\gamma(z,s)$. So, having $s$ fixed, from 9 we see under this assumption that

$$f^{p-1}(y+p^r s+p^{r+1}S) = y+p^r s+p^{r+1} \cdot (\gamma(z,s) + S \cdot (f'(y))^{p-1} + p^{r-1} \cdot \check{v}(s) + p^r \cdot \check{w}(S));$$

Thus, to satisfy condition (2) of Lemma 5.4, the following $\mathscr{B}$-function

$$G_{z,s}(S) = \gamma(z,s) + S \cdot (f'(y))^{p-1} + p^{r-1} \cdot \check{v}(s) + p^r \cdot \check{w}(S) \quad (11)$$

in variable $S$ must be ergodic on $\mathbb{Z}_p$.

However, $\mathscr{B}$-functions (in particular, polynomials with $p$-adic integer coefficients) that are ergodic on $\mathbb{Z}_p$ are completely characterized in [5].[9] We state the result as the following lemma.

**5.8 Lemma.** *A $\mathscr{B}$-function is ergodic on $\mathbb{Z}_p$ if and only if it is transitive modulo $p^3$ for $p \in \{2,3\}$, or modulo $p^2$, otherwise.*

Hence, if $r > 1$ in case $p > 3$, or if $r > 2$ in case $p \leq 3$, we conclude that the $\mathscr{B}$-function $G_{z,s}(S)$ of (11) is ergodic on $\mathbb{Z}_p$ if and only if the polynomial

$$L_{z,s}(S) = \gamma(z,s) + p^{r-1} \cdot \check{v}(s) + S \cdot (f'(y))^{p-1} \quad (12)$$

of degree 1 in variable $S$ is transitive modulo $p^2$ for $p > 3$, or modulo $p^3$ for $p \leq 3$.

Necessary and sufficient conditions providing the polynomial $\alpha + \beta \cdot x \in \mathbb{Z}_p[x]$ is transitive modulo $p^k$ for $k \geq 2$ are well known, see e.g. [18, Section 3.2.1]. We again state the result as the lemma.

---

[9] As for polynomials with integer coefficients, M. V. Larin was the first who gave the characterization in the beginning of 1980th. He used different terminology and techniques and published his result in [19] only in 2002. Also the characterization for polynomials over $\mathbb{Z}_p$ with odd $p$ could be derived from a general study of cycle structure of polynomial mappings in [13].

**5.9 Lemma.** *The polynomial $\alpha + \beta \cdot x \in \mathbb{Z}_p[x]$ is transitive modulo $p^k$ for some $k \geq 2$ (equivalently, for all $k = 1, 2, \ldots$)*[10] *if and only if the following conditions hold simultaneously:*
- $\alpha \not\equiv 0 \pmod{p}$;
- $\beta \equiv 1 \pmod{p}$ *for odd $p$, and $\beta \equiv 1 \pmod{4}$ for $p = 2$.*

From this lemma in view of (12) we immediately conclude that $f'(y) \not\equiv 0 \pmod{p}$. Now 8 immediately implies that to satisfy condition (1) of Lemma 5.4, the mapping $s \mapsto z + sf'(y) \pmod{p}$ must cyclically permute elements of the multiplicative group (i.e., the whole group of units) $(\mathbb{Z}/p\mathbb{Z})^*$ of the field $\mathbb{Z}/p\mathbb{Z}$. Hence, $z \equiv 0 \pmod{p}$ (that is, $f(y) \equiv y \pmod{p^{r+1}}$) since otherwise $s \mapsto 0 \pmod{p}$ for $s \equiv -\frac{z}{f'(y)} \pmod{p}$. From this moment we start considering the two cases $p = 2$ and $p > 2$ separately.

Case 1: $p > 2$. In this case the mapping $s \mapsto sf'(y) \pmod{p}$ cyclically permutes elements of $(\mathbb{Z}/p\mathbb{Z})^*$ if and only if $f'(y)$ is a primitive element of the field $\mathbb{Z}_p$ (that is, $f'(y)$ generates the cyclic group $(\mathbb{Z}/p\mathbb{Z})^*$).

Whenever this holds, each ball $y + p^r s + p^{r+1} \mathbb{Z}_p$, $s \in \{1, 2, \ldots, p-1\}$ is invariant under action of $f^{p-1}$ in view of (10). Moreover, since $z \equiv 0 \pmod{p}$, in case $f'(y)$ is primitive modulo $p$ we have that $\sigma(z, s) \equiv s \cdot (f'(y))^{p-1} \pmod{p^2}$ and whence $\gamma(z, s) \equiv bs \pmod{p}$, where $(f'(y))^{p-1} = 1 + pb$, $b \in \mathbb{Z}_p$ (see (10) and the text thereafter for the definition of $\sigma(z, s)$ and $\gamma(z, s)$).

Now, the polynomial (12) in variable $S$ is ergodic on $\mathbb{Z}_p$ (and so condition (2) of Lemma 5.4 is satisfied) if and only if $b \not\equiv 0 \pmod{p}$, see Lemma 5.9. Yet this means that $f'(y)$ must be a generator of the multiplicative group $(\mathbb{Z}/p^2\mathbb{Z})^*$.

Case 2: $p = 2$. In this case the sphere $S_{2^{-r}}(y) = y + 2^r + 2^{r+1}\mathbb{Z}_2$ is a ball, see (2). Moreover, the above condition $f'(y) \not\equiv 0 \pmod{p}$ means that $f'(y) \equiv 1 \pmod{2}$, and so the condition that the mapping $s \mapsto sf'(y) \pmod{p}$ is a single cycle permutation on the multiplicative group $(\mathbb{Z}/p\mathbb{Z})^*$, which just means that $z + f'(y) \equiv 1 \pmod{2}$ in this case, is automatically satisfied since we have already proved that $z \equiv 0 \pmod{p}$, (i.e., $z = pc$ for suitable $c \in \mathbb{Z}_p$) for any $p$.

Further, the condition that the polynomial $L_{z,s}(S)$ in variable $S$ is transitive modulo $p^3$ implies that $f'(y) \equiv 1 \pmod{4}$, see (12) and Lemma 5.9. That is, $f'(y) = 1 + 4b$ for some $b \in \mathbb{Z}_2$. Hence $\gamma(z, s) = c + 2b$ (see (10) and the text thereafter), so in view of (12) and Lemma 5.9, to provide the polynomial $L_{z,s}(S)$ is transitive modulo 8, must be $c \equiv 1 \pmod{2}$; that is, $f(y) = y + 2^r z = y + 2^{r+1} c \not\equiv y \pmod{2^{r+2}}$. This proves Theorem 5.7. □

The first important consequence of Theorem 5.7 is a solution of the problem of A. Khrennikov mentioned in Section 1:

---

[10] So in view of Theorem 1.2, the lemma states necessary and sufficient conditions providing a polynomial of degree 1 over $\mathbb{Z}_p$ is ergodic on $\mathbb{Z}_p$: It must be transitive modulo $p$ for odd $p$, or modulo 4 for $p = 2$.

**5.10 Proposition.** *The perturbed monomial mapping* $f: x \mapsto x^\ell + q(x)$, *where* $q(x) = p^{r+1}u(x)$ *for some function* $u \in \mathscr{B}$ (*e.g., for a polynomial* $u(x) \in \mathbb{Z}_p[x]$) *is ergodic on the sphere* $S_{p^{-r}}(1)$ (*where* $r > 1$) *if and only if* $\ell$ *is a generator of the multiplicative group* $(\mathbb{Z}/p^2\mathbb{Z})^*$.

*Proof.* Immediately follows from Theorem 5.7 with the only exception of the case $p = 3$ and $r = 2$. To handle this case, some extra efforts should be undertaken. Namely, for $p = 3$ in view of Theorem 5.5 one obtains

$$\begin{aligned} f^2(1+3^r s + 3^{r+1} S) &= f^2(1) + (3^r s + 3^{r+1} S) \cdot f'(f(1)) \cdot f'(1) \\ &+ \frac{1}{2}(3^r s + 3^{r+1} S)^2 (f''(f(1)) \cdot (f'(1))^2 + f'(f(1)) \cdot f''(1)) \\ &+ 3^{3r+1} \cdot \hat{w}(S), \end{aligned} \quad (13)$$

where $\hat{w}(S)$ is a $\mathscr{B}$-function in variable $S$. Now taking $f(x) = x^\ell + 3^{r+1} q(x)$, from (13) we obtain

$$\begin{aligned} f^2(1+3^r s + 3^{r+1} S) &= 1 + (\ell+1) 3^{r+1} u(1) + (3^r s + 3^{r+1} S) \cdot \ell^2 \\ &+ \frac{1}{2}(3^r s + 3^{r+1} S)^2 \cdot \ell^2 (\ell-1)(\ell+1) + 3^{2r+1} v(s) + 3^{2r+2} w(S), \end{aligned} \quad (14)$$

where $v$ and $w$ are $\mathscr{B}$-functions in variables $s$ and $S$, respectively. However, $\ell$ must be primitive modulo 3 (see Case 2 of the proof of Theorem 5.7); so $\ell \equiv 2 \pmod{3}$. Hence, $\ell^2 = 1 + 3b$ for a suitable $b \in \mathbb{Z}$. Also, $\ell(\ell-1)(\ell+1)$ is a multiple of 3; combining this altogether with (14) we obtain:

$$\begin{aligned} f^2(1+3^r s + 3^{r+1} S) &= 1 + 3^r s + 3^{r+1} \cdot (b + (\ell+1) \cdot u(1) + S\ell^2 \\ &+ 3^r \cdot \check{v}(s) + 3^{r+1} \cdot \check{w}(S)), \end{aligned} \quad (15)$$

for suitable $\mathscr{B}$-functions $\check{v}$ and $\check{w}$. Now we must check whether the $\mathscr{B}$-function

$$L(S) = b + (\ell+1) \cdot u(1) + S\ell^2 + 3^r \cdot \check{v}(s) + 3^{r+1} \cdot \check{w}(S)$$

is ergodic on $\mathbb{Z}_3$; cf. (11) where the residue term is $p^r \cdot \check{w}(S)$ rather than $3^{r+1} \cdot \check{w}(S)$ as in the case under consideration. The reason for this is that now extra factor 3 in the fourth term of 14 arises because of the multiplier $\ell(\ell-1)(\ell+1)$.

Applying Lemmas 5.8 and 5.9 to the $\mathscr{B}$-function $L$ in variable $S$ we see that $L$ is ergodic on $\mathbb{Z}_p$ if and only if $b \not\equiv 0 \pmod 3$ (since $(\ell+1)q(1) \equiv 0 \pmod 3$; we remind that $\ell \equiv 2 \pmod 3$). Thus, we finally conclude that $\ell$ must be primitive modulo $p^2$. □

There are some more consequences of Theorem 5.7. To start with, *Theorem 1.3, which is stated in Section 1, becomes now obvious in view of Theorem 5.7 and Corollary 5.2.*

Yet another immediate consequence follows:

**5.11 Corollary.** *Let* $y \in \mathbb{Z}_p$ *be a fixed point of the function* $f \in \mathscr{B}$, *and let $p$ be odd. Then, $f$ is ergodic on <u>all</u> spheres around $y$ of sufficiently small radii if and only if $f$ is ergodic on <u>some</u> sphere around $y$ of a sufficiently small radius.*

Some known results on ergodicity of polynomial mappings also follow from Theorem 5.7. For instance, [11] concerns ergodicity of simple polynomial mappings $M_{a,\ell}\colon z \mapsto az^\ell$ on spheres, where $\ell > 0$ is rational integer, $a \in \mathbb{Z}_p$. From Hensel's Lemma it follows that whenever $\ell \not\equiv 1 \pmod{p}$ and $a \in B_{p^{-1}}(1)$, the mapping $M_{a,\ell}$ has a unique fixed point $x_0 \in B_{p^{-1}}(1)$ (see [11, Lemma 8.2]). Under these assumptions, from Theorem 5.7 it immediately follows that $M_{a,\ell}$ is ergodic on $S_{p^{-r}}(x_0)$ (for $p$ odd) if and only if $a \cdot \ell$ is primitive modulo $p^2$, that is, *if and only if $\ell$ is primitive modulo $p^2$* since $a \equiv 1 \pmod{p}$ by the assumption; cf. [11, Theorem 8.4]. Similarly, the translation $T_{a,b}\colon z \mapsto az+b$, with $a,b \in \mathbb{Z}_p$, has a fixed point $y_0 = \frac{b}{1-a} \in \mathbb{Q}_p$ whenever $a \neq 1$. In case $y \in \mathbb{Z}_p$, Theorem 5.7 yields $T_{a,b}$ is ergodic on $S_{p^{-r}}(y)$ *if and only if $a$ is primitive modulo $p^2$*, cf. [11, Theorem 7.3].[11]

In view of Theorem 5.7 it is obvious that these results remain true in a 'perturbed form', that is, for mappings $z \mapsto M_{a,\ell}(z) + p^{r+1}v(z)$ and $z \mapsto T_{a,b} + p^{r+1}v(z)$, where $v$ is an arbitrary polynomial over $\mathbb{Z}_p$ (or even a $\mathcal{B}$-function), despite in this case $x_0$ (respectively, $y_0$) are not necessarily fixed points of the corresponding mappings.

Some important functions (for instance, some compatible integer-valued polynomials over $\mathbb{Q}_p$; i.e., those polynomials, which have not necessarily integer $p$-adic coefficients, that map $\mathbb{Z}_p$ into itself, and that satisfy Lipschitz condition with a constant 1 everywhere on $\mathbb{Z}_p$) do not lie in $\mathcal{B}$. However, they lie in a wider class $\mathcal{A}$, which is also introduced and studied in [5]: By the definition, the function $f\colon \mathbb{Z}_p \longrightarrow \mathbb{Z}_p$ lies in $\mathcal{A}$ if and only if $f$ is compatible (i.e., satisfies Lipschitz condition with a constant 1), and $p^n f \in \mathcal{B}$ for some non-negative rational integer $n$. It is important to note that *Theorem 5.7 remains true for $\mathcal{A}$-functions.*

Namely, since $f = \frac{1}{p^n}\bar{f}$ for a suitable $\mathcal{B}$-function $\bar{f}$ and suitable non-negative rational integer $n$, from Theorem 5.5 we immediately conclude that Taylor theorem for every $\mathcal{A}$-function $f$ holds in the following form:

**5.12 Theorem** (Taylor theorem for $\mathcal{A}$-functions). *For every $f \in \mathcal{A}$, $a, h \in \mathbb{Z}_p$ and $k = 1, 2, 3, \ldots$ the function $f(a + p^k h)$ in variable $h$ could be represented via convergent Taylor series*

$$f(a+p^k h) = f(a) + f'(a)\cdot p^k h + \frac{f''(a)}{2!}\cdot p^{2k}h^2 + \frac{f'''(a)}{3!}\cdot p^{3k}h^3 + \cdots. \qquad (16)$$

Note that $\frac{f^{(j)}(a)}{j!}$ are *not* necessarily $p$-adic integers now; however, in view of the second claim of Theorem 5.5, $\|\frac{f^{(j)}(a)}{j!}\|_p \leq p^n$ for all $j = 0, 1, 2, \ldots$. Moreover, $f'(a)$ is a $p$-adic integer. It is not difficult to prove that a derivative of a compatible function is a $p$-adic integer at any point the derivative exists, see e.g. [5].

---

[11] We note however that we prove not exactly the same results as in [11] since we impose conditions that are slightly different from the ones in [11].

Thus, we can re-write key equation 7 of Theorem 5.7 in the following form:

$$g(a+p^k h) = g(a) + g'(a) \cdot p^k h + p^{2k-n} \cdot h^2 \cdot \hat{g}(h), \qquad (17)$$

where $\hat{g} \in \mathscr{C}$ and $k$ is sufficiently large (to make $2k - n$ positive). Then from (8) we obtain (for a sufficiently large $r$) that

$$\begin{aligned} f(y+p^r s + p^{r+1}S) &= f(y) + (p^r s + p^{r+1}S) \cdot f'(y) + p^{2r-n} \cdot (s+pS)^2 \cdot \hat{w}(s+pS) \\ &= y + p^r z + p^r s \cdot f'(y) + p^{r+1}S \cdot f'(y) + p^{2r-n} \cdot v(s) + p^{2r+1-n} \cdot w(S), \quad (18) \end{aligned}$$

where $v$, $\hat{w}$ and $w$ are $\mathscr{C}$-functions in the respective variables. Now we assume that $r$ is so large that $2r - n \geq r + 3$ and finish the proof in the same way as in the one of Theorem 5.7. Note that now how small the sphere $S_{p^{-r}}(y)$ must be to satisfy the theorem depends not only on $p$ (as it is in case of Theorem 5.7) but also on $n$, i.e., on the function $f$.

Now in the same manner we could re-state the rest of results of the section for $\mathscr{A}$-functions rather than for $\mathscr{B}$-functions. We omit details.

We note in conclusion that Theorems 5.5 and 5.12 imply that despite a $\mathscr{B}$-function (or, an $\mathscr{A}$-function) $f$ may be non-analytic on $\mathbb{Z}_p$, it is analytic on every ball $a + p\mathbb{Z}_p$; that is, $f$ is locally analytic of order 1, in terminology of [22].

## 6. DISCUSSION

Main results of the paper are Theorem 1.2, which characterizes measure-preserving (or ergodic) transformations of the space of $p$-adic integers $\mathbb{Z}_p$, and Theorem 5.7, which characterizes ergodic transformations of a $p$-adic sphere, and which gives a solution to the problem of A. Khrennikov mentioned in the introduction. All the transformations are assumed to be compatible, that is, satisfying Lipschitz condition with a constant 1 (the latter class is denoted via $\mathscr{L}_1$).

To demonstrate the importance of Theorem 1.2 for the $p$-adic ergodic theory, we use for some time the already mentioned mappings, translations $T_{a,b} \colon z \mapsto az + b$ and simple polynomial mappings $M_{a,\ell} \colon z \mapsto az^\ell$ as running examples, since these mappings have seemingly attracted certain attention in the $p$-adic ergodic theory: We already have refer to [11] in this connection. Also, paper [12] considers the ergodicity of the mapping $M_a \colon z \mapsto az$ on the sphere $S_{p^{-1}}(0)$ in connection with a distribution modulo $p^n$ of Fibonacci numbers. In [21] ergodic decompositions of continuous automorphisms of the additive group $\mathbb{Z}_p$ were studied; the latter are of the form $M_a$ for $a \in S_1(0)$.

We see the role Theorem 1.2 (together with Note 5.3, with notes that precede Note 5.3, and with Lemma 5.4) plays in study of ergodicity of mappings on spheres and balls, as follows: *These results act like a bridge connecting together results from the p-adic ergodic theory with the number-theoretic results* concerning residue rings $\mathbb{Z}/p^n\mathbb{Z}$.

For instance, Theorem 1.2 implies that the translation $T_{a,b}$ is ergodic on $\mathbb{Z}_p$ if and only if the mapping $\bar{T}_{a,b} \colon z \mapsto az + b \pmod{p^n}$ is transitive for all $n = 1, 2, \ldots$. However, the latter mapping is the recurrence law of the so-called 'linear congruential generator', which is very well known to computer scientists, and which is often used in software to produce pseudorandom sequences, see [18, Section 3.2.1]. In the latter case it is important that the period of the sequence is a maximum possible, i.e., $p^n$. We already have quoted the corresponding criterion during the proof of Theorem 5.7, see Lemma 5.9 there; here we mention only that this Lemma is a 40-year old result of Hull and Dobell, see [18, Section 3.2.1, Theorem A].

Moreover, using the same approach (and Lemma 5.4) we immediately conclude that the translation $M_a$ is ergodic on the sphere $S_{p^{-r}}(0)$ if and only if $a$ is a generator of the multiplicative group $\mathbb{Z}/p^n\mathbb{Z}$ for all $n = 1, 2, \ldots$. Again, it is well-known (the result goes back to Gauss, see [18, Section 3.2.1, Theorem B, Exercise 12]) that this holds if and only if $a$ is primitive modulo $p$, and $a^p \not\equiv 1 \pmod{p^2}$; that is, if and only if $a$ is a generator of the cyclic group $(\mathbb{Z}/p^2\mathbb{Z})^*$. Now cf. [12, Theorem 1] and [11, Theorem 7.2].

Another use of Theorem 1.2 is that it brings a number of examples of ergodic transformations of balls and spheres, and moreover, invokes earlier results in order to obtain complete characterizations (in various forms) of ergodic transformations of the space $\mathbb{Z}_p$. Actually, in [3, 5, 7, 9, 8] we have proved a number of results on transitivity of compatible mappings modulo $p^n$ for all $n$. That is, in view of Theorem 1.2 these are statements about ergodicity of the mappings. Among them, the following results are of interest:

- **'Closed' form of ergodic functions:** For arbitrary $v \in \mathscr{L}_1$, the function $f(x) = 1 + x + p \cdot (v(x+1) - v(x))$ is ergodic on $\mathbb{Z}_p$; in case $p = 2$ the converse is true: Any ergodic function $f \in \mathscr{L}_1$ is of the form $f(x) = 1 + x + 2 \cdot (v(x+1) - v(x))$, for a suitable $v \in \mathscr{L}_1$.
- **Representation via Mahler's series:** For $p = 2$ the function $f: \mathbb{Z}_p \longrightarrow \mathbb{Z}_p$ is compatible[12] and ergodic on $\mathbb{Z}_p$ if and only if

$$f(x) = 1 + x + \sum_{i=1}^{\infty} c_i \cdot p^{\lfloor \log_p(i+1) \rfloor + 1} \binom{x}{i},$$

for suitable $c_i \in \mathbb{Z}_p$. For $p \neq 2$ the conditions remain sufficient, and not necessary.
- **Ergodicity of polynomials over $\mathbb{Q}_p$:** A polynomial $f(x) \in \mathbb{Q}_p[x]$ of degree $d$ with rational (and not necessarily integer) coefficients is integer-valued (i.e., $f(\mathbb{Z}_p) \subset \mathbb{Z}_p$) compatible, and ergodic on $\mathbb{Z}_p$ if and only if $f$ takes values in $\mathbb{Z}_p$ at the points $0, 1, \ldots, p^{\lfloor \log_p d \rfloor + 3} - 1$, and the mapping $z \mapsto f(z) \bmod p^{\lfloor \log_p d \rfloor + 3}$ is compatible and transitive on the residue ring $\mathbb{Z}/p^{\lfloor \log_p d \rfloor + 3}\mathbb{Z}$. Thus, to check whether the polynomial $f(x) \in \mathbb{Q}_p[x]$ is, simultaneously, integer-valued, satisfies Lipschitz condition with a constant 1, and is ergodic, it is enough to evaluate it at approximately $dp^3$ points.

Theorem 5.7 gives a complete description of $\mathscr{B}$-functions that are ergodic on a $p$-adic sphere. The class $\mathscr{B}$ (which is, loosely speaking, a closure in the sense of Stone-

---

[12] This means, we recall, that $f$ lies in $\mathscr{L}_1$.

Weierstrass theorem of the class $\mathscr{P}$ of all polynomials over $\mathbb{Z}_p$) contains the class $\mathscr{C}$ of all functions that could be represented by everywhere convergent power series over $\mathbb{Z}_p$ (thus, all $\mathscr{C}$-functions are analytic on $\mathbb{Z}_p$). However, $\mathscr{B}$ is wider than $\mathscr{C}$, a $\mathscr{B}$-function is not necessarily analytic on $\mathbb{Z}_p$.

With the use of Theorem 5.7 we immediately obtain a number of examples of various functions that are ergodic on a $p$-adic sphere: For instance, whenever a positive rational integer $\ell$ generates modulo $p^2$ the whole group of units of the residue ring $\mathbb{Z}/p^2\mathbb{Z}$, the functions $1 + \ell \cdot (-1 + x + p^2 \cdot v(x))$ and $\ell \cdot (ax + a^x - 2a) + 1$ are ergodic on all (sufficiently small) spheres around 1, for every $a \in 1 + p^2\mathbb{Z}_p$ and every $\mathscr{B}$-function $v$ (say, for $v$ being a polynomial over $\mathbb{Z}_p$); accordingly, the functions $\ell \cdot x + \ln_p(1 + p^2 x)$ and $\frac{\ell \cdot x}{1+p^2 x}$ are ergodic on all (sufficiently small) spheres around 0 (here $\ln_p$ stands for the $p$-adic logarithm).

With respect to the problem of A. Khrennikov on ergodicity of perturbed monomial mappings on spheres, it worth notice that in virtue of Theorem 5.7 the answer for the problem is affirmative if the perturbations are '$p$-adically small' $\mathscr{B}$-functions (and even $\mathscr{A}$-functions), and not only '$p$-adically small' polynomials over $\mathbb{Z}_p$, as in the original statement of the problem: e.g., $x^\ell + \frac{1}{p}(x^p - x)^2$.

Also, with the use of the above mentioned criterion of ergodicity for $\mathscr{B}$-functions on $\mathbb{Z}_p$ (see Lemma 5.8) we immediately conclude that the following functions are ergodic on $\mathbb{Z}_p$: $ax + a^x$ with $a \in 1 + p\mathbb{Z}_p$, $1 + x + \frac{p^3}{1+px}$, $1 + x + p^3 \cdot (1 + px)^{\frac{1}{1+px}}$, etc.

Some important functions (e.g., the above mentioned compatible and integer-valued polynomials over $\mathbb{Q}_p$) do not lie in $\mathscr{B}$. However, they lie in a wider class $\mathscr{A}$: By the definition, $f \in \mathscr{A}$ if and only if $f$ is compatible and $p^n f \in \mathscr{B}$ for some non-negative rational integer $n$. Theorem 5.7, as well as the consequences it implies, remain true for $\mathscr{A}$-functions. Here are examples of $\mathscr{A}$-functions (which are not $\mathscr{B}$-functions) that are ergodic on all sufficiently small spheres around 0 ($\ell$ is the same as above): $\ell \cdot x + \ln_p(1 + p^2 x) + \frac{1}{p}(x^p - x)^2$ and $\frac{\ell \cdot x}{1+p^2 x} + \frac{1}{p}(x^p - x)^2$.

We have demonstrated also that all $\mathscr{A}$-functions (whence, all $\mathscr{B}$-functions) are locally analytic of order 1, in terminology of [22]. Within this context it would be interesting to study whether it is possible to expand Theorem 5.7 to the class of all compatible functions that are locally analytic of order $n$, $n = 1, 2, \ldots$.

## ACKNOWLEDGMENTS

I am grateful to Professor Andrei Khrennikov for (a number of!) fruitful discussions, and also for his hospitality during my stay at the University of Växjö. I am indebted to Professor Franco Vivaldi for his stimulating questions, and to Professor Igor Volovich for his interest to my area of research. Last, but not the least, I wold like to express my admire with Professor Branko Dragovich for his really great work of organizing this excellent conference, the 2$^{nd}$ International Conference on $p$-adic Mathematical Physics.

# REFERENCES

1. Y. Amice, Interpolation $p$-adique, *Bull. Soc. Math. France* **92**, 117–180 (1964).
2. V. Anashin, A. Bogdanov, and I. Kizhvatov, ABC: A New Fast Flexible Stream Cipher, Version 2, Available from `http://crypto.rsuh.ru/papers/abc-spec-v2.pdf`, (2005).
3. V.S. Anashin, Uniformly distributed sequences of $p$-adic integers. *Mathematical Notes* **55**, no. 2, 109–133, (1994).
4. V.S. Anashin, Uniformly distributed sequences in computer algebra, or how to constuct program generators of random numbers, *J. Math. Sci.* **89**, no. 4, 1355–1390, (1998).
5. V.S. Anashin, Uniformly distributed sequences of $p$-adic integers, II. *Discrete Math. Appl.* **12**, no. 6, 527–590, (2002), `math.NT/0209407`.
6. V.S. Anashin, On finite pseudorandom sequences, In *Kolmogorov and contemporary mathematics*, Russian Academy of Sciences, Moscow State University, Abstracts of the Int'l Conference, pp. 382–383 (2003).
7. V.S. Anashin, Pseudorandom number generation by $p$-adic ergodic transformations, `cs.CR/0401030`.
8. V.S. Anashin, Pseudorandom number generation by $p$-adic ergodic transformations: An addendum, `cs.CR/0402060`.
9. V.S. Anashin, Uniformly distributed sequences over $p$-adic integers, In I. Shparlinsky A. J. van der Poorten and H. G. Zimmer, editors, *Number theoretic and algebraic methods in computer science, Proceedings of the Int'l Conference (Moscow, June–July, 1993)*, World Scientific, pp. 1–18 (1995).
10. V. Anashin, A. Bogdanov, and I. Kizhvatov, Increasing the ABC Stream Cipher Period, Technical report, ECRYPT, (2005), `http://www.ecrypt.eu.org/stream/papersdir/050.pdf`.
11. J. Bryk and C.E. Silva, Measurable dynamics of simple $p$-adic polynomials, *Amer. Math. Monthly* **112**, no. 3, 212–232, (2005).
12. Z. Coelho and W. Parry, Ergodicity of p-adic multiplications and the distribution of Fibonacci numbers, *Topology, Ergodic Theory, Real Algebraic Geometry*, number 2 in Amer. Math. Soc. Transl. Ser. 2, American Mathematical Society, Providence, pp. 51–70 (2001).
13. D.L. Desjardins and M.E. Zieve, On the structure of polynomial mappings modulo an odd prime power, `math.NT/0103046`.
14. M. Gundlach, A. Khrennikov, and K.O. Lindahl, Ergodicity on $p$-adic sphere, in *German Open Conference on Probability and Statistics*, March 21–24, University of Hamburg, 61 (2000).
15. A. Khrennikov and K.O. Lindahl, On ergodic behavior of $p$-adic dynamical systems, *Infinite Dimensional Analysis, Quantum Probability and Related Topics* **4**, no. 4, 569–577 (2001).
16. A.Yu. Khrennikov, K.O. Lindahl, and M. Gundlach, Ergodicity in the $p$-adic framework, S. Albeverio, N. Elander, W.N. Everitt, and P. Kurasov, editors, *Operator Methods in Ordinary and Partial Differential Equations (S.Kovalevski Symproium, Univ. of Stockholm, June 2000)*, Operator Methods: Advances and Applications Vol **132**, Birkhäuser, Basel-Boston-Berlin, 2002.
17. A.Yu. Khrennikov and M. Nilsson, *p-adic deterministic and random dynamics*, Kluver Academic Publ., Dordrecht etc., 2004.
18. D. Knuth *The Art of Computer Programming*, Vol **2**, Seminumerical Algorithms, Addison-Wesley, Third edition, 1998.
19. M.V. Larin, Transitive polynomial transformations of residue class rings, *Discrete Mathematics and Applications* **12**, no. 2, 141–154 (2002).
20. K. Mahler, *p-adic numbers and their functions*, Cambridge Univ. Press (2nd edition), 1981.
21. R. Oselies and H. Zieschang, Ergodische Eigenschaften der Automorphismen $p$-adischer Zahlen, *Arch. Math.* **26**, 144–153 (1975).
22. W.H. Schikhof, *Ultrametric calculus*, Cambridge University Press, 1984.

# $p$-Adic and Adelic Cosmology: $p$-Adic Origin of Dark Energy and Dark Matter

Branko Dragovich

*Institute of Physics, P.O.Box 57, 11001 Belgrade, SERBIA AND MONTENEGRO*
*email:* dragovich@phy.bg.ac.yu

**Abstract.** A brief review of $p$-adic and adelic cosmology is presented. In particular, $p$-adic and adelic aspects of gravity, classical cosmology, quantum mechanics, quantum cosmology and the wave function of the universe are considered.
$p$-Adic worlds made of $p$-adic matters, which are different from real world of ordinary matter, are introduced. Real world and $p$-adic worlds make the universe as a whole. $p$-Adic origin of the dark energy and dark matter are proposed and discussed.

**Keywords:** $p$-Adic Cosmology, Adelic Cosmology, Cosmic Acceleration, Dark Matter, Dark Energy, $p$-Adic Matter, $p$-Adic Worlds, Adelic Universe.
**PACS:** 02., 98.80.-k, 04.60.-m, 95.36.+x .

## 1. INTRODUCTION

Cosmology in its own right is a science devoted to the universe as a whole. It is based on the cosmological observational data and fundamental physical theories (especially: general theory of relativity, quantum theory and theory of elementary particles). So far many significant results have been obtained and the Standard Cosmological Model is established (see, e.g. [1]). According to this model at the very beginning the universe was very small, dense, hot and started to expand. This initial period of evolution should be described by quantum cosmology. According to general theory of relativity the universe is a (pseudo)Riemannian space, which is presently flat. From observational data follows that the universe has permanently expanded during all its history which is about 14 billion years. Since 1998, there are a lot of evidence that the universe is now in the stage of an accelerated expansion which began a few billion years ago (for a review, see [2, 3]). To explain this acceleration many models have been proposed but no one of them was generally accepted. A very natural and the most attractive is the approach with dark energy, which is a matter with a negative pressure and uniformly distributed in the space. About 70% of all energy content of the universe is related to the dark energy, while about 26% belongs to the dark matter and only 4% is made of baryons (protons and neutrons). However the nature of the dark energy is a big mystery, which presents one of the greatest problems and challenges of contemporary cosmology and theory of elementary particles.

It is well known that results of observational measurements, as well as of the experimental ones, are elements of the field $\mathbb{Q}$ of rational numbers. However theoretical models are constructed not over $\mathbb{Q}$ but traditionally over $\mathbb{R}$ (field of real numbers) or $\mathbb{C}$ (field of complex numbers), where $\mathbb{R}$ is completion of $\mathbb{Q}$ with respect to distance induced by the ordinary absolute value $|\cdot|_\infty$ and $\mathbb{C}$ is algebraic extension of $\mathbb{R}$. Thus it is worth noting that mathematical modeling of physical systems requires not only algebraic structure of $\mathbb{Q}$ but also its geometric properties, which are related to the possible norms on $\mathbb{Q}$. Besides $|\cdot|_\infty$ there are also infinitely many $p$-adic norms $|\cdot|_p$, i.e. there is one nontrivial and inequivalent norm for each prime number $p$ [4, 5].

Measurements represent comparison of a given quantity with respect to a fixed quantity of the same nature taken to be the unit one. We live in the world where our measurements are inherently connected with the (decimal) expansions of the real numbers. Namely, in the process of measurement one determines a finite number of digits in the decimal expansion. This number of digits can be enlarged using more precise tools, while all other digits remain hidden within measuring errors. According to the classical mechanics there are no in principle physical restrictions to measure all quantities of a system with arbitrary accuracy. However in quantum mechanics there is restriction as a consequence of the well known uncertainty relation $\Delta x \Delta k \geq \hbar/2$. When quantum mechanics is combined with gravity then there obtains strong restriction on measurement of very small distances in the form

$$\Delta x \geq \ell_0 = \sqrt{\frac{G\hbar}{c^3}} \sim 10^{-33}\,\text{cm}, \tag{1}$$

where $\ell_0$ is the Planck length. It follows that there is quite firm restriction to measure spatial distances with arbitrary accuracy. In other words, it is not possible to determine all digits related to positions in the space. Thus measurements of physical quantities up to the Planck scale give rational numbers with geometrical properties which can be described by the usual absolute value. Hence it has been natural to use real numbers and complex numbers to describe physical phenomena in the explored domain of the universe. It is not strange that just real analysis is used to describe these real data. However for a more profound physical theory there is a sense to employ also $p$-adic numbers [4, 6], which are completions of $\mathbb{Q}$ with respect to the $p$-adic norms.

If we cannot get as a result of direct measurement a rational number with $p$-adic norm properties, it does not mean that there is not any content of the universe which natural description is just by $p$-adic numbers. Suppose that such $p$-adic systems exist. Then real number, as a result of measurement, is a real measure of interaction between $p$-adic and real system, to which belong measuring instruments. In such case we have to use analysis with $p$-adic valued functions of $p$-adic argument to describe $p$-adic system itself and also to use analysis with real (complex) valued functions of $p$-adic argument to describe $p$-adic system from the real point of view. This is slightly similar to the employment of complex functions in quantum mechanics, where wave function $\psi(x) \in \mathbb{C}$ contains complete information on quantum system, but is not measurable quantity, while $|\psi(x)|^2 \in \mathbb{R}$ is related to the probability distribution which can be measured. Let us use terms real and $p$-adic to denote those aspects of the universe which can be naturally described by real and $p$-adic numbers, respectively. We conjecture here that the visible and dark sides of the universe are real and $p$-adic ones, respectively.

$p$-Adic strings were introduced in 1987 [7] and a nice adelic formula was obtained [8]. An effective theory with real numbers of $p$-adic strings was constructed [9, 10] and shown its importance in the context of the tachyon condensation [11]. Application of $p$-adic numbers in construction of various models for the first five years was presented in [12] and [6]. Current activity is reflected in the proceedings of the international conferences on $p$-adic mathematical physics [13, 14].

In Section 2 we give an introductory review of adeles as well as of $p$-adic and adelic analogs of classical cosmology. Sec. 3 is devoted to $p$-adic and adelic quantum mechanics and their employment in quantum cosmology. $p$-Adic matter and its cosmological aspects are considered in Sec. 4.

## 2. $p$-ADIC AND ADELIC CLASSICAL COSMOLOGY

Adelic classical cosmology is a generalization of the ordinary one (over real numbers) in such way that it employs all (real and $p$-adic) completions of the set of rational numbers $\mathbb{Q}$. It uses analysis based on adelic valued functions of adelic valued arguments.

Let us remind some basic properties of adeles and adelic valued functions. To consider real and $p$-adic numbers simultaneously and on equal footing one uses concept of adeles. An adele $x$ (see, e.g. [5]) is an infinite sequence

$$x = (x_\infty, x_2, \cdots, x_p, \cdots), \quad x_\infty \in \mathbb{R}, \; x_p \in \mathbb{Q}_p \tag{2}$$

with the restriction that for all but a finite set $\mathscr{P}$ of primes $p$ must be $x_p \in \mathbb{Z}_p$, where $\mathbb{Z}_p = \{y \in \mathbb{Q}_p | |y|_p \leq 1\}$ is the ring of $p$-adic integers. Componentwise addition and multiplication endow the ring structure to the set of all adeles $\mathbb{A}$, which is the union of direct products in the following form:

$$\mathbb{A} = \bigcup_{\mathscr{P}} \mathbb{A}(\mathscr{P}), \quad \mathbb{A}(\mathscr{P}) = \mathbb{R} \times \prod_{p \in \mathscr{P}} \mathbb{Q}_p \times \prod_{p \notin \mathscr{P}} \mathbb{Z}_p. \tag{3}$$

A multiplicative group of ideles $\mathbb{I}$ is a subset of $\mathbb{A}$ with elements $x = (x_\infty, x_2, \cdots, x_p, \cdots)$, where $x_\infty \in \mathbb{R}^* = \mathbb{R} \setminus \{0\}$ and $x_p \in \mathbb{Q}_p^* = \mathbb{Q}_p \setminus \{0\}$ with the restriction that for all but a finite set $\mathscr{P}$ one has $x_p \in \mathbb{U}_p$, where $\mathbb{U}_p = \{y \in \mathbb{Q}_p | |y|_p = 1\}$ is the multiplicative group of $p$-adic units. Thus the whole set of ideles is

$$\mathbb{I} = \bigcup_{\mathscr{P}} \mathbb{I}(\mathscr{P}), \quad \mathbb{I}(\mathscr{P}) = \mathbb{R}^* \times \prod_{p \in \mathscr{P}} \mathbb{Q}_p^* \times \prod_{p \notin \mathscr{P}} \mathbb{U}_p. \tag{4}$$

A principal adele (idele) is a sequence $(x, x, \cdots, x, \cdots) \in \mathbb{A}$, where $x \in \mathbb{Q}$ ($x \in \mathbb{Q}^* = \mathbb{Q} \setminus \{0\}$). $\mathbb{Q}$ and $\mathbb{Q}^*$ are naturally embedded in $\mathbb{A}$ and $\mathbb{I}$, respectively.

Let us define an ordering on the set $\mathbb{P}$, which consists of all finite sets $\mathscr{P}_i$ of primes $p$, by $\mathscr{P}_1 \prec \mathscr{P}_2$ if $\mathscr{P}_1 \subset \mathscr{P}_2$. It is evident that $\mathbb{A}(\mathscr{P}_1) \subset \mathbb{A}(\mathscr{P}_2)$ when $\mathscr{P}_1 \prec \mathscr{P}_2$. Spaces $\mathbb{A}(\mathscr{P})$ have natural Tikhonov topology and adelic topology in $\mathbb{A}$ is introduced by inductive limit: $\mathbb{A} = \lim \text{ind}_{\mathscr{P} \in \mathbb{P}} \mathbb{A}(\mathscr{P})$. A basis of adelic topology is a collection of open sets of the form $W(\mathscr{P}) = \mathbb{V}_\infty \times \prod_{p \in \mathscr{P}} \mathbb{V}_p \times \prod_{p \notin \mathscr{P}} \mathbb{Z}_p$, where $\mathbb{V}_\infty$ and $\mathbb{V}_p$ are open sets in $\mathbb{R}$ and $\mathbb{Q}_p$, respectively. Note that adelic topology is finer than the corresponding Tikhonov topology. A sequence of adeles $a^{(n)} \in \mathbb{A}$ converges to an adele $a \in \mathbb{A}$ if $(i)$ it converges to $a$ componentwise and $(ii)$ if there exist a positive integer $N$ and a set $\mathscr{P}$ such that $a^{(n)}, a \in \mathbb{A}(\mathscr{P})$ when $n \geq N$. In the analogous way, these assertions hold also for idelic spaces $\mathbb{I}(\mathscr{P})$ and $\mathbb{I}$.

Adelic valued functions of adelic arguments are maps $F_\mathbb{A} : U_\mathbb{A} \to V_\mathbb{A}$, where $U_\mathbb{A} \subset \mathbb{A}^n$, $V_\mathbb{A} \subset \mathbb{A}^m$ and have the form

$$F_\mathbb{A}(x) = (f_\infty(x_\infty), f_2(x_2), \cdots, f_p(x_p), \cdots), \quad f_\infty \in \mathbb{R}^m, \ f_p \in \mathbb{Q}_p^m, \qquad (5)$$

where for all but $p \in \mathscr{P}$ one has to satisfy $|f_p(x_p)|_p \leq 1$.

When $F_\mathbb{A}(x)$ is related to the same physical system it is natural to expect that $v$-adic $(v = \infty, 2, \cdots, p, \cdots)$ functions $f_v(x_v)$ have the same form of dependence on $x_v$. As an illustrative example one can take adelic valued exponential function

$$\exp_\mathbb{A} x = (\exp_\infty x_\infty, \exp_2 x_2, \cdots, \exp_p x_p, \cdots), \qquad (6)$$

where $v$-adic exponential functions are defined by the usual power series expansion

$$\exp_v x_v = \sum_{n=0}^\infty \frac{x_v^n}{n!}, \quad |x_p|_p \leq |2p|_p, \quad |\exp_p x_p|_p = 1. \qquad (7)$$

Similar situation is for functions $\sin_\mathbb{A} x, \cos_\mathbb{A} x, \sinh_\mathbb{A} x, \cosh_\mathbb{A} x$ and many other functions given by power series expansions with rational coefficients.

The Einstein gravitational field equations

$$R_{\mu\nu} - \frac{1}{2} R g_{\mu\nu} = \kappa T_{\mu\nu} - \Lambda g_{\mu\nu} \qquad (8)$$

can be also considered as $p$-adic ones if we take constants $\kappa = \frac{8\pi G}{c^4}$ and $\Lambda$ to be rational numbers. By this way the Einstein equations (8) become number field invariant and therefore more fundamental. A successful systematic study of $p$-adic gravity, especially $p$-adic differential geometry and gravitational field equations, started in 1991 [15].

For the sequel it is useful to introduce $\pi G = \bar{G}$ and the new Planck quantities

$$L_0 = \sqrt{\frac{h\bar{G}}{c^3}} = \sqrt{2}\pi \ell_0, \quad T_0 = \frac{L_0}{c} = \sqrt{2}\pi t_0, \quad M_0 = \sqrt{\frac{hc}{\bar{G}}} = \sqrt{2} m_0, \qquad (9)$$

where $\ell_0, t_0$ and $m_0$ are usual Planck length, time and mass, respectively. Let us take for the natural system of units these $L_0, T_0$ and $M_0$ instead of the standard $\ell_0, t_0$ and $m_0$. Then $\bar{G}, c$ and $h$ are rational numbers with unit values.

Constructing cosmological models it is useful to maximally exploit symmetrical properties of the universe. As a consequence one obtains dynamical system called minisuperspace cosmological model, which contains finite number degrees of freedom. Minisuperspace model may be regarded as a classical system given by Hamiltonian $H(q^i, k^i, t)$

or Lagrangian $L(q^i, \dot{q}^i, t)$, where $i = 1, 2, \cdots, n$. The corresponding adelic Hamiltonian is

$$H_{\mathbb{A}}(q^i, k^i, t) = (H_{\infty}(q^i_{\infty}, k^i_{\infty}, t_{\infty}), H_2(q^i_2, k^i_2, t_2), \cdots, H_p(q^i_p, k^i_p, t_p), \cdots), \quad (10)$$

where $H_{\infty} \in \mathbb{R}$ and $H_p \in \mathbb{Q}_p$ with restriction that $H_p \in \mathbb{Z}_p$ for all but $p \in \mathscr{P}$. Analogously one defines an adelic Lagrangian. $v$-Adic action of the minisuperspace model is

$$S_v[q] = \int_{t'}^{t''} dt \, N \left[ \frac{1}{2N^2} f_{\alpha\beta}(q) \dot{q}^{\alpha} \dot{q}^{\beta} - U(q) \right], \quad (11)$$

where $f_{\alpha\beta}$ is a metric on minisuperspace of some gravitational and matter field variables. The de Sitter minisuperspace cosmological model is a simple nontrivial, exactly solvable and instructive cosmological model. It is given by the Einstein-Hilbert action with cosmological constant $\Lambda$ ($c = 1$),

$$S[g_{\mu\nu}] = \frac{1}{16\pi G} \left[ \int_M d^4x \sqrt{-g}(R - 2\Lambda) + 2 \int_{\partial M} d^3x \sqrt{h} K \right] \quad (12)$$

and the Friedmann-Robertson-Walker (FRW) metric

$$ds^2 = -N^2 dt^2 + a^2(t) d\Omega_3^2, \quad (13)$$

where $N(t)$ is the lapse function and $d\Omega_3^2$ is the metric on the unit three-sphere. More complex models contain also additional action with matter fields. To simplify formalism and get quadratic Lagrangian one can take the Robertson-Walker metric in the form [17]

$$ds^2 = -\frac{N^2}{q(t)} dt^2 + q(t) d\Omega_3^2. \quad (14)$$

Using metric (14) in (12), one obtains

$$S[q] = \frac{1}{2} \int_{t'}^{t''} dt \, N \left( -\frac{\dot{q}^2}{4N^2} - \lambda q + 1 \right), \quad (15)$$

where $\lambda = \Lambda/3$. Choosing $N = 1$, the equation of motion is

$$\ddot{q} = 2\lambda. \quad (16)$$

Solution of the equation (16) which satisfies conditions $q'' = q(T)$ and $q' = q(0)$ is

$$q(t) = \lambda t^2 + \left( \frac{q'' - q'}{T} - \lambda T \right) t + q'. \quad (17)$$

Note that equations (16) and (17) are the same as for a particle with constant acceleration $a = 2\lambda$. The corresponding classical action (see, e.g. [16] and references therein) is

$$\bar{S}[q] = \frac{\lambda^2 T^3}{24} - [\lambda(q'' + q') - 2] \frac{T}{4} - \frac{(q'' - q')^2}{8T}. \quad (18)$$

The above consideration of the de Sitter model is performed for the real case but the expressions from (15) to (18) can be also regarded as $p$-adic valued.

Since parameter $\lambda = \frac{\Lambda}{3}$ has rational values we can write adelic Lagrangian in the form

$$L_{\mathbb{A}}(q,\dot{q}) = (L_\infty(q_\infty,\dot{q}_\infty), L_2(q_2,\dot{q}_2), \cdots, L_p(q_p,\dot{q}_p), \cdots), \tag{19}$$

where ($N = 1$)

$$L_v(q_v,\dot{q}_v) = \frac{1}{2}\left(-\frac{\dot{q}_v^2}{4} - \lambda q_v + 1\right). \tag{20}$$

It is evident that $|L_p(q_p,\dot{q}_p)|_p \leq 1$ when $|q_p|_p \leq 1$, $|\dot{q}_p|_p \leq 1$, $|\lambda|_p \leq 1$ and $p \neq 2$. Infinite sequence of actions

$$\bar{S}_{\mathbb{A}}[q] = (\bar{S}_\infty[q_\infty], \bar{S}_2[q_2], \cdots, \bar{S}_p[q_p], \cdots), \tag{21}$$

where $\bar{S}_v[q_v]$, ($v = \infty, 2, \cdots, p, \cdots$) are given by (18), becomes adelic if $T$ is principal idele and $q''$, $q'$ are adeles.

For some other $p$-adic and adelic quadratic classical cosmological models see [16].

## 3. $p$-ADIC AND ADELIC QUANTUM COSMOLOGY

$p$-Adic and adelic quantum cosmology [15, 18, 19, 16] is an appropriate application of $p$-adic [20] and adelic quantum mechanics [21, 22, 23] to the universe as a whole at its very early stage of evolution. Let us now first recall the basic properties of $p$-adic and adelic quantum mechanics.

### 3.1. $p$-Adic and Adelic Quantum Mechanics

It is remarkable that ordinary quantum mechanics on a real space can be generalized to quantum mechanics on $p$-adic spaces for any prime number $p$. There are two main approaches: with complex-valued [20] and $p$-adic valued [24] elements of the Hilbert space on $\mathbb{Q}_p^n$. $p$-Adic quantum mechanics with complex-valued wave functions is more suitable for connection with ordinary quantum mechanics, and in the sequel we will briefly review only this kind of $p$-adic quantum mechanics.

When wave functions are complex-valued and arguments are $p$-adic valued, one cannot construct a direct analog of the Schrödinger equation with a $p$-adic version of Heisenberg algebra. According to the Weyl quantization, canonical noncommutativity in $p$-adic case can be introduced by operators ($h = 1$)

$$\hat{Q}_p(\alpha)\psi_p(x) = \chi_p(-\alpha x)\psi_p(x), \quad \hat{K}_p(\beta)\psi_p(x) = \psi_p(x+\beta) \tag{22}$$

which satisfy

$$\hat{Q}_p(\alpha)\hat{K}_p(\beta) = \chi_p(\alpha\beta)\hat{K}_p(\beta)\hat{Q}_p(\alpha), \tag{23}$$

where $\chi_p(u) = \exp(2\pi i\{u\}_p)$ is additive character on the field $\mathbb{Q}_p$ and $\{u\}_p$ is the fractional part of $u \in \mathbb{Q}_p$.

Let $\hat{x}$ and $\hat{k}$ be operators of position $x$ and momentum $k$, respectively. Let us define operators $\chi_v(\alpha\hat{x})$ and $\chi_v(\beta\hat{k})$ by formulas

$$\chi_v(\alpha\hat{x})\chi_v(ax) = \chi_v(\alpha x)\chi_v(ax), \quad \chi_v(\beta\hat{k})\chi_v(bk) = \chi_v(\beta k)\chi_v(bk), \quad (24)$$

where index $v$ denotes real and any $p$-adic case, and $\chi_\infty(u) = \exp(-2\pi i u)$. These operators also act on a function $\psi_v(x)$, which has the Fourier transform $\tilde{\psi}(k)$, in the following way:

$$\chi_v(-\alpha\hat{x})\psi_v(x) = \chi_v(-\alpha\hat{x})\int \chi_v(-kx)\tilde{\psi}(k)d^n k = \chi_v(-\alpha x)\psi_v(x), \quad (25)$$

$$\chi_v(-\beta\hat{k})\psi_v(x) = \int \chi_v(-\beta k)\chi_v(-kx)\tilde{\psi}(k)d^n k = \psi_v(x+\beta), \quad (26)$$

where integration in $p$-adic case is with respect to the Haar measure $dk$ with the properties: $d(k+a) = dk$, $d(ak) = |a|_p dk$ and $\int_{|k|_p \leq 1} dk = 1$. Comparing (22) with (25) and (26) we conclude that $\hat{Q}_p(\alpha) = \chi_p(-\alpha\hat{x})$, $\hat{K}_p(\beta) = \chi_p(-\beta\hat{k})$. Instead of the Heisenberg relations one has

$$\chi_v(-\alpha_i\hat{x}_i)\chi_v(-\beta_j\hat{k}_j) = \chi_v(\alpha_i\beta_j\delta_{ij})\chi_v(-\beta_j\hat{k}_j)\chi_v(-\alpha_i\hat{x}_i), \quad (27)$$

$$\chi_v(-\alpha_i\hat{x}_i)\chi_v(-\alpha_j\hat{x}_j) = \chi_v(-\alpha_j\hat{x}_j)\chi_v(-\alpha_i\hat{x}_i), \quad (28)$$

$$\chi_v(-\beta_i\hat{k}_i)\chi_v(-\beta_j\hat{k}_j) = \chi_v(-\beta_j\hat{k}_j)\chi_v(-\beta_i\hat{k}_i). \quad (29)$$

One can introduce the Weyl operator

$$W_v(\alpha\hat{x},\beta\hat{k}) = \chi_v(\frac{1}{2}\alpha\beta)\chi_v(-\beta\hat{k})\chi_v(-\alpha\hat{x}), \quad (30)$$

which satisfies relation

$$W_v(\alpha\hat{x},\beta\hat{k})W_v(\alpha'\hat{x},\beta'\hat{k}) = \chi_v(\frac{1}{2}(\alpha\beta' - \alpha'\beta))W_v((\alpha+\alpha')\hat{x},(\beta+\beta')\hat{k}) \quad (31)$$

and is a unitary representation of the Heisenberg-Weyl group. Using $W_v(\alpha\hat{x},\beta\hat{k})$ one obtains generalized Weyl formula for quantization

$$\hat{f}_v(\hat{k},\hat{x}) = \int W_v(\alpha\hat{x},\beta\hat{k})\tilde{f}_v(\alpha,\beta)d^n\alpha d^n\beta. \quad (32)$$

As a basic instrument to treat dynamics of a $p$-adic quantum model is natural to take the kernel $\mathscr{K}_p(x'',t'';x',t')$ of the evolution operator $U_p(t'',t')$. This kernel obtains by generalization of its real analog, i.e.

$$\psi_v(x'',t'') = \int \mathscr{K}_v(x'',t'';x',t')\psi_v(x',t')d^n x', \quad (33)$$

where $\mathscr{K}_v(x'',t'';x',t')$ for quadratic Lagrangians can be defined by path integral

$$\mathscr{K}_v(x'',t'';x',t') = \int_{(x',t')}^{(x'',t'')} \chi_v\left(-\int_{t'}^{t''} L(\dot{q},q,t)\,dt\right) \mathscr{D}_v q. \tag{34}$$

In the Vladimirov-Volovich formulation [20], $p$-adic quantum mechanics is a triple

$$(L_2(\mathbb{Q}_p), W_p(z), U_p(t)), \tag{35}$$

where $W_p(z)$ corresponds to $W_p(\alpha\hat{x},\beta\hat{k})$ defined in (30).

Adelic quantum mechanics [18] is a natural generalization of the above formulation of ordinary and $p$-adic quantum mechanics: $(L_2(\mathbb{A}), W_\mathbb{A}(z), U_\mathbb{A}(t))$.

In complex-valued adelic analysis it is worth mentioning an additive character

$$\chi_\mathbb{A}(x) = \chi_\infty(x_\infty) \prod_p \chi_p(x_p), \tag{36}$$

a multiplicative character

$$|x|_\mathbb{A}^s = |x_\infty|_\infty^s \prod_p |x_p|_p^s, \ s \in \mathbb{C}, \tag{37}$$

and elementary functions of the form

$$\varphi_\mathscr{P}(x) = \varphi_\infty(x_\infty) \prod_{p \in \mathscr{P}} \varphi_p(x_p) \prod_{p \notin \mathscr{P}} \Omega(|x_p|_p), \tag{38}$$

where $\varphi_\infty(x_\infty)$ is an infinitely differentiable function on $\mathbb{R}$ and $|x_\infty|_\infty^n \varphi_\infty(x_\infty) \to 0$ as $|x_\infty|_\infty \to \infty$ for any $n \in \{0,1,2,\cdots\}$, $\varphi_p(x_p)$ are some locally constant functions with compact support, and

$$\Omega(|x_p|_p) = \begin{cases} 1, & |x_p|_p \leq 1, \\ 0, & |x_p|_p > 1. \end{cases} \tag{39}$$

All finite linear combinations of elementary functions (38) make the set $\mathscr{S}(\mathbb{A})$ of the Schwartz-Bruhat adelic functions. The Fourier transform of $\varphi(x) \in \mathscr{S}(\mathbb{A})$, which maps $\mathscr{S}(\mathbb{A})$ onto $\mathscr{S}(\mathbb{A})$, is

$$\tilde{\varphi}(y) = \int_\mathbb{A} \varphi(x)\chi_\mathbb{A}(xy)\,dx, \tag{40}$$

where $\chi_\mathbb{A}(xy)$ is defined by (36) and $dx = dx_\infty dx_2 dx_3 \cdots$ is the Haar measure on $\mathbb{A}$.

A basis of $L_2(\mathbb{A}(\mathscr{P}))$ may be given by the corresponding orthonormal eigenfunctions in a spectral problem of the evolution operator $U_\mathbb{A}(t)$, where $t \in \mathbb{A}$. Such eigenfunctions have the form

$$\psi_\mathscr{P}(x,t) = \psi_\infty(x_\infty, t_\infty) \prod_{p \in \mathscr{P}} \psi_p(x_p, t_p) \prod_{p \notin \mathscr{P}} \Omega(|x_p|_p), \tag{41}$$

where $\psi_\infty \in L_2(\mathbb{R})$ and $\psi_p \in L_2(\mathbb{Q}_p)$ are eigenfunctions in ordinary and $p$-adic cases, respectively. $\Omega(|x_p|_p)$ is an element of $L_2(\mathbb{Q}_p)$, defined by (39), which is invariant under transformation of an evolution operator $U_p(t_p)$ and provides convergence of the infinite product (41). Elements of $L_2(\mathbb{A})$ may be regarded as superpositions $\psi(x) = \sum_{\mathscr{P}} C(\mathscr{P}) \psi_{\mathscr{P}}(x)$, where $\psi_{\mathscr{P}}(x) \in L_2(\mathbb{A}(\mathscr{P}))$ (41) and $\sum_{\mathscr{P}} |C(\mathscr{P})|_\infty^2 = 1$.

Theory of $p$-adic generalized functions is presented in [6] and a theory of generalized functions on adelic spaces is in progress [25].

Adelic evolution operator $U_\mathbb{A}(t)$ is defined by

$$U_\mathbb{A}(t'') \psi(x'') = \int_\mathbb{A} \mathscr{K}_\mathbb{A}(x'',t'';x',t') \psi(x',t') dx'$$
$$= \prod_v \int_{\mathbb{Q}_v} \mathscr{K}_v(x_v'',t_v'';x_v',t_v') \psi_v(x_v',t_v') dx_v'. \quad (42)$$

The eigenvalue problem for $U_\mathbb{A}(t)$ reads

$$U_\mathbb{A}(t) \psi_{\mathscr{P}}(x) = \chi_\mathbb{A}(E_\alpha t) \psi_{\mathscr{P}}(x), \quad (43)$$

where $\psi_{\mathscr{P}}(x)$ are adelic eigenfunctions (41), and $E_\alpha = (E_\infty, E_2, ..., E_p, ...)$ is the corresponding adelic energy.

Adelic quantum mechanics takes into account ordinary as well as $p$-adic quantum effects and may be regarded as a starting point for construction of a more complete quantum cosmology [18], quantum field theory [26] and string/M-theory [27, 28]. In the limit of large distances adelic quantum mechanics effectively becomes the ordinary one [29].

Evaluation of $v$-adic kernel $\mathscr{K}_v(x'',t'';x',t')$ of the unitary evolution operator for one-dimensional systems with quadratic Lagrangians has the form [30, 31, 32]

$$\mathscr{K}_v(x'',t'';x',t') = \lambda_v \left(-\frac{1}{2h} \frac{\partial^2}{\partial x'' \partial x'} \bar{S}(x'',t'';x',t')\right)$$
$$\times \left|\frac{1}{h} \frac{\partial^2}{\partial x'' \partial x'} \bar{S}(x'',t'';x',t')\right|_v^{\frac{1}{2}} \chi_v\left(-\frac{1}{h} \bar{S}(x'',t'';x',t')\right), \quad (44)$$

where $\lambda_v$-functions are presented in [6].

## 3.2. $p$-Adic and adelic wave functions of the universe

The universe should be a quantum system, especially at the very beginning of its evolution. The main task of quantum cosmology is to provide formalism to specify quantum states and describe quantum dynamics of the universe as a whole. The quantum state of the universe at the beginning is of central importance, since it determines initial condition for its later behavior. Considering the universe as a quantum-mechanical system it has a quantum state which is encoded in the corresponding wave function.

This wave function is complex-valued and depends on some real quantities in standard approach. To include $p$-adic effects one has to reconsider its formulation. We maintain here the standard point of view that the wave function takes complex values, but we treat its arguments (space-time coordinates, gravitational and matter fields) to be not only real but also $p$-adic and adelic.

There is not $p$-adic generalization of the Wheeler - De Witt equation for cosmological models. Instead of differential approach, Feynman's path integral method in the Hartle-Hawking approach [33] was exploited [18] and minisuperspace cosmological models are also investigated by means of adelic quantum mechanics [16].

$p$-Adic and adelic minisuperspace quantum cosmology is an application of $p$-adic and adelic quantum mechanics to the cosmological models, respectively. In the path integral approach to standard quantum cosmology, the starting point is Feynman's path integral method. The amplitude to go from one state with intrinsic metric $h'_{ij}$ and matter configuration $\phi'$ on an initial hypersurface $\Sigma'$ to another state with metric $h''_{ij}$ and matter configuration $\phi''$ on a final hypersurface $\Sigma''$ is given by the path integral

$$\mathcal{K}_\infty(h''_{ij}, \phi'', \Sigma''; h'_{ij}, \phi', \Sigma') = \int \chi_\infty(-S_\infty[g_{\mu\nu}, \Phi]) \mathscr{D}_\infty g_{\mu\nu} \mathscr{D}_\infty \Phi \quad (45)$$

over all four-geometries $g_{\mu\nu}$ and matter configurations $\Phi$, which interpolate between the initial and final configurations. In (45) $S_\infty[g_{\mu\nu}, \Phi]$ is an Einstein-Hilbert action for the gravitational and matter fields. To perform $p$-adic and adelic generalization we make first $p$-adic counterpart of the action using form-invariance under change of real to the $p$-adic number fields. Then we generalize (45) and introduce $p$-adic complex-valued cosmological amplitude

$$\mathcal{K}_p(h''_{ij}, \phi'', \Sigma''; h'_{ij}, \phi', \Sigma') = \int \chi_p(-S_p[g_{\mu\nu}, \Phi]) \mathscr{D}_p g_{\mu\nu} \mathscr{D}_p \Phi. \quad (46)$$

The standard minisuperspace ground-state wave function in the Hartle-Hawking (no-boundary) proposal [33] is defined by functional integration in the Euclidean version of

$$\psi_\infty[h_{ij}] = \int \chi_\infty(-S_\infty[g_{\mu\nu}, \Phi]) \mathscr{D}_\infty g_{\mu\nu} \mathscr{D}_\infty \Phi, \quad (47)$$

over all compact four-geometries $g_{\mu\nu}$ which induce $h_{ij}$ at the compact three-manifold. This three-manifold is the only boundary of the all four-manifolds. Extending Hartle-Hawking proposal to the $p$-adic minisuperspace [15], an adelic Hartle-Hawking wave function is the infinite product

$$\psi_\mathbb{A}(q) = \prod_v \int \chi_v(-S_v[g_{\mu\nu}, \Phi]) \mathscr{D}_v g_{\mu\nu} \mathscr{D}_v \Phi, \quad (48)$$

where path integration must be performed over both, Archimedean and non-Archimedean geometries. If an evaluation of the corresponding functional integrals for a minisuperspace model yields $\psi(q_\alpha)$ in the form (41), then such cosmological model is a Hartle-Hawking adelic one.

Before to proceed in the above way it is worth mentioning another approach [15] which was the first one in *p*-adic quantum cosmology. The essence of this approach consists in the following *p*-adic proposal for the Hartle-Hawking type of the wave function:

$$\psi_\infty(q) = \sum_{a.m.} \prod_p \int \chi_p(-S_p[g_{\mu\nu}, \Phi]) \, \mathcal{D}_p g_{\mu\nu} \, \mathcal{D}_p \Phi, \qquad (49)$$

where summation is over algebraic manifolds. This proposal was illustrated on the above de Sitter model with the Euclidean version of $ds^2$ in (14).

It is shown [18] that the de Sitter minisuperspace model in $D = 4$ space-time dimensions is the Hartle-Hawking adelic one. Namely, according to the Hartle-Hawking proposal one has

$$\psi_\nu(q) = \int \mathcal{K}_\nu(q, T; 0, 0) \, dT, \quad \nu = \infty, 2, 3, \cdots, p, \cdots, \qquad (50)$$

where

$$\mathcal{K}_\nu(q'', T; q', 0) = \lambda_\nu(-8T) |4T|_\nu^{-\frac{1}{2}} \chi_\nu \left[ -\frac{\lambda^2 T^3}{24} + (\lambda q - 2)\frac{T}{4} + \frac{q^2}{8T} \right] \qquad (51)$$

is the kernel of the $\nu$-adic evolution operator. The functions $\lambda_\nu(a)$ have the properties [6]

$$|\lambda_\nu(a)|_\nu = 1, \; \lambda_\nu(b^2 a) = \lambda_\nu(a), \lambda_\nu(a)\lambda_\nu(b) = \lambda_\nu(a+b) \lambda_\nu(ab(a+b)). \qquad (52)$$

Employing the *p*-adic Gauss integral [6]

$$\int_{\mathbb{Q}_p} \chi_p(\alpha x^2 + \beta x) \, dx = \lambda_p(\alpha) |2\alpha|_p^{-\frac{1}{2}} \chi_p\left(-\frac{\beta^2}{4\alpha}\right), \quad \alpha \neq 0, \qquad (53)$$

one can rewrite *p*-adic version of (50) in the form

$$\psi_p(q) = \int_{\mathbb{Q}_p} dx \, \chi_p(qx) \int DT \chi_p \left[ -\frac{\lambda^2 T^3}{24} + \left(\frac{\lambda q}{4} - \frac{1}{2} - 2x^2\right) T \right]. \qquad (54)$$

Taking the region of integration to be $|T|_p \leq 1$ one obtains

$$\psi_p(q) = \int_{\mathbb{Q}_p} dx \, \chi_p(qx) \, \Omega\left(\left|\frac{\lambda q}{4} - \frac{1}{2} - 2x^2\right|_p\right), \quad \left|\frac{\lambda^2}{24}\right|_p \leq 1. \qquad (55)$$

An evaluation of the integral (55) yields

$$\psi_p(q) = \exp(i\pi \, \delta^1_{|q|_2} \delta^2_p) \, \Omega(|q|_p), \quad \left|\frac{\lambda^2}{24}\right|_p \leq 1, \qquad (56)$$

where $\delta^b_a$ is the Kronecker symbol. $\psi_\infty(q_\infty)$ was explored in [17] and the result depends on the contour of integration and has an exact solution

$$\psi_\infty(q_\infty) = \exp\left(\frac{1}{3\lambda}\right) Ai\left(\frac{1 - \lambda q_\infty}{(2\lambda)^{\frac{2}{3}}}\right), \qquad (57)$$

where $Ai(x)$ is the Airy function. Finally we obtain an adelic wave function for the de Sitter cosmological model in the form

$$\psi_{\mathbb{A}}(q) = \psi_\infty(q_\infty) \prod_p \exp(i\pi \, \delta^1_{|q|_2} \, \delta^2_p) \, \Omega(|q_p|_p), \quad |\frac{\lambda^2}{24}|_p \leq 1. \tag{58}$$

It is shown in [19, 16] that $p$-adic and adelic generalization of the minisuperspace cosmological models can be successfully performed in the framework of $p$-adic and adelic quantum mechanics [21, 22] without use of the Hartle-Hawking approach. The following cosmological models are investigated [16]: the de Sitter model, model with a homogeneous scalar field, anisotropic Bianchi model with three scale factors and some two-dimensional minisuperspace models. The necessary condition that a system can be regarded as the adelic one is the existence of $p$-adic ground state $\Omega(|q_\alpha|_p)$ ($\alpha = 1, 2, \cdots, n$) in the way

$$\int_{|q'_\alpha|_p \leq 1} \mathscr{K}_p(q''_\alpha, T; q'_\alpha, 0) \, dq'_\alpha = \Omega(|q''_\alpha|_p) \tag{59}$$

for all $p$ but a finite set $\mathscr{P}$. For the case of de Sitter model one obtains

$$\psi_p(q) = \begin{cases} \Omega(|q|_p), & |T|_p \leq 1, \; |\frac{\lambda^2}{24}|_p \leq 1, \; p \neq 2, \\ \Omega(|q|_2), & |T|_2 \leq \frac{1}{2}, \; |\frac{\lambda^2}{24}|_2 \leq 1, \; p = 2, \end{cases} \tag{60}$$

what is in a good agreement with the result (58) obtained by the Hartle-Hawking proposal. Some other forms of the $p$-adic wave functions are also found [16], e.g. $\psi_p(q) = \Omega(p^\mu |q|_p)$ and $\psi_p(q) = \delta(|q|_p - p^\mu)$ where $\mu \in \mathbb{Z}$.

Let us now consider the interpretation of an adelic wave function. Note that there is a general problem of the usual interpretation of the wave function of the universe. Here we are going to discuss only adelic aspects. It is natural to investigate adelic wave function on the rational values of its arguments. Without loss of generality we can use the above de Sitter cosmological model. For the adelic ground state (58) the density of probability in rational points is

$$|\psi_{\mathbb{A}}(q)|_\infty^2 = |\psi_\infty(q)|_\infty^2 \prod_p \Omega(|q|_p) = \begin{cases} |\psi_\infty(q)|_\infty^2, & q \in \mathbb{Z}, \\ 0, & q \in \mathbb{Q} \setminus \mathbb{Z}. \end{cases} \tag{61}$$

When (the density of) probability is equal (or close) to unity or zero it can be regarded as a certain event. From (61) it follows that $q$-space is discrete, because the density probability is nonzero only at integers. Note that here spacing is related to the Planck length but in some other adelic quantum models it will correspond to the characteristic length of the system. If we integrate (58) over $|q_p|_p \leq 1$ for every $p$ then we obtain ordinary wave function. In the case of adelic wave function of the form (41) with $\mathscr{P} \neq \emptyset$, $q$-space has not a sharp discreteness but fuzzyness.

As a result of $p$-adic effects, adelic quantum systems contain discreteness of the space with spacing equal to the own characteristic length. This kind of discreteness was obtained for the first time in the context of the Hartle-Hawking adelic de Sitter quantum model [18]. At the distances much larger than the characteristic length instead of an adelic wave function it is enough to use only the ordinary one over real space.

## 4. REAL AND *p*-ADIC WORLDS. ADELIC UNIVERSE

### 4.1. On Adelic Models

Let us introduce two distinct types of adelic models: (*i*) principal adelic models and (*ii*) non-principal adelic models.

In principal adelic models all parameters and constants, which characterize physical system, are principal ideles, i.e. they are nonzero rational numbers which are the same in real and all *p*-adic counterparts. Adelic product formulas can exist in these models, which connect real and *p*-adic aspects of the same quantity. A simple and illustrative such formula is

$$|x|_\infty^c \prod_p |x|_p^c = 1, \quad c \in \mathbb{C}, \quad x \in \mathbb{Q}\setminus\{0\}, \tag{62}$$

which is adelic multiplicative character on principal ideles. Another simple example is adelic additive character on principal adeles, i.e.

$$\chi_\infty(x) \prod_p \chi_p(x) = 1, \quad x \in \mathbb{Q}. \tag{63}$$

Employing $|x|_v^{\alpha-1}$ and $\chi_v(x)$ in some integrals one can obtain new product formulas. For instance (see [34] and references therein)

$$\Gamma_\infty(\alpha) \prod_p \Gamma_p(\alpha) = 1, \quad \Gamma_v(\alpha) = \int_{\mathbb{Q}_v} |x|_v^{\alpha-1} \chi_v(x) d_v x, \quad \alpha \neq 0, 1, \tag{64}$$

$$B_\infty(a,b) \prod_p B_p(a,b) = 1, \quad B_v(a,b) = \int_{\mathbb{Q}_v} |x|_v^{a-1} |1-x|_v^{b-1} d_v x, \quad a+b+c = 1. \tag{65}$$

While the formulas (62) and (63) are valid only for rational numbers, (64) and (65) have place for real and complex numbers. The product formula (65) connects real and *p*-adic amplitudes for scattering of two open string tachyons [8]. As a result of (65), complicated real amplitude can be expressed as product of inverse *p*-adic counterparts which are elementary functions. There are also some product formulas which include all *p*-adic counterparts but not the real one. Such an example is

$$\prod_p \Omega(|x|_p) = \begin{cases} 1, & x \in \mathbb{Z}, \\ 0, & x \in \mathbb{Q}\setminus\mathbb{Z}. \end{cases} \tag{66}$$

We used (66) in (61).

Adelic quantum mechanics [21, 22] and its generalizations to quantum field theory [26] and string theory [27] belong to the principal adelic models.

In non-principal adelic models parameters and constants are adeles but not principal ones, i.e. they are not rational numbers the same for real and $p$-adic counterparts. A non-principal adelic constant (or parameter) is an adele

$$c_\mathbb{A} = (c_\infty, c_2, c_3, \cdots, c_p, \cdots), \tag{67}$$

where at least two of components $c_v$ are not the same rational number. There are infinitely many possibilities to fix all $c_v$ in (67). One of them is the case when $c_p = p$ for all primes $p$. Another interesting case is when all but a finite number of $c_v$ are equal zero. In particular, all but one of $c_v$ can be zero. If only $c_\infty \neq 0$, adelic model is reduced to the real one. Analogously if only $c_p \neq 0$ for a fixed $p$, we have a p-adic model. In the sequel we will pay some attention to the case when $c_\infty \neq 0$ and $c_p \neq 0$ for a finite set $\mathscr{P}$ of primes $p$.

## 4.2. $p$-Adic Matter, Dark Energy and Dark Matter

Applying classical Einstein equations (8) and the FRW metric (13) (with $N = 1$, $c = 1$) to the universe, one obtains

$$\frac{\ddot{a}}{a} = -\frac{4\pi G}{3}\sum_i(\rho_i + 3p_i) + \frac{\Lambda}{3}, \quad \left(\frac{\dot{a}}{a}\right)^2 + \frac{k}{a^2} = \frac{8\pi G}{3}\sum_i \rho_i + \frac{\Lambda}{3}, \tag{68}$$

where $\rho_i$ and $p_i$ are $i$-th component of energy and pressure density, respectively, and $a$ is cosmological scale factor. The equation of state is $p_i = w_i \rho_i$. Since $\Lambda$ can be regarded as an energy density we can rewrite equations (68) in the following form

$$\frac{\ddot{a}}{a} = -\frac{4\pi G}{3}\sum_i(\rho_i + 3p_i), \tag{69}$$

$$\left(\frac{\dot{a}}{a}\right)^2 + \frac{k}{a^2} = \frac{8\pi G}{3}\sum_i \rho_i. \tag{70}$$

Recent observations (see reviews [2, 3]) show that

$$\Omega = \sum_i \frac{\rho_i}{\rho_{cr}} = \sum_i \Omega_i \approx 1, \quad \Omega_B \approx 0.04, \quad \Omega_{DM} \approx 0.26, \quad \Omega_{DE} \approx 0.70, \tag{71}$$

where $\Omega_B$, $\Omega_{DM}$ and $\Omega_{DE}$ are related to baryonic matter, dark matter and dark energy, respectively, and $\rho_{cr} = \frac{3}{8\pi G}H_0^2$ is critical energy density required for the universe to be spatially flat. $H_0 = \left(\frac{\dot{a}}{a}\right)_0$ is the rate of expansion at present time. While baryonic matter is partially luminous, the nature of dark matter and dark energy is presently unknown.

Dark matter was introduced (already in 1930's) to explain dynamics of galaxies in clusters as well as dynamics inside individual galaxies. It is nonbaryonic and clustered (attractive) matter. Many nonbaryonic particles (in particular, neutralino and axion) have been considered as candidates for dark matter.

Dark energy was introduced as an unclustered (repulsive) matter with negative pressure and responsible for accelerated expansion of the universe. Cosmic acceleration was discovered recently [35, 36] in observations of cosmologically distant Type Ia Supernovae (SNeIa) and independently verified by Cosmic Microwave Background and large scale structure investigations. Namely, to have an acceleration ($\ddot{a} > 0$), the RHS of eq. (69) has to be positive, i.e. $\sum_i(\rho_i + 3p_i) < 0$. Since $\rho_B + 3p_B > 0$ and $\rho_{DM} + 3p_{DM} > 0$ it follows that $\rho_{DE} + 3p_{DE} < 0$. Positivity of $\rho_{DE}$ yields $p_{DE} < -\frac{\rho_{DE}}{3} < 0$, i.e. $w < -1/3$. Current observational value of parameter $w$ (see [2] and references therein) is in the range $-1.61 < w < -0.78$.

There are many candidate models for the dark energy. The simplest model is the cosmological constant with $w = -1$ ($p = -\rho = const.$). Some other dynamical scalar field models are [2]: quintessence, braneworld, Chaplygin gas and Phantom dark energy (with $w < -1$). In spite of many interesting results, neither these nor other so far proposed models offer satisfactory answers to all questions around the dark energy.

I propose here $p$-adic origin of the dark energy and dark matter. In other words, dark energy and dark matter are two forms of the $p$-adic matter. If we take string theory seriously then there exist not only real strings but also $p$-adic strings. Real and $p$-adic matter consist of real and $p$-adic strings, respectively. Even without string theory, using $p$-adic quantum-mechanical models one can argue possible existence of $p$-adic matter.

An instructive model of $p$-adic matter is nonlocal and nonlinear effective scalar field theory [9, 10] which describes dynamics of $p$-adic tachyonic matter. The corresponding action is given by

$$S = \int d^d x \, \mathscr{L}_p, \quad \mathscr{L}_p = \frac{1}{g_p^2}\left[-\frac{1}{2}\phi\, p^{-\frac{1}{2}\Box}\phi + \frac{1}{p+1}\phi^{p+1}\right], \tag{72}$$

where

$$\frac{1}{g_p^2} = \frac{1}{g^2}\frac{p^2}{p-1}, \quad \Box = -\frac{\partial^2}{\partial t^2} + \frac{\partial^2}{\partial x_1^2} + \cdots + \frac{\partial^2}{\partial x_{d-1}^2}, \tag{73}$$

and the equation of motion is

$$p^{-\frac{1}{2}\Box}\phi = \phi^p. \tag{74}$$

This $p$-adic string theory was employed in the context of tachyon condensation [11] and for a recent review see [37]. Some mathematical aspects of time dependent solutions of (74) were analyzed in [38] (see also references therein). It is worth mentioning that this $p$-adic tachyon theory resembles dynamics of nonlocal real tachyon in string field theory, which is also under consideration as a candidate for cosmological dark energy [39]. It would be of great importance to find $p$-adic valued counterpart of Lagrangian (72) which reproduces the same complex-valued tachyon amplitudes at the tree level, as well as to establish a map between them.

According to this proposal baryonic matter is real one, while dark energy and dark matter have $p$-adic origin. Work on a concrete $p$-adic model of dark energy and dark matter is in progress.

## 4.3. Adelic Universe with Real and $p$-Adic Worlds

An example of a non-principal adelic physical system could be the universe as a whole. Namely, it seems natural to regard our universe as adelic one consisting of real and $p$-adic worlds which interact only gravitationally. Each of these worlds is made of its own matter and their main characteristics can be described by relevant real or $p$-adic numbers.

Even if we suppose that at the very beginning the universe was as a principal adelic system it could evolve to the non-principal one. Namely, according to string landscape (see recent discussion in [40]), effective potential in string theory has many local maxima and minima which determine values of various parameters. During evolution they can lead to different values of parameters in real and $p$-adic sectors and definitely to transformation of principal adelic strings to real and $p$-adic ones, which may be regarded as non-principal adelic strings.

## 5. CONCLUDING REMARKS

In an introductory way we have reviewed basic properties of $p$-adic and adelic cosmology. Proposal on $p$-adic origin of dark energy and dark matter, as well as on $p$-adic worlds of the universe, has been put forward.

It is worth mentioning some coincidences between $p$-adics and dimensionality of the space-time. A nonsingular form of an odd degree $r$ [41]

$$\sum_{i=1}^{n} a_i x_i^r, \quad a_i \in \mathbb{Q}, \quad r = 2k+1, \quad k \in \mathbb{N} \cup \{0\}, \tag{75}$$

represents all $p$-adic numbers in a nontrivial way in $\mathbb{Q}_p$ if $n \geq D = r^2 + 1$. Also the above form (75) nontrivially represents all rational numbers in $\mathbb{Q}$ when $n \geq D = r^2 + 1$. It follows that the lowest dimension $D$ for $r = 3$ is $D = 10$ and for $r = 5$, $D = 26$. It was noted [42] that these dimensions $D$ coincide with critical dimensions of space-time in string theory, i.e. $D = 10$ for superstrings and $D = 26$ for bosonic strings.

Another intriguing coincidence is related to quadratic algebraic extensions (of $\mathbb{R}$ and $\mathbb{Q}_p$) and dimensionality of M-theory. Recall [6] that in real, $p$-adic ($p \neq 2$) and 2-adic cases we have one, three and seven quadratic extensions, respectively. Their sum gives $1+3+7=11$, what is just number of space-time dimensions in M-theory.

$p$-Adic pseudoconstants [43] are $p$-adic valued functions $c_p(t)$ which have derivatives vanishing, i.e. $\dot{c}_p(t) = 0$ for $t \in \mathbb{Z}_p$. They give new cosmological possibilities comparing with real analysis and could be related [15] to stochastic inflation.

# ACKNOWLEDGMENTS

The work on this article has been supported by the Ministry of Science and Environmental Protection of the Republic of Serbia under projects No 1426 and No 144032. I would like to thank I.Ya. Aref'eva, V.S. Vladimirov and I.V. Volovich for inspirative and useful discussions I have had with them in various periods of work on some problems contained in this paper.

# REFERENCES

1. D. Scott, The Standard Cosmological Model, `astro-ph/0510731`.
2. V. Sahni, Dark Matter and Dark Energy, `astro-ph/0403324`.
3. T. Padmanabhan, Darker Side of the Universe, `astro-ph/0510492`.
4. W.H. Schikhof, *Ultrametric Calculus: An Introduction to p-Adic Analysis*, Cambridge Univ. Press, Cambridge, 1984.
5. I.M. Gel'fand, M.I. Graev and I.I. Pyatetski-Shapiro, *Representation Theory and Automorphic Functions*, Saunders, Philadelphia, 1969.
6. V.S. Vladimirov, I.V. Volovich and E.I. Zelenov, *p-Adic Analysis and Mathematical Physics*, World Scientific, Singapore, 1994.
7. I.V. Volovich, *Class. Quantum Grav.* **4**, L83–L87 (1987).
8. P.G.O. Freund and E. Witten, *Phys. Lett. B* **199**, 191–194 (1987).
9. L. Brekke, P.G.O. Freund, M. Olson and E. Witten, *Nucl. Phys. B* **302**, 365 (1988).
10. P.H. Frampton and Y. Okada, *Phys. Rev. D* **37**, 3077–3079 (1988).
11. D. Ghoshal and A. Sen, *Nucl. Phys. B* **584**, 300–312 (2000).
12. L. Brekke and P.G.O. Freund, *Phys.Rept.* **233**, 1–66 (1993).
13. *Selected Topics of p-Adic Mathematical Physics*, Proc. of the 1st Int. Conf. on *p*-Adic Math. Physics (Moscow, 2003), Proc V.A. Steklov Inst. Math. **245**, Nauka, Moscow, 2004.
14. *p-Adic Mathematical Physics 2005*, Proc. of the 2nd Int. Conf. on *p*-Adic Math. Physics (Belgrade, 2005), AIP Conference Proceedings, American Institute of Physics, New York, 2006.
15. I.Ya. Aref'eva, B. Dragovich, P.H. Frampton and I.V. Volovich, *Int. J. Mod. Phys. A* **6**, 4341–4358 (1991).
16. G.S. Djordjević, B. Dragovich, L.D. Nešić and I.V. Volovich, *Int. J. Mod. Phys. A* **17**, 1413–1433 (2002), `gr-qc/0105050`.
17. J.J. Halliwell and J. Louko, *Phys. Rev. D* **39**, 2206 (1989).
18. B. Dragovich, *Adelic Wave Function of the Universe*, in Proc. the Third A. Friedmann Int. Seminar on Grav. and Cosmology, St. Petersburg, pp. 311–321 (1995); *Adelic Wave Function of the de Sitter Universe*, in Proc. 7th Lomonosov Conf. on Elem. Particle Physics, Moscow, pp. 137–140 (1995); *Adelic Aspects of Quantum Cosmology*, in Proc. 8th Lomonosov Conf. on Elem. Particle Physics, Moscow, pp. 116–122 (1997).
19. B. Dragovich and Lj. Nešić, *Grav. Cosm.* **5**, 222 (1999), `gr-qc/0005103`.
20. V.S. Vladimirov and I.V. Volovich, *Commun. Math. Phys.* **123**, 659 (1989).
21. B. Dragovich, *Teor. Mat. Fizika* **101**, 349–359 (1994), `hep-th/0402193`.
22. B. Dragovich, *Int. J. Mod. Phys. A* **10**, 2349–2365 (1995), `hep-th/0404160`.
23. B. Dragovich, *p-Adic and Adelic Quantum Mechanics*, *Proc. V. A. Steklov Inst. Math.* **245**, 72–85, (2004), `hep-th/0312046`.
24. A. Yu. Khrennikov, *p-Adic Valued Distributions in Mathematical Physics*, Kluwer, Dordrecht, 1994.
25. B. Dragovich, *Integral Transforms and Spec. Funct.* **6**, 197–203 (1998), `math-ph/0404076`.
26. B. Dragovich, *Nucl. Phys. B* (Proc. Suppl.) **102/103**, 150–155 (2001).
27. B. Dragovich, *p-Adic and Adelic Strings*, in Proc. Int. Conf. Dedicated to the Memory of Prof. E. Fradkin: Quantization, Gauge Theory and Strings, Moscow, pp. 108–114 (2001).

28. B. Dragovich and A. Khrennikov, *p-Adic and Adelic Superanalysis*, hep-th/0512318.
29. G.S. Djordjević, B. Dragovich and Lj. Nešić, *Mod. Phys. Lett. A* **14**, 317–325 (1999), hep-th/0005216.
30. G.S. Djordjević and B. Dragovich, *On p-Adic Functional Integration*, in Proc. II Mathematical Conf. in Priština, ed. Lj. D. Kočinac, University of Priština, 1997, pp. 221 - 227; math-ph/0005025.
31. G.S. Djordjević and B. Dragovich, *Mod. Phys. Lett. A* **12**, 1455–1463 (1997), math-ph/0005026.
32. G.S. Djordjević, B. Dragovich and L. Nešić, *Inf. Dim. Analys. Quan. Probab. and Rel. Topics* **6**, 176–195 (2003), hep-th/0105030.
33. J. Hartle and S. Hawking, *Phys. Rev.* **28**, 2960 (1983).
34. V.S. Vladimirov, *Adelic Formulas for Four-Particle String and Superstring Tree Amplitudes in One-Class Quadratic Fields*, in Proc. of the 1st Int. Conf. on *p*-Adic Math. Physics (Moscow, 2003), *Proc V.A. Steklov Inst. Math.* **245**, Nauka, Moscow, pp. 9–28 (2004).
35. S.J. Perlmutter et al., *Astroph. J.* **517**, 565 1999.
36. A. Riess et al., *Astron. J.* **116**, 1009 1998.
37. P.G.O. Freund, *p-Adic Strings and Their Applications*, in these Proceedings, hep-th/0510192.
38. V.S. Vladimirov and Ya.I. Volovich, On the Nonlinear Dynamical Equation in the *p*-Adic String Theory, math-ph/0306018.
39. I.Ya. Aref'eva, Nonlocal String Tachyon as a Model for Cosmological Dark Energy, in these Proceedings.
40. S. Weinberg, Living in the Multiverse, hep-th/0511037.
41. Z.I. Borevich and I.R. Shafarevich, *Number Theory*, Academic Press, 1966.
42. B. Dragovich, *Mod. Phys. Lett. A* **6**, 2301–2307 (1991).
43. K. Mahler, *Introduction to p-Adic Numbers and Their Functions*, Cambridge Univ. Press, Cambridge, 1973.

# On the Chaotic Properties of Quadratic Maps Over Non-Archimedean Fields.

V. Dremov*, G. Shabat† and P. Vytnova**

*Moscow State University, Moscow, RUSSIA*
†*Moscow State University for Humanities, Moscow, RUSSIA* [1]
email: george@shabat.mccme.ru
**Moscow Independent University, Moscow, RUSSIA*
email: vytnova@mccme.ru

**Abstract.** We study dynamic properties of the quadratic maps over arbitrary non-archimedean fields. We find conditions under which these maps demonstrate the chaotic behavior. For the quadratic maps defined over a global field the chaos occurs only over a finite number of valuations.

**Keywords:** Discrete dynamics, non-archimedean norms, adelic chaos.
**MSC 2000:** 12J25, 37B40, 37B99.

## 1. INTRODUCTION

**1.0.** Consider a general discrete dynamical system on a *countable* set (=*phase space*). Formally it is a *deterministic* model of motion (we know *everything* about the orbit of any point) and there seems to be no context for the chaotic considerations.

However, if we are going to study and *describe* the orbits, we need some additional structures on the phase space.

First of all, we need some *language* to specify the points of the phase space. It can be formalized as a *recursive* structure, i.e. the distinguished class of numbering (= bijections with natural numbers) up to recursive renumberings.

For the most dynamical systems the *amount of information* needed to specify a point (it can be formalized in terms of *Kolmogorov complexity*) generically grows along the orbit. In most cases not all this information is valuable for describing the system qualitatively; e.g., if an orbit "goes to infinity" (in some sense) we might be not interested in the details of the positions of the points that are terribly far away.

Thus we impose some *topologies* on the phase space in order to be able to describe the orbits approximately. We emphasize the specific feature of the *nonclassical* discrete dynamics: it is not assumed that the phase space carries some distinguished topology; we rather consider the *set* of natural topologies. The product of the completions of the phase

---

[1] The work is partially supported by RFBR grant 04-01-00640.

space with respect to all these topologies is provided by a suitable *product topology*; the diagonal embedding of the phase space into this product should induce its true *discrete* topology.

The adelic dynamics provides a perfect framework for this approach, the phase space being global number fields; the topologies are defined by their non-archimedean valuations.

In the present paper we consider the simplest non-linear model of this kind — the iterations of quadratic maps. Conceptually our main result is the Theorem 5, according to which the system demonstrates the chaotic behavior only over the finite number of valuations — precisely over those ones over which the quadratic map is in some sense *averagely expanding* in the fixed points.

The results of the paper generalize the earlier results of two of the authors Shabat [2] and Dremov [1]. The similar results over $p$-adic fields with $p \neq 2$ were obtained considerably earlier in Thiran at al. [3].

**1.1.** The paper is organized as follows. Sections 1 and 2 are devoted to certain elementary properties of the quadratic maps over non-archimedean fields. Sections 3 and 4 are technical: under some assumptions the preimages of 0 and of a "large disc" around it are described. In the section 5 the filled Julia sets for all the quadratic maps over all the non-archimedean local fields are described. In the section 6 under the assumptions of the section 3 the isomorphism between the quadratic dynamics on the filled Julia set and some sequence dynamics (Bernoulli shift on the left-infinite sequences) is established. In the section 7 the main results are formulated; the 2-adic case is considered separately. In the section 8 some adelic interpretation of our results is suggested.

**1.2.** Some of the notations we use are not quite standard.

For a map $T: X \mapsto X$ and for $n \in \mathbf{N}$ we denote by $T^{n\circ}$ its $n$th iterate and by $T^{-n\circ}$ its $n$ inverse iterate (possibly multivalued). By $T^{\mathbf{N}\circ}(x)$ we denote the $T$-orbit of $x \in X$; finally, for $Y \subseteq X$ denote by $T^{-\mathbf{N}\circ}Y := \bigcup_{n \in \mathbf{N}} T^{-n\circ}Y$ and $T^{-\infty\circ}Y := \bigcap_{n \in \mathbf{N}} T^{-n\circ}Y$.

When $X$ is a metric space denote by $\mathscr{FJ}(T)$ the *filled Julia set*, i.e. the set of elements of $X$ with bounded $T$-orbits.

For an alphabet (=finite set of characters) $A$ denote by $A^{-\mathbf{N}} = \{\ldots a_2 a_1 a_0\}$ (where $a_0, a_1, a_2 \cdots \in A$) the set of sequences of elements of $A$, infinite *to the left*. For a finite sequence $\varepsilon$ we denote its length by $|\varepsilon|$.

For a field $\mathbf{k}$ denote its set of squares by $\mathbf{k}^{2\cdot} := \{x^2 \mid x \in \mathbf{k}\}$.

For a field $\mathbf{k}$ with the norm $\|\cdot\|$ for $a \in \mathbf{k}$ and $r \in \mathbf{R}_{>0}$ denote the open and closed discs by

$$D(a,r) := \{x \in \mathbf{k} \mid \|x-a\| < r\}$$
$$D[a,r] := \{x \in \mathbf{k} \mid \|x-a\| \leq r\}$$

## 2. CANONICAL FORMS OF QUADRATIC MAPS

**2.0.** We fix a field $\mathbf{k}$ with char $\mathbf{k} \neq 2$ and consider the general quadratic map

$$q\colon \mathbf{A}^1(\mathbf{k}) \mapsto \mathbf{A}^1(\mathbf{k})$$

defined by

$$q(x) = Ax^2 + Bx + C$$

with $A, B, C \in \mathbf{k}$ and $A \neq 0$.

**2.1.** The dynamical properties of the above $q$ depend only on the similarity class of $q$; it means that we consider the action of the group of affine transformation of argument

$$x \mapsto L(x) := mx + n \text{ with } m \in \mathbf{k}^{\bullet}, n \in \mathbf{k}$$

on the set of quadratic transformations. This action is defined by

$$L \bullet q = L \circ q \circ L^{-1\circ};$$

$q$ and thus defined $L \bullet q$ are called *similar*. The problem is to find the simplest (and traditional) representatives of similarity classes of the quadratic map.

**2.2.** It is easy to check that in all the cases the transformation

$$L(x) := Ax + \frac{B}{2}$$

sends

$$q(x) = Ax^2 + Bx + C$$

to

$$[L \bullet q](y) = y^2 + c$$

with

$$c = AC - \frac{B^2}{4} + \frac{B}{2}$$

thus the standard form of the quadratic map

$$x \mapsto x^2 + c$$

is universal, and we are going to stick to it in this paper.

The invariant meaning of $c$ is as follows. Denote by $\mathrm{Fix}(q)$ the (generally 2-element) set of fixed points of $q$, i.e., the set of solutions of the quadratic equation

$$Ax^2 + Bx + C = x.$$

It belongs to $\mathbf{k}$ or to its quadratic extension depending on whether or not the discriminant of the above equation

$$(B-1)^2 - 4AC$$

45

is a square in **k**. But one checks that

$$c := AC - \frac{B^2}{4} + \frac{B}{2} = \frac{1}{4} \prod_{x \in \text{Fix}(q)} q'(x)$$

is always in **k**. We'll see that in the case when **k** is equipped with a (usually non-archimedean) metric the dynamical properties of $q$ depend drastically on the norm of $c$; in particular, $q$ generates the chaotic behavior iff $\|c\| > 1$, i.e., when $q$ is *averagely expanding in the fixed points*. We are not aware of any reasonable generalization of this observation.

**2.3.** The map $q$ is not always similar to another standard form (the *logistic* map)

$$[L \bullet q](y) = \lambda y(1-y).$$

(hence the results of this paper are a bit stronger than those in Shabat [2] even in the case $\mathbf{k} = \mathbf{Q}_p$). The obvious necessary condition is the existence of fixed points of $q$ defined over **k**. It is easy to show that this condition is sufficient as well.

## 3. BEHAVIOR OF NORMS ALONG THE ORBITS

**3.0.** We fix a field **k** with the non-archimedean norm $\|\cdot\|$ and for any element $c \in \mathbf{k}$ consider the quadratic map

$$T_c \colon \mathbf{A}^1(\mathbf{k}) \mapsto \mathbf{A}^1(\mathbf{k})$$

defined by

$$T_c(x) := x^2 + c$$

**3.1.** Every $x \in \mathbf{k}$ defines a sequence $\|T_c^{n\circ}(x)\|$. In most cases the behavior of the norm is quite simple.

**Theorem 1.** *According to the values of $\|c\|$ and $\|x\|$ the following statements hold:*

|  | $\|c\| < 1$ | $\|c\| = 1$ | $\|c\| > 1$ |
|---|---|---|---|
| $\|x\| < 1$ | $\lim_{n \to \infty} \|T_c^{n\circ}(x)\| = \|c\|$, | No general statement | $\lim_{n \to \infty} \|T_c^{n\circ}(x)\| = \infty$ |
| $\|x\| = 1$ | $\|T_c^{n\circ}(x)\| \equiv 1$ | No general statement | $\lim_{n \to \infty} \|T_c^{n\circ}(x)\| = \infty$ |
| $\|x\| > 1$ | $\lim_{n \to \infty} \|T_c^{n\circ}(x)\| = \infty$ | $\lim_{n \to \infty} \|T_c^{n\circ}(x)\| = \infty$ | $\|T_c^{n\circ}\|$ is either constant or $\to \infty$ |

*Proof.* All the statements about existing limits and about the norms $\|T_c^{n\circ}\|$ being constant are obvious. In the case $\|c\| = \|x\| = 1$ the $\lim_{n \to \infty} \|T_c^{n\circ}(x)\|$ can exist. E.g., in any field where $\|2\| = 1$, $x = -1$ is a fixed point of $x \mapsto x^2 - 2$. But it is possible as well that $\|c\| = \|x\| = 1$, but $\lim_{n \to \infty} \|T_c^{n\circ}(x)\|$ does not exist. Over any field the map

$$x \mapsto x^2 - 1$$

provides a cycle that gives a sequence of norms $0, 1, 0, 1, \ldots$

In the case $\|c\| > 1, \|x\| > 1$ the trajectories generally tend to $\infty$. E.g., for $\mathbf{k} = \mathbf{Q}_3$ and $x = c = \frac{1}{3}$ we have the orbit

$$\frac{1}{3} \longrightarrow \frac{4}{9} \longrightarrow \frac{43}{81} \longrightarrow \cdots$$

with the sequence of norms $3, 9, 81, \ldots$ But in some special cases (which are the most interesting from the viewpoint of the present paper) the norms along the orbits are constant. E.g., over $\mathbf{k} = \mathbf{Q}_5$ the map

$$x \to x^2 - \frac{1}{25}$$

has two fixed points $\frac{1}{2} \pm \frac{\sqrt{21}}{16} \in \mathbf{Q}_5$ of the norm 5. □

## 4. THE PREORBIT OF 0.

**4.0.** We fix the triple $\mathbf{k} \supset \mathcal{O} \supset \mathcal{M}$ consisting of a local field, its valuation ring and its maximal ideal; let $p = \mathrm{char}(\mathcal{O}/\mathcal{M})$. We fix the non-archimedean norm $\|\cdot\|$ on $\mathbf{k}$, normalized by the condition $\|p\| = \frac{1}{p}$ and the element $c \in \mathbf{k} \setminus \mathcal{O}$ (i.e. $\|c\| > 1$; this is the only case we'll need). Our goal is to describe the set $T_c^{-N\circ}(0)$.

**4.1.** Informally,

$$T_c^{-1\circ}(0) = \{x \mid x^2 + c = 0\} = \pm\sqrt{-c},$$

$$T_c^{-2\circ}(0) = \{x \mid x^2 + c \in T_c^{-1\circ}(0)\} = \{x \mid x^2 = -c \pm \sqrt{-c}\} = \pm\sqrt{-c \pm \sqrt{-c}}$$

and so on. We should is to give the precise sense to the expressions with nested roots

$$\pm\sqrt{\ldots \pm \sqrt{-c \pm \sqrt{-c \pm \sqrt{-c}}}}$$

(continued recursively to the *left*).

Note that if the roots do not belong to the corresponding fields our notations would be just the convenient names of the elements of their quadratic extensions; however, we are most interested in the case where these roots belong to $\mathbf{k}$ and we are going rather to provide for our nested roots certain *analytic* sense.

**4.2 Proposition.** *The following statements are equivalent:*

(a) $-c \in \mathbf{k}^{2\cdot}$;

(b) $T_c^{-1\circ}(0)$ *is non-empty*

(c) *For any positive natural n the set* $T_c^{-n\circ}(0)$ *is non-empty and, moreover,*

$$\#\{T_c^{-n\circ}(0)\} = 2^n.$$

*Proof.* Implications (a) $\iff$ (b) $\impliedby$ (c) are trivial; concentrate on (a) $\implies$ (c). The assumption (a) implies $c = -a^2$ for some $a \in \mathbf{k}$ with $\|a\| > 1$. In fact, we have *arbitrarily* attributed the signs to $\pm\sqrt{-c}$. Further,

$$\pm\sqrt{-c \pm \sqrt{-c}} = \pm\sqrt{a^2 \pm a} = \pm a(1 \pm \frac{1}{a})^{\frac{1}{2}} :=$$

$$= \pm a\left[1 + \frac{\frac{1}{2}}{1!}\left(\pm\frac{1}{a}\right) + \frac{\frac{1}{2}(\frac{1}{2}-1)}{2!}\left(\pm\frac{1}{a}\right)^2 + \frac{\frac{1}{2}(\frac{1}{2}-1)(\frac{1}{2}-2)}{3!}\left(\pm\frac{1}{a}\right)^3 + \ldots\right],$$

and this series converges $p$-adically (we use $p \neq 2$); see Lemma 1 below.

The longer expressions with nested roots are also defined by the convergent series; see the next subsection. A similar description in terms of *dichotomic variables* can be found in Thiran at al. [3]. $\square$

**4.3. Notations of the elements of** $T_c^{-N\circ}(0)$**.** We assume $c = -a^2$ for all $a \in \mathbf{k}$ and introduce recursively the numbers $b_\varepsilon \in \mathbf{k}$ labeled by the strings $\varepsilon$ of +'s and -'s

$$b := 0,$$
$$b_\pm := \pm a,$$
$$\cdots\cdots\cdots$$
$$b_{\pm\varepsilon} := \{ \text{ solution of } x^2 - a^2 = b_\varepsilon\}.$$

In order to choose the signs for $b_{\pm\varepsilon}$ we introduce recursively the following Laurent series $B_\varepsilon \in \mathbf{Q}((\frac{1}{A}))$ :

$$B_\pm := \pm A,$$

$$B_{\pm\varepsilon} := \pm\sqrt{A^2 + B_\varepsilon} := \pm A\left(1 + \frac{B_\varepsilon}{A^2}\right)^{\frac{1}{2}} = \pm A\left[1 + \frac{\frac{1}{2}}{1!}\frac{B_\varepsilon}{A^2} + \frac{\frac{1}{2}(\frac{1}{2}-1)}{2!}\left(\frac{B_\varepsilon}{A^2}\right)^2 + \ldots\right],$$

and it makes sense since one proves inductively that

$$B_\varepsilon \in \pm A + \mathbf{Z}[\frac{1}{2}]\left[\left[\frac{1}{A}\right]\right]$$

We check that after substituting the free variable $A$ by $a \in \mathbf{k}$ all the $B_\varepsilon$ ' s converge in $\|\cdot\|$-norm and hence define $b_\varepsilon \in \mathbf{k}$.

## 5. LARGE DISC AND THE INVERSE DYNAMICS ON IT

**5.0.** We keep the same notations, including $c = -a^2$. Besides, for any $S \subset \mathbf{k}$ we denote by $\sqrt{S}$ the set $\{x \in \mathbf{k} \mid x^2 \in S\}$.

**Lemma 1** ((Effective openness of the set of squares)). *Let $x_0 \in \mathbf{k}^{2\cdot}$. Then $B(x_0, \|x_0\|) \subset \mathbf{k}^{2\cdot}$.*

*Proof.* Let $y \in \mathbf{k}$ be such that $y^2 = x_0$. By Taylor formula for any $x$ with $\|x\| < \|x_0\|$

$$(y^2 + x)^{1/2} = y\left(1 + \frac{x}{y^2}\right)^{1/2} = y \sum_{n=0}^{\infty} \frac{1(-1)(-3)\ldots(3-2n)}{2^n n!} \cdot \left(\frac{x}{y^2}\right)^n.$$

In order to prove the convergence of this series estimate the norm of its general term. Using

$$-\log_p \|n!\|_p = \left[\frac{n}{p}\right] + \left[\frac{n}{p^2}\right] + \ldots \sim \frac{n}{p} \cdot \frac{1}{1 - 1/p} = \frac{n}{p-1}$$

We see that $\sqrt[n]{\|n!\|_p} \sim p^{-\frac{1}{(p-1)}}$, $\sqrt[n]{\|(2n-1)!!\|_p} = \sqrt[n]{\left\|\frac{(2n)!}{2^n n!}\right\|_p} \sim p^{-\frac{1}{(p-1)}}$. Then $n$th root of general term satisfies

$$\sqrt[n]{\left\|y \frac{(-1)(-3)\ldots(3-2n)}{2^n n!} \cdot \left(\frac{x}{y^2}\right)^n\right\|} = \sqrt[n]{\left\|y \frac{(2n-3)!!}{2^n n!}\right\| \cdot \left\|\frac{x}{y^2}\right\|} \sim \sqrt[n]{\left\|\frac{(2n-1)!!}{n!}\right\|_p} \left\|\frac{x}{x_0}\right\| < 1$$

$\square$

**5.1.** By definition, for all $\varepsilon \in \bigsqcup_{n=0}^{\infty} \{\pm\}^{\{-n\ldots 0\}}$

$$D_\varepsilon := D\left[b_\varepsilon; \frac{1}{\|a\|^{|\varepsilon|-1}}\right].$$

In particular, the one marked by the empty word is

$$D = D[0, \|a\|].$$

**Theorem 2.** *For any $n \in \mathbf{N}$*

$$T_{-a^2}^{-n\circ}(D) = \bigsqcup_{|\varepsilon|=n} D_\varepsilon.$$

**Lemma 2.** *Let $a \in \mathbf{k}$ and $r \in \mathbf{R}_{>0}$ satisfy $\|a\| > 1$ and $D[a^2, r^2] \subset \mathbf{k}^{2\cdot}$. Then*

$$\sqrt{D[a^2, r^2]} = D\left[a, \frac{r^2}{\|a\|}\right] \sqcup D\left[-a, \frac{r^2}{\|a\|}\right]$$

*Proof.* First of all note that $\|a\| > r$, since $D[a^2, r^2] \subset \mathbf{k}^{2\cdot}$.

We are going to show that $\sqrt{D(a^2,r^2)} \supseteq D\bigl[a,\frac{r^2}{\|a\|}\bigr] \sqcup D\bigl[-a,\frac{r^2}{\|a\|}\bigr]$. Let $x \in D\bigl[a,\frac{r^2}{\|a\|}\bigr] \sqcup D\bigl[-a,\frac{r^2}{\|a\|}\bigr]$, then $\|x\| = \|a\|$, as $\|x-a\| < \|a\|$ or $\|x+a\| < \|a\|$. For one of the choices of the sign $\|x \mp a\| = \max(\|x\|,\|a\|) = \|a\|$. Then $\|x \pm a\| < \|a\|$, and

$$\|x^2 - a^2\| = \|x \mp a\| \cdot \|x \pm a\| \leq \frac{r^2}{\|a\|} \cdot \|a\| \leq r^2.$$

Hence $x^2 \in D[a^2, r^2]$.

Now show that $\sqrt{D[a^2,r^2]} \subseteq D\bigl[a,\frac{r^2}{\|a\|}\bigr] \sqcup D\bigl[-a,\frac{r^2}{\|a\|}\bigr]$. Let $x \in \sqrt{D[a^2,r^2]}$, then (as in the previous case), $\|x\| = \|a\|$. Therefore $\|x \mp a\| = \|a\|$. Hence $\frac{\|a^2\|}{\|a^2-x^2\|} = \frac{\|a\|}{\|x \pm a\|}$. Therefore $\|a \pm x\| = \frac{\|a^2-x^2\|}{\|a\|} \leq \frac{r^2}{\|a\|}$. So $x \in D\bigl[a,\frac{r^2}{\|a\|}\bigr] \sqcup D\bigl[-a,\frac{r^2}{\|a\|}\bigr]$. □

Now we prove the Theorem 2 by the induction in $n$. It follows from the effective openness of $\mathbf{k}^{2\cdot}$ that for the disc $D[a^2, \|a\|]$ belongs to $\mathbf{k}^{2\cdot}$. Therefore by Lemma 2 $T_{-a^2}^{-1\circ} D[0,\|a\|] = \sqrt{D[a^2,\|a\|]} = D[a,1] \sqcup D[-a,1] = \bigsqcup_{|\varepsilon|=1} D_\varepsilon$. Since $\pm a \in D[0,\|a\|]$, we have

$$T_{-a^2}^{-1\circ} D[0,\|a\|] \subset D[0,\|a\|].$$

So for any $n$

$$T_{-a^2}^{-n\circ} D[0,\|a\|] = \bigsqcup_{|\varepsilon|=n} D_\varepsilon \subset D[0,\|a\|],$$

and the Lemma 2 is applicable to every disk it is used for. The Theorem 2 is proved.

**Corollary 1.**

$$T_{-a^2}^{-\infty}(D) = \bigcap_{n=0}^{\infty} \bigsqcup_{|\varepsilon|=n} D_\varepsilon$$

# 6. THE FILLED JULIA SETS

Keep the notations of the previous section (with the exception of $c$ that now is arbitrary).

**Theorem 3.** *If* $\|c\| \leq 1$, *then* $\mathscr{FJ}(T_c) = \mathscr{O} = D[0,1]$. *If* $\|c\| > 1$, *then*
(a) *if* $-c \notin \mathbf{k}^{2\cdot}$, *then* $\mathscr{FJ}(T_c) = \emptyset$;
(b) *if* $-c \in \mathbf{k}^{2\cdot}$, *i.e.* $c = -a^2$ *for some* $a \in \mathbf{k}$, *then*

$$\mathscr{FJ}(T_{-a^2}) = T^{-\infty} D[0,\|a\|].$$

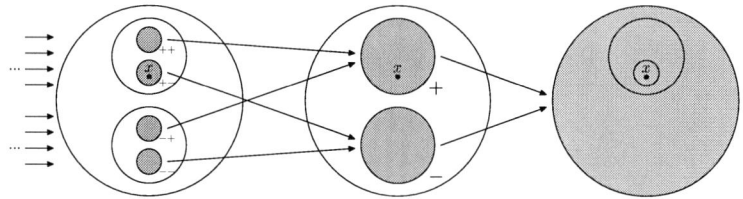

*Proof.* The statement in the case $\|c\| \leq 1$ follows from the properties of the norm sequence for $T^{n\circ}(x)$, see section 2.

In the case $\|c\| > 1$ we see that if $\|x\| > \sqrt{\|c\|}$, then $\|T^{n\circ}(x)\| = \|x\|^{2^n} \to \infty$ and if $\|x\| < \sqrt{\|c\|}$, then $\|T^{n\circ}(x)\| = \|c\|^{2^{n-1}} \to \infty$. Hence the $\mathscr{FJ}$ lies on the circle defined by $\|x\| = \sqrt{\|c\|}$.

Consider the case *(a)*. The assumption $-c \notin \mathbf{k}^{2\cdot}$ for any $x$ satisfying $\|x\| = \sqrt{\|c\|}$ implies $\|x^2 + c\| \geq \|c\|$. Indeed, if $\|x^2 + c\| < \|c\|$, then $-c \in D(x^2, \|x^2\|) \subset \mathbf{k}^{2\cdot}$ by the effective openness of squares. Hence $\|T^{n\circ}(x)\| \geq \|c\|^{2^{n-1}} \to \infty$.

In the case *(b)* we just use our construction of indexed discs:

$$\mathscr{FJ} \subset D = D[0, \|a\|].$$

Then $\mathscr{FJ} \subseteq T^{-n\circ}(D) = \bigsqcup_{|\varepsilon|=n} D_\varepsilon$, so $\mathscr{FJ} \subseteq \bigcap_{n=0}^{\infty} T^{-n\circ}(D) = T^{-\infty}D[0, \|a\|]$

The opposite inclusion $\mathscr{FJ} \supseteq T^{-\infty}D[0, \|a\|]$ is obvious. □

## 7. ISOMORPHISM WITH THE SEQUENCE DYNAMICS

Keep the notations of the section 4. Consider the space $\{\pm\}^{-\mathbf{N}} := \{\ldots \varepsilon_2, \varepsilon_1, \varepsilon_0 \mid \varepsilon_n \in \{+,-\}\}$ of sequences of pluses and minuses infinite *to the left* endowed with Tikhonov topology. Denote by

$$\sigma \colon \{\pm\}^{-\mathbf{N}} \mapsto \{\pm\}^{-\mathbf{N}} \colon \ldots \varepsilon_2 \varepsilon_1 \varepsilon_0 \mapsto \ldots \varepsilon_3 \varepsilon_2 \varepsilon_1$$

the *Bernoulli shift*.

**Theorem 4.** *For any $a$ satisfying $\|a\| > 1$ there is an isomorphism of dynamical systems (i.e. compacts with continuous endomorphisms)*

$$(\mathscr{FJ}(T_{-a^2}), T_{-a^2}) \simeq (\{\pm\}^{-\mathbf{N}}, \sigma).$$

*Proof.* For any $x \in \mathscr{FJ}(T_{-a^2})$ there exists a unique sequence of embedded discs.

$$D_{\varepsilon_0\varepsilon_1\varepsilon_2} \subset D_{\varepsilon_0\varepsilon_1} \subset D_{\varepsilon_0} \subset D$$

such that $\{x\} = \ldots \cap D_{\varepsilon_0\varepsilon_1} \cap D_{\varepsilon_0} \cap D$ and $\{T(x)\} = \ldots \cap D_{\varepsilon_1} \cap D \cap T(D)$. This construction defines

$$I \colon \mathscr{FJ}(T_{-a^2}) \mapsto \{\pm\}^{\mathbf{N}} \colon x \mapsto \ldots \varepsilon_2\varepsilon_1\varepsilon_0,$$

and it is easy to check that $I$ is a homeomorphism satisfying $I \circ T_{-a^2} = \sigma \circ I$. $\square$

## 8. CHAOTIC PROPERTIES OF QUADRATIC MAPS

Restore the notations $\mathbf{k} \supset \mathscr{O} \supset \mathscr{M}$ (a local field, its valuation ring and its maximal ideal); $p := \operatorname{char}(\mathscr{O}/\mathscr{M})$. Extend the polynomial maps we consider from $\mathbf{A}^1(\mathbf{k})$ to the projective line $\mathbf{P}^1(\mathbf{k})$, sending infinity to infinity.

Here are the main results of the paper.

**Theorem 5.** *If $p \neq 2$, then the map*

$$T_c \colon \mathbf{P}^1(\mathbf{k}) \to \mathbf{P}^1(\mathbf{k}) \colon x \mapsto x^2 + c$$

*has positive topological entropy iff $\|c\| > 1$ and $-c \in \mathbf{k}^{2\cdot}$.*

*Proof.* Follows from the Theorem 4 and the results of Nitecki [4] and Adlet et al. [5]. See details in Shabat [2]. $\square$

**Theorem 6.** *If $p = 2$, then the map*

$$\mathbf{P}^1(\mathbf{k}) \to \mathbf{P}^1(\mathbf{k}) \colon x \mapsto x^2 + c$$

*has positive topological entropy iff $\|4c\| > 1$ and $(1-4c) \in \mathbf{k}^{2\cdot}$.*

*Proof.* We formulate and outline the proofs of the analogues of our main statements for $p = 2$.

Consider the case $\|c\| \leq \|1/4\|$. Denote the roots of $T_c(x) - x$ by $x_1$ and $x_2$. We have $\mathbf{K} := \mathbf{k}[x_1] = \mathbf{k}[x_2]$, with $(\mathbf{K} : \mathbf{k}) \in \{1, 2\}$. Our norm can be extended to the field $\mathbf{K}$. Then $\|2x_1\| \leq 1$, $\|2x_2\| \leq 1$ and moreover $\|x_1 - x_2\| = \|\sqrt{1-4c}\| \leq 1$. So $D[x_1, 1] = D[x_2, 1]$. Now prove the formula $\mathscr{FJ}(T_c) = \mathbf{k} \cap D_{\mathbf{K}}[x_1, 1]$. For $t := x - x_1$ we obtain $\|T(x) - x_1\| = \|(x_1 + t)^2 + c - x_1\| = \|t(2x_1 + t)\|$. Hence for $\|t\| \leq 1$ we have $\|T_c^{n\circ}(x) - x_1\| \leq 1$ and for $\|t\| > 1$ we have $\|T_c^{n\circ}(x) - x_1\| = \|t\|^{2^n}$.

For any two points $x, y \in \mathscr{FJ}(T_c)$ we have

$$\|T_c(x) - T_c(y)\| = \|(x-y)(x+y)\| \leq \|x-y\|\|2x_1 + (x-x_1) + (y-x_1)\| \leq \|x-y\|.$$

52

Hence if $\|c\| \leq 1/4$, then the topological entropy of $T_c$ equals zero.

Consider the case $\|c\| > 1/4$. Now we have two distinct disks $D[x_1,1]$ and $D[x_2,1]$, with $\|x_1\| = \|x_2\| = \sqrt{\|c\|}$ and $\|x_1 - x_2\| = \sqrt{\|4c\|}$. We introduce $b_\pm := x_{1,2}$, and construct the $b_\varepsilon$'s and $D_\varepsilon$ as in Subsections **4.3, 5.1** (excluding the empty word). We argue similarly to the case $p \neq 2$, but have to introduce some modifications.

As in the case $p \neq 2$, $\|T_c(x) - x_1\| = \|(x_1+t)^2 + c - x_1\| = \|t(2x_1+t)\|$.

For $x_1 \notin \mathbf{k}$ we have $\|T_c^{no}(x) - x_1\| = \|t\|^{2^n}$ for $\|t\| > \|2x_1\|$ and $\|T_c(x) - x_1\| = = \|2x_1\| \cdot \|x - x_1\| > \|x - x_1\|$ for $0 < \|t\| \leq \|2x_1\|$. Hence the filled Julia set is empty and the entropy is equals zero.

But for $x_1 \in \mathbf{k}$ we have $x_2 = 1 - x_1 \in \mathbf{k}$ and moreover all the discs $D_\varepsilon$ lie within $\mathbf{k}$ since Lemma 1 holds for the disks $D(x_0, \|4x_0\|)$.

Lemma 2 is replaced by the statement $\sqrt{D[a^2, r^2]} = D[a, \frac{r^2}{\|2a\|}] \sqcup D[-a, \frac{r^2}{\|2a\|}]$ for all the discs $D[a^2, r^2]$ with $r^2 < \|4a^2\|$ (in particular, for all the shifted disks in the proof of the Theorem 2). Hence for $D_\varepsilon$ we obtain the formula $D_\varepsilon = D[b_\varepsilon, \|2a\|^{1-|\varepsilon|}]$.

So we prove that on $\mathscr{FJ}(T_c)$ our dynamical system is equivalent to the Bernoulli shift as in the Theorem 4. Its topological entropy is positive. $\square$

## 9. ADELIC INTERPRETATION

Let $\mathscr{K}$ be a global number field, $(\mathscr{K}:\mathbf{Q}) < \infty$. Consider $c \in \mathscr{K}$ and

$$T_c: \mathscr{K} \longrightarrow \mathscr{K}: x \mapsto x^2 + c.$$

For any $c$ there is only a finite number of $v$'s such that $T_c: \mathscr{K}_v \mapsto \mathscr{K}_v$ demonstrates chaotic behavior. For any non-archimedean valuation

$$v: \mathscr{K} \longrightarrow \mathbf{Z} \sqcup \{\infty\}$$

we extend $T_c$ to

$$T_c: \mathscr{K}_v \longrightarrow \mathscr{K}_v.$$

According to the Theorems 5 and 6 we can introduce the quantitative measure of global chaos:

$\mathrm{chao}(c) := \#\{v \in \mathrm{val}(\mathscr{K}) \mid T_c: \mathscr{K}_v \longrightarrow \mathscr{K}_v \text{ is chaotic}\} = \#\{v \in \mathrm{val}(\mathscr{K}) \mid \|c\|_v > 1, c \in \mathscr{K}_v^{2\cdot}\}$.

Perhaps, it deserves further study.

# REFERENCES

1. V.A. Dremov, *On a certain p-adic set* (in russian), *Uspekhi Mat. Nauk* **58**, no. 6 (354), (2003).
2. G.B. Shabat, *p-adic entropies of logistic maps*, *Tr. Mat. Inst. Steklova*, Izbr. Vopr. *p*-adich. Mat. Fiz. i Anal., **245** 257–263 (2004); translation in *Proc. Steklov Inst. Math.*, no. **2** (245), 243–249 (2004).
3. E. Thiran, D. Verstegen and J. Weyers, *p-adic dynamics*, *J. Statist. Phys.* **54**, no. 3-4, 893–913 (1989).
4. Z.H. Nitecki, *Topological entropy and the preimage structure of maps*, *Real Anal. Exchange* **29**, no. 1, 9–41 (2003/04).
5. R.G. Adler, A.G. Konheim and M. H. McAndrew, *Topological entropy*, *Trans. Am. Math. Soc.* **114** (1965).

# On Ultrametricity and a Symmetry Between Bose-Einstein and Fermi-Dirac Systems

Alexandr A. Ezhov[*] and Andrei Yu. Khrennikov[†]

[*]*Troitsk Institute for Innovation and Fusion Research. 142190, Troitsk, Moscow Region, RUSSIA*
[†]*International Center for Mathematical Modelling in Physics and Cognitive Sciences,
MSI, University of Växjö, S-35195, SWEDEN
email: Andrei.Khrennikov@msi.vxu.se*

**Abstract.** We present the results of computer simulations which give the evidence for the existence of an interesting symmetry in a multi-agent model demonstrating, in special cases, both Bose-Einstein and Fermi-Dirac statistics. This symmetry is expressed in the coincidence of the degree of ultrametricity and the fraction of isosceles of the sets of agents memories coded by two different information loss coding schemes.

**Keywords:** Ultrametricity, p-adic probability, quantum statistics, multi-agent model.
**PACS:** 05.10.Ln, 02.50.Le, 02.50.Cw.

## 1. INTRODUCTION

Take-off of interest in ultrametricity in statistical physics is attributed to Mézard et al. [1] – it has been demonstrated that the distribution of pure states in spin glasses is ultrametric. As it has been noted by Parisi and Ricci-Tersenghi [2], ultrametricity implies that the distance between different states is such that they can be put in a taxonomic tree. Since ultrametricity is closely related to hierarchical structures which can characterize many sorts of data sets including time series, linguistic constructions etc. [3, 4], there is a keen interest in this topic in many fields far from physics.

In a series of papers (see [5] for references) Murtagh proposed to consider ultrametricity as a fingerprint of some special hidden feature of a phenomenon.

By studying ultrametricity of statistical data we can distinguish two different phenomena. Similar idea was presented in the framework of so called *p*-adic probability theory (see [6]–[10]). Here *p*-adic probability was also considered as a fingerprint of some special feature of a random phenomenon (see [6], Chapter 7)

Here we argue that the search for surprisingly new roles that ultrametricity plays in physics can be fruitful. We report that this property can be very useful in quantum statistics, and can be used not only to distinguish but also to find what different data have in common. For this we performed a numerical modelling of some recently proposed multi-agent system demonstrating both types of quantum statistics [11]. This model demonstrates Bose-Einstein and Fermi-Dirac statistics in the populations of *friendly* right brain hemisphere dominant agents (*"Rights"*) and also in the population of *competitive* left brain hemisphere dominant agents (*"Lefts"*) (see [11] for details). Using computer simulations we found that it can also demonstrate an unusual form of symmetry between bosons and fermions. This symmetry is intrinsically connected with the degree of ultrametricity of the set of agent *memories* coded using two different *information loss schemes*. Note, that there are many other systems, both quantum and classic, whose state of equilibrium is described with quantum statistical distributions. For example, Evans found the Bose-Einstein condensation (BEC) while solving the heterogeneous transport problem [12]. Bianconi and Barabasi [13] demonstrated that the Bose-Einstein statistics describes the growing Internet. Staliunas [14] adduced arguments that the BEC can arise in classic systems far from thermal equilibrium due to the system coherent dynamics, or due to the equivalent autocatalytic dynamics in a system momentum space. In addition, Bianconi found that a growing Cayley-tree with a different number of nodes and thermal noise is described by Fermi-Dirac statistics [15]. Derrida and Lebowitz [16] found both the Fermi-Dirac and Bose-Einstein distributions when studying fully asymmetric exclusion processes at a ring containing $N$ sites and $p$ particles.

After a brief description of the model presented in detail in [11] we give its simple physical analog related to particle transport and introduce two different coding schemes of particle histories (memories). Finally, we present the results of computer simulations which reveal a surprising coincidence of ultrametric properties of the memory sets of bosons and fermions.

## 2. THE MODEL

The model considers a *world* consisting of $n$ *cells* which can be occupied by $N$ agents. Every agent has two kinds of *resources*: *a physical* and *a mental* ones, which have to be held positive at any time. The first resource degrades in time but can be compensated by consuming the *food* which randomly appears in the world cells from their environment. The second resource decreases each time agent changes its cell. The appearance of the first resource can be described by probability $f_i$ and by cell's energy $\varepsilon_i = -\theta \log f_i$, where parameter $\theta$ characterizes *temperature* of the environment (see also [13]). We interpret appearance of food in the cell being free of a specific agent as the *environmental proposal* to enhance the first kind of resource accompanied by a decrease of the second kind of resource. Let Boolean variable, $a$, denotes this proposal, and $a = 0$, if the environment offers to change the cell. The appearance of the unit of the first resource in the cell occupied by the given agent, can be considered as a proposal for the agent *to preserve its second resource* and to enhance its first resource. Let $a = 1$, if the environment offers the agent to keep its cell. It is suggested in [11] that every agent can *accept* (A) or *reject* (R) such a proposal interacting with a randomly chosen partner (including itself). Specifically, it is suggested that if food is offered to agent $\alpha$, *this agent regards that it is also offered to agent $\beta$*. In accordance with Lefebvre [17], we also suggest that an agent can consider two types of relations with another agent i.e. friendly and competitive ones. The decision of agent $\alpha$ depends both on the environment proposal $a$ to agent $\alpha$ and also on its proposal $b$ to agent $\beta$; *from the point of view of*

**TABLE 1.** The decisions of an agent taking into account the proposal to its friend and to its enemy.

| a | b | Right's friend | Left's enemy |
|---|---|---|---|
| 0 | 0 | 1 | 0 |
| 0 | 1 | 0 | 1 |
| 1 | 0 | 1 | 1 |
| 1 | 1 | 1 | 1 |

*agent $\alpha$ the same unit of the first resource is offered to agent $\beta$.* Now an intention of agent $\alpha$ becomes a function of two variables, $\psi = \psi(a,b)$. The intentions of the two types of agents can be considered as a function of two variables: $a$ and $b$. As it was argued in [11] there are two reasonable functions of this type which can be named as *right brain dominant* and *left brain dominant* ones. In fact, for the right brain agent $\alpha$ which considers *other* agent $\beta$ as its friend, we obtain the function values presented in the third column of Table 1. The right brain agent accepts the environment proposal to consume the food in a new cell only if the environment does not demand to do this of its friend (with which agent $\alpha$ interacts). Enemies *do not influence* intentions of the right brain agent at all. This is also true for the left brain agent which interacts with a friend. On the contrary, for the left brain agent which takes into account the situation with an enemy, the decisions are presented in the fourth column of Table 1. Focusing on the second raw of this column, we conclude that the left brain agent does not accept the environment proposal to consume the food in another cell, unless the environment demands of the enemy to change its cell (so that the food is offered just in the cell which occupies a random enemy, with which agent $\alpha$ interacts). In the case of agent self-interaction (when it randomly chooses itself for interaction) the situation changes. Indeed, the right brain and left brain agents act identically unconditionally accepting the environment proposal. It was demonstrated in [11] that these dynamic rules lead to the Bose-Einstein distribution in the population of friendly right brain agents

$$\langle N_i(\varepsilon_i) \rangle = \frac{1}{e^{(\varepsilon_i - \mu)/\theta} - 1}, \qquad (1)$$

where $\theta$ is a temperature and $\mu$ is a chemical potential. In the population of competitive left brain agents the equilibrium distribution

$$\langle N_i(\varepsilon_i) \rangle = \frac{N}{e^{(\varepsilon_i - \mu)/\theta} + 1}. \qquad (2)$$

is the Fermi-Dirac one with high ($G = N$) degeneracy of energy levels.

*HISTORIES CODING.* It is convenient to reformulate our model in physical terms as follows. Suppose that we use Monte Carlo code to model particle transport in the media consisting of very heavy nuclei which are able to scatter them both elastically (with conservation of particle energy after collision) or inelastically (with change of particle energy). We also suggest that inelastically scattered particles can both lose and also increase their energy. It is possible to consider the discrete time intervals $\Delta t$ in which the particle path is $\Delta l = \sqrt{2\varepsilon/m}\Delta t$. In this case the proposal of food in an energy cell occupied by an agent will correspond to the proposal to the particle to continue its free motion after passing the interval $\Delta l$. In Monte Carlo code it means that the random value $\gamma \in [0,1]$ used for choice of free path length is less than $\exp(-\Sigma_{tot}\Delta l)$, where $\Sigma_{tot}$ is the total cross section of particle-nuclei interaction. Surely, every particle will unconditionally *accept* this proposal despite the type of possible pair interaction with the other particle. Suppose, that collision means that the environment proposes to the particle to change its energy in inelastic scattering. If such a particle *rejects* this proposal it undergoes elastic scattering holding its energy (cell in a multi-agent model). Note, that in this formulation particles interact both with nuclei (environment) and also, if collision takes place, with the other particle (other agent). Using this interpretation we can describe the history (or agent memory) of a particle using the following complete symbolic code (see Figure 1 – top left): 1) $A_f$ – the environment proposes the particle to continue its free ($f$) motion and the particle accepts ($A$) this proposal unconditionally (in our cellular model it corresponds to $a = 1$); 2) $A_i$ – the environment proposes the particle to scatter inelastically ($i$) with a change of its energy, and the particle accepts ($A$) this proposal; 3) $R$ – the environment proposes the particle to scatter inelastically with a change of its energy, but the particle rejects ($R$) this proposal. Two last cases correspond to $a = 0$ in the multi-agent model.

We can describe the same trajectory using two *alternative three letter codes* which describe whether the particle stays in the previous energy state ($S$) or goes to another one ($G$) (see Figure 1 – top right): 1) $S_f$ – the environment proposes the particle to continue its free ($f$) motion and the particle stays ($S$) in the previous energy state; 2) $G$ – the environment proposes the particle to scatter inelastically with a change of its energy, and the particle goes ($G$) to a new energy state; 3) $S_e$ – the environment proposes the particle to scatter inelastically with a change of its energy, but the particle scatters elastically ($e$) holding its energy. Further let us consider two *binary* coding schemes which can be derived from the schemes defined above *by omitting* the low indexes in the notation of events (see Figure 1 – bottom). These two schemes can be considered as two projections of the full three-character codes to the two orthogonal planes. They give an **incomplete description** of the particle history (agent memory). The first information loss scheme uses the letters $A$ and $R$ to describe whether a particle accepts or rejects the environment proposal with the loss of information about *what kind of proposal has been accepted* (to continue free motion or to scatter inelastically). On the other hand, the second information loss scheme uses the letters $S$ and $G$ to describe whether a particle stays in a previous energy state (*regardless if it moves freely or scatters elastically*) or goes to a new energy state.

These two coding schemes are evidently non-equivalent. It means that it is impossible to reconstruct the particle history in a binary $(A - R)$ representation if it is given in $(S - G)$ representation *and vice versa*.

At last, we can use well known *urn model* to illustrate coarse coding scheme (and also to obtain the final result). In this model the probability for the randomly chosen ball to go to the urn $i$ can be calculated as $c(a + N_i)$, where $c$ and $a$ are constants. Bose-Einstein statistics corresponds to the case $a = 1$, while the Fermi-Dirac distribution (2) corresponds to the case $a = -N$. The final statistics does not depend crucially on the

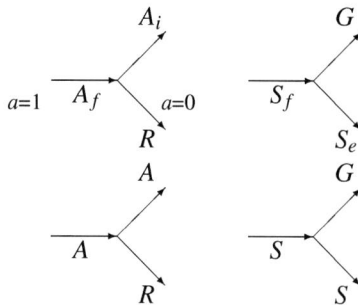

**FIGURE 1.** *Top left*: Full coding based on the acceptance or rejection of environment proposal (a). For $a = 1$ the particle unconditionally accepts the proposal to continue its free motion ($A_f$). For $a = 0$ it either accepts the proposal to change its energy ($A_i$), or rejects it (R). *Top right*: Full coding which suggests that the particle stays in the previous state of free motion ($S_f$) when the environment proposes this $a = 1$, or rejects the proposal to change its state ($S_i$). In the last case accepting the proposal means that the particle goes (G) to other state. *Bottom left and right*: Two incomplete coding schemes follows from the full ones by omitting code subscripts.

constant $c$, but the urn model gives the result analogous to the agent and particle model in the case $c = N^{-1}$ for Bose-Einstein statistics and in the case $c = -N^{-1}$ for the Fermi-Dirac statistics with no limit of number of particles in one urn.

So, for any interpretation of basic model we can perform its Monte Carlo simulation which will generate for each agent (particle, ball) the history of the form as follows: AARARRAARRR.. or SGSGSSGSSSS....

*ULTRAMETRICAL PROPERTIES.* It is well known, that high dimensional, sparsely populated data spaces can be characterized in terms of ultrametric topology. Ultrametric spaces are characterized by strong triangular inequality $d(x,z) \leq \max\{d(x,y), d(y,z)\}$ for any triplet $x, y, z$. From this one of the most prominent features of ultrametric spaces follows: any triangle formed by any triplet is *isosceles* with two large sides equal, or is *equilateral*. It is reasonable to study how well the set of agent memories (particle histories) can be embedded in an ultrametric topology. We analyze agent memory sets in the space of vectors with binary-valued components (corresponding to the two information loss schemes defined above) and with Hamming metrics. In this sense the question is to quantify how ultrametric given metric space is. Using two different non-complete memory codings described in the previous section we calculate the degree of ultrametricity of the memory sets arisen in different statistical tests.

# 3. MONTE CARLO MODELLING

Here we summarize in details the computational scheme used to obtain the final result.

- The statistics (Bose-Einstein or Fermi-Dirac with no limitation on the number of agents in a cell) is chosen.
- The cellular world consists of $N_{cells}$ cells (in simulations $N_{cells} = 10$). Each cell has an energy value $\varepsilon_i$, which grows linearly with the cell index $i$ (from zero to one in further simulations).
- The number of the agents populated the world equals to $N$ ($N = 50$).
- Initial distribution of the agents $N_i$ in the cells $i$ ($i = 1, \ldots, N_{cells}$) is random. Initially the history of each agent $h_i$ contains no items – $m_i = 0$.
- The dynamics of the agent population is simulated in discrete time points $t = 0, \ldots$.
- Each time $t$ a random test agent $\alpha$ which occupies the cell $i$, a random partner agent $\beta$ located in the cell $j$ and a questionable destination cell $k$ are chosen. The probability that $k$th cell is a destination one is equal to

$$p_k = \frac{\exp(-\varepsilon_k/\theta)}{\sum_{l=1}^{N_{cells}} \exp(-\varepsilon_l/\theta)}, \qquad (3)$$

(in simulations $\theta = 0.2$).

- If the destination cell $k$ coincides with the cell $i$ occupied by the test agent then the last one keeps this cell unconditionally and its memory grows $m \to m+1$ by adding the memory component $A$ (accept) for $(A-R)$ coding or $S$ (stay) for $(S-G)$ coding.
- If $k \neq i$ then the test agents goes to the destination cell if:
  - the partner agent $\beta$ coincides with the test agent $\alpha$ (self-interaction);
  - the partner agent $\beta$ occupies the destination cell $k$ and the agents are bosons;
  - the partner agent $\beta$ does not occupy the destination cell $k$ and the agents are fermions.

  The test agent memory grows ($m \to m+1$) by adding the memory component $A$ (accept) for $(A-R)$ coding or $G$ (go) for $(S-G)$ coding.
  Otherwise the test agent keeps its previous cell its memory grows ($m \to m+1$) by adding the memory component $R$ (reject) for $(A-R)$ coding or $S$ (go) for $(S-G)$ coding.

The process continues until an equilibrium state is reached. At this moment all the memories of the agents are emptied in order to exclude the description of transition process and to start the memorization of the histories from $m = 0$ again. The process is further continued up to the moment when memory length of each agent reaches prescribed value, $m$ (up to 100).

The set of vectors $\mathbf{h}^{(\alpha)} = (h_1^{(\alpha)}, \ldots, h_m^{(\alpha)})$, $\alpha = 1, \ldots, N$, with binary components $h_l^{(\alpha)} \in \{A, R\} (\{S, G\})$ are investigated for ultrametricity.

*MAIN OBSERVATION.* In contrast with the sophisticated approaches to the definition of the parameter proposed, e.g., in [18, 19, 20], we use just the fraction of triplets satisfying the strong triangle inequality as a measure of ultrametricity. We also add one more parameter – the fraction of improper isosceles triangles to the total number of isosceles *and* proper ones, which we call *fraction of isosceles*. In the ultrametric topology this parameter reflects the form of a data tree structure. The most surprising result of the modelling was the discovery that both of these parameters – the degree of ultrametricity and the fraction of isosceles *seem to be identical for the memory sets of agents obeying Bose-Einstein and Fermi-Dirac statistics if the former one uses (A − R) coding, while the latter one – (S − G) coding*. This observation can indicate to both the possible importance of degree of ultrametricity and fraction of isosceles as informative parameters of particle dynamics in quantum statistics and also to some new possible form of supersymmetry. In Figures 2,3 we present the dependence of these two parameters for thermodynamically equilibrium ensembles of 50 friendly right brainers and also competitive left brainers on the length of history, $m$ in interval $m \in [10, 100]$ (we found that the system reaches equilibrium after approximately 20 updates of each agent cell location). It follows from this figure that average values of the degree of ultrametricity and of the fraction of isosceles calculated by averaging the results of 50 tests are completely different for different incomplete coding schemes, but **are not statistically different** for $(A − R)$ coding of bosons and $(S − G)$ coding of fermions (see Table 2) in the whole interval of memory lengths. The coincidence of the degree of ultrametricity and the fraction of isosceles for the right and left brainer statistics for $(A − R)$ and $(S − G)$ binary coding, correspondingly, is not trivial because the first order statistics of e.g. "A" and "S" is different. Namely, a $t$-test reveals a highly significant difference ($t = 19.4, p < 0.001$) between a mean fraction of $A$ and $S$ in history ($p(A) = 0.412, p(S) = 0.393$) with standard deviations 0.0067 and 0.0028, correspondingly. So, the memory sets in these two cases are not statistically equivalent.

Note, that the phenomenon observed depends not only on the equllibrium statistical distributions (Bose-Einstein and Fermi-Dirac), but also on the concrete dynamical law determining the agent interactions. Just this law determines the rate of agents movement in the cell world, and, finally, their memory. There are many laws which leads to the same final distributions which obey equivalent relation of the detailed balance.

## 4. CONCLUSION

It has been shown recently, that the action functions corresponding to Bose-Einstein and Fermi-Dirac statistics fulfill Riccati equation which is in the core of supersymmetric quantum mechanics [21]. We guess if the observed phenomenon is also a specific manifestation of supersymmetry in quantum statistics? Anyway, it seems very intriguing to study further ultrametric properties of the space of the histories which characterize the

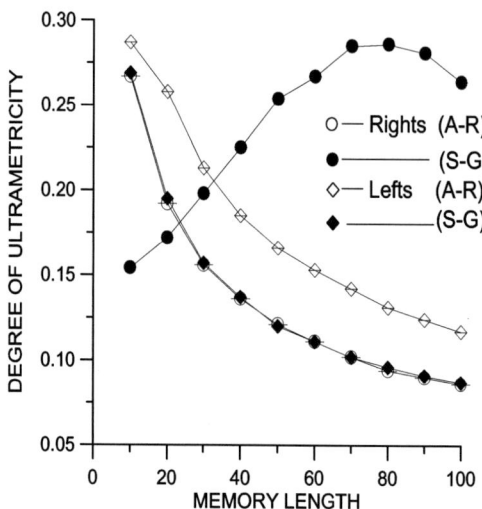

**FIGURE 2.** Mean values of the degree of ultrametricity versus the length of memory ($m$) averaged over 50 trials for the ensembles of friendly right brain dominant agents (Rights) and of competitive left brain dominant agents (Lefts) for two different coding schemes: $(A - R)$ and $(S - G)$.

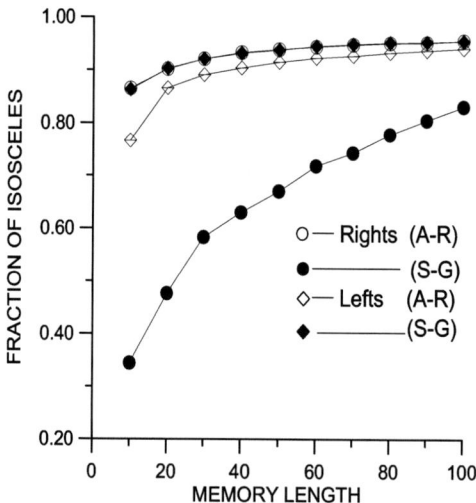

**FIGURE 3.** Mean values of the fraction of isosceles versus the length of memory ($m$) averaged over 50 trials for the ensembles of friendly right brain dominant agents (Rights) and of competitive left brain dominant agents (Lefts) for two different coding schemes: $(A - R)$ and $(S - G)$.

TABLE 2. Mean degree of ultrametricity ($U$) and fraction of isosceles ($I$) vs memory length ($m$) for right brain dominant agents (bosons) obeying Bose-Einstein statistics in Accept-Reject ($A-R$) binary coding, and also for left brain dominant agents (fermions) obeying Fermi-Dirac statistics for Stay-Go ($S-G$) coding. Populations consisting of 50 agents are used, inverse temperature $\beta = 5$. Total number of $N = 50$ trials is used. The values of $t$-parameter reveals no difference in means ($p > 0.05$).

| $m$ | 10 | 30 | 50 | 70 | 90 |
|---|---|---|---|---|---|
| $U$ bosons (A-R) | 0.267 | 0.156 | 0.121 | 0.102 | 0.090 |
| – fermions (S-G) | 0.269 | 0.157 | 0.120 | 0.102 | 0.091 |
| $t_{50}$ | -0.808 | -0.422 | 0.345 | 0.516 | -0.236 |
| $m$ | 10 | 30 | 50 | 70 | 90 |
| $I$ bosons (A-R) | 0.865 | 0.921 | 0.939 | 0.948 | 0.952 |
| – fermions (S-G) | 0.863 | 0.921 | 0.938 | 0.949 | 0.953 |
| $t_{50}$ | 0.892 | -0.304 | 0.855 | -0.958 | -0.702 |

particles obeying Bose-Einstein and Fermi-Dirac statistics in general. In conclusion, by computer modelling we found that ensemble averaged ultrametric properties of the set of memories coded using two non-equivalent information loss binary schemes for agents obeying quantum statistics can be used for revealing the symmetry between bosons and fermions at least in the case of the considered model. This work was supported by the grant of the Swedish Royal Academy of Science on the collaboration with scientists of the former Soviet Union; Profile Mathematical Modelling of Växjö University, EU-network on Quantum Probability and Applications.

# REFERENCES

1. M. Mézard, G. Parisi, N. Sourlas, G. Toulouse, M.A. Virasoro, *Phys. Rev. Lett.* **52**, 1156–1159 (1984).
2. G. Parisi, F. Ricci-Tersenghi, *J. Physics. A* **33**, 113–129 (2000).
3. R.N. Mantegna, *European Physical Journal B* **11**, 193–197 (1999).
4. F. Murtagh, *European Physical Journal B* **43**, 573–579 (2005).
5. F. Murtagh, *Identifying the Ultrametricity of Networks*, Technical report, (2005), http://www.cs.rhul.ac.uk/home/fionn/papers/fm36.pdf.
6. A.Yu. Khrennikov, *p-adic Valued Distributions in Mathematical Physics*, Kluwer Academic Press, 1994.
7. A.Yu. Khrennikov, *Interpretation of probability*, VSP Int. Publ., Utrecht, 1999.
   A.Yu. Khrennikov, *Non-Archimedian Analysis: Quantum Paradoxes, Dynamical Systems and Biological Models*, Kluwer Academic Publishers, Dordrecht, 1997.
8. A.Yu. Khrennikov, "*p*-adic probability distribution of hidden variables", *Physica A* **215**, 577–587 (1995).
9. A.Yu. Khrennikov, "Non-Kolmogorov probabilistic models with *p*-adic probabilities and foundations of quantum mechanics", *Stochastic Analysis and Related Topics* **4**, 275–304 (1998).
10. A.Yu. Khrennikov, "*p*-adic probability predictions of correlations between particles in the two slit and neutron interferometry experiments", *Il Nuovo Cimento* B **113**, 751–760 (1998).

11. A.A. Ezhov and A.Yu. Khrennikov, *Phys. Rev. E* **71**, 016138(1-9) (2005).
12. M.R. Evans, *Europhys. Lett.* **36**, 13–18 (1996).
13. G. Bianconi and A.-L. Barabasi, *Phys. Rev. Lett.* **86**, 5632–5635 (2001).
14. K. Staliunas, `e-print cond-mat/0001347`, (2000).
15. G. Bianconi,*Phys. Rev. E* **66**, 036116(1-5), (2002).
16. B. Derrida and J.L. Lebowitz, *Phys. Rev. Lett.* **80**, 209–213 (1998).
17. V.A. Lefebvre, *Algebra of Conscience*, Kluwer Academic Publisher, 2001.
18. R. Rammal, G. Toulouse, and M.A. Virasoro, *Reviews of Modern Physics* **58**, 765–788 (1986).
19. I.C. Lerman, *Classification et Analyse Ordinale de Donneés* Paris, Dunod, 1981.
20. F. Murtagh, *J. of Classification* **21**, 167–184 (2004).
21. H.C. Rosu and F.A. de la Cruz, *Physica Scripta* **65**, 377–382 (2002).

# *p*-Adic Strings and Their Applications

Peter G. O. Freund

*Enrico Fermi Institute and Department of Physics, University of Chicago, Chicago, IL 60637, USA*
*email: freund@theory.uchicago.edu*

**Abstract.** The theory of *p*-adic strings is reviewed along with some of their applications, foremost among them to the tachyon condensation problem in string theory. Some open problems are discussed, in particular that of the superstring in 10 dimensions as the end-stage of the 26-dimensional closed bosonic string's tachyon condensation.

## 1. INTRODUCTION

Though, within errors, the values yielded by measurement of any observable can, of necessity, always be expressed as rational numbers, it is useful to treat the "true" spectra of many observables as continuous, i.e. endow them with a topology. This requires a completion of the field **Q** of rational numbers. Normally, it is considered evident that this completion be Archimedean and one is thus led to the field $\mathbf{R} = \mathbf{Q}_\infty$ of real numbers. But just as space, originally viewed as evidently commutative and Euclidean, turned out to offer interesting and even realistic new possibilities when these constraints were relaxed, so we can ask what would happen were we to entertain the possibility of a non-Archimedean (*p*-adic) [1] completion of **Q**.

As it turns out, the resulting theories are extremely simple and related to the Archimedean theory in a well-defined manner [2, 3, 4, 5, 6]. One main advantage of this simplicity is that we can express the effective tachyon lagrangian in closed form, and therefore address directly the issue of tachyon condensation [7, 8].

I will first review the main ideas and results of the theory of *p*-adic strings. Then I shall show how *p*-adic strings are used in understanding tachyon condensation. Surprisingly, some open issues in tachyon condensation are related to some long-open issues in *p*-adic string theory. I will therefore consider to what extent this relation can be used to further our understanding of both.

## 2. REVIEW OF *p*-ADIC AND ADELIC STRINGS

To complete the field **Q** of rational numbers, we need Cauchy sequences and this requires a norm on **Q**. Remarkably, all norms on **Q** are known. There is of course the familiar "absolute value" norm, which is Archimedean (i.e. for any pair of rational numbers $x$ and $y$, with $|x| < |y|$, there exists an integer $N$, such that $|Nx| > |y|$). For our purposes it will be convenient to denote this norm as $|x|_\infty$, rather than $|x|$. Besides this Archimedean norm, we have an infinity of non-Archimedean norms $|x|_p$ labeled by the prime numbers $p$. These $p$-adic norms measure the divisibility of the rational number $x$ by the prime $p$ (the more divisible, the smaller). According to Ostrowski's famous theorem [1], the Archimedean norm $|x|_\infty$ along with the $p$-adic norms $|x|_p$ yield all the possible norms on the field **Q**.

In other words, there are infinitely many non-equivalent ways to complete the rationals **Q** to a field endowed with a topology. Of these completions precisely one is Archimedean and the infinite set of non-Archimedean completions are unambiguously labeled by the primes.

Using these $p$-adic norms to construct Cauchy sequences, we obtain locally compact completions $\mathbf{Q_p}$ of **Q**. These are the fields of $p$-adic numbers. Like $\mathbf{R} \equiv \mathbf{Q}_\infty$, all these $\mathbf{Q_p}$ are continua. Being locally compact, they have additive and multiplicative Haar measures [9], and as such we can integrate over these fields, just as we can over the reals. Moreover, these fields, just like the field of real numbers, have [9] additive and multiplicative $\mathbf{C}^*$-valued characters.

These observations allow us to construct all the $p$-adic string amplitudes. Indeed, the ordinary Archimedean Koba-Nielsen tree amplitudes are multiple integrals over $\mathbf{Q}_\infty$ of integrands built entirely of multiplicative characters of $\mathbf{Q}_\infty$. Replacing these characters of $\mathbf{Q}_\infty$ with characters of $\mathbf{Q_p}$, and replacing at the same time the integrals over $\mathbf{Q}_\infty$ by integrals over $\mathbf{Q_p}$, we obtain the tree amplitudes of a theory of $p$-adic strings. That these $p$-adic integrals do indeed represent the tree amplitudes of a string theory, is readily confirmed, by verifying that they possess the following features: they are a) meromorphic, b) Möbius-invariant, c) crossing symmetric, d) free of pairs of incompatible poles and e) they factorize.

Unlike for the ordinary Archimedean string amplitudes, for these $p$-adic string amplitudes the integrations can be fully carried out [5, 10, 11]and the result expressed [5] in terms of elementary functions.

The $p$-adic string, unlike its Archimedean counterpart, has no Hagedorn spectrum. Rather, its full spectrum is given by a single spin-zero tachyon. Therefore, without having to integrate out any higher modes, the $p$-adic amplitudes themselves can be used to generate the effective *target-space* tachyon action $S_p$ with the result,

$$S_p = \int d^d x L_p, \qquad (1)$$

with the lagrangian $L_p$ given as,

$$L_p = \frac{p}{p-1}[-\frac{1}{2}\sigma p^{-\partial^2-1}\sigma + \frac{1}{g^2}\frac{p}{p+1}(1+\frac{g\sigma}{p})^{p+1} - \frac{\sigma}{g} - \frac{1}{g^2}\frac{p}{p+1}]+C_p. \qquad (2)$$

Here $\sigma$ is the real scalar tachyon field propagating in $d$-dimensional Archimedean space-time (the familiar value $d = 26$ of the critical dimension of Archimedean bosonic strings can be shown, using adelic methods, to be needed in the *p*-adic case as well) and $g$ is the (*p*-dependent) coupling constant. Notice that the kinetic term is exponential in derivatives, making this lagrangian non-local. This non-locality has the consequence that along with the tachyon pole at $m^2 = -2$, the *p*-adic string amplitudes also have equally spaced *complex* poles at $m^2 = -2 + i4\pi N/\ln p$ for all integer $N$. The, in a lagrangian irrelevant, additive constant $C_p$ has been introduced here for future convenience. It is important to emphasize again that this lagrangian gets integrated over the full *real d*-dimensional target space. It remembers its *p*-adic origins through its non-locality and through the explicit appearance of the prime $p$ in the lagrangian. Momenta and target-space coordinates remain real (or complex upon suitable continuation), in other words Archimedean. Only the worldsheet is *p*-adic.

The field-redefinition,

$$\phi = 1 + g\sigma/p \tag{3}$$

allows us to rewrite this lagrangian as

$$L_p = \frac{1}{g^2}\frac{p^2}{p-1}\left[-\frac{1}{2}\phi p^{-\partial^2/2}\phi + \frac{1}{p+1}\phi^{p+1}\right]. \tag{4}$$

Notice that this $L_p$ vanishes for $\phi = 0$. This has been achieved by setting the arbitrary constant

$$C_p = -\frac{1}{2g^2}\frac{p^2}{p+1}, \tag{5}$$

This will make the discussion of tachyon condensation particularly transparent.

Exact solitonic solutions of the field equation

$$p^{-\partial^2/2}\phi = \phi^p \tag{6}$$

can be found. These solutions are important in understanding tachyon condensation, as will be shown in the next section.

What is the physical meaning of the *p*-adic strings? For an open string, the worldsheet's boundary is no longer the real line as in the ordinary Archimedean case, but is now postulated to be the *p*-adic line. What kind of worldsheet has such a boundary? The answer to this question is a certain *p*-adic symmetric space [12], a Bruhat-Tits tree. It is as if we were discretizing the worldsheet, for after all a Bruhat-Tits tree is a Bethe lattice. There is here an intrinsic difference between the target space and the worldsheet. We will return to this problem in the following sections.

Finally, before concluding our very brief review of *p*-adic strings, let us notice that any *p*-adic string requires the choice of the prime $p$. Now, even if the mentioned non-locality of the theory were not a major problem, were we trying to use *p*-adic strings for phenomenology, we would be in the preposterous position of having to ask our experimental colleagues to *measure* the prime number $p$ which underlies the theory. It would be much more satisfactory if somehow we could treat all primes on an equal footing, and this includes the prime "at infinity" which corresponds to the usual Archimedean strings. This can be achieved by considering the so-called *adelic* strings [3, 4, 6]. Adelic strings establish such a relation between all $p$-adic strings and the Archimedean strings. Specifically, by taking the (suitably regularized) product of the 4-point amplitudes of *all* $p$-adic strings and then multiplying the result with the ordinary Veneziano amplitude, we obtain a constant. In the limit in which all Archimedean and non-Archimedean string couplings are set equal to one, the value of this constant itself turns out to be equal to one. Strictly speaking this result holds at the tree level, but its generalization to the full amplitudes has also been proposed [6].

## 3. TACHYON CONDENSATION

A tachyon in the spectrum of a quantum system is a signal that we are expanding it around a false vacuum. It is therefore important to understand how the system finds its true vacuum. The bosonic string — whether Archimedean or $p$-adic — has a tachyon in its spectrum and as such we know that it is expanded around a false vacuum. This expansion corresponds to small $\sigma$ field, and indeed $\sigma = 0$, or equivalently $\phi = 1$ is a solution of the field equation (6). Being constant over all of target space, it corresponds to a target space volume-filling D-brane, specifically a D-(d-1)-brane. As expected this D-brane is the one around which we quantized the string and found the tachyon. The energy density of this D-brane can be read off Equation (4),

$$T_{d-1}^p = -L_p(\phi = 1) = \frac{1}{2g^2}\frac{p^2}{p+1}. \tag{7}$$

The linearized field equation around $\phi = 1$ has plane wave solutions with momentum constrained to the open string tachyon mass-shell.

The configuration $\phi = 0$ is also a solution of the field equation (6). With the choice (5) for $C_p$, we had $L_p(\phi = 0) = 0$, so the energy of this solution vanishes. There is no D-brane in this case, and there are therefore no plane wave solutions with finite momentum squared, in other words no open-string-like perturbative excitations around this $\phi = 0$ vacuum. All that is left are the more fundamental closed string excitations, but these can not be readily found in this not yet second-quantized picture.

Finally [5], the field equation (6) admits interesting soliton solutions. Specifically, the configuration,

$$\phi(x) = \Pi_{j=q+1}^{j=d-1} f(x^j) := F^{d-q-1}(x_\perp), \tag{8}$$

with

$$f(\eta) = p^{\frac{1}{2(p-1)}} \exp(-\frac{1}{2}\frac{p-1}{p\ln p}\eta^2), \ x_\perp = (x^{q+1},...,x^{d-1}), \tag{9}$$

solves the field equation (6), and its energy is localized around the hyperplane $x_\perp = 0$. Call $T_q$ the energy per unit $q$-volume of this solution. It is finite, exactly calculable, and has the remarkable property [7, 8] that the ratio $T_q/T_{q-1}$ is independent of $q$. This is precisely what happens for the ratio of the tensions of an ordinary D-$q$-brane and an ordinary D-$(q-1)$-brane in Archimedean bosonic string theory. Moreover these ordinary D-branes are also of solitonic nature. It then stands to reason to interpret [7, 8] the solitons of Eqs. (8), (9) as the D-$q$-branes of the $p$-adic string. In support of this interpretation, one can show [7, 8] that the spectrum and dynamics of fluctuations around these D-$q$-branes coincides with what one expects for the open string quantized on the D-$q$-brane.

This interpretation is indirect in the following sense. The whole argument was made at the level of the effective tachyon action. This involves the theory of a *real* scalar field in ordinary Archimedean space-time, with the $p$-adic features encoded, as explained above, through the non-locality of the lagrangian, and the explicit appearance of the prime $p$ in it. The fluctuations around the D-brane are identified with $p$-adic strings also at the tachyon lagrangian level. This description lacks the intuitive picture of a string moving through space-time with its ends attached to the D-brane. This is just as good, since in the $p$-adic case, the world lines of the string's ends are $p$-adic lines and as such could not be fit into a D-brane all of whose coordinates are valued in the *Archimedean* field of real numbers.

The just described features of the effective tachyon action which were readily obtained in the $p$-adic string case, correspond to some very general conjectures about tachyon condensation [8]. The advantage in studying $p$-adic strings is precisely that one can easily do analytic calculations and verify these conjectures.

It is interesting to note that in the (local) limit $p \to 1$ the p-adic string's effective action has been found [13], [14] to approximate up to terms with two derivatives, the effective action of the Archimedean bosonic string calculated using boundary string field theory.

In the usual *Archimedean* bosonic string, when $N$ D-branes coincide, a $U(N)$ symmetry with the attendant Chan-Paton (CP) rules is known to emerge. Does something similar happen in the $p$-adic case as well? A possible theory of $p$-adic strings with CP rules has been proposed [6]. It had the undesirable feature that for $N = 1$ its amplitudes did not reduce to the those of the theory without CP rules described above. This problem can be corrected by constructing an alternative $p$-adic string with CP rules as follows. In the lagrangian (4) let $\phi$ denote not a unique real scalar field, but an $N \times N$ matrix of such fields, with $L_p$ obtained from Eq. (4) by taking the matrix trace. Then the theory clearly reduces to what it should for $N = 1$. The question that remains to be answered is: what is the worldsheet description of the $p$-adic theory whose effective tachyon action is given by this matrix-alternative construction?

In the Archimedean case, unstable D-branes decay into closed strings, so that the open string theory contains the whole dynamics of D-branes and even much of the closed string dynamics. This is the content of Sen's far-reaching *open string completeness conjecture*. Closed bosonic strings also have a tachyon, but here the tachyon condensation is a much more complex issue. One might at first think that also for closed strings the $p$-adic case will allow for an exact treatment. Unfortunately this is not the case. The main difference between the open and closed bosonic Archimedean strings is that to obtain the tree amplitudes one integrates over the field $\mathbf{R} = \mathbf{Q}_\infty$ for the open strings and over the field $\mathbf{C}$ of complex numbers for the closed strings. The field $\mathbf{C}$ being a quadratic extension of the field $\mathbf{R}$, at the $p$-adic level one can try to use a quadratic extension of the field $\mathbf{Q_p}$. But the Archimedean and non-Archimedean situations are different. The field $\mathbf{C}$ of complex numbers is algebraically closed, whereas quadratic extensions of the field $\mathbf{Q_p}$ of $p$-adic numbers – just like the quadratic extensions of the "global" field $\mathbf{Q}$ of rational numbers — are neither unique nor algebraically closed.

If one uses these quadratic extensions of $\mathbf{Q_p}$ anyway, one obtains sets of consistent $n$-point amplitudes, with the corresponding 4-point amplitudes in just as nice an adelic product relationship with the Archimedean *closed* string amplitudes, as was the case for $\mathbf{Q_p}$ and the open string 4-point amplitudes. This may lead us to think that these are the proper closed $p$-adic strings. But this is not necessarily so. First of all we have to ask what would have happened, had we taken not a quadratic, but a higher algebraic extension of the field $\mathbf{Q_p}$. These would be adelically related not to open or closed string Archimedean 4-point amplitudes, but rather to products of various powers of open and closed string amplitudes, the precise values of these powers depending on precisely which of the infinitely many possible algebraic extensions is being considered. A product of $n_r$ open and $n_i$ closed string amplitudes, needed in such an adelic product does not seem to make any physical sense when $n_r > 1$, or $n_i > 1$, or $n_r n_i = 1$. It appears that we are dealing in all these cases with ever more exotic forms of open string amplitudes. This is most easily confirmed by noticing that for all these cases one can construct an effective tachyon action, with the corresponding field equations admitting various solitonic D-brane solutions, a clear sign that one is dealing with — admittedly more exotic kinds of — open strings. To get to what are "true" closed strings, one may have to switch from the boundaries of Bruhat-Tits trees to the boundaries of infinite graphs that are no longer trees, but contain loops as well [15]. Unfortunately effective tachyon actions are no longer available in this case.

There remains the question as to the nature of the closed bosonic string's vacuum. If anything like what happened for the open bosonic string, happens here as well, then in this vacuum there should be no closed *bosonic* strings. In other words, there should be no 26-dimensional graviton! Then all of 26-dimensional space should disappear as well. What is *it* replaced by? Obviously by a system without a tachyon. But to eliminate tachyons we must end up with something closely resembling supersymmetry, as has been shown some time ago by Kutasov and Seiberg [16]. The most natural outcome would then be the 10-dimensional space of the superstring, with its graviton. This is in line with an old idea of the superstring as a vacuum of the bosonic string [17, 18, 19]. This way, not only would the five types of 10-dimensional superstring theories be unified, but the odd-man-out, the 26-dimensional bosonic string would be unified with its supersymmetric partners as well. We could start with bosons only, and then fermions would be forced upon us, so as to get rid of the tachyons.

As in the Archimedean case, in the $p$-adic case as well, an antisymmetric rank-2 tensor field $B$ can be included [20],with a resultant non-commutative target space geometry underlying the effective tachyon field action. This takes some doing and involves $p$-adic sign-functions in a manner similar to that used in [5] when introducing CP rules. There, as was already mentioned, for $N = 1$ one did not recover the ordinary theory without CP rules. Here a similar problem arises, for if one sets $B = 0$ one does not recover the ordinary string amplitudes without $B$ field. It would be interesting to see whether there is a connection between these singular behaviors observed when introducing CP rules and non-vanishing $B$ field.

Finally, although not directly related to the issue of tachyon condensation, I should mention one more, potentially important, adelic feature of string theory. In the compactifications of various string theories one encounters [21, 22] scalar fields which live on the so-called Narain moduli spaces (similar spaces were already encountered [23] in toroidally compactified supergravities). All these Narain moduli spaces are double coset spaces of the form $G(\mathbf{Z})\backslash G(\mathbf{R})/K_\mathbf{R}$, with $G(\mathbf{R})$ a non-compact real form of some Lie group, $K_\mathbf{R}$ its maximal compact subgroup, and $G(\mathbf{Z})$, which ultimately functions as duality group, the suitably defined discrete group $G$ over the integers. Typical cases are $G = O(22,6;\mathbf{R})$ for heterotic string compactifications and $E_{7,7}$ for certain type II compactifications. The remarkable feature is that these double coset spaces can, on the basis of the strong approximation theorem [9], also be represented as the *adelic* double coset spaces $G(\mathbf{Q})\backslash G(\mathbf{A})/K_\mathbf{A}$, with $G(\mathbf{A})$ the group $G$ over the adeles, $K_\mathbf{A}$ *its* maximal compact subgroup, and $G(\mathbf{Q})$, the group $G$ over the rationals. (this equivalence was used in the theory of scattering on the hyperbolic plane [24]). This observation suggests that a proper understanding of string dualities should involve an adelic formulation. At the same time it should be mentioned, that at the $p$-adic level string compactifications have not yet been explored.

## 4. OUTLOOK

Originally, the possibility of the $p$-adic and adelic strings reviewed above, was recognized, once the importance of algebraic geometry in string theory became clear. After all, the worldsheets of closed strings are compact Riemann surfaces, all of which are algebraic curves (one *complex* dimension), as was known already to Riemann and then rigorously proved by Weyl. On the other hand, algebraic geometry is intimately tied in with number theory, as had been recognized by mathematicians for over a century. This immediately suggested the generalization of strings to the $p$-adic and adelic case. The most ambitious possibility is that in nature we are really dealing with adelic strings, and Archimedean strings are only an approximation to these adelic strings, say at low energies. This far-reaching point of view, though still tenable, has not led to any major progress. Part of the reason for this is that $p$-adic strings in their current form do not seem to take well to supersymmetry. The other difficulty is that adelic product formulas have only been obtained for 4-point tree amplitudes. Though ways of going beyond tree level have been suggested [6], they are more complicated and have not led to any major insights.

Then came the realization [7] that the remarkable intuition [8] about tachyon condensation can be easily checked in the $p$-adic case. In this context non-Archimedean strings have regained some popularity, but essentially as "toy models."

If we insist on a fundamental role for $p$-adic and adelic strings, then we first have to master the supersymmetric case and higher orders in perturbation theory.

There is also the issue whether it is consistent to go *p*-adic only on the worldsheet, while leaving the coordinates of target space (and therefore their conjugate momenta) Archimedean. We have seen how the effective tachyon actions of *p*-adic strings handle this issue in a rather indirect and non-intuitive manner. But could one entertain the possibility of a world, which is *p*-adic both on the worldsheet and in target space? This possibility has been advocated [25, 26], but within its context even the strings, where everything started, have not yielded a meaningful theory.

I am aware of the fact that what I have just written has a pessimistic tinge to it. To end on a positive note, let me stress that the adelic string product formulas and the solitonic D-brane solutions of the field equations derived by extremizing the effective tachyon action, are both mathematically beautiful and physically interesting. As far as Physics is concerned, they have contributed a useful laboratory where one can test tachyon condensation ideas. I simply cannot think of all this as a sequence of amusing accidents and dismiss it as a "toy model." I think more work on this is both needed and worthwhile. After all, to put things in perspective, the main *known* virtues of the whole edifice of *Archimedean* string theory are *its* mathematical richness and *its* role as a useful laboratory where one can test ideas about black holes.

## ACKNOWLEDGMENTS

I wish to thank Prof. Branko Dragovich for inviting me to participate in the organization of and to speak at the Belgrade Conference on *p*-adic Mathematical Physics.

## REFERENCES

1. N. Koblitz, *p-adic Numbers, p-adic Analysis and Zeta Functions*, Springer, Berlin,1984.
2. P.G.O. Freund and M. Olson, *Phys. Lett.* **B 191**, 186 (1987).
3. P.G.O. Freund and E. Witten, *Phys. Lett.* **B 199**, 191 (1987).
4. Yu.I. Manin, Talk at the Poiana Braşov Workshop, 1987.
5. L. Brekke, P.G.O. Freund, M. Olson and E. Witten, *Nucl. Phys.* **B 302**, 365 (1988).
6. For a review, see L. Brekke and P.G.O. Freund, *Phys. Rep.* **233**, 1 (1993).
7. D. Ghoshal and A. Sen, *JHEP* **0011**, 021 (2000), hep-th/0009191.
8. The current status of the important tachyon condensation conjectures made in 1998 by A. Sen is reviewed by him in, hep-th/0410103. The older review of W. Taylor and B. Zwiebach, in *Boulder 2001, Strings, Branes and Extra Dimensions*, hep-th/0311017 discusses also the early work of K. Bardakci and M.B. Halpern on tachyon condensation in string theory.
9. I.M. Gel'fand, M.I. Graev and I.I. Pyatetskii-Shapiro, *Representation Theory and Automorphic Functions*, Saunders, London 1966.
10. Z. Hlousek and D. Spector, *Phys. Lett.* **B 214** 19 (1988); *Ann. Phys.* **189**, 370 (1989).
11. P.H. Frampton and H. Nishino, *Phys. Lett.* **B 242**, 354 (1990).
12. A.V. Zabrodin, *Mod. Phys. Lett.* **A 4**, 367 (1989); *Commun. Math. Phys.* **123**, 463 (1989).
13. A.A. Gerasimov and S. Shatashvili, *JHEP* **0011**, 034 (2000), hep-th/0009103.
14. D. Ghoshal, *JHEP* **0409**, 041 (2004), hep-th/0406259.

15. L.O. Chekhov, A.D. Mironov and A.V. Zabrodin, *Commun. Math. Phys.* **125**, 675 (1989).
16. D. Kutasov and N. Seiberg, *Nucl. Phys.* **B 358**, 600 (1991).
17. P.G.O. Freund, *Phys. Lett.* **B 151**, 387 (1984).
18. A. Casher, F. Englert, H. Nicolai and A. Taormina, *Phys. Lett.* **B 162** 121 (1985); F. Englert, in *Fifty years of Yang Mills theory*, G. 't Hooft, editor, World Scientific, Singapore, 2005, `hep-th/0406162`.
19. The possible relevance of the work [17,18] on superstrings as vacua of the bosonic string, in the context of the recent work on tachyon condensation in the case of the closed bosonic string, was independently realized also by M. Roček (private communication).
20. D. Ghoshal and T. Kawano, `hep-th/0409311`.
21. K.S. Narain, *Phys. Lett.* **B 169**, 41 (1986).
22. For a review see J. Schwarz, *Nucl. Phys. Proc. Suppl.* **55**, 1 (1997), `hep-th/9607201`.
23. B. Julia and E Cremmer, *Nucl. Phys.* **B 159**, 141 (1979).
24. P.G.O. Freund, *Phys. Lett.* **B 257**, 119 (1991).
25. I.V. Volovich, *Class. Quant. Grav.* **4** L83, (1987).
26. I.Ya. Aref'eva, B. Dragovich, P.H. Frampton and I.V. Volovich, *Int. J. Mod. Phys.* **A 6**, 4341 (1991).

# Proof of the Kurlberg-Rudnick Rate Conjecture

Shagmar Gurevich* and Ronny Hadani*

*School of Mathematical Sciences, Tel Aviv University, Tel Aviv, ISRAEL
emails: shamgar@math.tau.ac.il , nogaporat@hotmail.com

**Abstract.** In this paper we present a proof of the *Hecke quantum unique ergodicity conjecture* for the Berry-Hannay model, a model of quantum mechanics on a two dimensional torus. This conjecture was stated in Z. Rudnick's lectures at MSRI, Berkeley, 1999 and ECM, Barcelona, 2000.

## 1. INTRODUCTION

**Hannay-Berry model.** In 1980 the physicists Sir M.V. Berry and J. Hannay [1] explore a model for quantum mechanics on the two dimensional symplectic torus $(\mathbf{T}, \omega)$.

**Quantum chaos.** Consider the ergodic discrete dynamical system on the torus, which is generated by an hyperbolic automorphism $A \in \mathrm{SL}_2(\mathbb{Z})$. Quantizing the system, we replace: the classical phase space $(\mathbf{T}, \omega)$ by a Hilbert space $\mathcal{H}_\hbar$, classical observables, i.e., functions $f \in C^\infty(\mathbf{T})$, by operators $\pi_\hbar(f) \in \mathrm{End}(\mathcal{H}_\hbar)$ and classical symmetries by a unitary representation $\rho_\hbar : \mathrm{SL}_2(\mathbb{Z}) \longrightarrow \mathrm{U}(\mathcal{H}_\hbar)$. A fundamental meta-question in the area of quantum chaos is to *understand* the ergodic properties of the quantum system $\rho_\hbar(A)$, at least in the semi-classical limit as $\hbar \to 0$.

**Hecke quantum unique ergodicity.** This question was addressed in a paper by Kurlberg and Rudnick [6]. In this paper they formulated a rigorous definition of quantum ergodicity for the case $\hbar = \frac{1}{p}$. The basic observation is that the representation $(\rho_\hbar, \mathcal{H}_\hbar)$ is finite dimensional and factors through the quotient group $\mathrm{SL}_2(\mathbb{F}_p)$. We denote by $\mathrm{T}_A \subset \mathrm{SL}_2(\mathbb{F}_p)$ the centralizer of the element $A$, now considered as an element of the quotient group $\mathrm{SL}_2(\mathbb{F}_p)$. The group $\mathrm{T}_A$ is called (cf. [6]) the *Hecke* torus corresponding to the element $A$. The Hecke torus acts semisimply on $\mathcal{H}_\hbar$. Therefore we have a decomposition:

$$\mathcal{H}_\hbar = \bigoplus_{\chi : \mathrm{T}_A \to \mathbb{C}^*} \mathcal{H}_\chi$$

where $\mathcal{H}_\chi$ is the Hecke eigenspace corresponding to the character $\chi$. Considering a unit vector $v \in \mathcal{H}_\chi$, one defines the *Wigner* distribution $\mathcal{W}_\chi : C^\infty(\mathbf{T}) \longrightarrow \mathbb{C}$ by the formula $\mathcal{W}_\chi(f) := \langle v | \pi_\hbar(f) v \rangle$. The main statement in [6] asserts about an explicit bound of the semi-classical asymptotic of $\mathcal{W}_\chi(f)$:

$$\left| \mathcal{W}_\chi(f) - \int_\mathbf{T} f\omega \right| \leq \frac{C_f}{p^{1/4}}$$

where $C_f$ is a constant that depends only on the function $f$. In Rudnick's lectures at MSRI, Berkeley 1999 [7] and ECM, Barcelona 2000 [8] he conjectured that a stronger bound should hold true, namely:

**Conjecture 1.1 (Rate Conjecture).** *The following bound holds:*

$$\left| \mathcal{W}_\chi(f) - \int_{\mathbf{T}} f \omega \right| \le \frac{C_f}{p^{1/2}}.$$

The basic *clues* suggesting the validity of this stronger bound come from two main sources. The first source is *computer* simulations [5] accomplished over the years to give extremely precise bounds for considerably large values of $p$. A more mathematical argument is based on the fact that for special values of $p$, in which the Hecke torus *splits*, namely $T_A \simeq \mathbb{F}_p^*$, one is able to compute explicitly the eigenvector $v \in \mathcal{H}_\chi$ and as a consequence to give an explicit *formula* for the Wigner distribution [4]. More precisely, in case $\xi \in \mathbf{T}^\vee$, i.e., a character, the distribution $\mathcal{W}_\chi(\xi)$ turns out to be equal to an exponential sum very much similar to the Kloosterman sum:

$$\frac{1}{p} \sum_{a \in \mathbb{F}_p^*} \psi\left(\frac{a+1}{a-1}\right) \sigma(a)\chi(a)$$

where $\sigma$ and $\psi$ denote the Legendre character and the standard additive character correspondingly. In this case the classical Weil bound [10] yields the result.

**Geometric approach.** The basic observation to be made is that the theory of quantum mechanics on the torus, in case $\hbar = \frac{1}{p}$, can be equivalently recast in the language of representation theory of finite groups in characteristic $p$. Consider the quotient $\mathbb{F}_p$ vector space $V = \mathbf{T}^\vee / p\mathbf{T}^\vee$, where $\mathbf{T}^\vee$ is the lattice of characters on $\mathbf{T}$. We denote by $H = H(V)$ the Heisenberg group. The group $\mathrm{SL}_2(\mathbb{F}_p)$ is naturally identified with the group of linear symplectomorphisms of V. We have an action of $\mathrm{SL}_2(\mathbb{F}_p)$ on H. The Stone-von Neumann theorem states that there exists a unique irreducible representation $\pi : H \longrightarrow \mathrm{GL}(\mathcal{H})$, with the non-trivial central character $\psi$, for which its isomorphism class is fixed by $\mathrm{SL}_2(\mathbb{F}_p)$. This is equivalent to saying that $\mathcal{H}$ is equipped with a compatible projective representation $\rho : \mathrm{SL}_2(\mathbb{F}_p) \longrightarrow \mathrm{PGL}(\mathcal{H})$. Noting that H and $\mathrm{SL}_2(\mathbb{F}_p)$ are the sets of rational points of corresponding algebraic groups, it is natural to *ask* whether there exists an algebro-geometric object that underlies the pair $(\pi, \rho)$? The answer to this question is *positive*. The construction is proposed in an unpublished letter of Deligne to Kazhdan [2]. In one sentence, the content of this letter is a construction of *Representation Sheaves* $\mathcal{K}_\pi$ and $\mathcal{K}_\rho$ on the algebraic varieties $\mathbb{H}$ and $\mathbb{SL}_2$ respectively. One obtains, as a consequence, the following general principle:

(*) **Motivic principle**: All quantum mechanical quantities in the Berry-Hannay model are motivic in nature.

By this we mean that every quantum-mechanical quantity $\mathcal{Q}$, is associated with a vector space $V_\mathcal{Q}$ endowed with a Frobenius action $\mathrm{Fr} : V_\mathcal{Q} \longrightarrow V_\mathcal{Q}$ s.t. $\mathcal{Q} = \mathrm{Tr}(\mathrm{Fr}_{|V_\mathcal{Q}})$. The *main contribution* of this paper is to implement this principle. In particular we show that there exists a two dimensional vector space $V_\chi$, endowed with an action $\mathrm{Fr} : V_\chi \longrightarrow V_\chi$

s.t. $\mathscr{W}_\chi(\xi) = \mathrm{Tr}(\mathrm{Fr}_{|V_\chi})$. This, combined with a bound on the modulus of the eigenvalues of Frobenius, i.e., $|\mathrm{e.v}(\mathrm{Fr}_{|V_\chi})| \leq \frac{1}{p^{1/2}}$, completes the proof of the rate conjecture.

## ACKNOWLEDGMENTS

We thank our Ph.D. adviser J. Bernstein for his interest and guidance in this project. We thank P. Kurlberg and Z. Rudnick who discussed with us their papers and explained their results. We would like to thank David Kazhdan for sharing his thoughts about the possible existence of canonical Hilbert spaces. Finally, we would like to thank P. Deligne for letting us publish his ideas about the geometrization of the Weil representation which appeared in a letter he wrote to David Kazhdan in 1982.

## 2. CLASSICAL TORUS

Let $(\mathbf{T}, \omega)$ be the two dimensional symplectic torus. Together with its linear symplectomorphisms $\Gamma \simeq \mathrm{SL}_2(\mathbb{Z})$ it serves as a simple model of classical mechanics (a compact version of the phase space of the harmonic oscillator). More precisely, let $\mathbf{T} = W/\Lambda$ where W is a two dimensional real vector space and $\Lambda$ is a rank two unimodular lattice in W. We denote by $\Lambda^* \subseteq W^*$ the dual lattice, i.e., $\Lambda^* = \{\xi \in W^* \mid \xi(\Lambda) \subset \mathbb{Z}\}$. The lattice $\Lambda^*$ is identified with the lattice of characters of $\mathbf{T}$ by the map $\xi \in \Lambda^* \longmapsto e^{2\pi i \langle \xi, \cdot \rangle} \in \mathbf{T}^\vee$, where $\mathbf{T}^\vee := \mathrm{Hom}(\mathbf{T}, \mathbb{C}^*)$.

**Classical mechanical system.** We consider a very simple discrete mechanical system. An hyperbolic element $A \in \Gamma$, i.e., $|\mathrm{Tr}(A)| > 2$, generates an ergodic discrete dynamical system on $\mathbf{T}$.

## 3. QUANTIZATION OF THE TORUS

**The Weyl quantization model.** The Weyl quantization model works as follows. Let $\mathscr{A}_\hbar$ be a one parameter deformation of the algebra $\mathscr{A}$ of trigonometric polynomials on the torus. This algebra is known in the literature as the Rieffel torus [9]. The algebra $\mathscr{A}_\hbar$ is constructed by taking the free algebra over $\mathbb{C}$ generated by the symbols $\{s(\xi) \mid \xi \in \Lambda^*\}$ and quotient out by the relation $s(\xi + \eta) = e^{\pi i \hbar \omega(\xi,\eta)} s(\xi) s(\eta)$. Here $\omega$ is the form on $W^*$ induced by the original form $\omega$ on W. The algebra $\mathscr{A}_\hbar$ contains as a standard basis the lattice $\Lambda^*$. Therefore, one can identify the algebras $\mathscr{A}_\hbar \simeq \mathscr{A}$ as vector spaces. Hence, every function $f \in \mathscr{A}$ can be viewed as an element of $\mathscr{A}_\hbar$. For a fixed $\hbar$ a representation $\pi_\hbar : \mathscr{A}_\hbar \longrightarrow \mathrm{End}(\mathscr{H}_\hbar)$ serves as a quantization protocol.

**Equivariant Weyl quantization of the torus.** The group $\Gamma$ acts on the lattice $\Lambda^*$, therefore it acts on $\mathscr{A}_\hbar$. For an element $B \in \Gamma$, we denote by $f \longmapsto f^B$ the action of

$B$ on an element $f \in \mathscr{A}_\hbar$. Let $\Gamma_p \simeq \mathrm{SL}_2(\mathbb{F}_p)$ denotes the quotient group of $\Gamma$ modulo $p$.

**Theorem 3.1 (Canonical equivariant quantization).** *Let $\hbar = \frac{1}{p}$, where $p$ is an odd prime. There exists a unique (up to isomorphism) pair of representations $\pi_\hbar : \mathscr{A}_\hbar \longrightarrow \mathrm{End}(\mathscr{H}_\hbar)$ and $\rho_\hbar : \Gamma \longrightarrow \mathrm{GL}(\mathscr{H}_\hbar)$ satisfying the compatibility condition (Egorov identity) $\rho_\hbar(B)\pi_\hbar(f)\rho_\hbar(B)^{-1} = \pi_\hbar(f^B)$, where $\pi_\hbar$ is an irreducible representation and $\rho_\hbar$ is a representation of $\Gamma$ that factors through the quotient group $\Gamma_p$.*

**Quantum mechanical system.** Let $(\pi_\hbar, \rho_\hbar, \mathscr{H}_\hbar)$ be the canonical equivariant quantization. Let $A$ be our fixed hyperbolic element, considered as an element of $\Gamma_p$. The element $A$ generates a quantum dynamical system. For every (pure) quantum state $v \in S(\mathscr{H}_\hbar) = \{v \in \mathscr{H}_\hbar : \|v\| = 1\}$, $v \longmapsto v^A := \rho_\hbar(A)v$.

## 4. HECKE QUANTUM UNIQUE ERGODICITY

Denote by $\mathrm{T}_A$ the centralizer of $A$ in $\Gamma_p \simeq \mathrm{SL}_2(\mathbb{F}_p)$. We call $\mathrm{T}_A$ the *Hecke torus* (cf. [6]). The precise statement of the **Kurlberg-Rudnick conjecture** (cf. [7, 8]) is given in the following theorem:

**Theorem 4.1 (Hecke Quantum Unique Ergodicity).** *Let $\hbar = \frac{1}{p}$, $p$ an odd prime. For every $f \in \mathscr{A}_\hbar$ and $v \in S(\mathscr{H}_\hbar)$, we have:*

$$\left| \mathrm{Av}_{\mathrm{T}_A}(<v|\pi_\hbar(f)v>) - \int_{\mathbb{T}} f\omega \right| \leq \frac{C_f}{\sqrt{p}}, \qquad (1)$$

*where* $\mathrm{Av}_{\mathrm{T}_A}(<v|\pi_\hbar(f)v>) := \sum_{B \in \mathrm{T}_A} <v|\pi_\hbar(f^B)v>$ *is the average with respect to the group* $\mathrm{T}_A$ *and $C_f$ is an explicit constant depending only on $f$.*

## 5. PROOF OF THE HECKE QUANTUM UNIQUE ERGODICITY CONJECTURE

It is enough to prove the conjecture for the case when $f$ is a non-trivial character $\xi \in \Lambda^*$ and $v$ is an Hecke eigenvector with eigencharacter $\chi : \mathrm{T}_A \longrightarrow \mathbb{C}^*$. In this case Theorem 4.1 can be restated in the form:

**Theorem 5.1 (Hecke Quantum Unique Ergodicity (Restated)).** *Let $\hbar = \frac{1}{p}$, where $p$ is an odd prime. For every $\xi \in \Lambda^*$ and every character $\chi : \mathrm{T}_A \longrightarrow \mathbb{C}^*$ the following holds:*

$$\left| \sum_{B \in \mathrm{T}_A} \mathrm{Tr}(\rho_\hbar(B)\pi_\hbar(\xi))\chi(B) \right| \leq 2\sqrt{p}.$$

**The trace function.** Denote by $F$ the function $F : \Gamma \times \Lambda^* \longrightarrow \mathbb{C}$ defined by $F(B,\xi) = \mathrm{Tr}(\rho(B)\pi_h(\xi))$. We denote by $V := \Lambda^*/p\Lambda^*$ the quotient vector space, i.e., $V \simeq \mathbb{F}_p^2$. The symplectic form $\omega$ specializes to give a symplectic form on $V$. The group $\Gamma_p$ is the group of linear symplectomorphisms of $V$, i.e., $\Gamma_p = \mathrm{Sp}(V, \omega)$. Set $Y_0 := \Gamma \times \Lambda^*$ and $Y := \Gamma_p \times V$. We have a natural quotient map $Y_0 \longrightarrow Y$.

**Lemma 5.2.** *The function $F : Y_0 \longrightarrow \mathbb{C}$ factors through the quotient $Y$.*

From now on $Y$ will be considered as the default domain of the function $F$. The function $F : Y \longrightarrow \mathbb{C}$ is invariant with respect to the action of $\Gamma_p$ on $Y$ given by the following formula:

$$\begin{aligned}\Gamma_p \times Y &\xrightarrow{\alpha} Y, \\ (S,(B,\xi)) &\longmapsto (SBS^{-1}, S\xi).\end{aligned} \qquad (2)$$

**Geometrization (Sheafification).** Next, we will phrase a geometric statement that will imply Theorem 5.1. Moving into the geometric setting, we replace the set $Y$ by an algebraic variety and the functions $F$ and $\chi$ by sheaf theoretic objects, also of a geometric flavor.

**Step 1.** The set $Y$ is the set of rational points of an algebraic variety $\mathbb{Y}$ defined over $\mathbb{F}_p$. To be more precise, $\mathbb{Y} \simeq \mathbb{S}p \times \mathbb{V}$. The variety $\mathbb{Y}$ is equipped with an endomorphism $\mathrm{Fr} : \mathbb{Y} \longrightarrow \mathbb{Y}$ called Frobenius. The set $Y$ is identified with the set of fixed points of Frobenius $Y = \mathbb{Y}^{\mathrm{Fr}} = \{y \in \mathbb{Y} : \mathrm{Fr}(y) = y\}$. Finally, we denote by $\alpha$ the algebraic action of $\mathbb{S}p$ on the variety $\mathbb{Y}$ (cf. (2)).

**Step 2.** The following theorem proposes an appropriate sheaf theoretic object standing in place of the function $F : Y \longrightarrow \mathbb{C}$. Denote by $\mathscr{D}^b_{c,w}(\mathbb{Y})$ the bounded derived category of constructible $\ell$-adic Weil sheaves on $\mathbb{Y}$.

**Theorem 5.3 (Geometrization Theorem).** *There exists an object $\mathscr{F} \in \mathscr{D}^b_{c,w}(\mathbb{Y})$ satisfying the following properties:*

1. *(Function) It is associated, via the sheaf-to-function correspondence, to the function $F : Y \longrightarrow \mathbb{C}$, i.e., $f^{\mathscr{F}} = F$.*
2. *(Weight) It is of weight $w(\mathscr{F}) \leq 0$.*
3. *(Equivariance) For every element $S \in \mathbb{S}p$ there exists an isomorphism $\alpha_S^* \mathscr{F} \simeq \mathscr{F}$.*
4. *(Formula) On introducing coordinates $\mathbb{V} \simeq \mathbb{A}^2$ we identify $\mathbb{S}p \simeq \mathbb{SL}_2$. Then there exists an isomorphism $\mathscr{F}_{|\mathbb{T}\times\mathbb{V}} \simeq \mathscr{L}_{\psi(\frac{1}{2}\lambda\mu\frac{a+1}{a-1})} \otimes \mathscr{L}_{\sigma(a)}$.[1]*

   Here $\mathbb{T} := \{\begin{pmatrix} a & 0 \\ 0 & a^{-1} \end{pmatrix}\}$ stands for the standard torus, $(\lambda, \mu)$ are the coordinates on $\mathbb{V}$ and $\mathscr{L}_\psi$, $\mathscr{L}_\sigma$ the Artin-Schreier and Kummer sheaves.

**Geometric statement.** Fix an element $\xi \in \Lambda^*$ with $\xi \neq 0$. We denote by $i_\xi$ the inclusion map $i_\xi : \mathbb{T}_A \times \xi \longrightarrow \mathbb{Y}$. Going back to Theorem 5.1 and putting its content in a functorial

---

[1] By this we mean that $\mathscr{F}_{|\mathbb{T}\times\mathbb{V}}$ is isomorphic to the extension of the sheaf defined by the formula in the right-hand side.

notation, we write the following inequality:

$$\left| pr_!(i_\xi^*(F) \cdot \chi) \right| \leq 2\sqrt{p}.$$

In words, taking the function $F : Y \longrightarrow \mathbb{C}$ and restricting $F$ to $\mathrm{T}_A \times \xi$ and get $i_\xi^*(F)$. Multiply $i_\xi^* F$ by the character $\chi$ to get $i_\xi^*(F) \cdot \chi$. Integrate $i_\xi^*(F) \cdot \chi$ to the point, this means to sum up all its values, and get a scalar $a_\chi := pr_!(i_\xi^*(F) \cdot \chi)$. Here $pr$ stands for the projection $pr : \mathrm{T}_A \times \xi \longrightarrow pt$. Then Theorem 5.1 asserts that the scalar $a_\chi$ is of an absolute value less than $2\sqrt{p}$.

Repeat the same steps in the geometric setting. We denote again by $i_\xi$ the closed imbedding $i_\xi : \mathbb{T}_A \times \xi \longrightarrow \mathbb{Y}$. Take the sheaf $\mathscr{F}$ on $\mathbb{Y}$ and apply the following sequence of operations. Pull-back $\mathscr{F}$ to the closed subvariety $\mathbb{T}_A \times \xi$ and get the sheaf $i_\xi^*(\mathscr{F})$. Take the tensor product of $i_\xi^*(\mathscr{F})$ with the Kummer sheaf $\mathscr{L}_\chi$ and get $i_\xi^*(\mathscr{F}) \otimes \mathscr{L}_\chi$. Integrate $i_\xi^*(\mathscr{F}) \otimes \mathscr{L}_\chi$ to the point and get the sheaf $pr_!(i_\xi^*(\mathscr{F}) \otimes \mathscr{L}_\chi)$ on the point.

Recall $w(\mathscr{F}) \leq 0$. Knowing that the Kummer sheaf has weight $w(\mathscr{L}_\chi) \leq 0$ we deduce that $w(i_\xi^*(\mathscr{F}) \otimes \mathscr{L}_\chi) \leq 0$.

**Theorem 5.4 (Deligne, Weil II [3]).** *Let $\pi : \mathbb{X}_1 \longrightarrow \mathbb{X}_2$ be a morphism of algebraic varieties. Let $\mathscr{L} \in \mathscr{D}_{c,w}^b(\mathbb{X}_1)$ be a sheaf of weight $w(\mathscr{L}) \leq w$ then $w(\pi_!(\mathscr{L})) \leq w$.*

Using Theorem 5.4 we get $w(pr_!(i_\xi^*(\mathscr{F}) \otimes \mathscr{L}_\chi)) \leq 0$.

Now, consider the sheaf $\mathscr{G} := pr_!(i_\xi^*(\mathscr{F}) \otimes \mathscr{L}_\chi)$. It is an object in $\mathscr{D}_{c,w}^b(pt)$. The sheaf $\mathscr{G}$ is associated by *Grothendieck's Sheaf-To-Function correspondence* to the scalar $a_\chi$:

$$a_\chi = \sum_{i \in \mathbb{Z}} (-1)^i \mathrm{Tr}(\mathrm{Fr}|_{\mathrm{H}^i(\mathscr{G})}). \tag{3}$$

Finally, we can give the geometric statement about $\mathscr{G}$, which will imply Theorem 5.1.

**Lemma 5.5 (Vanishing Lemma).** *Let $\mathscr{G} = pr_!(i_\xi^*(\mathscr{F}) \otimes \mathscr{L}_\chi)$. All cohomologies $\mathrm{H}^i(\mathscr{G})$ vanish except for $i = 1$. Moreover, $\mathrm{H}^1(\mathscr{G})$ is a two dimensional vector space.*

Theorem 5.1 now follows easily. By Lemma 5.5 only the first cohomology $\mathrm{H}^1(\mathscr{G})$ does not vanish and it is two dimensional. Having that $w(\mathscr{G}) \leq 0$ implies that the eigenvalues of Frobenius acting on $\mathrm{H}^1(\mathscr{G})$ are of absolute value $\leq \sqrt{p}$. Hence, using formula (3) we get $|a_\chi| \leq 2\sqrt{p}$.

**Proof of the Vanishing Lemma. Step 1.** All tori in $\mathrm{Sp}$ are conjugated. On introducing coordinates, i.e., $\mathbb{V} \simeq \mathbb{A}^2$, we make the identification $\mathrm{Sp} \simeq \mathrm{SL}_2$. In these terms there exists an element $S \in \mathrm{SL}_2$ conjugating the *Hecke* torus $\mathbb{T}_A \subset \mathrm{SL}_2$ with the standard torus $\mathbb{T} = \{\begin{pmatrix} a & 0 \\ 0 & a^{-1}\end{pmatrix}\} \subset \mathbb{SL}_2$, namely $S\mathbb{T}_A S^{-1} = \mathbb{T}$. The situation is displayed in the following diagram:

$$\begin{array}{ccc} \mathbb{SL}_2 \times \mathbb{A}^2 & \xrightarrow{\alpha_S} & \mathbb{SL}_2 \times \mathbb{A}^2 \\ i_\xi \uparrow & & i_\eta \uparrow \\ \mathbb{T}_A \times \xi & \xrightarrow{\alpha_S} & \mathbb{T} \times \eta \\ pr \downarrow & & pr \downarrow \\ pt & = & pt \end{array}$$

where $\eta = S \cdot \xi$ and $\alpha_S$ is the restriction of the action $\alpha$ to the element $S$.

**Step 2.** Using the equivariance property of the sheaf $\mathscr{F}$ (see Theorem 5.3, property 3) we see that it is *sufficient* to prove the Vanishing Lemma for the sheaf $\mathscr{G}_{st} := pr_!(i_\eta^* \mathscr{F} \otimes \alpha_{S!}\mathscr{L}_\chi)$.

**Step 3.** The Vanishing Lemma holds for the sheaf $\mathscr{G}_{st}$. We write $\eta = (\lambda, \mu)$. By Theorem 5.3 Property 4 we have $i_\eta^* \mathscr{F} \simeq \mathscr{L}_{\psi(\frac{1}{2}\lambda\mu\frac{a+1}{a-1})} \otimes \mathscr{L}_{\sigma(a)}$, where $a$ is the coordinate of the standard torus $\mathbb{T}$ and $\lambda \cdot \mu \neq 0^2$. The sheaf $\alpha_{S!}\mathscr{L}_\chi$ is a character sheaf on the torus $\mathbb{T}$. A direct computation proves the Vanishing Lemma. □

## REFERENCES

1. Hannay J.H. and Berry M.V., Quantization of linear maps on the torus - Fresnel diffraction by a periodic grating, *Physica D1*, 267–291 (1980).
2. Deligne P., Metaplectique, *A letter to Kazhdan* (1982).
3. Deligne P., La conjecture de Weil II, *Publ. Math. I.H.E.S* **52**, 313–428 (1981).
4. Degli Esposti M., Graffi S. and Isola S. Classical limit of the quantized hyperbolic toral automorphisms, *Comm. Math. Phys.* **167**, no. 3, 471–507 (1995).
5. Kurlberg P., *private communication, Chalmers University, Gothenburg, Sweden* (September, 2003).
6. Kurlberg P. and Rudnick Z., Hecke theory and equidistribution for the quantization of linear maps of the torus, *Duke Math. Jour.* **103**, 47–78 (2000).
7. Rudnick Z., The quantized cat map and quantum ergodicity, *Lecture at the MSRI conference "Random Matrices and their Applications"*, Berkeley, June 7–11, 1999.
8. Rudnick Z., On quantum unique ergodicity for linear maps of the torus, *European Congress of Mathematics*, Vol. II (Barcelona, 2000), Progr. Math., **202**, Birkhäuser, Basel, pp. 429–437 (2001).
9. Rieffel M.A., Non-commutative tori—a case study of non-commutative differentiable manifolds, *Contemporary Math.* **105**, 191–211 (1990).
10. Weil A., *Sur les courbes algébriques et les variétés qui s'en déduisent*, Hermann et Cie., Paris 1948.

---

[2] This is a direct consequence of the fact that $A \in SL_2(\mathbb{Z})$ is an hyperbolic element and does not have eigenvectors in $\Lambda^*$.

# p-Adic Description of Hierarchical Systems Dynamics

K. Lukierska-Walasek[*] and K. Topolski[†]

[*]*Institute of Physics, University of Zielona Góra, ul. Z. Szafrana 4a,
65-516 Zielona Góra, POLAND
email:* klukie@proton.if.uz.zgora.pl

[†]*Institute of Mathematics, Wrocław University, pl. Grunwaldzki 2/4,
50-384 Wrocław, POLAND*

**Abstract.** We show that $p$-adic analysis provides a quite natural basis for the description of relaxation in hierarchical systems. For our purposes, we specify the Markov stochastic process considered by S. Albeverio and W. Karwowski. As a result we have obtained a random walk on the $p$-adic integer numbers, which provides the generalization of Cayley tree proposed by Ogielski and Stein. The temperature-dependent power-law decay and the Kohlrausch law are derived.

**Keywords:** Spin glass, dynamics of relaxation, $p$-adics, Markov processes.
**PACS:** 67.40.Fd, 75.10.Nr .

## 1. INTRODUCTION

The growing interest in spin glassy dynamics has stimulated intense research in experimental [1, 2, 3, 4] as well as theoretical domain [5, 6, 7, 8, 9, 10, 11, 12, 13, 14, 15, 16, 17]. The theoretical studies have been successfully developed in the following two directions:

**(i)** scaling theory for growing domains and droplets [5, 6, 7]

**(ii)** the hierarchical structures in ultrametric spaces for relaxation dynamics [8, 9, 10, 11, 12, 13, 14, 15, 16, 17].

The goal of this paper is to investigate the second item (ii), using $p$-adic analysis. The temperature cycle experiments [1, 2, 3, 4] are usually interpreted as the existence of a continuous hierarchy in spin glass dynamics. Hierarchical structure of metastable states corresponds to such structure expressed in terms of pure states in the sense of Parisi solution [17]. If we lower the temperature the energy barriers separating states are becoming higher, even infinite.

The dominant process in the relaxation dynamics has been described as the hopping between the physical states [4]. Such mechanism leads to the studies of the models with hierarchy-constrained dynamics with ultrametric topology. These models did appear very useful for the description of complex systems, called glassy systems, with highly degenerated metastable states.

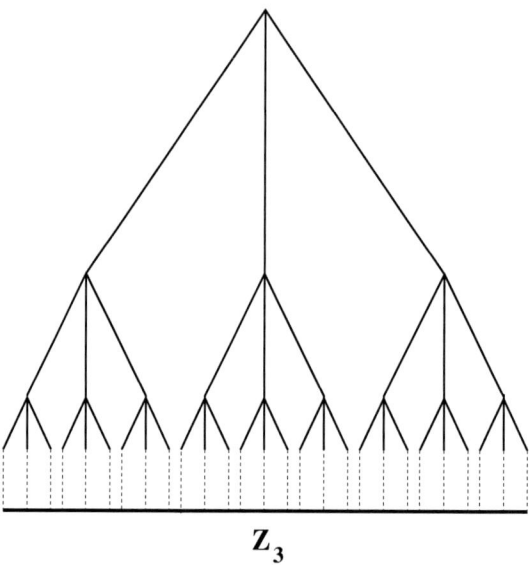

**FIGURE 1.** Cayley tree with p=3. The bottom represents $\mathbf{Z}_3$.

In this paper we consider dynamics for glassy systems using a random walk on the *p*-adic numbers.

There are two main approaches to the construction of the random walks on the *p*-adic numbers. First one is probabilistic approach developed in the papers of Albeverio and Karwowski [18, 19] and the second one which is based on an analytical approach developed in the papers of Avetisov, Bikulov, Kozyrev and Osipov [20, 21, 22, 23]. The reader may find the relation between different construction of the random walks on the *p*-adic numbers in [24].

In our paper we follow the probabilistic approach and for the description of dynamics in an ultrametric space we specify the Markov process, considered in [18]. As the state space of the Markov process we use the space of *p*-adic integers, which for our purposes is an appropriate example of the ultrametric space. We should notice that the *p*-adic integers may be represented by the bottom of the regular, infinite the Cayley tree. Let us recall that in the Cayley tree of the order *p* every branch at each level splits into *p* other branches. For the illustration we present on Fig. 1 the Cayley tree with $p = 3$ from which we obtain, after infinite numbers of splitting, the 3-adic integers described by the bottom of the tree.

We can introduce a random walk in the *p*-adic integers $\mathbf{Z_p}$ analogously as a random walk in the ultrametric space represented by leaves of the regular, finite Cayley tree. The pure physical states of the system are represented by these leaves of the infinite Cayley tree. In order to construct the Markov process on the *p*-adic integers first we consider Markov process on the *p*-adic balls with the same finite radius and finally we contract the ball radius to zero. In this way we obtain transition probabilities in the space of the *p*-adic integers which correspond to the ones considered by Ogielski and Stein [9], in the case of finite ultrametric space.

The plan of the paper is following. In Section 2 we introduce basic notions and describe a Markov process on the $p$-adic integers. In Section 3 we pass to the physical application. We consider the thermally activated random walk on the ultrametric space described as a Markov process on the space of $p$-adic balls. The temperature-dependent power-law decay and the Kohlrausch law are derived. Let us notice finally that recently one observes an increasing interest in the application of $p$-adic numbers in mathematical physics [25]. The $p$-adic analysis was used to study stochastic processes [26, 25], especially the ultrametric jump diffusions [20, 21, 22].

## 2. CONSTRUCTION OF MARKOV PROCESS ON $p$-ADIC INTEGERS

Before we consider a Markov process with $p$-adic integers as a state space let us first introduce some notation and basic properties of $p$-adic numbers which we will be use. More details can be found in e.g. [27], [25].

Let $p$ be an arbitrary prime number and let $\mathbf{Z_p}$ denote the set of $p$-adic integers with $p$–adic norm $||\cdot||_p$. We consider $p$-adic integer as a *formal series* $\sum_{i \geq 0} a_i p^i$ with coefficients $a_i$ satisfying $0 \leq a_i \leq p-1$. With this definition, a $p$-adic integer $a = \sum_{i \geq 0} a_i p^i$ can be identified with the sequence $(a_i)_{i \geq 0}$ of its coefficients.

In terms of the $p$-adic norm $||\cdot||_p$ we introduce the $p$-adic metrics $d_p(x,y) = ||x-y||_p$, which fulfils the ultrametric condition

$$d_p(x,y) \leq \max\{d_p(x,z), d_p(z,y)\}. \tag{1}$$

For $M \geq 0$ and $a \in \mathbf{Z_p}$ we define a closed $p$-adic ball $B(a,M)$ with center $a$ and radius $p^{-M}$

$$B(a,M) = \{x \in \mathbf{Z_p} : d_p(x,a) \leq p^{-M}\}. \tag{2}$$

If $a$ center of the ball has $p$-adic representation $a = \sum_{j=0}^{\infty} a_j p^{-j}$, then the ball $B(a,M)$ is completely determined by

$$\{a\}_M \equiv a_M, a_{M-1}, \ldots, a_0. \tag{3}$$

We will use the following property of the $p$-adic balls:

**(A)** *each ball $B(a, M)$ of radius $p^{-M}$ may be represented as a finite union of* **disjoint** *balls $B(a_i, M+1)$ of radius $p^{-(M+1)}$*

$$B(a, M) = \bigcup_{i=0}^{p-1} B(a_i, M+1), \tag{4}$$

Notice that property (**A**) implies that $\mathbf{Z_p}$ may be represented as a union of **disjoint** balls $B(a_i, M)$ of radius $p^{-M}$

$$\mathbf{Z_p} = \bigcup_{i=0}^{p^M-1} B(a_i, M). \tag{5}$$

Let us finally define the distance between two arbitrary balls $B_1$ and $B_2$ as

$$d_p(B_1, B_2) = \inf\{d_p(x,y) : x \in B_1, y \in B_2\}. \tag{6}$$

With such definition two balls with representations $\{a\}_M \equiv a_M, a_{M-1}, \ldots, a_k, a_{k-1}, \ldots, a_0$ and $\{b\}_M \equiv b_M, b_{M-1}, \ldots, b_k, a_{k-1}, \ldots, a_0$, differ for the first time at $k-th$ position, we have

$$d_p(\{a\}_M, \{b\}_M) = p^{-k}. \tag{7}$$

Let $\{X(t), t \geq 0\}$ be a Markov process with the state space $\mathbf{Z_p}$ and transition rates between states depending only on $p$-adic distance between these states. To define such a process we follow general construction given in [18]. In order to do this we represent $\mathbf{Z_p}$ as a finite union of disconnected balls $B_i^M$ with the radius $p^{-M}$

$$\mathbf{Z_p} = \bigcup_{i=0}^{p^M-1} B_i^M \tag{8}$$

For a fixed $M$ we may consider a finite state space Markov process $\{X_M(t), t \geq 0\}$ with the state space $E_M = \{B_0^M, B_1^M, \ldots, B_{p^M-1}^M\}$, the set of $p$-adic balls with the radius $p^{-M}$, and transition probabilities from ball $B_i^M$ at time 0 to ball $B_j^M$ at time $t$, $P_{i,j}^{(M)}(t)$, defined as

$$P_{i,j}^{(M)}(t) \equiv P\left(X_M(t) = B_j^M \mid X_M(0) = B_i^M\right)$$

$$= P\left(X(t) \in B_j^M \mid X(0) \in B_i^M\right). \tag{9}$$

Transition probabilities $P_{i,j}^{(M)}(t)$, $(i,j = 0, 1, \ldots, p^M - 1)$ are solutions of the system of Kolmogorov equations

$$\frac{d}{dt} P_{i,j}^{(M)}(t) = -q_i^{(M)} P_{i,j}^{(M)}(t) + \sum_{\substack{0 \leq k \leq p^M-1 \\ k \neq i}} q_{ik}^{(M)} P_{k,j}^{(M)}(t), \tag{10}$$

with the initial condition $P_{i,j}(0) = \delta_{ij}$, where $q_{ij}^{(M)}$ the infinitesimal transition probability and $q_i^{(M)}$ the intensity of stay in state $i$ of the process are defined for any pair $i, j$ of different states represented by balls $B_i^M$, $B_j^M$ with radius $p^{-M}$, by

$$q_{ij}^{(M)} = \lim_{h \downarrow 0} \frac{P_{i,j}^{(M)}(h)}{h}, \tag{11}$$

and for any state $i$, by

$$q_i^{(M)} = \lim_{h \downarrow 0} \frac{1 - P_{i,j}^{(M)}(h)}{h}. \tag{12}$$

Observe, that each ball $B_i^M$ is a union of disjoint balls $B_{ik}^{M+1}$, $(k = 1, \ldots, p)$ of radius $p^{-(M+1)}$. We stress here, that $q_{ij}^{(M)}$ depend only on $p$-adic distance between balls $B_i^M$ and $B_j^M$. For this reason we may represent transition probabilities of the process $X_M(t)$ by appropriate transition probabilities of the process $X_{M+1}(t)$

$$\begin{aligned} P_{i,j}^{(M)}(t) &= P\left(X_M(t) = B_j^M \mid X_M(0) = B_i^M\right) \\ &= P\left(X(t) \in \bigcup_{k=1}^p B_{jk}^{M+1} \mid X(0) \in \bigcup_{k=1}^p B_{ik}^{M+1}\right) \\ &= pP\left(X_{M+1}(t) = B_j^{M+1} \mid X_{M+1}(0) = B_i^{M+1}\right) \\ &= pP_{i,j}^{(M+1)}(t)) \end{aligned} \tag{13}$$

This equality leads to the recurrence relation for the local characteristics $q_{i,j}^{(M)}$ and $q_{i,j}^{(M+1)}$ of the processes $\{X_M(t)\}$ and $\{X_{M+1}(t)\}$. Let $dist_p(B_i^M, B_j^M) = p^{-n}$, then using the notation

$$q_{i,j}^{(M)} \equiv u(-M, M-n), \tag{14}$$

we may notice that (13) implies that for any $M > 0$ and $0 < m \leq M$

$$u(-M+1, m-1) = pu(-M, m). \tag{15}$$

Finally, taking into account that $u(-M, m)$ represents probability intensity transition of the Markov process, we get from the recurrence relations (15) that

$$u(-M, m) = \frac{p^{-m+1}}{p-1}(a(-M+m-1) - a(-M+m)), \tag{16}$$

where $\{a(-n)\}$, $n = 0, 1, 2, \ldots$. is a sequence of positive numbers such that

$$\begin{aligned} &(i) \quad a(-n) \geq a(-n+1) \tag{17} \\ &(ii) \quad \lim_{n \to \infty} a(-n) = W, \quad \text{where } W \text{ is a positive number or } +\infty. \end{aligned}$$

Proceeding in the similar way as in [18] we obtain the solution of the Kolmogorov equations for the Markov process with local characteristics $q_{i,j}^{(M)}$ given by (14) and (15) in the form

$$P_{i,i}^{(M)}(t) = \frac{p-1}{p} \sum_{i=0}^{M} p^{-i} \exp\left\{\frac{[pa(-M+i) - a(-M+i+1)]t}{1-p}\right\}, \tag{18}$$

if $dist_p(B_i, B_j) = p^{-M+m}$ then

$$P_{i,j}^{(M)}(t) = \frac{p-1}{p^{m+1}} \sum_{i=0}^{M-m} p^{-i} \exp\left\{\frac{[pa(-M+m+i) - a(-M+m+i+1)]t}{1-p}\right\} \\ - \frac{1}{p^m} \exp\left\{\frac{[pa(-M+m-1) - a(-M+m)]t}{1-p}\right\} \quad (19)$$

Now we define the transition probabilities of the Markov process on $\mathbf{Z_p}$ in the following way.

For any $x \in \mathbf{Z_p}$ let

$$P_t(x; B(a,M)) \equiv P(X_M(t) \in B(a,M) | X_M(0) \in B(x,M)), \quad (20)$$

By arguments similar to those in [18] we may prove that there exists continuous time Markov stochastic process $\{X(t), t \geq 0\}$ with the state space $\mathbf{Z_p}$ and transition probabilities $P_t(x; B)$ given by (20).

Observe that (20) gives us direct connection between the Markov processes $\{X_M(t), t \geq 0\}$ on $p$-adic balls with radius $p^{-M}$, and the Markov process $\{X(t), t \geq 0\}$ on $\mathbf{Z_p}$. Moreover the properties of the process $\{X(t), t \geq 0\}$ can be described in the natural way in terms of the corresponding sequence $\{a(-n)\}_{n=0}^{\infty}$, see e.g. [28, 29]. In fact, the process defined in such a way is an additive rotation-invariant, i.e., invariant under multiplication by any elements of norm 1. Moreover in a similar way it is possible to construct a rotation-invariant, additive process on any local field as it has been presented in [29].

## 3. DYNAMICS AS THERMAL HOPPING IN $p$-ADIC SPACE

In this section we shall apply the mathematical considerations from Section 2 to the physical system with the hierarchy of states which can be linked with Cayley tree structure. By studying the dynamics of such systems the temperature-dependent power law decay and the Kohlrausch law are derived.

We consider the systems in which transitions between states are thermally activated. The height of the energy barriers, $\Delta_k$ $k = 1, 2, \ldots$, which the system overcomes can be ordered in the increasing sequence $\Delta_1 < \Delta_2 < \ldots < \Delta_k \ldots$. The time evolution of the system is described by a random walk on the space of states. The simplest model of such dynamics is the model proposed by Ogielski and Stein [9]. They consider the regular Cayley tree with $M$ levels and fixed branching ratio $p$. The total number of leaves, points on the bottom of the tree, is $n = p^M$. The natural ultrametric distance $d(k,l)$ between leaves $k$ and $l$, is defined as equal to the height $m$ ($m = 0, 1, \ldots, M$) of their closest common ancestor. Now, identifying the states $x$ and $y$ separated by the energy barrier $\Delta_m$ with leaves $k$ and $l$, the probability of moving from state $x$ to state $y$ may be defined as equal to transition probability from leaf $k$ to leaf $l$ separated by the ultrametric distance $m$. Thus dynamics in the space of states separated by the energy barrier may be studied in terms of an appropriate Markov process involving the end points of the Cayley tree, as the space of states. It is a nontrivial observation that the probability transition intensities between states are depending on their ultrametric distance. Due to hierarchical structure of the state space a probability intensities matrix of the process has Parisi matrix structure. Parisi matrix has regular form, and for illustration we present the case $p = 2$.

$$\begin{bmatrix} \varepsilon_0 & \varepsilon_1 & & & & & & \\ \varepsilon_1 & \varepsilon_0 & & \mathbf{E_1} & & & \mathbf{E_2} & \\ & & \varepsilon_0 & \varepsilon_1 & & & & \\ \mathbf{E_1} & & \varepsilon_1 & \varepsilon_0 & & & & \\ & & & & \varepsilon_0 & \varepsilon_1 & & \\ & & & & \varepsilon_1 & \varepsilon_0 & \mathbf{E_1} & \\ \mathbf{E_2} & & & & & \varepsilon_0 & \varepsilon_1 \\ & & & & \mathbf{E_1} & & \varepsilon_1 & \varepsilon_0 \\ & & & & \cdots & & & \end{bmatrix}, \quad (21)$$

where $\mathbf{E_i}$ is the matrix with all elements equal to $\varepsilon_i$.

One can observe that end points of the regular Cayley tree with $M$ levels and fixed branching ratio $p$ may be represented as a set of disconnected balls $\{B_0^M, \ldots, B_{p^M-1}^M\}$ with radius $p^{-M}$ covering $\mathbf{Z_p}$.

Let us consider now a special case of the Markov process on the $p$-adic integers. We assume that the transition probability intensities of the process $\{X(t), t \geq 0\}$ depend on $p$-adic distance only. For this process we have a corresponding Markov chain $\{X_M(t), t \geq 0\}$ with the set of disconnected balls $\{B_0^M, \ldots, B_{p^M-1}^M\}$ with the radius $p^{-M}$ covering $\mathbf{Z_p}$, as a state space. If we enumerate these balls in such a way that $dist_p(B_0^M, B_i^M)$ increase with $i$ then $\mathbf{Q} = [q_{ij}]$, $(0 \leq j \leq p^M - 1, 0 \leq i \leq p^M - 1)$, the matrix of transition probability intensities of the process $\{X_M(t), t \geq 0\}$ has Parisi matrix form. By appropriate choice of $u(-M, k)$, we obtain process studied in [9].

Let for $m = 1, 2, \ldots, M - 1$

$$a(-M+m) = (p-1) \sum_{i=m}^{M-1} p^i \varepsilon_{i+1}. \quad (22)$$

Such a choice of a sequence $\{a(-n)\}$ gives us that for $k = 1, 2, \ldots, M-1$

$$\varepsilon_k = u(-M, k) \equiv q_{ij}^{(M)}, \quad (23)$$

where $q_{ij}^{(M)}$ is transition probability intensity of a jump from the ball $B_i^{(M)}$ to the ball $B_j^{(M)}$, separated by $p$-adic distance $p^{-M+k}$.

From (18) we see that process which at time 0 starts from the ball $B_0^M$ will be found at this ball at time $t$ with probability

$$P_t(B_0^M, B_0^M) = \frac{p-1}{p} \sum_{i=0}^{M} p^{-i} \exp\left\{-t[p\sum_{k=i}^{M-1} \varepsilon_{k+1}p^k - \sum_{k=i+1}^{M-1} \varepsilon_{k+1}p^k]\right\} \quad (24)$$

which after some algebra gives

$$P_t(B_0^M, B_0^M) = \frac{p-1}{p}\left(p^{-M} + \sum_{i=0}^{M-1} p^{-i}\exp\left\{-t[\varepsilon_{i+1}p^{i+1} + (p-1)\sum_{k=i+2}^{M}\varepsilon_k p^{k-1}]\right\}\right) \quad (25)$$

For a special case $p = 2$ equation (25) has the same form as corresponding equation (6) from [9]

$$P_t(B_0^M, B_0^M) = 2^{-(M+1)} + \sum_{i=0}^{M-1} 2^{-(i+1)}\{\exp(-t[2a_{i+1} + \sum_{k=i+2}^{M} a_k]\} \quad (26)$$

where $a_k = 2^{k-1}\varepsilon_k$ represents the probability intensity of a jump an ultrametric distance $k$ from a starting sit at the Cayley tree, while in our case $a_k$ represents the probability transition intensity of a jump to any ball of radius $2^{-M}$ at 2–adic distance $2^{-M+k}$. Finally we contract the ball radiuses to zero and performing the procedure analogical to [9] and specifying the form of energy barriers and probability intensities of crossing these barriers we are able to compare different scenario. The simplest case is a sequence of linearly growing barriers $\Delta_k = \Delta k$ for some positive constant $\Delta$, and $a_k = e^{-\Delta k/T}$ for some fixed temperature $T$. In this case, for large time $t$, the probability $P_t(B_0^M, B_0^M)$ fulfils the temperature-dependent power law

$$\lim_{M\to\infty} P_t(B_0^M, B_0^M) \sim t^{-T\ln 2/\Delta}. \quad (27)$$

For a sequence of energy barriers which grows in slower way, i.e $\Delta_k = \Delta \ln k$ for some positive constant $\Delta$, and $a_k = e^{-\Delta \ln k/T}$ for some fixed temperature $T$, the probability $P_t(B_0^M, B_0^M)$ fulfils, for large $t$, the Kohlrausch law

$$\lim_{M\to\infty} P_t(B_0^M, B_0^M) \sim \exp(-t^{T/\Delta}). \quad (28)$$

## 4. CONCLUDING REMARKS

In our paper we use a random walk framework and we show that $p$-adic analysis is a very natural tool to describe the relaxation process in glassy systems. The $p$-adic space has the ultrametric topology which we apply in this paper.

For the description of dynamics of the hierarchical systems, apart from the random walks on the $p$-adic numbers, one can alternatively use the processes of jump diffusions on the

$p$-adic numbers as it has been presented by Avetisov, Bikulov, Kozyrev and Osipov in the series of papers [20, 21, 22, 23]. In these papers the authors write down the $p$-adic counterpart to the diffusion equation, description of which employs the pseudodifferential operators [28] in the $p$-adic space, and investigate the Cauchy problem for this equation. Solutions of this equations define the transition probabilities of the stochastic processes with the $p$-adic numbers as a state space. In our paper to construct a continuous time random walk on $\mathbf{Z_p}$, the integer $p$-adic numbers, we proceed a similar way as in [18, 19]. First we determine a suitable Markov process with the set of $p$-adic balls of fixed radius as the state space then shrinking the radius of the balls to zero we were able to obtain transition probabilities for a process with $\mathbf{Z_p}$ as the state space. In this way the Markov process on $\mathbf{Z_p}$ is obtained by the usual Kolmogorov construction. The process obtained by this standard construction can be identified with the one constructed in [20, 21, 22, 23]. Our results are in the full agreement with these obtained in [9] moreover from the approach presented in our paper it is clearly seen that the processes on the $p$-adics can be looked upon as the appropriate processes on the regular infinite Cayley tree and this allowed us to find the connection between physical models on the Cayley tree and the correspondent models on the $p$-adic integers space.

It seems that the $p$-adic analysis is a good tool not only for the description of the dynamics of the glassy systems, but also for other hierarchical processes like the evolution of fractals [30], the avalanches [31], protein folding [32] etc.. It is interesting to notice that the $p$-adic space inherently includes the natural hierarchy: $p$-adic balls can be represented as the union of smaller, disjoint balls. The hierarchy of nested balls corresponds to the hierarchy of the scales of the configuration rearrangements, as it is seen from the scaling theory for the growing domains and droplets [5–7]. The droplets may be broken into smaller ones when the temperature decreases.

It is also interesting to notice that construction of the Markov processes presented in our paper can be extended to all local fields, and helps to investigate some of their properties.

In conclusion one can comment that the memory effects in spin glassy systems can be well described by the $p$-adic topology.

## REFERENCES

1. L. Lundgren, P. Svedlindh, P. Nordblad, O. Beckman, *Phys. Rev. Lett.* **51**, 911 (1983).
2. M. Lederman, R. Orbach, J.M. Hammann, M. Ocio, E. Vincent, *Phys. Rev. B.* **44**, 7403 (1991).
3. F. Lefloch, R. Orbach, J.M. Hammann, M. Ocio, E. Vincent, *Europhys. Lett.* **18**, 647 (1992).
4. E. Vincent, J.M. Hammann, M. Ocio, J.P. Bouchand, L.F. Cugliandolo, *Proceedings of Sitges Conference on Glassy Systems*, ed. Rubi, Springer, Berlin (1996).
5. A.J. Bray, M.A. Moore, *J. Phys. C.* **17**, L613 (1984).
6. D.S. Fisher, D.A. Huse, *Phys. Rev. B.* **38**, 373 (1988); **38**, 386 (1988).
7. G. Koper, H. Hilhorst, *J. Phys. (France)* **49**, 429 (1988).
8. R.G. Palmer, D.L. Stein, E. Abrahams, P.W. Anderson, *Phys. Rev. Lett.* **53**, 958 (1984).
9. A.T. Ogielski, D.L. Stein, *Phys. Rev. Lett.* **55**, 1634 (1985).
10. G. Paladin, M. Mezard, C. De Domimics, *J. Phys. (France)* **46**, L985 (1985).
11. M. Schreckenberg, *Z. Phys. B* **60**, 483 (1985).

12. S. Grossmann, F. Wegner, K.H. Hoffmann, *J. Phys. Lett. (France)* **46**, L575 (1985).
13. P. Sibani, K.H. Hoffmann, *Phys. Rev. Lett.* **63**, 2853 (1989).
14. P. Sibani, *Phys. Rev. B.* **34**, R3555 (1986).
15. J.P. Bouchaud, D.S. Dean, *J. Phys. (France)* **5**, 265 (1995).
16. R. Rammal, G. Toulouse, A. Virasoro, *Rev. Mod. Phys.* **58**, 80 (1986).
17. M. Mezard, G. Parisi, A. Virasoro, *Spin glasses theory and beyoud*, World Scientific, Singapore, 1987.
18. S. Albeverio, W. Karwowski, *Stochastic. Process. Appl.* **53**, 1 (1994).
19. S. Albeverio, W. Karwowski, X. Zhao, *Stochastic. Process. Appl.* **83**, 39 (1999).
20. V.A. Avetisov, A.H. Bikulov, S.V. Kozyrev, *J. Phys. A: Math. Gen.* **34**, 8785 (1999).
21. V.A. Avetisov, A.H. Bikulov, S.V. Kozyrev, V.A. Osipov, *J. Phys. A: Math. Gen.* **35**, 177 (2002).
22. V.A. Avetisov, A.H. Bikulov, V.A. Osipov, *J. Phys. A: Math. Gen.* **24**, 3985 (2003).
23. S.V. Kozyrev, V.A. Osipov, V.A. Avetisov, cond-mat/0403440.
24. S. Albeverio, X. Zhao, *Markov Process, Related Fields* **6**(2), 239 (2000).
25. V.S. Vladimirov, I.V. Volovich, Y.I. Zelenov, *p-adic analysis and mathematical physics*, World Scientific, Singapore, 1994.
26. S.N. Evans, *J. Theoret.Probab.* **2**, 209 (1989).
27. N. Koblitz, *p-adic numbers, p-adic analysis and zeta functions*, Springer-Verlag, New York, 1977.
28. A.N. Kochubei, *Pseudo-differential equations and stochastics over non-Archimedean fields*, Monographs and Textbooks in Pure and Applied Mathematics, Vol. **244**, Marcel Dekker Inc., New York, 2001.
29. K. Yasuda, *J. Math. Sci. Univ. Tokyo* **3**, 629 (1996).
30. H.O. Peitgran, H. Jürgens, D. Saupe, *Introduction to Fractal and Chaos*, Springer-Verlag, New-York, 1992.
31. S. Boettcher, *Physica A* **266**, 330 (1999).
32. J.D. Bringelson and P.G. Wolynes, *J. Phys. Chem* **93**, 6902 (1989).

# Capacities and Function Spaces on the Local Field

Hiroshi Kaneko

*Department of Mathematics, Tokyo University of Science,*
*26 Wakamiya, Shinjuku, 162-8601 Tokyo, JAPAN*
*email:* kaneko@home.email.ne.jp

**Abstract.** In this article, we will compare non-linear capacities with Hausdorff measure and present a trace theorem on Besov space over local field. These studies play important roles in existing fractal analysis on the Euclidean space.

**Keywords:** Besov space, Hausdorff measure, non-linear capacity, p-adic number field, trace theorem.

**PACS:** 02.10.De, 02.30.Em, 02.50.-r.

## 1. INTRODUCTION

A probabilistic counterpart of the Sobolev space was introduced by Fukushima and the author in [13], and it has been investigated by many researchers. For instance, we can look at recent development of the theory in the articles by Farkas, Hirsh, Hoh, Jacob, Kazumi, Schilling, Shigekawa and Song [11, 12, 17, 18, 19, 20 and 28]. In those articles, we see many aspects which are tightly related to the non-linear capacity theory associated with the ordinary Sobolev space. The study of potential theory on the field $Q_p$ of p-adic numbers has been developed with keeping tight relationship with some difference operators on the field. For instance, some potential theoretic features such as Harnack inequalities, equilibrium measure and $\alpha$-capacity based on Riesz potential were explored by Haran ([15]). In [25], Yasuda and the author found some properties of non-linear capacity on an infinite extension over $Q_p$ introduced in [29 and 36], after some preliminary observation on non-linear capacity on finite extensions over $Q_p$.

The first objective of this article is to look at the relationship of non-linear capacity in [13] with Hausdorff measure based on a counterpart $\mathcal{V}_r^\alpha$ of Bessel kernel of the $\alpha$-stable process as a preparation for Besov space. This can be viewed in [24] as applications of Hardy-Littlewood-Wiener theorem and Whitney decomposition on a local field. On the other hand, Kochubei showed some features of Hausdorff measure on the infinite extension of $Q_p$ in [30].

These observations could be related to some results in the fractal analysis. This is mainly because we see that if two extensions $K', K$ of $Q_p$ with $K' \subset K$ are given, then $K$ contains $K'$ as a $d$-set with respect to the Haar measure $\mu_{K'}$ on $K'$ and that this inclusion is regarded as a counterpart of $d$-set in the Euclidean space as observed in [24].

The second objective of this paper is to present a trace theorem for the Besov space on local field, which is different from the one presented by Haran ([16]). The proof of our trace theorem will be performed by recalling some properties of the function $\mathscr{V}_r^{(\alpha)}$, the Whitney decomposition and by establishing a counterpart of the Besov space. In this article, $K' \subset K$ is said to be a $d$-set, if there exists a Radon measure $\mu'$ on $K'$ and positive number $c$ and $r_0$ satisfying $\mu'(B(x,r) \cap K') \leq cr^d$ for any $x \in K$ and $r$ with $0 < r \leq r_0$, where $B(x,r)$ stands for the ball centered at $x$ with radius $r$.

In what follows, we will denote a fixed prime number by $p_0$ and the field of $p_0$-adic numbers by $Q_{p_0}$. The non-linear capacities will be denoted more specifically as $(r,p)$-capacity.

## 2. A COUNTERPART OF BESSEL KERNEL

Let $K$ be a finite extension of the field $Q_{p_0}$ of $p_0$-adic numbers. Then, $K$ admits a norm characterized as a unique extension of the norm on $Q_{p_0}$. The maximal ideal $P = \{x \in K | \, \|x\| < 1\}$ of the ring $R = \{x \in K | \, \|x\| \leq 1\}$ has an element $\pi$ of maximal norm so that $\pi R = P$ is satisfied.

Sine the residue field $R/P$ is a finite extension of $F_{p_0} = Z/Z_{p_0}$, one can choose a family $\{s_i\}_{i=1}^{f_K} \subset R$ so that their natural images in the residue field $R/P$ are the basis over the finite field $F_{p_0}$. In what follows, $p_0^{f_K}$ will be denoted by $q_K$ and the extension degree of $K$ over $Q_{p_0}$ by $m_K$. Then, the normalized Haar measure $\mu$ on $K$ is characterized by $\mu(B(x,q_K^{\ell/m_K})) = q_K^\ell$ for any integer $\ell$.

We introduce the following conditions on sequence $A = \{a(m)\}_{m=-\infty}^{\infty}$:

(1) $\quad a(m) \geq a(m+1)$,

(2) $\quad \lim\limits_{m \to \infty} a(m) = 0 \quad \text{and} \quad \lim\limits_{m \to -\infty} a(m) > 0 \text{ or } = \infty.$

We denote the family of all sequences with these two properties by $\mathscr{A}_{A\text{-}K\text{-}Y}$.

Denoting $\frac{q_K-1}{q_K} \sum_{i=0}^{\infty} q_K^{-i} \exp\left(-\frac{q_K a(N+i) - a(N+i+1)}{q_K-1}t\right)$ by $P_t(N)$, we have an explicit expression

$$P_t^{(A)} 1_{B(x,q_K^{N/m_K})}(y) = \begin{cases} P_t(N), & y \in B(x, q_K^{N/m_K}), \\ \dfrac{q_K^{1-\ell}}{q_K - 1}\left(P_t(N+\ell) - P_t(N+\ell-1)\right), \\ \qquad y \in B(x, q_K^{(N+\ell)/m_K}) \setminus B(x, q_K^{(N+\ell-1)/m_K}) \end{cases}$$

of the transition probability semi-group kernels $\{P_t^{(A)}\}$ of Albeverio, Karwowski and Yasuda's random walk associated with $A = \{a(m)\} \in \mathscr{A}_{A\text{-}K\text{-}Y}$ (cf [35]). For any $r \geq 1$, the kernel $V_r^{(A)}$ is defined in [13] by

$$V_r^{(A)} = \frac{1}{\Gamma(r/2)} \int_0^\infty t^{r/2-1} e^{-t} P_t^{(A)} dt.$$

Especially, the image of the map $V_1^{(A)} : L^2(K;\mu) \to L^2(K;\mu)$ gives the domain $\mathscr{F}^{(A)}$ of the Dirichlet form $\mathscr{E}^{(A)}$ determined by the sequence $A = \{a(m)\}$.

We restate some fundamental facts on $V_r^{(A)}$ demonstrated in [24].

**Lemma 1.** *If $A = \{a(m)\}_{m=-\infty}^\infty \in \mathscr{A}_{A\text{-}K\text{-}Y}$ satisfies $\lim_{m \to -\infty} a(m) = \infty$, then $V_r^{(A)}(x, \{x\}) = 0$ for any $x \in K$.*

In what follows, we will focus only on the important class of $\alpha$-stable processes which are materialized by taking $A$ as $\{q_K^{-m\alpha/m_K}\} \in \mathscr{A}_{A\text{-}K\text{-}Y}$ with some $\alpha > 0$ ([35]). Thanks to Lemma 1, we can introduce the density function $\mathscr{V}_r^{(\alpha)}(x)$ of the kernel $V_r^{(\alpha)}(0, dx)$ with respect to the Haar measure $\mu$ on $K$. A fundamental fact on the convolution shows $\mathscr{V}_r^{(\alpha)} * f \in L^p(K;\mu)$ whenever $f \in L^p(K;\mu)$. The $(r,p)$-capacity $\mathrm{Cap}_{r,p}(O)$ of open set $O$ is defined in [13].

**Lemma 2.** *If $A = \{a(m)\}_{m=-\infty}^\infty$ satisfies $\lim_{m \to -\infty} a(m) = \infty$,*

$$\mathrm{Cap}_{r,p}(O) = \inf\{\|f\|_p^p \mid f \in L^p(K;\mu), \mathscr{V}_r^{(\alpha)} * f(x) \geq 1 \text{ for all } x \in O\},$$

*for any open set $O$ in $K$ and $p > 1$.*

*Proof.* The continuity of $\mathscr{V}_r^{(\alpha)} * f$ at any point $x$ satisfying $\mathscr{V}_r^{(\alpha)} * f(x) < \infty$ proved in [24, Lemma 2 (ii)] shows that the condition $\mathscr{V}_r^{(\alpha)} * f(x) \geq 1$ $\mu$-a.e. $x \in O$ implies $\mathscr{V}_r^{(\alpha)} * f(x) \geq 1$ for all $x \in O$. □

This capacity is extended so as to be an outer capacity. Therefore, we can drive from general theory of non-linear capacity in [1] that $\mathrm{Cap}_{r,p}(E) = \sup\{\nu(E) \mid \nu$ is a Radon measure satisfying $\mathrm{supp}[\nu] \subset E$ and $\|\mathscr{V}_r^{(\alpha)} * \nu\|_{p'} \leq 1\}$ for any compact set $E$ in $K$.

**Lemma 3.** *There exist some positive constants $C_1(\alpha, r, q_K)$ and $C_2(\alpha, r, q_K)$ such that $\mathscr{V}_r^{(\alpha)}(x) \leq C_1(\alpha, r, q_K) q_K^{-k(1-r\alpha/2m_K)}$, for any $x \in K$ satisfying $\|x\| = q_K^{k/m_K}$ with some non-positive integer $k$, and $\mathscr{V}_r^{(\alpha)}(x) \leq C_2(\alpha, r, q_K) q_K^{-k(1+\alpha/m_K)}$, for any $x \in K$ satisfying $\|x\| = q_K^{k/m_K}$ with some positive integer $k$.*

*Proof.* Firstly, we note that

$$\mathscr{V}_r^{(\alpha)}(x) = \sum_{i=0}^\infty q_K^{-(k+i)} \left(\frac{1}{\left(\frac{q_K q_K^{-(k+i)\alpha/m_K} - q_K^{-(k+i+1)\alpha/m_K}}{q_K - 1} + 1\right)^{r/2}}\right.$$

$$-\frac{1}{\left(\frac{q_K q_K^{-(k+i-1)}\alpha/m_K - q_K^{-(k+i)}\alpha/m_K}{q_K - 1} + 1\right)^{r/2}}\right).$$

Therefore, the first inequality in the assertion is derived from the following estimate validated by taking some positive constant $C_1(\alpha, r, q_K)$:

$$\mathcal{V}_r^{(\alpha)}(x) \leq \sum_{i=0}^{\infty} q_K^{-(k+i)} \frac{1}{\left(\frac{q_K q_K^{-(k+i)\alpha/m_K} - q_K^{-(k+i+1)\alpha/m_K}}{q_K - 1}\right)^{r/2}}$$

$$= \sum_{i=0}^{\infty} q_K^{-(k+i)} q_K^{(k+i)r\alpha/2m_K} \frac{1}{\left(\frac{q_K^{\alpha/m_K} - 1}{q - 1}\right)^{r/2}}.$$

On the other hand, the second inequality is derived from the following estimate validated by taking some positive constant $C_2(\alpha, r, q_K)$:

$$\mathcal{V}_r^{(\alpha)}(x) \leq C_2(\alpha, r, q) \sum_{i=0}^{\infty} q_K^{-(k+i)} \left( \frac{1}{1 + \frac{r}{2} q_K^{-k\alpha/m_K} \frac{q_K q_K^{-i\alpha/m_K} - q_K^{-(i+1)\alpha/m_K}}{q-1}} \right.$$

$$\left. - \frac{1}{1 + \frac{r}{2} q_K^{-k\alpha/m_K} \frac{q_K q_K^{-(i-1)\alpha/m_K} - q_K^{-i\alpha/m_K}}{q-1}} \right)$$

$$\leq C_2(\alpha, r, q_K) \sum_{i=0}^{\infty} q_K^{-(k+i)} \frac{r}{2} q_K^{-(k+i)\alpha/m_K} \left( \frac{q_K q_K^{\alpha/m_K} - 1}{q_K - 1} - \frac{q_K - q_K^{-\alpha/m_K}}{q_K - 1} \right). \quad \square$$

## 3. FUNDAMENTAL FACTS

The following three facts have already been justified on $K$ over $Q_{p_0}$ as in [24].

**Proposition 1 (Hardy-Littlewood-Wiener theorem).** *For any Radon measure $\nu$ with $\mathrm{supp}[\nu] \subset B(0,1)$,*

$$\mu(\{x \in K | M\nu(x) > \lambda\}) \leq \frac{1}{\lambda} \nu(B(0,1)) \quad \text{for all} \quad \lambda > 0,$$

*where* $M\nu(x) = \sup_{r>0} \frac{\nu(B(x,r))}{\mu(B(x,r))}$.

**Proposition 2 (Whitney decomposition).** *For any non-empty closed set $F$ in $K$, there exists at most countable family $\{B_i\}$ of balls such that*
(a)  $\cup B_i = F^c$,
(b)  $\{B_i\}$ *are mutually disjoint,*

(c) $q_K^{1/m_K} \text{diam}(B_i) = \text{dist}(B_i, F)$ for all $n$.

**Theorem 1 (Wolff's inequality).** *For any $r > 1$ and $p > 1$ with $0 < rp\alpha/2 < m_K$, there exists a positive constant $C(r, p, q_K)$ such that*

$$C(r,p,q_K)^{-1}\int_K |\mathcal{V}_r^{(\alpha)} v(x)|^{p'} d\mu(x) \leq \int_K W_{r,p}^v(x) dv(x) \leq C(r,p,q_K)\int_K |\mathcal{V}_r^{(\alpha)} v(x)|^{p'} d\mu(x),$$

*where $W_{r,p}^v(x) = \sum_{k=0}^{\infty} (q_K^{k(m_K - r\alpha p/2)/m_K} v(B(x, q_K^{-k/m_K})))^{p'-1}$ with the real number $p'$ satisfying $1/p + 1/p' = 1$.*

## 4. HAUSDORFF CONTENT AND NON-LINEAR CAPACITIES

We present some results on the relationship between non-linear capacities and Hausdorff content (for the definition, see [1]).

**Theorem 2.** *If $0 < rp\alpha/2 < m_K$ and if an increasing function $g$ on $[0, \infty)$ with $g(0) = 0$ satisfies*

$$\sum_{k=-\infty}^{0} \left(\frac{g(q_K^{k/m_K})}{q_K^{k(m_K - r\alpha p/2)/m_K}}\right)^{1/(p-1)} < \infty,$$

*then there exists a constant $C_{g,E}$ dependent on $g$ and compact set $E$ in $K$ such that*

$$H_g^{(\infty)}(E) \leq C(r,p,q_K)^{p-1} C_{g,E}^{p-1} \text{Cap}_{r,p}(E),$$

*where $H_g^{(\infty)}$ stands for the Hausdorff content associated with the function $g$.*

*Proof.* We can take a Radon measure $v$ satisfying $\text{supp}[v] \subset E$, $v(B(x, q_K^{\ell/m_K})) \leq g(q_K^{\ell/m_K})$ for any $\ell \in \mathbb{Z}$ and the inequalities $\Lambda_g^{-1} H_g^{(\infty)}(E) \leq v(E) \leq H_g^{(\infty)}(E)$ with some positive constant $\Lambda_g$, where $H_g^{(\infty)}$ stands for the Hausdorff content with respect to function $g$. On the other hand, Theorem 1 shows that

$$\int_K |\mathcal{V}_r^{(\alpha)} * v(x)|^{p'} d\mu(x) \leq C(r,p,q_K) \int_K W_{r,p}^v(x) dv(x).$$

The assertion is legitimized by setting

$$C_{g,E} = \sum_{k=-\infty}^{\ell} \left(\frac{g(q_K^{k/m_K})}{q_K^{k(1-r\alpha p/2m_K)}}\right)^{p'-1} + H_g^{(\infty)}(E)^{p'-1} \sum_{k=\ell+1}^{0} \left(\frac{1}{q_K^{k(1-r\alpha p/2m_K)}}\right)^{p'-1},$$

where $\ell$ is maximal integer enjoying $g(q_K^{\ell/m_K}) \leq H_g^{(\infty)}(E)$. This is because we have

$$\mathrm{Cap}_{r,p}(E)^{1/p} \geq \frac{v(E)}{\|\mathscr{V}_r^{(\alpha)} * v\|_{L^{p'}(K;\mu)}} \geq \frac{v(E)^{1-1/p'}}{C(r,p,q_K)^{1/p'} C_{g,E}^{1/p'}} \geq \frac{\Lambda_g^{-1/p} H_g^{(\infty)}(E)^{1/p}}{C(r,p,q_K)^{1/p'} C_{g,E}^{1/p'}}.$$

□

By some specific choice of function $g$, we can prove the following assertion:

**Corollary.** For any compact set $E$ in $K$ and $p_1, p_2 > 1$ satisfying $0 < r_2 \alpha p_2/2 \leq r_1 \alpha p_1/2 \leq m_K$ with some positive numbers $r_1, r_2$, there exists a constant $C_g$ independent of $E$ such that

(i) $H_g^{(\infty)}(E)^{m_K - r_1 p_1 \alpha/2} \leq C_g \mathrm{Cap}_{r_1, p_1}(E)^{m_K - r_2 p_2 \alpha/2}$, if $r_2 p_2 \alpha/2 < r_1 p_1 \alpha/2 < m_K$,

(ii) $\left(1 + \log_+ \frac{1}{H_g^{(\infty)}(E)}\right)^{1-p_1} \leq C_g \mathrm{Cap}_{r_1, p_1}(E)$, if $r_2 p_2 \alpha/2 < r_1 p_1 \alpha/2 = m_K$,

(iii) $H_g^{(\infty)}(E)^{p_1 - 1} \leq C_g \mathrm{Cap}_{r_1, p_1}(E)^{p_2 - 1}$, if $r_2 p_2 \alpha/2 = r_1 p_1 \alpha/2 = m_K$ and $p_1 < p_2$,

where $g(q_K^{k/m_K}) = q_K^{k(1 - r_2 p_2 \alpha/2 m_K)}$ in (i) and (ii) and $g(q_K^{k/m_K}) = (\log_+ \frac{2}{q_K^{k/m_K}})^{1-p_2}$

in (iii).

## 5. CHARACTERIZATION OF THE BESOV SPACE

For the definition of the Besov space on a finite separable extension $K$ of $Q_{p_0}$, we introduce the sequence $\{\varphi_i\}$ of functions on $K$ given by

$$\varphi_i = \begin{cases} 1_{B(0,1)} & \text{if } i = 0, \\ 1_{B(0,\theta^{-i}) \setminus B(0,\theta^{-i+1})} & \text{if } i = 1, 2 \cdots, \end{cases}$$

where $\theta = \|\pi\|$. Then, the Fourier transform of integrable function $f$ on $K$ is defined by $\hat{f}(\xi) = \int_K \chi_K(x\xi) f(x) \mu(dx)$, where the additive character $\chi_K$ of $K$ is involved. The inverse Fourier transform of integrable function $\varphi$ on $K$ is given by $\check{\varphi}(x) = \int_K \chi_K(-x\xi) \varphi(\xi) \mu(d\xi)$, then the original integrable function $f$ is obtained by taking the inverse Fourier transform of $\hat{f}$.

**Definition.** For $p > 1$, the Besov space $B_r^{p,p}(K)$ is defined by

$$B_r^{p,p}(K) = \{f \in \mathscr{S}' \mid \sum_{j=0}^{\infty} \theta^{-rpj} \int_K |(\varphi_j \hat{f})^{\vee}(x)|^p \mu(dx) < \infty\},$$

whose norm is given by $\|f\|_{B^{p,p}_r(K)} = \left(\sum_{j=0}^{\infty} \theta^{-rpj} \int_K |(\varphi_j \hat{f})^{\vee}(x)|^p \mu(dx)\right)^{1/p}$, where $\mathscr{S}' = \{f \in L^1_{loc}(K;\mu)| \sup \frac{|f*\phi(x)|}{1+\|x\|^k} < \infty$ with some positive integer $k$ for any locally constant function $\phi$ with compact support$\}$.

By denoting $\frac{1}{\mu(B(0,\theta^j)\setminus B(0,\theta^{j+1}))} \int_{B(0,\theta^j)\setminus B(0,\theta^{j+1})} (\chi_K(-h\xi) - 1)\mu(dh)$ by $\psi_j$ for $j = 0,1,2,\cdots$, we obtain $\theta^{m_K+1}\varphi_j + (1 - \theta^{m_K+1})\varphi_{j+1} = \psi_j - \psi_{j-1}$. From the basic estimate $\left(\int_K |(\varphi_i \hat{f})^{\vee}(x)|^p \mu(dx)\right)^{1/p} \leq c_1 \|f\|_{L^p(K;\mu)}$ with some positive constant $c_1$, we can derive

$$\begin{aligned}
\|(\varphi_j \hat{f})^{\vee}\|_{L^p(K;\mu)} &\leq \|(\varphi_j(((\theta^{m_K+1}\varphi_{j-1} + (1-\theta^{m_K+1})\varphi_j) \\
&\quad + \theta^{m_K+1}\varphi_j + (1-\theta^{m_K+1})\varphi_{j+1})\hat{f}))^{\vee}\|_{L^p(K;\mu)} \\
&\leq c_1 \Big(\|((\theta^{m_K+1}\varphi_{j-1} + (1-\theta^{m_K+1})\varphi_j)\hat{f})^{\vee}\|_{L^p(K;\mu)} \\
&\quad + \|((\theta^{m_K+1}\varphi_j + (1-\theta^{m_K+1})\varphi_{j+1})\hat{f})^{\vee}\|_{L^p(K;\mu)}\Big) \\
&\leq c_1 \Big(\|(\psi_{j-1}\hat{f})^{\vee}\|_{L^p(K;\mu)} + \|(\psi_j \hat{f})^{\vee}\|_{L^p(K;\mu)} \\
&\quad + \|(\psi_{j+1}\hat{f})^{\vee}\|_{L^p(K;\mu)} + \|(\psi_{j+2}\hat{f})^{\vee}\|_{L^p(K;\mu)}\Big).
\end{aligned}$$

Accordingly, from this inequality, we can derive

$$\left(\sum_{j=1}^{\infty} \theta^{-rpj} \int_K |(\varphi_j \hat{f})^{\vee}(x)|^p \mu(dx)\right)^{1/p}$$

$$\leq c_1 \left(\sum_{j=1}^{\infty} \theta^{-rpj} \sum_{\ell=-1}^{2} \int_K |(\psi_{j+\ell}\hat{f})^{\vee}(x)|^p \mu(dx)\right)^{1/p}$$

$$\leq \frac{4c_1}{\theta^r} \left(\sum_{j=1}^{\infty} \int_{\{\|h\|=\theta^j\}} \int_K |\Delta_h f(x)|^p \mu(dx) \frac{1}{\|h\|^{(m_K+rp)}} \mu(dh)\right)^{1/p}$$

$$\leq \frac{4c_1}{\theta^r} \left(\int_{B(0,1)\setminus\{0\}} \int_K \frac{|\Delta_h f(x)|^p}{\|h\|^{(m_K+rp)}} \mu(dx)\mu(dh)\right)^{1/p}$$

$$\leq \frac{4c_1}{\theta^r} \left(\sum_{j=1}^{\infty} \theta^{-rpj} \int_K |(\psi_j \hat{f})^{\vee}(x)|^p \mu(dx)\right)^{1/p}$$

$$\leq \frac{8c_1}{\theta^r} \left(\sum_{j=1}^{\infty} \theta^{-rpj} \int_K |(\varphi_j \hat{f})^{\vee}(x)|^p \mu(dx)\right)^{1/p}$$

$$\quad + \frac{4c_1}{\theta^r} \left(\sum_{j=1}^{\infty} \theta^{-rpj} \int_K |(\psi_{j+1}\hat{f})^{\vee}(x)|^p \mu(dx)\right)^{1/p}$$

$$\leq \frac{8c_1}{\theta^r} \left(\sum_{j=1}^{\infty} \theta^{-rpj} \int_K |(\varphi_j \hat{f})^{\vee}(x)|^p \mu(dx)\right)^{1/p}$$

$$+\frac{4c_1}{\theta^r}\theta^r\left(\sum_{j=1}^{\infty}\theta^{-rpj}\int_K|(\psi_j\hat{f})(x)|^p\mu(dx)\right)^{1/p}$$

$$\leq \frac{8c_1}{\theta^r}\frac{1}{1-\theta^r}\left(\sum_{j=1}^{\infty}\theta^{-rpj}\int_K|(\varphi_j\hat{f})(x)|^p\mu(dx)\right)^{1/p}.$$

Consequently, we can pay attention to the following simple inequalities:

$$\|(\varphi_0\hat{f})\|_{L^p(K;\mu)} \leq \sum_{j=0}^{\infty}\|(\varphi_j\hat{f})\|_{L^p(K;\mu)}$$

$$\leq \left(\sum_{j=0}^{\infty}\theta^{rqj}\right)^{1/q}\left(\sum_{j=1}^{\infty}\theta^{-rpj}\int_K|(\varphi_j\hat{f})(x)|^p\mu(dx)\right)^{1/p}$$

$$\leq \|f\|_{B_r^{p,p}(K)}.$$

As a result, it turns out that the Besov space $B_r^{p,p}(K)$ admits an equivalent norm

$$\|f\|_{L^p(K;\mu)} + \int_{B(0,1)\setminus\{0\}}\int_K \frac{|\Delta_h f(x)|^p}{\|h\|^{(m_K+rp)}}\mu(dx)\mu(dh).$$

## 6. TRACE THEOREM

The aim of this section is to show the counterpart of trace theorem for a closed $d$-set $K'$ in $K$. Firstly we note that for the real number $s$ and $p$ satisfying $0 < s < 1$, $1 \leq p < \infty$ and any fixed non-positive integer $L$, $B_s^{p,p}(K')$ is defined as a family of functions $u$ with finite value of norm

$$\|u\|_{L^p(K';\mu')} + \left(\sum_{\ell=j}^{\infty}\theta^{-\ell(sp+d)}\int\int_{\|x-y\|\leq\theta^{\ell+L}}|u(x)-u(y)|^p\mu'(dx)\mu'(dy)\right)^{1/p},$$

for any fixed integer $j$, where $\mu'$ stands for the Radon measure in the definition of $d$-set.

For the description of the counterpart, we take a Whitney decomposition $\{B_i\}$ for the non-trivial closed $d$-set $K'$ established in Proposition 2 and denote the ball containing $B_i$ with diameter $\theta^{-1}\text{diam}(B_i)$ by $\bar{B}_i$. When $K'$ is compact, we can find a ball $B_{i_0}$ satisfying $K' \subset \bar{B}_{i_0}$. In this case, we exclude all balls whose diameters are greater than $\text{diam}(B_{i_0})$ from the family $\{B_i\}$ and denote the redefined family of the balls again by $\{B_i\}$.

**Theorem 3.** Let $K'$ be a $d$-set in $K$ with $0 < d \leq m_K$. Then for any $p$ with $1 < p < \infty$ and any positive numbers $r, s$ satisfying $s = r - (m_K - d)/p$,

(i) the restriction of any function in $B_r^{p,p}(K)$ to $K'$ is an element of $B_s^{p,p}(K')$,

(ii) if $K'$ is compact, any function in $B_s^{p,p}(K')$ admits an extension which coincides with some element of $B_r^{p,p}(K)$.

The assertion consists of two parts. One is a statement for the restriction of function in $B_r^{p,p}(K)$ and the other is for the extension of function in $B_s^{p,p}(K')$. We start with preliminaries for the assertion on restriction.

**Lemma 4.** *For any $u = \mathscr{V}_r^{(\alpha)} * f$ with some $f \in L^p(K;\mu)$, $\|u\|_{L^p(K';\mu')} \leq c\|f\|_{L^p(K;\mu)}$ with some positive constant $c$.*

*Proof.* Since $\mathscr{V}_r^{(\alpha)}$ is the density function of a probability distribution on $K$, we have $\int_K \mathscr{V}_r^{(\alpha)}(x-y)\mu(dx) = 1$. The basic inequality

$$\int_{B(x,\theta^{-k})} \|x-z\|^{-\ell}\mu'(dz) = \sum_{j=-k}^{\infty} \theta^{j(-\ell)}\mu'(B(x,\theta^j)\setminus B(x,\theta^{j+1})) \leq c\theta^{-k(d-\ell)}$$

for any real number $\ell < d$ and any integers $k$ shows that $\int_{K'} \mathscr{V}_r^{(\alpha)}(x-y)\mu'(dx) < \infty$. The assertion is derived from the general theory for convolution kernels. □

*Proof of Theorem 3* (i). Take $r_1$ and $r_2$ satisfying $r_1 \neq n, r_2 \neq n$ and $r_1 < r < r_2$. Then we introduce $s_i = r_i - (m_K - d)/p$ for $i = 1, 2$. Lemma 4 shows that taking the restriction to $K'$ is an bounded operator from $\mathscr{F}_{r_i,p}$ to $L^p(K';\mu')$. This restriction gives the bounded operator from the intermediate space $(\mathscr{F}_{r_1,p}, \mathscr{F}_{r_2,p})_\zeta$ to the intermediate space $(L^p(K';\mu'), L^p(K';\mu'))_\zeta$ for $0 < \zeta < 1$ satisfying $r = (1-\zeta)r_1 + \zeta r_2$.

Here, we note that the basic method in the intermediate space presented such as in [21] is valid to prove that $\mathscr{F}_{r_i,p}$ is embedded into $B_{r_i}^{p,\infty}(K)$ and that the intermediate space of $B_{r_1}^{p,\infty}(K)$ and $B_{r_2}^{p,\infty}(K)$ is imbedded into $B_r^{p,p}(K)$. Combining this with the fact that the latter intermediate space is $L^p(K';\mu')$, it turns out that $\|u\|_{L^p(K';\mu')} \leq c_0 \|u\|_{B_r^{p,p}(K)}$ with some positive constant $c_0$.

By introducing $B = L^p(K' \times K'; \frac{1}{\|x-y\|^d}\mu'(dx)\mu'(dy))$ and the space $\ell_p^s(B)$ of $B$-valued sequence $\{a_\ell\}$ with norm $\|\{a_\ell\}\|_{\ell_p^s(B)} = \sum_{\ell=0}^{\infty}(\theta^{-sp\ell}\|a_\ell\|_{L^p(K'\times K';\frac{1}{\|x-y\|^d}\mu'(dx)\mu'(dy))}^p)^{1/p}$, we can define the map $Tf = (T_\ell f)_{\ell=0}^{\infty}$

$$T_\ell f(x,y) = \begin{cases} f(x) - f(y), & \text{if } \|x-y\| = \theta^{-\ell}, \\ 0, & \text{otherwise.} \end{cases}$$

Then, we see that $T$ is a map from $\mathscr{F}_{r,p}$ to $\ell_\infty^s(B)$, where $\ell_\infty^s(B)$ stands for the space of $B$-valued sequence $\{a_\ell\}$ with norm $\|\{a_\ell\}\|_{\ell_\infty^s(B)} = \sup_\ell(\theta^{-s\ell}\|a_\ell\|_{L^p(K'\times K';\frac{1}{\|x-y\|^d}\mu'(dx)\mu'(dy))})$. This is because we can derive a similar result to Lemma C in [21].

Since $T$ is bounded from $(\mathscr{F}_{r_1,p}, \mathscr{F}_{r_2,p})_\zeta$ to $(\ell_\infty^{s_1}(B), \ell_\infty^{s_2}(B))_\zeta$, by applying a similar method in [21] we can conclude that the map $T$ is from $B_r^{p,p}(K)$ to $\ell_p^s(B)$. As the result, the restriction part of the trace theorem has been proved. □

Based on the family of balls $\{\overline{B}_i\}$ associated with compact $d$-set $K'$, we have the family of functions

$$\phi_j(x) = \left(\frac{\operatorname{diam}(B_j)^{m_K}}{N_j(x)} 1_{\overline{B}_j}(x)\right) \bigg/ \left(\sum_j \frac{\operatorname{diam}(B_j)^{m_K}}{N_j(x)}\right) \quad (j=1,2,\cdots),$$

where $N_j(x)$ stands for the number of balls in the family $\{\overline{B}_i\}$ containing $x$ with $\operatorname{diam}(B_j) = \operatorname{diam}(B_i)$. We have obtained the partition of the unity $\{\phi_i\}$ for the open set $\cup_i \overline{B}_i$. For any function $u$ in $B_s^{p,p}(K')$, we define the function $u_{\mathscr{E}}$ defined outside $K'$ by

$$u_{\mathscr{E}}(x) = \sum_i \frac{1}{\mu'(\overline{B}_i)} \phi_i(x) \int_{\overline{B}_i} u(y)\mu'(dy)$$

and define $\overline{\Delta}_\ell$ as the union of all balls $\{\overline{B}_{n_\lambda}\}$ in $\{\overline{B}_i\}$ satisfying $\operatorname{diam}(\overline{B}_{n_\lambda}) = \theta^{\ell-1}$.

**Lemma 5.** Let $\Delta_\ell$ be the union of all balls $\{B_{n_\lambda}\}$ in $\{B_i\}$ satisfying $\operatorname{diam}(B_{n_\lambda}) = \theta^\ell$ and let $h$ be a non-negative function defined on $K'$.

(i) For any non-positive integer $L$, $g(x) = \int_{B(x,\theta^{\ell+L})} h(w)\mu'(dw)$ enjoys
$\int_{\Delta_\ell \cap B(y,\theta^m)} g(x)\mu(dx) \leq \theta^{\ell m_K} \int_{\|w-y\| \leq \max\{\theta^m,\theta^{\ell+L}\}} h(w)\mu'(dw)$.

(ii) In particular, if $K'$ is compact, then $g(x) = \sum_{x \in \overline{B}_n} \int_{\overline{B}_n} h(w)\mu'(dw)$ enjoys $\int_{\Delta_\ell} g(x)\mu(dx) \leq c_2 \int_{\overline{\Delta}_\ell} h(w)\mu'(dw)$ with some positive constant $c_2$.

*Proof.* (i) For the set $N_{y,m} = \{n_\lambda \mid \operatorname{diam}(B_{n_\lambda}) = \theta^\ell, B_{n_\lambda} \cap B(y,\theta^m) \neq \emptyset\}$, we have

$$\int_{\Delta_\ell \cap B(y,\theta^m)} g(x)\mu(dx) \leq \sum_{n_\lambda \in N_{y,m}} \int_{B_{n_\lambda}} g(x)\mu(dx)$$

$$\leq \sum_{n_\lambda \in N_{y,m}, B_{n_\lambda} \subset B(z,\theta^{\ell+L})} \theta^{\ell m_K} \int_{B(z,\theta^{\ell+L})} h(w)\mu'(dw)$$

$$\leq \theta^{\ell m_K} \int_{\|w-y\| \leq \max\{\theta^m,\theta^{\ell+L}\}} h(w)\mu'(dw).$$

(ii) For the set $N_\ell = \{n_\lambda \mid \operatorname{diam}(B_{n_\lambda}) = \theta^\ell\}$, we see

$$\int_{\Delta_\ell} g(y)\mu(dy) \leq \sum_{n_\lambda \in N_\ell} \int_{B_{n_\lambda}} g(y)\mu(dy)$$

$$\leq \theta^{-m_K} \sum_{n_\lambda \in N_\ell} \operatorname{diam}(B_{n_\lambda})^{m_K} \int_{\overline{B}_{n_\lambda} \cap K'} h(w)\mu'(dw)$$

$$\leq \theta^{-m_K} \operatorname{diam}(B_{n_\lambda})^{m_K} \sum_{n_\lambda \in N_\ell} \int_{\overline{B}_{n_\lambda} \cap K'} h(w)\mu'(dw)$$

$$= \theta^{-m_K} \operatorname{diam}(B_{n_\lambda})^{m_K} \int_{\overline{\Delta}_\ell} h(w)\mu'(dw).$$

Since the compactness of $K'$ implies

$$\sup_{\ell} \sup_{n_\lambda \in N_\ell, \overline{B}_{n_\lambda} \cap K' \neq \emptyset} \operatorname{diam}(B_{n_\lambda}) < \infty,$$

the assertion has been proved. □

**Proposition 4.** *If $K'$ is compact, $u_{\mathscr{E}} \in B_r^{p,p}(K)$.*

*Proof.* We previously see that an equivalent norm of $B_r^{p,p}(K)$ is given by

$$\|u\|_{L^p(K;\mu)} + \int_{\|h\| \leq \theta^L} \frac{|\Delta_h u|_{L^p(K;\mu)}^p}{\|h\|^{m_K + rp}} \mu(dh) \qquad \text{for any non-positive integer } L,$$

therefore it suffices to show that

$$\|u_{\mathscr{E}}\|_{L^p(K;\mu)} + \int_{\|h\| \leq \theta^L} \frac{\|\Delta_h u_{\mathscr{E}}\|_{L^p(K;\mu)}^p}{\|h\|^{m_K + rp}} \mu(dh) \leq \|u\|_{B_r^{p,p}(K')}.$$

Firstly, we see $|u_{\mathscr{E}}(x)|^p \leq \sum_{x \in \overline{B}_n} \mu'(\overline{B}_n)^{-1} \int_{\overline{B}_n} |u(w)|^p \mu'(dw)$ for any $x \notin K'$. By the assumption of Theorem 3 (ii), we can derive from Lemma 5 (ii) that

$$\|u_{\mathscr{E}}(x)\|_{L^p(K;\mu)}^p \leq \sum_{\ell=0}^{\infty} \int_{\Delta_\ell} |u_{\mathscr{E}}(x)|^p \mu(dx) \leq c_2 \sum_{\ell=0}^{\infty} \int_{\Delta_\ell} |u(w)|^p \mu'(dw) \leq c_2 \|u\|_{L^p(K';\mu')}^p,$$

with some positive constant $c_2$. The right-hand side is dominated by constant times $\|u\|_{B_s^{p,p}(K')}^p$. Secondly, we see

$$\int\int_{\|h\| < \theta^j} \|h\|^{-m_K - rp} |\Delta_h u_{\mathscr{E}}(x)|^p \mu(dh) \mu(dx)$$

$$= \sum_{\ell=j}^{\infty} \int_{\Delta_\ell} \int_{\|h\| < \theta^\ell} \|h\|^{-m_K - rp} |\Delta_h u_{\mathscr{E}}(x)|^p \mu(dh) \mu(dx)$$

$$+ \sum_{\ell=j}^{\infty} \int_{\Delta_\ell} \int_{\theta^{\ell+1} \leq \|h\| < \theta^j} \|h\|^{-m_K - rp} |\Delta_h u_{\mathscr{E}}(x)|^p \mu(dh) \mu(dx)$$

$$= \sum_{\ell=j}^{\infty} \sum_{m=j}^{\ell} \int_{\Delta_\ell} \int_{\theta^{m+1} \leq \|h\| < \theta^m} \|h\|^{-m_K - rp} |\Delta_h u_{\mathscr{E}}(x)|^p \mu(dh) \mu(dx)$$

$$= \sum_{m=j}^{\infty} \sum_{k=m}^{\infty} \int_{\theta^{m+1} \leq \|h\| < \theta^m} \int_{\Delta_k} \|h\|^{-m_K - rp} |\Delta_h u_{\mathscr{E}}(x)|^p \mu(dh) \mu(dx)$$

$$= \sum_{m=j}^{\infty} \int_{\theta^{m+1} \leq \|h\| < \theta^m} \|h\|^{-m_K - rp} \int_{F_m} |\Delta_h u_{\mathscr{E}}(x)|^p \mu(dh) \mu(dx),$$

where $F_m = \cup_{j=m}^{\infty} \Delta_j$. On the other hand, we obtain

$$\int_{F_m} \int_{\|h\|<\theta^j} \|h\|^{-m_K-rp} |\Delta_h u_{\mathcal{E}}(x)|^p \mu(dh)\mu(dx)$$
$$\leq c_3 \sum_{m=j}^{\infty} \theta^{-m(m_K+rp)} \int \int_{\|x-y\|<\theta^m, x,y\in F_m} |u_{\mathcal{E}}(x) - u_{\mathcal{E}}(y)|^p \mu(dy)\mu(dx).$$

with some constant $c_3$. Accordingly, by combining Lemma 5 (i) and a basic inequality

$$|u_{\mathcal{E}}(x) - u_{\mathcal{E}}(y)|^p \leq \sup_{x\in \overline{B}_i, y\in \overline{B}_j} \mu'(\overline{B}_i)^{-1}\mu'(\overline{B}_j)^{-1} \int_{\overline{B}_i}\int_{\overline{B}_j} |u(w) - u(z)|^p \mu'(dz)\mu'(dw),$$

one sees that

$$\int_{\|x-y\|<\theta^m, x\in\Delta_j, y\in\Delta_N} |u_{\mathcal{E}}(x) - u_{\mathcal{E}}(y)|^p \mu(dx)\mu(dy)$$
$$\leq c_4 \theta^{-j(d-m_K)} \theta^{-N(d-m_K)} \int\int_{\|w-z\|<\theta^{m-1}} |u(w) - u(z)|^p \mu'(dz)\mu'(dw)$$

for any $j \geq m$ and $y \in \Delta_N$ with $N \geq m$. This is because $\sum_{m=j}^{\infty} \theta^{-m(m_K+rp)}$ $\int_{\|x-y\|<\theta^m, x,y\in F_m} |u_{\mathcal{E}}(x) - u_{\mathcal{E}}(y)|^p \mu(dx)\mu(dy)$ is dominated by

$$c_4 \sum_{m=j}^{\infty} \theta^{-m(d-m_K)} \theta^{-m(d-m_K)} \theta^{-m(m_K+rp)} \int\int_{\|w-z\|<\theta^{m-1}} |u(w) - u(z)|^p \mu'(dz)\mu'(dw)$$

with some positive constant $c_4$. Therefore, the assertion follows from the following estimates:

$$\sum_{m=j}^{\infty} \theta^{-m(d-m_K)} \theta^{-m(d-m_K)} \theta^{-m(m_K+sp)} \int\int_{\|w-z\|<\theta^{m-1}} |u(w) - u(z)|^p \mu'(dz)\mu'(dw)$$
$$\leq \sum_{m=j}^{\infty} \theta^{-m(d+sp)} \int\int_{\|w-z\|<\theta^{m-1}} |u(w) - u(z)|^p \mu'(dz)\mu'(dw) \leq \|u\|_{B_s^{p,p}(K')}. \quad \square$$

*Proof of Theorem 3 (ii).* Since $u_{\mathcal{E}} \in B_r^{p,p}$, it suffices to show that $u_{\mathcal{E}}|_F = u$. Since we have $|u_{\mathcal{E}}(x) - u(x_0)|^p \leq c_5 \sup_{x\in \overline{B}_n} \frac{1}{\mu'(\overline{B}_n)} \big(\int_{\overline{B}_n} |u(w) - u(x_0)|^p \mu'(dw)\big)$, with some positive constant $c_5$. we can derive from Lemma 5 (i) that

$$\int_{\Delta_\ell \cap B(x,\theta^m)} |u_{\mathcal{E}}(w) - u(x_0)|^p \mu(dw) \leq c_5 \theta^{-\ell d + \ell m_K} \int_{\|w-x\|\leq \max\{\theta^m, \theta^{\ell-1}\}} |u(w) - u(x_0)|^p \mu'(dw)$$

for any $\ell > m$. Therefore, we obtain

$$\int_{B(x,\theta^m)} |u_\mathcal{E}(w) - u(x_0)|^p \mu(dw) \leq c_5 \theta^{m(m_K-d)} \int_{B(x,\theta^{m-1})} |u(w) - u(x_0)|^p \mu'(dw)$$

$$\leq c_5 \theta^{m(m_K+sp)} \int_{B(x,\theta^{m-1})} \frac{|u(w)-u(x_0)|^p}{\|w-x_0\|^{d+sp}} \mu'(dw).$$

As a result, the assertion is derived from the following estimate:

$$|\mu(B(x_0,\theta^m))^{-1} \int_{B(x_0,\theta^m)} u_\mathcal{E}(w) \mu(dw) - u(x_0)|$$
$$\leq c_5 \left( \theta^{-mm_K} \int_{B(x_0,\theta^m)} |u_\mathcal{E}(w) - u(x_0)|^p \mu(dw) \right)^{1/p}.$$

$\square$

## ACKNOWLEDGMENTS

The author sends his thanks to the chief organizers and the local organizers of the 2<sup>nd</sup> International Conference on *p*-Adic Mathematical Physics, Belgrade (2005).

## REFERENCES

1. D.R. Adams and L.I. Hedberg, *Function Spaces and Potential Theory*, Springer-Verlag, Berlin, 1996.
2. S. Albeverio and W. Karwowski, "Diffusion on *p*-adic numbers," in *Gaussian Random Fields*, edited by K. Itô and H. Hida, World Scientific, Singapore, pp. 86–99 (1991).
3. S. Albeverio and W. Karwowski, *Stochastic. Process. Appl.* **53**, 1–22 (1994).
4. S. Albeverio and W. Karwowski, "Real time random walks on *p*-adic numbers," in *Proceeding of Conference dedicated to Ludwig Streit on the Occasion of His 60th Birthday*, 1999.
5. S. Albeverio, W. Karwowski and X. Zhao, *Stochastic Process. Appl.* **83**, 39–59 (1999).
6. S. Albeverio, A. Khrennikov, S. De Smedt, and B. Tirozzi, *Theor. Math. Phys.* **114**, 349–365 (1998).
7. S. Albeverio and X. Zhao, *Ann. Probab.*, **28**, 1680–1710 (2000).
8. D. Aldous and S. Evans, *J. Theoret. Probab.* **12**, 839–857 (1999).
9. P. Baldi, E. Casadio-Tarabusi and A. Figà-Talamanca, *Stable laws on local fields and the real line arising from hitting distributions on homogeneous trees and the hyperbolic half–plane*. Preprint, Università di Roma, 1998.
10. S. Evans, *J. Theoret. Probab.* **2**, 209–259 (1989).
11. W. Farkas, N. Jacob and R.L. Schilling, *Forum Math.* **13**, 51–90 (2001).
12. W. Farkas, N. Jacob and R.L. Schilling, *Diss. Math.*, **CCCXCIII** 1–62 (2001).
13. M. Fukushima and H. Kaneko, "On $(r,p)$-capacities for general Markov semigroups," in *Infinite dimensional analysis and stochastic processes Proc. USP-meeting at Bielfeld 1983*, edited by S. Albeverio, Research Notes in Math. **124**, Pitman, Boston-London, , pp.41–47 (1985).
14. M. Fukushima, Y. Oshima, and M. Takeda, *Dirichlet Forms and Symmetric Markov Processes*, Walter de Gruyter, Berlin, 1994.
15. S. Haran, *Ann. Inst. Fourier* **43**, 905–944 (1993).
16. S. Haran, *Ann. Inst. Fourier* **43**, 997–1053 (1993).

17. F. Hirsch, and S. Song, *Probab. Theory. Relat. Fields.* **103**, 45–71 (1996).
18. W. Hoh, and N. Jacob, *Towards an $L^p$ potential theory for sub-Markovian semigroups: variational inequalities and balayage theory*, to appear in *J. Evol. Equ.*
19. N. Jacob, *Pseudo differential operators and Markov processes. Vol. II.*, Imperial College Press, London, 2002.
20. N. Jacob and R. L. Schilling, *Towards an $L^p$ potential theory for sub-Markovian semigroups: kernels and capacities*, preprint.
21. A. Jonnson and H. Wallin, *Function spaces on Subsets of $R^n$*, Harwood Academic Pub., London, 1984.
22. H. Kaneko, *Osaka J. Math.* **23**, 325–336 (1986).
23. H. Kaneko, *Stochastic Process. Appl.* **88**, 161–174 (2000).
24. H. Kaneko, *$(r,p)$-capacity and Hausdroff measure on a local field*, to apper in *Indag. Math.*
25. H. Kaneko and K. Yasuda, *Capacities associated with Dirichlet space on an infinite extension of a local field*, to apper in *Forum Math.*
26. H. Kaneko and X. Zhao, *Forum Math.* **16**, 69–95 (2004).
27. W. Karwowski and R. Vilela-Mendes, *J. Math. Phys.* **35**, 4637–4650 (1994).
28. T. Kazumi and I. Shigekawa, "Measures of finite $(r,p)$-energy and potentials on a separable metric space," in *Séminaire de Probabilité XXVI*, edited by J. Azéma, P.A. Meyer and M. Yor, Lecture Notes Math. **1526**, Springer, Berlin, pp. 415–444 (1992).
29. A. Kochubei, *Potential Analysis.* **10**, 305–325. (1999).
30. A. Kochubei, *J. Theoret. Probab.* **15**, 951–972. (2002).
31. W.H. Schikhof, *Ultrametric calculus*, Cambridge University Press, Cambridge, 1984.
32. H. Triebel, *Theory of function spaces*, Birkhäuser, Basel – Boston – Stuttgart, 1983.
33. V.S. Vladimirov, *Russian Math. Surveys* **43**, 19–64 (1988).
34. V.S. Vladimirov, I.V. Volovich, and E.I. Zelenov, *p-adic numbers in mathematical physics*, World Scientific, Singapore, 1993.
35. K. Yasuda, *J. Math. Sci. Univ. Tokyo* **3**, 629–654 (1996).
36. K. Yasuda, *Osaka J. Math.* **37**, 967–985 (2000).

# $p$-Adic Probability Theory and Its Generalizations

Andrei Khrennikov[1]

*International Center for Mathematical Modeling in Physics, Engineering and Cognitive science MSI, Växjö University, S-35195, SWEDEN*
email: Andrei.Khrennikov@msi.vxu.se

**Abstract.** This is a review about $p$-adic valued probabilities, applications to quantum physics and cognitive sciences. We also do the next step in development of non-Kolmogorovian models and consider an analogue of probability theory for probabilities taking values in topological groups. The main attention is paid to statistical interpretation of $p$-adic probabilities as well as more general probabilities with values in topological groups.

**Keywords:** Probability, frequency, measure-theoretic, negative, complex, $p$-adic, taking values in a topological group, significance level and neighborhood.
**PACS:** 02.10.De, 02.50.-r, 03.65.Ca.

## 1. INTRODUCTION

Since the creation of the modern probabilistic axiomatics by A. N. Kolmogorov [1] in 1933, probability theory was merely reduced to the theory of normalized $\sigma$-additive measures taking values in the segment [0,1] of the field of real numbers **R**. In particular, the main competitor of Kolmogorov's measure-theoretic approach, von Mises' frequency approach to probability [2] totally disappeared from the probabilistic arena. On one hand, this was a consequence of difficulties with von Mises' definition of randomness (via place selections), see e.g., [3]-[5].[2] On the other hand, von Mises' approach (as many others) could not compete with precisely and simply formulated Kolmogorov's theory.

We mentioned von Mises' approach not only, because its attraction for applications, but also because von Mises' model with frequency probabilities played the important role in the process of formulation of the conventional axiomatics of probability theory. If one opens Kolmogorov's book [1], he will see numerous remarks about von Mises' theory. Andrei Nikolaevich Kolmogorov used properties of the frequency probability to justify his choice of the axioms for probability. In particular, Kolmogorov's probability belongs

---

[1] This paper was partially supported by EU-network "Quantum Probability and Applications" and the grant of Swedish Royal Academy of Science for collaboration with states of former Soviet Union.
[2] However, see also [5], where von Mises' approach was simplified, generalized, and then fruitfully applied to theoretical physics.

to the segment [0,1] of the real line **R**, because the same takes place for von Mises' frequency probability (relative frequencies $v_N = n/N$ as well as their limits - probabilities – always belong to the segment [0,1] of the real line **R**). In the same way Kolmogorov's probability is additive, because the frequency probability is additive: the limit of the sum of two frequencies equals to the sum of limits. And so on... Thus by using *THEOREMS* of von Mises' frequency theory Kolmogorov justified AXIOMATIZATION of probability as a normalized finite-additive measure taking values in [0,1]. Finally, he added the condition of $\sigma$-additivity.

We would like to mention that Kolmogorov's (as well as von Mises') assumptions were also based on a fundamental, but hidden, assumption:

*Limiting behavior of relative frequencies is considered with respect to one fixed topology on the field of rational numbers* **Q**, *namely, the real topology.*

In particular, consideration of this asymptotic behaviour implies that probabilities belong to the field real numbers **R**. In fact, additivity of the probability is a consequence of the fact that **R** is an *additive topological group*. We also remark that Bayes' formula

$$\mathbf{P}(B|A) = \frac{\mathbf{P}(A \cap B)}{\mathbf{P}(A)}, \mathbf{P}(A) \neq 0.$$

is also a theorem in von Mises' theory, see, e.g., [2], [5]. It is derived as a consequence of the fact that $\mathbf{R} \setminus \{0\}$ is a *multiplicative topological group*.

However, it is possible to study asymptotic behavior of relative frequencies (which are always rational numbers) in other topologies on field of rational numbers **Q**. In this way we derive another probability-like structure that recently appeared in theoretical physics. This is so called *p-adic probability*. We recall that *p*-adic numbers are applied intensively in different domains of physics – quantum logic, string theory, cosmology, quantum mechanics, quantum foundations, see, e.g., [6]–[18], dynamical systems [19]–[24], [13], biological and cognitive models [13], [24]–[26].

In this paper we shall concentrate our study to probabilistic models that could be obtained through changing the range of values of probabilities. Thus our "generalized probabilities" do not more belong to the segment [0,1] of **R**, cf. with [27]-[32]: there were considered negative and complex "probabilities."

We consider natural generalizations of properties of probability that are obtained through the transition from **R** to an arbitrary topological group. We consider **R** as a topological group (with respect to addition) and extract the main properties of Kolmogorov's measure-theoretic or von Mises' frequency probability corresponding to the group structure (algebraic and topological) on **R**. Then we use generalizations of these properties to define generalized probabilities that take values in an arbitrary topological group $G$.

Before developing the general axiomatics, we will pay more attention to the *p*-adic valued probabilities [12], [13], [33]–[35], [5]. In fact, it was the first example of the mathematically rigorous formalism for probabilities that take values in a topological group $G$ which is different from **R**.

## 2. *p*-ADIC PROBABILITY THEORY

We recall that the field of real numbers **R** is constructed as the completion of the field of rational numbers **Q** with respect to the metric $p(x,y) = |x-y|$, where $|\cdot|$ is the usual valuation given by the absolute value. The fields of *p*-adic numbers $\mathbf{Q}_p$ are constructed in a corresponding way, but by using *p*-adic valuations, see [26] for detail. Write $U_r(a) = \{x \in \mathbf{Q}_p : |x-a|_p \leq r\}$, where $r = p^n$ and $n = 0, \pm 1, \pm 2, \ldots$ These are the "closed" balls in $\mathbf{Q}_p$ while the sets $\mathbf{S}_r(a) = \{x \in \mathbf{Q}_p : |x-a|_p = r\}$ are the spheres in $\mathbf{Q}_p$ of such radii $r$.

As in the ordinary probability theory [2], the first *p*-adic probability model was the frequency one, [12], [13], [33]–[35], [5]. This model was based on the simple remark that relative frequencies $v_N = \frac{n}{N}$ always belong to the field of rational numbers **Q**. And **Q** can be considered as a (dense) subfield of **R** as well as $\mathbf{Q}_p$ (for each prime number *p*). Therefore behaviour of sequences $\{v_N\}$ of (rational) relative frequencies can be studied not only with respect to the real topology on $Q$, but also with respect to any *p*-adic topology on $Q$. Roughly speaking a *p*-adic probability (as real von Mises' probability) is defined as:

$$\mathbf{P}(\alpha) = \lim_N v_N(\alpha). \qquad (1)$$

Here $\alpha$ is some label denoting a result of a statistical experiment. Denote the set of all such labels by the symbol $\Omega$. In the simplest case $\Omega = \{0, 1\}$. Here $v_N(\alpha)$ is the relative frequency of realization of the label $\alpha$ in the first $N$ trials. The $\mathbf{P}(\alpha)$ is the frequency probability of the label $\alpha$.

The main *p*-adic lesson is that it is impossible to consider, as we did in the real case, limits of the relative frequencies $v_N$ when the $N \to \infty$. Here the point "$\infty$" belongs, in fact, to the real compactification of the set of natural numbers. So $|N| \to \infty$, where $|\cdot|$ is the real absolute value. The set of natural numbers **N** is bounded in $\mathbf{Q}_p$ and it is densely embedded into the ring of *p*-adic integers $\mathbf{Z}_p$ (the unit ball of $\mathbf{Q}_p$). Therefore sequences $\{N_k\}_{k=1}^\infty$ of natural numbers can have various limits $m = \lim_{k \to \infty} N_k \in \mathbf{Z}_p$.

In the *p*-adic frequency probability theory we proceed in the following way to provide the rigorous mathematical meaning for the procedure (1), see [34], [5]. We fix a *p*-adic integer $m \in \mathbf{Z}_p$ and consider the class, $L_m$, of sequences of natural numbers $s = \{N_k\}$ such that $\lim_{k \to \infty} N_k = m$ in $\mathbf{Q}_p$.

Let us consider the fixed sequence of natural numbers $s \in L_m$. We define a *p*-adic *s*-probability

$$\mathbf{P}(\alpha) = \lim_{k \to \infty} v_{N_k}(\alpha), s = \{N_k\}.$$

This is the limit of relative frequencies with respect to the fixed sequence $s = \{N_k\}$ of natural numbers. For any subset $A$ of the set of labels $\Omega$, we define its *s*-probability as

$$\mathbf{P}(A) = \lim_{k \to \infty} v_{N_k}(A), s = \{N_k\},$$

where $v_{N_k}(A)$ is the relative frequency of realization of labels $\alpha$ belonging to the set $A$ in the first $N$ trials. As $\mathbf{Q}_p$ is an additive topological semigroup (as well as **R**), we obtain that the *p*-adic probability is additive:

**Theorem 1.** $\quad\mathbf{P}(A_1 \cup A_2) = \mathbf{P}(A_1) + \mathbf{P}(A_2), \ A_1 \cap A_2 = \emptyset.$ (2)

As $\mathbf{Q}_p$ is even an additive topological group (as well as $\mathbf{R}$), we get that

**Theorem 2.** $\quad\mathbf{P}(A_1 \setminus A_2) = \mathbf{P}(A_1) - \mathbf{P}(A_1 \cap A_2).$ (3)

Trivially, for any sequence $s = \{N_k\}$, $\mathbf{P}(\Omega) = \lim_{k\to\infty} v_{N_k}(\Omega) = 1$, as $v_N(\Omega) = \frac{N}{N} = 1$ for any N. As $\mathbf{Q}_p$ is a multiplicative topological group (as well as $\mathbf{R}$), we get (see von Mises [2] for the real case and [5] for the $p$-adic case) Bayes' formula for conditional probabilities:

**Theorem 3.** $\quad\mathbf{P}(A|B) = \lim_{k\to\infty} \frac{v_{N_k}(A \cap B)}{v_{N_k}(A)} = \frac{\mathbf{P}(A \cap B)}{\mathbf{P}(A)}, \ \mathbf{P}(A) \neq 0.$ (4)

As we know, frequency probability played the crucial role in conventional probability theory in determination of the range of values (namely, the segment [0,1]) of a probabilistic measure, see remarks on von Mises' theory in Kolmogorov's book [1]. Frequencies always lie between zero and one. Thus their limits (with respect to the real topology) belong to the same range.

In the $p$-adic case we can proceed in the same way. Let $r \equiv r_m = \frac{1}{|m|_p}$ (where $r = \infty$ for $m = 0$). We can easily get, see [34], [5], that for the $p$-adic frequency $s$-probability, $s \in L_m$, the values of $\mathbf{P}$ always belong to the $p$-adic ball $U_r(0) = \{x \in \mathbf{Q}_p : |x|_p \leq r\}$. In the $p$-adic probabilistic model such a ball $U_r(0)$ plays the role of the segment [0,1] in the real probabilistic model.

As in the real case, the structure of an additive topological group of $\mathbf{Q}_p$ induces the main properties of probability that can be used for the axiomatization in the spirit of Kolmogorov, [1]. Let us fix $r = p^{\pm l}, l = 0, 1, \ldots$, or $r = \infty$.

**Axiomatics 1.** *Let $\Omega$ be an arbitrary set (a sample space) and let F be a field of subsets of $\Omega$ (events). Finally, let $\mathbf{P} : F \to U_r(0)$ be an additive function (measure) such that $\mathbf{P}(\Omega) = 1$. Then the triple $(\Omega, F, \mathbf{P})$ is said to be a p-adic r-probabilistic space and $\mathbf{P}$ p-adic r-probability.*

Following to Kolmogorov we should find some technical mathematical restriction on $\mathbf{P}$ that would induce fruitful integration theory and give the possibility to define averages. Kolmogorov (by following Borel, Lebesque, Lusin, and Egorov) proposed to consider the $\sigma$-additivity of measures and the $\sigma$-structure of the field of events. Unfortunately, in the $p$-adic case the situation is not so simple as in the real one. One could not just copy Kolmogorov's approach and consider the condition of $\sigma$-additivity. There is, in fact, a No-Go theorem, see, e.g., [36], [37]:

**Theorem 4.** *All $\sigma$-additive p-adic valued measures defined on $\sigma$-fields are discrete.*

Here the difficulty is not induced by the condition of $\sigma$-additivity, but by an attempt to extend a measure from the field $F$ to the $\sigma$-field generated by $F$. Roughly speaking there exist σ-additive "continuous" $\mathbf{Q}_p$-valued measures, but they could not be extended from the field $F$ to the σ-field generated by $F$. Therefore it is impossible to choose the σ-additivity as the basic integration condition in the p-adic probability theory.

The first important condition (that was already invented in the first theory of non-Archimedian integration of Monna and Springer [38]) is *boundedness*:

$$||A||_\mathbf{P} = \sup\{|\mathbf{P}(A)|_p : A \in F\} < \infty.$$

Of course, if $\mathbf{P}$ is a p-adic r-probability with $r < \infty$, then this condition is fulfilled automatically. It is nontrivial only if the range of values of a p-adic probability is unbounded in $\mathbf{Q}_p$.[3] We pay attention to one important particular case in that the condition of boundedness alone implies fruitful integration theory. Let $\Omega$ be a compact zero-dimensional topological space.[4] Then the integral

$$E\xi = \int_\Omega \xi(\omega)\mathbf{P}(d\omega)$$

is well defined for any continuous function $\xi : \Omega \to \mathbf{Q}_p$. For example, this theory works well for the following choice: $\Omega$ is the ring of q-adic integers $Z_q$, and $\mathbf{P}$ is a bounded p-adic r-probability, $r < \infty$. The integral is defined as the limit of Riemannian sums [36]–[38].

But in general boundedness alone does not imply fruitful integration theory. We should consider another condition, namely *continuity* of $\mathbf{P}$. The most general continuity condition was proposed by A. van Rooij [37].[5]

**Definition 1.** *A p-adic valued measure that is bounded, continuous, and normalized is called p-adic probability measure.*

Everywhere below we consider p-adic probability spaces endowed with p-adic probability measures.

Let $(\Omega, F, \mathbf{P})$ be a p-adic probabilistic space. *Random variables* $\xi : \Omega \to \mathbf{Q}_p$ are defined as $\mathbf{P}$-integrable functions.

As the frequency p-adic probability theory induces, see [5], (as a Theorem) Bayes' formula for conditional probability, we can use (4) as the definition of conditional probability in the p-adic axiomatic approach (as it was done by Kolmogorov in the real case).

**Example 1.** (*p-adic valued uniform distribution on the space of q-adic sequences*). Let $p$ and $q$ be two prime numbers. We set $X_q = \{0, 1, \ldots, q-1\}, \Omega_q^n = \{x = (x_1, \ldots, x_n) : x_j \in X_q\}, \Omega_q^\star = \bigcup_n \Omega_q^n$ (the space of finite sequences), and

$$\Omega_q = \{\omega = (\omega_1, \ldots, \omega_n, \ldots) : \omega_j \in X_q\}$$

---

[3] In the frequency formalism this corresponds to considering of p-adic (frequency) s-probabilities for $s \in L_0$; e.g., $s = \{N_k = p^k\}$. In this case $m = \lim_{k \to \infty} p^k = 0$.
[4] There exists a basis of neighborhoods that are open and closed at the same time.
[5] We remark that in many cases continuity coincides with σ-additivity.

(the space of infinite sequences). For $x \in \Omega_q^n$, we set $l(x) = n$. For $x \in \Omega_q^\star, l(x) = n$, we define a cylinder $U_x$ with the basis $x$ by $U_x = \{\omega \in \Omega_q : \omega_1 = x_1, \ldots, \omega_n = x_n\}$. We denote by the symbol $F_{\text{cyl}}$ the field of subsets of $\Omega_q$ generated by all cylinders. In fact, the $F_{\text{cyl}}$ is the collection of all finite unions of cylinders

First we define the uniform distribution on cylinders by setting $\mu(U_x) = 1/q^{l(x)}, x \in \Omega_q^\star$. Then we extend $\mu$ by additivity to the field $F_{\text{cyl}}$. Thus $\mu : F_{\text{cyl}} \to \mathbf{Q}$. The set of rational numbers can be considered as a subset of any $\mathbf{Q}_p$ as well as a subset of $\mathbf{R}$. Thus $\mu$ can be considered as a $p$-adic valued measure (for any prime number $p$) as well as the real valued measure. We use symbols $\mathbf{P}_p$ and $\mathbf{P}_\infty$ to denote these measures. The probability space for the uniform $p$-adic measure is defined as the triple

$$\mathscr{P} = (\Omega, F, \mathbf{P}), \text{ where } \Omega = \Omega_q, F = F_{\text{cyl}} \text{ and } \mathbf{P} = \mathbf{P}_p.$$

The $\mathbf{P}_p$ is called a *uniform $p$-adic probability distribution*.

The uniform $p$-adic probability distribution is a probabilistic measure iff $p \neq q$. The range of its values is a subset of the unit $p$-adic ball.

**Remark 1.** Values of $\mathbf{P}_p$ on cylinders coincide with values of the standard (real-valued) uniform probability distribution (Bernoulli measure) $\mathbf{P}_\infty$. Let us consider, the map $j_\infty(\omega) = \sum_{j=0}^\infty \frac{\omega_j}{2^{j+1}}$. The $j_\infty$ maps the space $\Omega_q$ onto the segment $[0, 1]$ of the real line $\mathbf{R}$ (however, $j_\infty$ is not one to one correspondence). The $j_\infty$-image of the Bernoulli measure is the standard Lebesque measure on the segment $[0,1]$ (the uniform probability distribution on the segment $[0,1]$).

**Remark 2.** The map $j_q : \Omega_q \to \mathbf{Z}_q, j_q(\omega) = \sum_{j=0}^\infty \omega_j q^j$, gives (one to one!) correspondence between the space of all $q$-adic sequences $\Omega_q$ and the ring of $q$-adic integers $\mathbf{Z}_q$. The field $F_{\text{cyl}}$ of cylindrical subsets of $\Omega_q$ coincides with the field $B(\mathbf{Z}_q)$ of all clopen (closed and open at the same time) subsets of $\mathbf{Z}_q$. If $\Omega_q$ is realized as $\mathbf{Z}_q$ and $F_{\text{cyl}}$ as $B(\mathbf{Z}_q)$, then $\mu_p$ is the $p$-adic valued Haar measure on $\mathbf{Z}_q$. The use of the topological structure of $\mathbf{Z}_q$ is very fruitful in the integration theory (for $p \neq q$). In fact, the space of integrable functions $f : \mathbf{Z}_q \to \mathbf{Q}_p$ coincides with the space of continuous functions (random variables) $C(\mathbf{Z}_q, \mathbf{Q}_p)$, see [36]–[38], [5].

## 3. $p$-ADIC LAW OF LARGE NUMBERS

Everywhere in this section $p$ is a prime number distinct from 2. We start with considering the classical Bernoulli scheme (in the conventional probabilistic framework) for random variables $\xi_j(\omega) = 0, 1$ with probabilities $1/2, j = 1, 2, \ldots$. First we consider a finite number $n$ of random variables: $\xi_1(\omega), \ldots, \xi_n(\omega)$. A sample space corresponding to these random variables can be chosen as the space $\Omega_2^n = \{0, 1\}^n$. The probability of an event $A$ is defined as

$$\mathbf{P}^{(n)} = \frac{|A|}{|\Omega_2^n|} = \frac{|A|}{2^n},$$

where the symbol $|B|$ denotes the number of elements in a set $B$. The typical problem of ordinary probability theory is to find the asymptotic behavior of the probabilities $\mathbf{P}^{(n)}(A), n \to \infty$. It was the starting point of the theory of limit theorems in conventional probability theory.

But the probabilities $\mathbf{P}^{(n)}(A)$ belong to the field of rational numbers $\mathbf{Q}$. We may study behavior of $\mathbf{P}^{(n)}(A)$, not only with respect to the usual real metric $\rho_\infty(x,y)$ on $\mathbf{Q}$, but also with respect to an arbitrary metric $\rho(x,y)$ on $\mathbf{Q}$. We have studied the case of the $p$-adic metric on $\mathbf{Q}$, see [39], [40]. We remark that $\mathbf{P}^{(n)}(A) = \sum_{x \in A} \mu(U_x)$, where $\mu$ is the uniform distribution on $\Omega_2^n$. By realizing $\mu$ as the (real valued) probability distribution $\mathbf{P}_\infty$ we use the formalism of conventional probability theory. By realizing $\mu$ as the $p$-adic valued probability distribution $\mathbf{P}_p$ we use the formalism of $p$-adic probability theory.

What kinds of events $A$ are naturally coupled to the $p$-adic metric? Of course, such events must depend on the prime number $p$. As usual, we consider the sums

$$S_n(\omega) = \sum_{k=1}^{n} \xi_n(\omega).$$

We are interested in the following question. Does $p$ divide the sum $S_n(\omega)$ or not? Set $A(p,n) = \{\omega \in \Omega_2^n : p \text{ divides the sum} S_n(\omega)\}$. Then $\mathbf{P}^{(n)}(A(p,n)) = L(p,n)/2^n$, where $L(p,n)$ is the number of vectors $\omega \in \Omega_2^n$ such that $p$ divides $|\omega| = \sum_{j=1}^{n} \omega_j$. As usual, denote by $\bar{A}$ the complement of a set $A$. Thus $\bar{A}(p,n)$ is the set of all $\omega \in \Omega_q^n$ such that $p$ does not divide the sum $S_n(\omega)$. We shall see that the sets $A(p,n)$ and $\bar{A}(p,n)$ are asymptotically symmetric from the $p$-adic point of view:

$$\mathbf{P}^{(n)}(A(p,n)) \to \frac{1}{2} \text{ and } \mathbf{P}^{(n)}(\bar{A}(p,n)) \to \frac{1}{2} \quad (5)$$

in the $p$-adic metric when $n \to 1$ in the same metric. Already in this simplest case we shall see that the behavior of sums $S_n(\omega)$ depends crucially on the choice of a sequence $s = \{N_k\}_{k=1}^{\infty}$ of natural numbers. A limit distribution of the sequence of random variables $S_n(\omega)$, when $n \to \infty$ in the ordinary sense, does not exist. We have to describe all limiting distributions for different sequences $s$ converging in the $p$-adic topology.

Let $(\Omega, F, \mathbf{P})$ be a $p$-adic probabilistic space and $\xi_n : \Omega \to \mathbf{Q}_p (n = 1, 2, \ldots)$ be a sequence of equally distributed independent random variables, $\xi_n = 0, 1$ with probability $1/2$.[6] We start with the following result that can be obtained through purely combinatorial considerations (behavior of binomial coefficients $C_m^r$ in the $p$-adic topology).

**Theorem 5.** *Let $m = 0, 1, \ldots, p^s - 1 (s = 1, 2, \ldots), r = 0, \ldots, m$, and $l \geq s$. Then*

$$\lim_{n \to m} \mathbf{P}(\omega : S_n(\omega) \in U_{1/p^l}(r)) = \frac{C_m^r}{2^m}.$$

Formally this theorem can be reformulated as the following result for the convergence of probabilistic distributions:

---

[6] Here $1/2$ is considered as a $p$-adic number. In the conventional theory $1/2$ is considered as a real number.

The limiting distribution on $\mathbf{Q}_p$ of the sequence of the sums $S_n(\omega)$, where $n \to m$ in $\mathbf{Q}_p$, is the discrete measure $\kappa_{1/2,m} = 2^{-m} \sum_{r=0}^{m} C_m^r \delta_m$.

We consider the event $A(p,n,r) = \{\omega : S_n(\omega) = pi + r\}$ for $r = 0, 1, \ldots, p-1$. This event consists of all $\omega$ such that the residue of $S_n(\omega)$ mod $p$ equals to $r$. Note that the set $A(p,n,r)$ coincides with the set $\{\omega : S_n(\omega) \in U_{1/p}(r)\}$.

**Corollary 1.** *Let $n \to m$ in $\mathbf{Q}_p$, where $m = 0, 1, \ldots, p-1$. Then the probabilities $\mathbf{P}^{(n)}(A(p,n,r))$ approach $C_m^r / 2^m$ for all residues $r = 0, \ldots, m$.*

In particular, as $A(p,n) \equiv A(p,n,0)$, we get (5). What happens in the case $m \geq p$? We have only the following particular result:

**Theorem 6.** *Let $n \to p$ in $\mathbf{Q}_p$ and $r = 0, 1, 2, \ldots, p$. Then*

$$\lim_{n \to p} \mathbf{P}(\omega : S_n(\omega) \in U_{1/p^s}(r)) = \frac{C_p^r}{2^p},$$

*where $s \geq 2$ for $r = 0, p$ and $s \geq 1$ for $r = 1, \ldots, p-1$.*

**Remark 3.** (Bernard-Letac asymtotics) In [41] J. Bernard and G. Letac have studied $p$-adic asymptotic of multi binomial coefficients. Although they did not consider the $p$-adic probabilistic terminology (at that moment there were no physical motivations to consider the $p$-adic generalization of probability), their results may be interpreted as a kind of a limit theorem for $p$-adic probability.

We now study the general case of dichotomic equally distributed independent random variables: $\xi_n(\omega) = 0, 1$ with probabilities $q$ and $q' = 1 - q, q \in \mathbf{Z}_p$. We shall study the weak convergence of the probability distributions $\mathbf{P}_{S_{N_k}}$ for the sums $S_{N_k}(\omega)$. We consider the space $C(\mathbf{Z}_p, \mathbf{Q}_p)$ of continuous functions $f : \mathbf{Z}_p \to \mathbf{Q}_p$. We will be interested in convergence of integrals

$$\int_{\mathbf{Z}_p} f(x) d\mathbf{P}_{S_{N_k}}(x) \to \int_{\mathbf{Z}_p} f(x) d\mathbf{P}_S(x), f \in C(\mathbf{Z}_p, \mathbf{Q}_p),$$

where $\mathbf{P}_S$ is the limiting probability distribution (depending on the sequence $s = \{N_k\}$). To find the limiting distribution $\mathbf{P}_S$, we use the method of characteristic functions. We have for characteristic functions

$$\phi_{N_k}(z,q,a) = \int_{\Omega} \exp\{z S_{N_k}(\omega)\} d\mathbf{P}(\omega) = (1 + q'(e^z - 1))^{N_k}.$$

Here $z$ belong to a sufficiently small neighborhood of zero in the $\mathbf{Q}_p$; see [13] for detail about the $p$-adic method of characteristic functions. Let $a$ be an arbitrary number from $\mathbf{Z}_p$. Let $s = \{N_k\}_{k=1}^{\infty}$ be a sequence of natural numbers converging to $a$ in the $\mathbf{Q}_p$. Set $\phi(z,q,a) = (1 + q'(e^z - 1))^a$. This function is analytic for small $z$. It is easy to see that the sequence of characteristic functions $\{\phi_{N_k}(z,q,a)\}$ converges (uniformly on every ball of a sufficiently small radius) to the function $\phi(z,q,a)$. Unfortunately, we could not prove (or disprove) a $p$-adic analogue of Levy's theorem. Therefore in the general case the convergence of characteristic functions does not give us anything. However, we shall see that we have Levy's situation in the particular case under consideration: There exists a bounded probability measure distribution, denoted by $\kappa_{q,a}$, having the characteristic function $\phi(z,q,a)$ and, moreover, $\mathbf{P}_{S_{N_k}} \to \mathbf{P}_S = \kappa_{q,a}, N_k \to a$.

We start with the first part of the above statement. Here we shall use Mahlers integration theory on the ring of $p$-adic integers, see e.g., [36]–[38], [12], [13]. We introduce a system of binomial polynomials: $C(x,k) = C_x^k = \frac{x(x-1)...(x-k+1)}{k!}$ (that are considered as functions from $\mathbf{Z}_p$ to $\mathbf{Q}_p$). Every function $f \in C(\mathbf{Z}_p, \mathbf{Q}_p)$ is expanded into a series (a Mahler expansion, see [36]) $f(x) = \sum_{k=0}^{\infty} a_k C(x,k)$. It converges uniformly on $\mathbf{Z}_p$. If $\mu$ is a bounded measure on $\mathbf{Z}_p$, then

$$\int_{\mathbf{Z}_p} f(x) \mu(dx) = \sum a_k \int_{\mathbf{Z}_p} C(x,n) \mu(dx).$$

Therefore to define a $p$-adic valued measure on $\mathbf{Z}_p$ it suffices to define coefficients $\int_{\mathbf{Z}_p} C(x,n) \mu(dx)$. A measure is bounded iff these coefficients are bounded. Using the Mahler expansion of the function $\phi(z,q,a)$, we obtain

$$\lambda_m(q,a) = \int_{\mathbf{Z}_p} C(x,m) \kappa_{q,a}(dx) = (1-q)^m C(a,m).$$

As $|C(a,m)|_p \leq 1$ for $a \in \mathbf{Z}_p$, we get that the distribution $\kappa_{q,a}$ (corresponding to $\phi(z,q,a)$) is bounded measure on $\mathbf{Z}_p$. Set $\lambda_{mn}(q,a) = \int_{\Omega} C(S_n(\omega),m) dP(\omega)$. We find

$$\lambda_{mN_k}(q,a) = (1-q)^m C_{N_k}^m.$$

Thus $\lambda_{mN_k}(q,a) \to \lambda_m(q,a), N_k \to a$. This implies the following limit theorem.

**Theorem 7.** (*p*-adic Law of Large Numbers.) *The sequence of probability distributions* $\{\mathbf{P}_{S_{N_k}}\}$ *converges weakly to* $\mathbf{P}_S = \kappa_{q,a}$, *when* $N_k \to a$ *in* $\mathbf{Q}_p$.

## 4. GROUP VALUED PROBABILITY

Let $G$ be a commutative (additive) topological group. In general, it can be nonlocally compact.[7] Let us choose a fixed subset $\Delta$ of the group $G$.

**Axiomatics 2.** *Let* $\Omega$ *and F be as in Axiomatics 1. Let* $\mathbf{P} : F \to \Delta$ *be an additive function (measure). The triple* $(\Omega, F, \mathbf{P})$ *is said to be a G-probabilistic space (with the $\Delta$-range of probability).*

We also have to add an integration condition. Such a condition depends on the topological structure of $G$. It seems to be impossible to propose a general condition providing fruitful integration theory.

---

[7] In principle, we could proceed in the same way in the non-commutative case.

The reader might say that our definition of a $G$-probabilistic measure is too general. Moreover, our real probabilistic intuition would protest against disappearance of the *unit probability* from consideration. We shall discuss this problem in section 5.

We now consider a modification of the above axiomatics that includes a kind of 'unit probability'. Let $E = \mathbf{P}(\Omega)$ be a nonzero element in $G$. Let $G$ be metrizable (with the metric $\rho$). The additional "unit-probability axiom" should be of the following form:

$$\sup_{A \in F} \rho(0, \mathbf{P}(A)) = \rho(0, E). \tag{6}$$

A $G$-probabilistic space in that (6) holds is called a $G$-probabilistic space with *unit probability axiom*. Of course, the consideration of such probabilistic spaces seems to be more natural from the standard probabilistic viewpoint. Therefore it would be natural to start with consideration of such models. However, for many important $G$-probabilistic spaces the unit probability axiom does not hold true. At the moment we know a few examples of $G$-probabilities having applications:

1) $G = \mathbf{R}$ and $\Delta = [0, 1]$ (the conventional probability theory);
2) $G = \mathbf{R}$ and $\Delta = \mathbf{R}$ ("negative probabilities", see, e.g., [27]–[29], they are realized as signed measures, charges).
3) $G = \mathbf{C}$ and $\Delta = \mathbf{C}$ ("complex probabilities", see, e.g., [29], [31], they are realized as $\mathbf{C}$-valued measures).
4) $G = \mathbf{Q}_p$ and $\Delta$ is a ball in $\mathbf{Q}_p$ ("$p$-adic probabilities", [5], [12], [13], [33]–[34], they are realized as $\mathbf{Q}_p$-valued measures).

The $p$-adic model can be essentially (and rather easily) generalized. Let $K$ be an arbitrary complete non-Archimedian field with the valuation (absolute value) $|\cdot|$. We can define $K$-valued probabilistic measures by using the same integration conditions as in the $p$-adic case, namely boundedness and continuity.

We note that in all considered examples the additive group $G$ has the additional algebraic structure, namely the field structure. The presence of such a field structure gives the possibility to develop an essentially richer probabilistic calculus than in the general case. Here we can introduce conditional probability by using Bayes' formula and define the notion of independence of events.

The following slight generalization gives the possibility to consider a few new examples. Let $G$ be a non-Archimedian normed ring. To simplify considerations, we again consider the commutative case. Here:

1) $||x|| \geq 0, ||x|| = 0 \leftrightarrow x = 0$;
2) $||x|| \, ||y|| \leq ||x|| \, ||y||$ and $||x+y|| \leq \max(||x||, ||y||)$.

We set, for $A \in F, ||A||_\mathbf{P} = \sup\{||\mathbf{P}(B)|| : B \in F, B \subset A\}$. We define a $G$-probabilistic measure as a normalized $G$-valued measure satisfying to the conditions of boundedness and continuity. Corresponding integration theory is developed in the same way as in the case of a non-Archimedian field. One of the most important examples of non-Archimedian normed rings is a ring of $m$-adic numbers $\mathbf{Q}_m$, where $m \neq p^k, p - prime$.

It is a locally compact ring. We can present numerous examples of non-Archimedian normed rings by considering various functional spaces of $\mathbf{Q}_p$ (or $\mathbf{Q}_m$)-valued functions. For a ring $G$, we can define averages for $G$-valued random variables, $\xi : \Omega \to G$. In particular, we can represent the probability distribution of the sum $\eta = \xi_1 + \xi_2$ of two $G$-valued random variables as the convolution of corresponding probability distributions. Here we define the convolution of two $G$-valued measures on $G$ as:

$$\int_G f(x) M_1 \star M_2(dx) = \int_{G \times G} f(x_1 + x_2) M_1(dx_1) M_2(dx_2),$$

where $f : G \to G$ is a "sufficiently good" function. If $G$ is a ring and $A \in F$ is such that $\mathbf{P}(A)$ is invertible, then we can define conditional probabilities by using Bayes' formula.

We obtain a large class of new mathematical problems related to $G$-probabilistic models. We emphasize that, despite a rather common opinion, probability theory is not just a part of functional analysis (measure theory). Probability theory has also its own ideology. The probabilistic ideology induces its own problems. Such problems would be impossible to formulate in the framework of functional analysis (of course, methods of functional analysis can be essentially used for the investigation of these problems).

One of the most important problems is to find analogues of limit theorems, compare, e.g., with [42].

**Open problem 1.** Let $s = \{N_k\}$ be a sequence of natural numbers and let

$$M_{11}, M_{12}, M_{1n_1};$$

$$M_{21}, M_{22}, \ldots, M_{2n_2};$$

$$M_{k1}, M_{k2}, \ldots, M_{kN_k};$$

be $G$-probabilistic measures. As usual, we have to study behavior of convolutions:

$$\alpha_k = M_{k1} \star M_{k2} \star \ldots \star M_{kN_k}$$

to find analogues of limit theorems. For example, an analogue of the law of large numbers could be formulated in the following way. Let $d$ be a nonzero element of a topological additive group $G$. Let $s = \{N_k\}_{k=1}^\infty$ be a sequence of natural numbers. Suppose that the corresponding sequence $\{N_k d\}_{k=1}^\infty$ of elements of $G$ converges to some element $a \in G$ or to $a = \infty$. The latter has the standard meaning: for each neighborhood $U$ of zero in $G$ there exists $t$ such that $N_k d \notin U$ for all $k \geq t$.

Let $(\Omega, F, \mathbf{P})$ be a $G$-probabilistic space. Let $\xi_n(\omega) = 0, d$, with $G$-probabilities $q$, $q' = E - q$ where $E = \mathbf{P}(\Omega)$, be a sequence of independent random variables. Let $S_n(\omega)$ be the sum of $n$ first variables.

**Open Problem 2.** *Does the sequence of probability distributions $\mathbf{P}_{S_{N_k}}$ converge weakly to some probability distribution $\mathbf{P}_S$, when $k \to \infty$?*

# 5. INTERPRETATION OF PROBABILITY

In fact, Kolmogorov's probability theory has two (more or less independent) counterparts: (a) axiomatics (a mathematical representation); (i) interpretation (rules for application). The first part is the measure-theoretic formalism. The second part is a mixture of frequency and ensemble interpretations: "... we may assume that to an event $A$ which has the following characteristics: (a) one can be practically certain that if the complex of conditions $\Sigma$ is repeated a large number of times, $N$, then if $n$ be the number of occurrences of event $A$, the ratio $n/N$ will differ very slightly from $\mathbf{P}(A)$; (b) if $\mathbf{P}(A)$ is very small, one can be practically certain that when conditions $\Sigma$ are realized only once the event $A$ would not occur at all", [1].

As we have already noticed, (a) and (i) are more or less independent. Therefore Kolmogorov's measure-theoretic formalism, (a), is used successfully, for example, in the subjective probability theory.

In practice we apply Kolmogorov's (conventional) interpretation, (i), in the following way. First of all we have to fix $0 < \varepsilon < 1$, *significance level*. If the probability $\mathbf{P}(A)$ of some events $A$ is less than $\varepsilon$, this event is considered as practically impossible. We now generalize the conventional interpretation of probability to the case of $G$-valued probabilities. First of all we have to fix some neighborhood of zero, $V$, *significance neighborhood*.

If the probability $\mathbf{P}(A)$ of some event $A$ belongs to $V$, this event is considered as practically impossible.

If a group $G$ is metrizable, then the situation is even more similar to the standard (real) probability. We choose $\varepsilon > 0$ and consider the ball $V_\varepsilon = \{x \in G : \rho(0,x) < \varepsilon\}$. If $\rho(0,\mathbf{P}(A)) < \varepsilon$, then the event $A$ is considered as practically impossible.

Let us borrow some ideas from statistics. We are given a certain sample space $\Omega$ with an associated distribution $\mathbf{P}$. Given an element $\omega \in \Omega$, we want to test the hypothesis "$\omega$ belongs to some reasonable majority." A reasonable majority $\mathscr{M}$ can be described by presenting *critical regions* $\Omega^{(\varepsilon)}(\in F)$ of the significance level $\varepsilon, 0 < \varepsilon < 1 : \mathbf{P}(\Omega^{(\varepsilon)}) < \varepsilon$. The complement $\bar{\Omega}^{(\varepsilon)}$ of a critical region $\Omega^{(\varepsilon)}$ is called $(1-\varepsilon)$ confidence interval. If $\omega \in \Omega^{(\varepsilon)}$, then the hypothesis '$\omega$ belongs to majority $\mathscr{M}$' is rejected with the significance level $\varepsilon$. We can say that $\omega$ fails the test to belong to $\mathscr{M}$ at the level of critical region $\Omega^{(\varepsilon)}$.

$G$-statistical machinery works in the same way. The only difference is that, instead of significance levels $\varepsilon$, given by real numbers, we consider significance levels $V$ given by neighborhoods of zero in $G$. Thus we consider critical regions $\Omega^{(V)}(\in F)$:

$$\mathbf{P}(\Omega^{(V)}) \in V.$$

If $\omega \in \Omega^{(V)}$, then the hypothesis "$\omega$ belongs to majority $\mathscr{M}$" (represented by the statistical test $\{\Omega^{(V)}\}$) is rejected with the significance level $V$. If $G$ is metrizable, then we have even more similarity with the standard (real) statistics. Here $V = V_\varepsilon, \varepsilon > 0$.

Of course, the strict mathematical description of the above statistical considerations can be presented in the framework of Martin-Löf [3]–[5] statistical tests. We remark that such a *p*-adic framework was already developed in [5]. In the *p*-adic case (as in the real case) it is possible to enumerate effectively all *p*-adic tests for randomness. However, a universal *p*-adic test for randomness does not exist [5]. If the group $G$ is metrizable we can proceed in the same way as in the real and *p*-adic case [5] and define $G$-random sequences, namely sequences $\omega = (\omega_1, \ldots, \omega_N, \ldots), \omega_j = 0, 1$, that are random with respect to a $G$-valued probability distribution. However, if $G$ is not metrizable, then the notion of a recursively enumerable set would not be more the appropriative basis for such a theory. In any case we have an interesting

**Open problem 3.** *Development of randomness theory for an arbitrary topological group.*

The general scheme of the application of $G$-valued probabilities is the same as in the ordinary case: 1) we find initial probabilities; 2) then we perform calculations by using calculus of $G$-valued probabilities; 3) finally, we apply the above interpretation to resulting probabilities.

The main question is "How can we find initial probabilities?" Here the situation is more or less similar to the situation in the ordinary probability theory. One of possibilities is to apply the frequency arguments (as R. von Mises). We have already discussed such an approach for *p*-adic probabilities. Another possibility is to use subjective approach to probability. I think that everybody agrees that there is nothing special in segment $[0, 1]$ as the set of labels for the measure of belief in the occurrence of some event. In the same way we can use, for example, the segment $[-1, 1]$ (signed probability) or the unit complex disk (complex probability) or the set of *p*-adic integers $\mathbf{Z}_p$ (*p*-adic probability). If $G$ is a field we can apply the machinery of Bayesian probabilities and, finally, use our interpretation of probabilities to make a statistical decision. The third possibility is to use symmetry arguments, Laplacian approach. For example, by such arguments we can choose (in some situations) the uniform $\mathbf{Q}_p$-valued distribution.

We now turn back to the role of the unit probability and, in particular, the axiom (6). In fact, by considering the interpretation of probability based on the notion of the significance level we need not pay the special attention to the probability $E = \mathbf{P}(\Omega)$. It is enough to consider $V$-impossible events, $V = V(0)$. If $V$ is quite large and $\mathbf{P}(A) \notin V$, then an event $A$ can be considered as practically definite.

**Example 2.** (A *p*-adic statistical test) Theorem 5 implies that, for each *p*-adic sphere $\mathbf{S}_{1/p^l}(r)$, where $l, r, m$ were done in Theorem 5:

$$\lim_{k \to \infty} \mathbf{P}(\{\omega \in \Omega_2 : S_{N_k}(\omega) \in \mathbf{S}_{1/p^l}(r)\}) = 0,$$

for each sequence $s = \{N_k\}, N_k \to m, k \to \infty$. We can construct a statistical test on the basis of this limit theorem (as well as any other limit theorem). Let $s = \{N_k\}, N_k \to m$, be a fixed sequence of natural numbers. For any $\varepsilon > 0$, there exists $k_\varepsilon$ such that, for all $k \geq k_\varepsilon$,

$$|\mathbf{P}(\{\omega \in \Omega_2 : S_{N_k}(\omega) \in \mathbf{S}_{1/p^l}(r)\})|_p < \varepsilon.$$

We set $\Omega^{(\varepsilon)} = \bigcup_{k \geq k_\varepsilon} \{\omega \in \Omega_2 : S_{N_k}(\omega) \in \mathbf{S}_{1/p^l}(r)\}$. We remark that

$$|\mathbf{P}(\Omega^{(\varepsilon)})|_p < \varepsilon.$$

We now define reasonable majority of outcomes as sequences that do not belong to the sphere $\mathbf{S}_{\frac{1}{p^l}}(r)$, "nonspherical majority." Here the set $\Omega^{(\varepsilon)}$ is the critical region on the significance level $\varepsilon$.

Suppose that a sequence $\omega$ belongs to the set $\Omega^{(\varepsilon)}$. Then the hypothesis "$\omega$ belongs nonspherical majority" must be rejected with the significance level $\varepsilon$. In particular, such a sequence $\omega$ is not random with respect to the uniform $p$-adic distribution on $\Omega_2$. If, for some sequence of 0 and 1, $\omega = (\omega_j)$ we have $\omega_1 + \ldots + \omega_{N_k} - r = \alpha$ mod $p^l$, $\alpha = 1, \ldots, p-1$, for all $k \geq k_\varepsilon$, then it is rejected.

The simplest test is given by $m = 1, r = 0, N_k = 1 + p^k$ and $\omega_1 + \ldots + \omega_{N_k} = \alpha$ mod $p$, $\alpha = 1, \ldots, p-1$.

## 6. GENERALIZED FREQUENCY MODELS

We proposed a theory of $G$-valued probabilities in the measure-theoretic framework.[8] In principle, we can also proceed in the frequency framework. However, an arbitrary topological group is too general base for such frequency probabilities. We have to start with a topological field $T$ that contains the field of rational numbers as a dense subfield. We proceed in the same way as in the $p$-adic case. Let $s = \{N_k\}$ be a sequence of natural numbers converging in $T$ to some $m$. A $T$-valued $s$-probability is defined as the limit $\mathbf{P} = \lim_{k \to \infty} v_{N_k} \in T$ (if it exists).

As $T$ is a topological field, we get additivity (2), formula (3) and Bayes' formula (4). The range of values of such a frequency probability depends on the sequence $s$ and the topology on $T$.

In fact, Kolmogorov's theory does not look extremely anti-frequencist due to the presence of the strong law of large numbers. Of course, this law was strongly criticized from the frequency point of view, see e.g., von Mises [2]. The main critical argument is that we could not say anything about behaviour of frequencies for a concrete sequence of trials. Nevertheless, there are no problems in average. Therefore, by obtaining a kind of law of large numbers (of course, in the case when the field of rational numbers $Q$ is dense in $T$) we could strongly improve the measure-theoretic approach for $T$-valued probabilities. A kind of such a law we have in the $p$-adic case.

---

[8] However, we use the frequency experience of the real and $p$-adic probabilities to find reasonable properties of 'probability'.

# ACKNOWLEDGMENTS

I would like to thank L. Accardi, S. Albeverio, A. Bendikov, B. Dragovich, W. Hazod, H. Heyer, G. Letac, V. Maximov, D. Neuenschwander, A. Shiryaev, Yu. Prohorov, V. Maslov, A. Bulinski, S. Gudder, S. Kozyrev, T. Hida, V. Vladimirov and I. Volovich for fruitful discussions. I also use this occasion to congratulate Branko Dragovich with his 60th birthday and wish him new great achievements in application of $p$-adic and adelic numbers to theoretical physics and especially to discrete models of space-time on Planck distance.

# REFERENCES

1. A.N. Kolmogorov, *Grundbegriffe der Wahrscheinlichkeitsrechnung*, Springer Verlag, Berlin, 1933; reprinted: *Foundations of the Probability Theory*, Chelsey Publ. Comp., New York, 1956.
2. R. Von Mises, *The Mathematical Theory of Probability and Statistics*, Academic, London, 1964.
3. A.N. Kolmogorov, Three approaches to the quantitative definition of information, *Problems Inform. Transmition* **1**, 1–7 (1965).
4. P. Martin-Löf, On the concept of random sequences, *Theory of Probability Appl.* **11**, 177–179 (1966).
5. A.Yu. Khrennikov, *Interpretation of probability*, VSP Int. Publ., Utrecht, 1999.
6. E. Beltrametti, and G. Cassinelli, Quantum mechanics and $p$-adic numbers, *Found. of Physics* **2**, 1–7 (1972).
7. I.V. Volovich, $p$-adic string, *Class. Quant. Grav.* **4**, 83–87 (1987).
8. P.G.O. Freund and E. Witten, Adelic string amplitudes, *Phys. Lett. B* **199**, 191–195 (1987).
9. B. Dragovic, On signature change in $p$-adic space-time, *Mod. Phys. Lett.* **6**, 2301–2307 (1991).
10. I.Ya. Aref'eva, B. Dragovic, I.V. Volovich, On the $p$−adic summability of the anharmonic oscillator, *Phys. Lett. B* **200**, 512–514 (1988).
11. V.S. Vladimirov, I.V. Volovich, and E.I. Zelenov, *p-adic Analysis and Mathematical Physics*, World Scientific Publ., Singapore, 1993.
12. A.Yu. Khrennikov, *p-adic Valued Distributions in Mathematical Physics*, Kluwer Academic Publishers, Dordrecht, 1994.
13. A.Yu. Khrennikov, *Non-Archimedian Analysis: Quantum Paradoxes, Dynamical Systems and Biological Models*, Kluwer Academic Publishers, Dordrecht 1997.
14. A.Yu. Khrennikov, $p$-adic probability distribution of hidden variables, *Physica A* **215**, 577–587 (1995).
15. A.Yu. Khrennikov, Non-Kolmogorov probabilistic models with $p$-adic probabilities and foundations of quantum mechanics, *Stochastic Analysis and Related Topics* **4**, 275–304 (1998).
16. A.Yu. Khrennikov, $p$-adic probability predictions of correlations between particles in the two slit and neuron interferometry experiments, *Il Nuovo Cimento B* **113**, 751–760 (1998).
17. S. Albeverio, A.Yu. Khrennikov, A regularization of quantum field Hamiltonians with the aid of p-adic numbers, *Acta Appl. Math.* **50**, 225–251 (1998).
18. A.Yu. Khrennikov, Description of experiments detecting $p$-adic statistics in quantum diffraction experiments, *Doklady Mathematics* **58**, 478–480 (1998).
19. E. Thiran, D. Verstegen, J. Weyers, $p$-adic dynamics, *J. Stat. Phys.* **54**, 893–913 (1989).
20. D.K. Arrowsmith, and F. Vivaldi, Some $p$adic representations of the Smale horseshoe, *Phys. Lett. A* **176**, 292–294 (1993).
21. D.K. Arrowsmith, and F. Vivaldi, Geometry of $p−$adic Siegel discs, *Physica D* **74**, 222–236 (1994).
22. V.S. Anashin, Uniformly distributed sequences over $p$-adic numbers, *Mat. Zametki* **55**, 3–46 (1994); English translation in *Math. Notes* **55**, 109–133 (1994).
23. V.S. Anashin, Uniformly distributed sequences of $p$-adic integers, *Diskret. Mat.* **14**, 3–64 (2002); English translation in *Discrete Math. Appl.* **12**, 527–590 (2002).

24. A.Yu. Khrennikov, M. Nilsson, *p-adic Deterministic and Random Dynamical Systems*, Kluwer Academic Publishers, Dordrecht, 2004.
25. D. Dubischar, V.M. Gundlach, O. Steinkamp, A.Yu. Khrennikov, Attractors of random dynamical systems over *p*-adic numbers and a model of noisy cognitive processes, *Physica D* **130**, 1–12 (1999).
26. A.Yu. Khrennikov, *Information Dynamics in Cognitive, Psychological and Anomalous Phenomena*, Kluwer Academic Publishers, Dordrecht, 2004.
27. P.A.M. Dirac, The physical interpretation of quantum mechanics, *Proc. Roy. Soc. London* **A 180**, 1–39 (1942).
28. E. Wigner, Quantum-mechanical distribution functions revisted, *Perspectives in quantum theory*, edited by W. Yourgrau, and A. van der Merwe, MIT Press, Cambridge MA, pp. 121–146 (1971).
29. W. Muckenheim, A review on extended probabilities, *Phys. Reports* **133**, 338–401 (1986).
30. A.Yu. Khrennikov, *p*-adic description of Dirac's hypothetical world with negative probabilities, *Int. J. Theor. Phys.* **34**, 2423–2434 (1995).
31. P.A.M. Dirac, On the analogy between classical and quantum mechanics, *Rev. of Modern Phys.* **17**, 195–199 (1945).
32. E. Prugovecki, Simultaneous measurement of several observables, *Found. of Physics* **3**, 3–18 (1973).
33. A.Yu. Khrennikov, *p*-adic probability and statistics, *Dokl. Akad. Nauk USSR* **322**, 1075–1079 (1992).
34. A.Yu. Khrennikov, An extension of the frequency approach of R. von Mises and the axiomatic approach of N.A. Kolmogorov to the *p*-adic theory of probability, *Theory of Probability and Appl.* **40**, 458–463 (1995).
35. A.Yu. Khrennikov, Interpretation of probability and their *p*-adic extensions, *Theory of Probability and Appl.* **46**, 311–325 (2001).
36. W. Schikhof, *Ultrametric Calculus*, Cambridge Univ. Press, Cambridge, 1984.
37. A. van Rooij, *Non-Archimedian Functional Analysis*, Marcel Dekker, Inc., New York, 1978.
38. A. Monna, and T. Springer, Integration non-Archimédienne, *Indag. Math.* **25**, 634–653 (1963).
39. A.Yu. Khrennikov, Laws of large numbers in non-Archimedian probability theory, *Izvestia Akademii Nauk, Math.* **64**, 211–223 (2000).
40. A.Yu. Khrennikov, Limit behaviour of sums of independent random variables with respect to the uniform *p*-adic distribution, *Statistics and Probability Lett.* **51**, 269–276 (2001).
41. J. Bernard, and G. Letac, Construction d'evenements equiprobables et coefficients multinomiaux modulo $p^n$, *Illinois J. Math.* **17**, 317–332 (1997).
42. H. Heyer, *Probability Measures on Locally Compact Groups*, Springer-Verlag, Berlin-Heidelberg-New York, 1977.

# Ultrametric Analysis and Interbasin Kinetics

S. V. Kozyrev

*Steklov Mathematical Institute, Russian Academy of Sciences,
Gubkin st. 8, Moscow, 119991, RUSSIA
email:* kozyrev@mi.ras.ru

**Abstract.** We discuss ultrametric pseudodifferential operators and wavelets and applications to models of interbasin kinetics. We show, that, using the language of ultrametric pseudodifferential operators, it is possible to describe interbasin kinetics for general complex landscape.

**Keywords:** Interbasin kinetics, ultrametric analysis.
**PACS:** 02.60.Nm,82.40.Qt.

## 1. INTRODUCTION

In the present paper we discuss theory of ultrametric pseudodifferential operators and applications to models of interbasin kinetics. The basic example of ultrametric pseudodifferential operators is the Vladimirov operator of $p$–adic fractional differentiation [1].

Interbasin kinetics approximates random walks on complex landscapes by system of differential (kinetic) equations, which describe transitions between group of states of the system. The mentioned group of states, called basins, in the interbasin kinetic approach form some hierarchy (i.e. larger basins consist of smaller basins). Different basins are separated by potential barriers, and the larger basins are separated by higher barriers.

Different models of interbasin kinetics and hierarchical dynamics were studied in many works, see for example [2], [3], [4], [5]. Interbasin kinetic models were discussed in relation to dynamics of proteins [6], [7]. $p$–Adic models of interbasin kinetics were discussed in [8], [9], [10].

Models of complex landscapes and relaxation behavior were discussed in many works, see for example [11], [12]. In [13], [14] $p$–adic diffusion were discussed in relation to relaxation in spin glasses.

In the present paper we apply theory of general ultrametric pseudodifferential operators to investigation of interbasin kinetic models for general landscapes, where hierarchy of basins can be arbitrary. We show, that in this case models of interbasin kinetics can be described by ultrametric pseudodifferential equations of simple form. This shows that the language of ultrametric pseudodifferential operators gives a simple description to interbasin kinetics on complex landscapes. Moreover, for some particular cases the obtained pseudodifferential equations may be solved analytically with the help of the ultrametric wavelet transform.

These results show that theory of ultrametric pseudodifferential operators has wide range of applications, from theory of functions [17] to spin glasses [8], [19], [20], [21], dynamics of macromolecules and string theory [22], [23], [24].

## 2. ULTRAMETRIC ANALYSIS

In this Section we put the results on ultrametric analysis, which mainly may be found in [15], [16]. We discuss here ultrametric wavelet analysis and analysis of ultrametric pseudodifferential operators (PDO).

**Definition 1.** *An ultrametric space is a metric space with the ultrametric $|xy|$ (where $|xy|$ is called the distance between x and y), i.e. the function of two variables, satisfying the properties of positivity and non degeneracy*

$$|xy| \geq 0, \qquad |xy| = 0 \implies x = y;$$

*symmetricity*

$$|xy| = |yx|;$$

*and the strong triangle inequality*

$$|xy| \leq \max(|xz|, |yz|), \qquad \forall z.$$

We say that the ultrametric space X has the analytic type if it satisfies the following properties:

1) *The set of all the balls of nonzero diameter in X is no more than countable;*
2) *For any decreasing sequence of balls $\{D^{(k)}\}$, $D^{(k)} \supset D^{(k+1)}$, diameters of the balls tend to zero;*
3) *Any ball is a finite union of maximal subballs.*

Remind that a directed set is a partially ordered set, where for any pair of elements there exists the unique supremum with respect to the partial order.

Denote $\mathcal{T}(X)$ the set of balls of nonzero diameter in analytic ultrametric space $X$. Consider the set $X \bigcup \mathcal{T}(X)$. This set is directed. The direction if defined by inclusion of balls and inclusion of points into balls. In particular, the supremum

$$\sup(x, y) = I$$

of points $x, y \in X$ is the minimal ball $I$, containing the both points. This directed set has the natural structure if a tree.

Consider a Borel $\sigma$–additive measure $\nu$ with a countable basis on analytic ultrametric space $X$, such that for arbitrary ball $D$ its measure $\nu(D)$ is a positive number (i.e. is not equal to zero).

Consider a basis of ultrametric wavelets in the space $L^2(X, \nu)$ of quadratically integrable with respect to the measure $\nu$ functions. This is a generalization of basis of $p$–adic wavelets [17]. Generalization of $p$–adic wavelets onto the family of abelian locally compact groups was performed by J.J. Benedetto and R.L. Benedetto [18].

Denote $V_I$ the space of functions on the absolute, generated by characteristic functions of the maximal subballs in the ball $I$ of nonzero radius. Correspondingly, $V_I^0$ is the subspace of codimension 1 in $V_I$ of functions with zero mean with respect to the measure $\nu$. Spaces $V_I^0$ for the different $I$ are orthogonal. Dimension of the space $V_I^0$ is equal to $p_I - 1$.

We introduce in the space $V_I^0$ some orthonormal basis $\{\psi_{Ij}\}$, $j = 1, \ldots, p-1$. The next theorem shows how to construct the orthonormal basis in $L^2(X, \nu)$, taking the union of bases $\{\psi_{Ij}\}$ in spaces $V_I^0$ over all non minimal $I$.

**Theorem 2.** *1) Let the ultrametric space $X$ contains an increasing sequence of embedded balls with infinitely increasing measure. Then the set of functions $\{\psi_{Ij}\}$, where $I$ runs over all non minimal vertices of the tree $\mathcal{T}$, $j = 1, \ldots, p_I - 1$ is an orthonormal basis in $L^2(X, \nu)$.*
*2) Let for the ultrametric space $X$ there exists the supremum of measures of the balls, which is equal to $A$. Then the set of functions $\{\psi_{Ij}, A^{-\frac{1}{2}}\}$, where $I$ runs over all non minimal vertices of the tree $\mathcal{T}$, $j = 1, \ldots, p_I - 1$ is an orthonormal basis in $L^2(X, \nu)$.*

The introduced in the present theorem basis we call the basis of ultrametric wavelets.

We study the ultrametric pseudodifferential operator (or the PDO) of the form considered in [16], [15]

$$Tf(x) = \int T(\sup(x,y))(f(x) - f(y))d\nu(y)$$

Here $T(I)$ is some nonnegative function on the tree $\mathcal{T}(X)$. Thus the structure of this operator is determined by the direction on $X \bigcup \mathcal{T}(X)$. This kind of ultrametric PDO we call *the sup–operators* (or operators of the sup–type).

The next theorem shows that the basis of ultrametric wavelets is the basis of eigenvectors for ultrametric PDO of the sup–type.

**Theorem 3.** *Let the following series converge:*

$$\sum_{J>R} T(J)(\nu(J) - \nu(J-1, R)) < \infty \qquad (1)$$

*where $R$ is some ball in $X$. Then the operator*

$$Tf(x) = \int T(\sup(x,y))(f(x) - f(y))d\nu(y)$$

*is self–adjoint, has the dense domain in $L^2(X,\nu)$, and is diagonal in the basis of ultrametric wavelets from the theorem 2:*

$$T\psi_{Ij}(x) = \lambda_I \psi_{Ij}(x) \qquad (2)$$

with the eigenvalues:

$$\lambda_I = T(I)v(I) + \sum_{J>I} T(J)(v(J) - v(J-1,I))  \qquad (3)$$

Here $(J-1,I)$ is the maximal subball in $J$ which contains $I$.
Also the operator $T$ kills constants.

An important example of measure on analytic ultrametric space is the homogeneous measure $\mu$ (an analogue of the Haar measure), defined as follows: for any ball $I$ measures of all maximal subballs $I_j$ in $I$ are equal: $\mu(I_j) = \mu(I_{j'})$.

## 3. INTERBASIN KINETICS AS ULTRAMETRIC DIFFUSION

Let us discuss application of ultrametric analysis to models of interbasin kinetics, which describe approximation of dynamics on complex landscapes.

Consider the system of kinetic equations of the form:

$$\frac{d}{dt} f(x,t) = -\sum_y [T(x,y)f(x,t) - T(y,x)f(y,t)] v(y) \qquad (4)$$

Here $x$ enumerates states of the system, $T(x,y) \geq 0$ is the transition probability rate for transitions from $x$ to $y$, $v(y) > 0$ are positive numbers.

We say that the system above describes interbasin kinetics, if all the states $x$ can be separated into nonintersecting subsets, called basins, and the transition probability rate $T(x,y)$ for states in some basins depends only on basins and does not depend on the choice of states inside basins.

Then we separate the set of basins into set of basins of the second order (or superbasins) with the analogous condition on the transition probability rate $T(x,y)$, and iterate the procedure. The matrix $T(x,y)$ described by this procedure will be a block matrix with large number of equal matrix elements.

Hierarchy of nested basins define some directed tree $\mathscr{T}$. Vertices of the tree are basins, and the order is defined by inclusion of basins. States of the system are the minimal vertices of $\mathscr{T}$. We introduce measure $v$ on the space $X(\mathscr{T})$. Distance in $X$ we define $|xy|$, basins are balls with respect to this distance.

Transition probability rate $T(x,y)$ in this language is the nonnegative function of two variables $x$, $y$ from $X(\mathscr{T})$. The block structure for $T(x,y)$ takes the form of the local constancy conditions

$$T(x,y) = T(x,z), \quad |yz| < |xy| \qquad (5)$$

$$T(x,y) = T(z,y), \quad |xz| < |xy| \qquad (6)$$

We have the following theorem.

**Theorem 4.** *If the system (4) describes interbasin kinetics (i.e. the function $T(x,y)$ satisfies (5), (6)), this system is equivalent to the following ultrametric pseudodifferential equation:*

$$\frac{d}{dt}f(x,t) = -\int_X (T(x,y)f(x,t) - T(y,x)f(y,t))d\nu(y) \tag{7}$$

$$T(x,y) \geq 0, \quad T(x,y) = T(x,z), \quad |yz| < |xy|$$

$$T(x,y) = T(z,y), \quad |xz| < |xy|$$

*Here $|xy|$ is the natural ultrametric on $X = X(\mathcal{T})$, where $\mathcal{T}$ is the directed tree of basins. Measure of a minimal $y$ in $X(\mathcal{T})$ equals to $\nu(y) > 0$.*

## 4. INTERBASIN KINETICS FOR RANDOM WALK ON LANDSCAPE

Consider dynamics of complex system described by random walk of energy landscape $\Phi(x)$, where $x$ takes values in some subset in $R^d$.

There is a procedure of approximation of complex landscape by directed tree of basins by F.H. Stillinger, T.A. Weber [2], [3], see also [6] (we not discuss here in details this procedure). Let us build an interbasin kinetics approximation for random walk on complex landscape, using described in the previous section language of ultrametric pseudodifferential equations.

Random walk on the landscape $\Phi$ is described by the transition probability rate between two neighbor infinitesimal volumes $dV \to dV'$, situated at $\xi$, $\xi + d\xi$ correspondingly:

$$e^{-\beta(\Phi(\xi+d\xi)-\Phi(\xi))\theta(\Phi(\xi+d\xi)-\Phi(\xi))}dVdV'$$

where $\beta$ is the inverse temperature. Here $\theta(x)$ is the $\theta$–function, which is equal to 1 for positive and 0 for negative arguments.

In general it is not possible to investigate random walk on complex landscape. For approximate description of this random walk we consider discretization of the landscape — we divide all the landscape on some domains, and consider transitions between the domains. The discretization of the landscape will be a graph, vertices of the graph correspond to the domains, and the neighbor domains are connected by edges. The landscape will generate function $\Phi(I)$ on vertices of the graph. We take the procedure of discretization which generate the mentioned above directed tree of basins. The corresponding domains will be vicinities of local minima and transition areas between the local minima. The dynamics on the landscape will be approximated by the random walk on the tree with transition probability rates between neighbor vertices $J$ and $I$, proportional to

$$e^{-\beta(\Phi(J)-\Phi(I))\theta(\Phi(J)-\Phi(I))}$$

We consider the regime of low temperatures, when the system is concentrated on minimal vertices of the tree. In this regime random walk on the tree is approximated by ultrametric diffusion.

The transition probability rate for transitions between minimal vertices $x$ and $y$ is the product of entropic and energetic (enthalpic) factors.

Energetic factor is equal to
$$e^{-\beta(\Phi(\sup(x,y))-\Phi(x))}$$
which is equal to the product of the Boltzmann factors along increasing part of the path in the tree between $x$ and $y$.

Entropic factor in the simplest case can be estimated as the product of the branching indices in the degree minus one along the path between $x$ and $y$:
$$\prod_{J\in xy} p_J^{-1}$$
(here we take only the inner vertices of this path). This product is equal to
$$\frac{\mu(x)}{\mu(\sup(x,y))}\frac{\mu(y)}{\mu(\sup(x,y))} \qquad (8)$$
where $\mu$ is the homogeneous measure. This estimate is based on the idea, that the main contribution to the transition probability rate gives the unique path without self intersections between $x$ and $y$, and at any vertex of this path the system can perform transitions in the directions of any edges incident to the vertex. There exist more accurate estimates, which justify the above formula, but we will not discuss these estimates here.

Thus for the density $f(x)$ of occupation of the minimal basins of the landscape (i.e. the occupation of the basin $x$ is equal to $f(x)\mu(x)$), we get the system of kinetic equations
$$\frac{d}{dt}f(x,t)+\sum_y \frac{e^{-\beta\Phi(\sup(x,y))}}{\mu^2(\sup(x,y))}\left[e^{\beta\Phi(x)}f(x)-e^{\beta\Phi(y)}f(y)\right]\mu(y)=0$$

This system satisfies the conditions of detailed balance, and the equilibrium state $f(x) = e^{-\beta\Phi(x)}$ is the unique stationary state.

Using the theorem 4, this system of differential equations can be put into the form of pseudodifferential equation
$$\frac{d}{dt}f(x,t)+\int \frac{e^{-\beta\Phi(\sup(x,y))}}{\mu^2(\sup(x,y))}\left[e^{\beta\Phi(x)}f(x)-e^{\beta\Phi(y)}f(y)\right]d\mu(y)=0 \qquad (9)$$
which describes the ultrametric diffusion in potential on ultrametric space of general form.

Note that in the case, when the energies of the lowest states in the tree of basins are constant (i.e. $\Phi(x) = \text{Const}$, where $x$ are local minima, or equivalently the elements of ultrametric space), the equation above can be solved analytically with the help of the ultrametric wavelet transform.

## ACKNOWLEDGMENTS

The author would like to thank V.A. Avetisov, A.Kh. Bikulov, A.Yu. Khrennikov and I.V. Volovich for fruitful discussions and valuable comments. The author has been partly supported by DFG Project 436 RUS 113/809/0-1, The Russian Foundation for Basic Research (project 05-01-00884-a), by the grant of the President of Russian Federation for the support of scientific schools NSh 1542.2003.1, by the Program of the Department of Mathematics of Russian Academy of Science "Modern problems of theoretical mathematics".

## REFERENCES

1. V.S. Vladimirov, I.V. Volovich, E.I. Zelenov, *p–Adic analysis and mathematical physics*, World Scientific, Singapore, 1994 (See also Nauka, Moscow, 1994, in Russian).
2. F.H. Stillinger, T.A. Weber, Hidden Structure in Liquids, *Phys. Rev. A* **25**, 978–989 (1982).
3. F.H. Stillinger, T.A. Weber, Packing Structures and Transitions in Liquids and Solids, *Science* **225**, 983–989 (1984).
4. K.H. Hoffmann, P. Sibani, Diffusion in Hierarchies, *Phys. Rev. A* **38**, 4261–4270 (1988).
5. F.H. Stillinger, Relaxation behavior in atomic and molecular glasses, *Phys. Rev. B* **41**, 2409–Ű2416 (1990).
6. O.M. Becker, M. Karplus, The Topology of Multidimensional Protein Energy Surfaces: Theory and Application to Peptide Structure and Kinetics, *J. Chem. Phys.* **106**, 1495–1517 (1997).
7. H. Frauenfelder, S.G. Sligar, P.G. Wolynes, The Energy Landscape and Motions of Proteins. *Science* **254**, 1598–1603 (1991).
8. V.A. Avetisov, A.H. Bikulov, S.V. Kozyrev, Application of *p*–adic analysis to models of spontaneous breaking of the replica symmetry, *J. Phys. A: Math. Gen.* **32**, no. 50, 8785–8791 (1999), cond-mat/9904360.
9. V.A. Avetisov, A.H. Bikulov, S.V. Kozyrev, V.A. Osipov, *p*–Adic Models of Ultrametric Diffusion Constrained by Hierarchical Energy Landscapes, *J. Phys. A: Math. Gen.* **35**, no. 2, 177–189 (2002), cond-mat/0106506.
10. V.A. Avetisov, A.H. Bikulov, V.A. Osipov, *p*–Adic description of characteristic relaxation in complex systems, *J. Phys. A: Math.and Gen.* **36**, no. 15, 4239–4246 (2003), cond-mat/0210447.
11. H. Yoshino, Hierarchical diffusion, aging and multifractality, *J. Phys. A* **30**, 1143 (1997), cond-mat/9604033.
12. A.T. Ogielski, D.L. Stein, Dynamics on ultrametric spaces, *Phys. Rev. Lett.* **55**, no. 15, 1634–1637 (1985).
13. L. Brekke, M. Olson, *p*–Adic diffusion and relaxation in glasses, Preprint UTTG-16-89, EFI-89-23.
14. L. Brekke, P.G.O. Freund, *p*–Adic numbers in physics, *Phys. Rept.* **233**, no. 1, 1–66 (1993).
15. A.Yu. Khrennikov, S.V. Kozyrev, Pseudodifferential operators on ultrametric spaces and ultrametric wavelets, *Russian Math. Izv.* **69**, no. 5, (2005), math-ph/0412062.
16. A.Yu. Khrennikov, S.V. Kozyrev, Wavelets on ultrametric spaces, *Applied and Computational Harmonic Analysis* **19**, 61–76 (2005).

17. S.V. Kozyrev, Wavelet analysis as a $p$–adic spectral analysis, *Russian Math. Izv.* **66**, no. 2, 367 (2002), math-ph/0012019.
18. J.J. Benedetto, R.L. Benedetto, A wavelet theory for local fields and related groups, *The Journal of Geometric Analysis* **14**, no. 3, 423–456 (2004).
19. G. Parisi, N. Sourlas, $p$–Adic numbers and replica symmetry breaking, *European Phys. J. B* **14**, 535–542 (2000), cond-mat/9906095.
20. A.Yu. Khrennikov, S.V. Kozyrev, Replica symmetry breaking related to a general ultrametric space I: replica matrices and functionals, *Physica A* **359**, 222–240 (2006).
21. A.Yu. Khrennikov, S.V. Kozyrev, Replica symmetry breaking related to a general ultrametric space II: RSB solutions and the $n \to 0$ limit, *Physica A* **359**, 241-266 (2006).
22. I.V. Volovich, $p$–Adic String, *Class. Quantum Gravity* **4** L83–L87 (1987).
23. I.Ya. Aref'eva, B. Dragovich, I.V. Volovich, On the adelic string amplitudes, *Phys. Lett. B* **209**, 445–450 (1988).
24. P.G.O. Freund, E. Witten, Adelic string amplitudes, *Phys. Lett. B* **199**, 191–194 (1987).

# Critical Exponents in $p$-Adic $\varphi^4$-Model

Moukadas D. Missarov* and Roman G. Stepanov*

*Department of Computing Mathematics and Cybernetics,
Kazan State University, Kazan, RUSSIA
emails: Moukadas.Missarov@ksu.ru, Roman.Stepanov@ksu.ru

**Abstract.** We consider $\varphi^4$-model with $O(N)$-symmetry in $d$-dimensional $p$-adic space using the approach of renormalized projection Hamiltonians. Critical exponents $\nu$ and $\eta$ are calculated up to three orders of perturbation theory using two types of expansions: $(4-d)$-expansion and $(\alpha - 3/2d)$-expansion, where $\alpha$ is a renormalization group parameter. Some resemblances and differences between the Euclidean and $p$-adic models are discussed.

**Keywords:** Renormalization group, critical exponents, $p$-adic space.
**PACS:** 05.10.Cc, 02.10.De.

## 1. INTRODUCTION

In recent years a considerable amount of work has been done in $p$-adic models of string models, quantum mechanics, statistical physics, dynamical systems etc [1, 4, 5, 6, 7, 8, 16, 17]. Particularly, it was shown that $p$-adic quantum field models can be considered as a continuous versions of the hierarchical models which were introduced in statistical physics by F. Dyson [9, 11]. Dyson's models are defined on the hierarchical lattices and especially suitable for the block-spin renormalization group analysis [3]. It is noteworthy that femionic hierarchical model is exactly solvable and gives a lot of interesting non-perturbative information and new comprehension of functional integral, ultraviolet divergences and renormalization procedure [10, 12, 13, 14, 15]. One of the interesting problem is to investigate and explain the similarities and distinctions between properties of $p$-adic and the Euclidean models.

In this paper we give renormalization group description of $p$-adic $N$ - component bosonic $\varphi^4$-model. To emphasize the similarity between $p$-adic and the Euclidean cases we use Wilson renormalization group transformation in momentum space representation. The method of renormalized projection Hamiltonians, elaborated for the Euclidean models [2], is adaptable for the p-adic case. Though Wilson renormalization group (RG) in $p$-adic case is discrete semigroup, in the frameworks of $\varepsilon$-expansions we can locally embed RG-transformation in continuous semigroup and use its generator. One of the advantages of the $p$-adic models is a theoretical possibility of computation of Feynman amplitudes and its vertex parts in any order of perturbation theory. In this paper we describe two variants of $\varepsilon$-expansions and compute critical exponents up to the 3-rd order of perturbation theory and compare these results with Euclidean ones.

Let $\mathbb{Q}_p^d$ denote the $d$-dimensional $p$-adic space: $\mathbb{Q}_p^d = \{(q_1, \ldots, q_d) : q_i \in \mathbb{Q}_p\}$. The norm on $\mathbb{Q}_p^d$ is defined as $|q|_p = \max\{|q_1|_p, \ldots, |q_d|_p\}$, where $|\cdot|_p$ is a $p$-adic norm, $p$ is some fixed prime number.

Let the $N$–component Gibbsian field $\sigma(q) = (\sigma_1(q), \ldots, \sigma_N(q))$ is defined in $d$–dimensional ball $\Omega_R = \{q \in \mathbb{Q}_p^d : |q| \leq R\}$, $R = p^k$ for some integer $k$.

The action of Wilson renormalization group $r_{\tau,\alpha}$ is defined as

$$r_{\tau,\alpha}\sigma(q) = (s_{\tau,\alpha}\sigma(q))\chi(q), \qquad s_{\tau,\alpha}\sigma(q) = p^{-\tau\alpha/2}\sigma(qp^\tau),$$

where $\tau \in \mathbb{N}$ is a scaling parameter, $\alpha$ is a parameter of RG, $\chi(q)$ is an indicator function of the ball $\Omega_R$. In fact, $r_{\tau,\alpha}$ is an operator of restriction of the field $s_{\tau,\alpha}\sigma(q)$ defined in a ball $\Omega_{p^\tau R}$ into the ball $\Omega_R$.

The $O(N)$-symmetric Gibbsian field $\sigma(q)$ is given by a Hamiltonian of the form

$$H(\sigma) = \sum_{k=1}^{\infty} \sum_{i_1, \ldots, i_{2k}=1}^{N} F_{i_1,\ldots,i_{2k}} \int_{\Omega_R^{2k}} h_k(q_1, \ldots, q_{2k}) \delta(q_1 + \cdots + q_{2k}) \prod_{s=1}^{2k} \sigma_{i_s}(q_s) dq, \qquad (1)$$

where $h_k(q_1, \ldots, q_{2k})$ are coefficient functions invariant under permutations of arguments,

$$F_{i_1,\ldots,i_{2k}} = \frac{1}{(2k-1)!!} \sum_{\gamma} \prod_{(m_1,m_2) \in \gamma} \delta_{i_{m_1}, i_{m_2}}, \qquad (2)$$

the summation $\sum_\gamma$ runs over all partitions of the set $\{1, \ldots, 2k\}$ on pairs. Particularly,

$$F_{i_1,i_2} = \delta_{i_1,i_2}, \quad F_{i_1,i_2,i_3,i_4} = \frac{1}{3}\left(\delta_{i_1,i_2}\delta_{i_3,i_4} + \delta_{i_1,i_3}\delta_{i_2,i_4} + \delta_{i_1,i_4}\delta_{i_2,i_3}\right).$$

The $O(N)$–symmetry of the field defined by (1) is a consequence of the fact that the Hamiltonian depends on $\sigma(q)$ via scalar products

$$<\sigma(q_1), \sigma(q_2)> = \sum_{i_1,i_2=1}^{N} \delta_{i_1,i_2}\sigma_{i_1}(q_1)\sigma_{i_2}(q_2),$$

because we can write

$$\sum_{i_1,\ldots,i_{2k}=1}^{N} F_{i_1,\ldots,i_{2k}} \prod_{s=1}^{2k} \sigma_{i_s}(q_s) = \frac{1}{(2k-1)!!} \sum_{\gamma} \prod_{(m_1,m_2) \in \gamma} <\sigma(q_{m_1}), \sigma(q_{m_2})>.$$

In the space of Hamiltonians the renormalization group transformation $R_{\tau,\alpha}$ is a composition of two transformations

$$R_{\tau,\alpha} = P_\tau S_{\tau,\alpha},$$

where the scaling transformation $S_{\tau,\alpha}$ is induced by the transformation $s_{\tau,\alpha}$. Its action on coefficient functions if defined as

$$h'_k(q_1, \ldots, q_{2k}) = p^{\tau(\alpha k - 2kd + d)} h_k(q_1 p^\tau, \ldots, q_{2k} p^\tau).$$

Projection transformation $P_\tau$ is induced by the operation of restricting the field $\sigma'(q) = s_{\tau,\alpha}\sigma(q)$ that is defined in the ball $\Omega_{p^\tau R}$ into the ball $\Omega_R$. Let $\sigma_0(q) = \chi(q)\sigma'(q)$, $\sigma_1(q) = (1-\chi(q))\sigma'(q)$. The field $\sigma_0(q)$ is defined in the ball $\Omega_R$, the field $\sigma_1(q)$ is defined in the ring $\Omega_{p^\tau R} \setminus \Omega_R$, and $\sigma = \sigma_0 + \sigma_1$. If $H'(\sigma')$ is a Hamiltonian of the field $\sigma'$ then the Hamiltonian $P_\tau H'$ is a Hamiltonian of the field $\sigma_0$ and is given by the functional integral

$$P_\tau H'(\sigma_0) = -\ln \frac{\int D\sigma_1 \exp(-H'(\sigma_0+\sigma_1))}{\int D\sigma_0 \int D\sigma_1 \exp(-H'(\sigma_0+\sigma_1))}.$$

Let $H(\sigma)$ is a Hamiltonian of a field $\sigma(q)$ defined in the whole space $\mathbb{Q}_p^d$. Denote by $PH$ the projection of $H$ into the ball $\Omega_R$:

$$PH(\sigma_0) = -\ln \frac{\int D\sigma_1 \exp(-H(\sigma_0+\sigma_1))}{\int D\sigma_0 \int D\sigma_1 \exp(-H(\sigma_0+\sigma_1))},$$

where $\sigma_0(q) = \sigma(q)\chi(q)$ is a field in the ball $\Omega_R$, $\sigma_1 = \sigma(q)(1-\chi(q))$ is a field in $\mathbb{Q}_p^d \setminus \Omega_R$.

It is not difficult to show that the following equation holds:

$$R_{\tau,\alpha} P = P S_{\tau,\alpha}. \tag{3}$$

The only way to calculate the above functional integrals is to use the technique of Feynman diagrams.

Let $G$ is some connected Feynman diagram with $n$ vertices, and all vertices are uniquely labeled by numbers between 1 and $n$. The number of lines connected to a vertex $v \in \{1,\ldots,n\}$ is called the *degree of vertex* $v$. Denote the degree of vertex $v$ by $2k_v$ (it is always an even number). Let the lines connected to a vertex $v$ are uniquely labeled by numbers between 1 and $2k_v$.

Let the diagram $G$ has $2k$ external lines. Every external line is identified with a pair $(v,m)$, where $v$ is a serial number of vertex connected to the line, $m$ is a serial number of the line among all lines of the vertex.

Every internal line may by identified with four numbers $(v_1,m_1,v_2,m_2)$. We assume that the line $(v_1,m_1,v_2,m_2)$ puts together vertices $v_1, v_2$ and has a serial number $m_1$ for vertex $v_1$, and a serial number $m_2$ for vertex $v_2$.

Let $L(G) = \{(v_1,m_1,v_2,m_2)\}$ be a collection of internal lines of $G$, $E(G) = \{(v,m)\}$ is a collection of external lines of $G$.

**Definition.** *Symmetry factor* of diagram $G$ is the following expression:

$$S_N(G) = N^{-k} \sum_{\substack{i_1^1,\ldots,i_{2k_1}^1=1 \\ i_1^n,\ldots,i_{2k_n}^n=1}}^{N} \left(\prod_{v=1}^{n} F_{i_1^v,\ldots,i_{2k_v}^v}\right) \left(\prod_{(v_1,m_1,v_2,m_2)\in L(G)} \delta_{i_{m_1}^{v_1},i_{m_2}^{v_2}}\right). \tag{4}$$

**Example.** If we consider the Feynman diagram

>—<      then we have $n = 2, k_1 = k_2 = k = 2$,

$$L(G) = \{(1,1,2,1),(1,2,2,2)\}, \qquad E(G) = \{(1,3),(1,4),(2,3),(2,4)\},$$

$$S_N(G) = N^{-2} \sum_{i_1^1 \ldots i_4^2 = 1}^{N} F_{i_1^1,i_2^1,i_3^1,i_4^1}^1 F_{i_1^2,i_2^2,i_3^2,i_4^2}^2 \delta_{i_1^1,i_1^2} \delta_{i_2^1,i_2^2} = \frac{N+8}{9}.$$

## 2. $(\alpha - 3/2d)$ – EXPANSION

Denote

$$H_{0,\alpha,R}(\sigma) = \sum_{i=1}^{N} \int_{q_1, q_2 \in \Omega_R} \delta(q_1 + q_2) |q_1|^{\alpha - d} \sigma_i(q_1) \sigma_i(q_2) \, dq,$$

$$H_k(\sigma) = \sum_{i_1,\ldots,i_{2k}=1}^{N} F_{i_1,\ldots,i_{2k}} \int_{q_1,\ldots,q_{2k} \in R^d} \delta(q_1 + \cdots + q_{2k}) \prod_{s=1}^{2k} \sigma_{i_s}(q_s) \, dq,$$ (5)

It is easy to see that the Hamiltonian $\frac{1}{2}H_{0,\alpha,R}$ is a Gaussian fixed point of the transformation $R_{\tau,\alpha}$.

The analysis of the differential of RG on the Gaussian branch of fixed points shows that the value $\alpha = 3/2d$ is a bifurcation value. A parameter $\varepsilon = \alpha - 3/2d$ may be used as a small parameter of expansions.

Non-Gaussian Hamiltonians invariant under RG transformation are constructed in a space of projection Hamiltonians of the form

$$P\left(\frac{1}{2}H_{0,\alpha,\infty} + u_1 H_1 + u_2 H_2\right).$$

Let $\mathscr{F}_G(q)$ is a Feynman amplitude of graph $G$ with a propagator

$$<\sigma(p), \sigma(q)> = \delta(p+q)\theta_1(q), \qquad \theta_1(q) = |q|^{d-\alpha}(1-\chi(q)).$$

For some natural numbers $k_1, \ldots, k_n$ we denote by $<H_{k_1} H_{k_2} \ldots H_{k_n}>_{\theta_1}^c$ the connected Wick ordering according to Gaussian field with above propagator:

$$<H_{k_1} H_{k_2} \ldots H_{k_n}>_{\theta_1}^c = \sum_{G} \sum_{\substack{i_1^1,\ldots,i_{2k_1}^1=1 \\ \vdots \\ i_1^n,\ldots,i_{2k_n}^n = 1}}^{N} \left(\prod_{v=1}^{n} F_{i_1^v,\ldots,i_{2k_v}^v}\right) \times$$

$$\times \left(\prod_{(v_1,m_1,v_2,m_2)\in L(G)} \delta_{i_{m_1}^{v_1}, i_{m_2}^{v_2}}\right) \int \mathscr{F}_G(q) \prod_{(v,m) \in E(G)} \sigma_{i_m^v}(q_m^v) \, dq, \quad (6)$$

where $N$ is a number of components of the field, the summation $\Sigma_G$ runs over all connected non-vacuum Feynman graphs $G$ with $n$ vertices of degrees $2k_1,\ldots,2k_n$. Integration extends over subspace of impulses $q_m^v \in \mathbb{Q}_p^d$, $(v,m) \in E(G)$, satisfying the law of momentum conservation:

$$\sum_{(v,m)\in E(G)} q_m^v = 0.$$

For an arbitrary power series $F(x_1,\ldots,x_n)$ a value $< F(H_1,\ldots,H_n) >_{\theta_1}^c$ may be defined by linearity.

The operation $P$ is related to the operation $<\cdot>_{\theta_1}^c$ in the following way:

$$P\left(\frac{1}{2}H_{0,\alpha,\infty} + u_1 H_1 + u_2 H_2\right) = \frac{1}{2}H_{0,\alpha,R} - < \exp(-u_1 H_1 - u_2 H_2) >_{\theta_1}^c . \tag{7}$$

The Feynman amplitudes that define the coefficient functions of projection Hamiltonians (7) diverge when $\varepsilon \to 0$. To make Hamiltonian finite at $\varepsilon = 0$ we need a procedure of renormalization. The most natural one is a procedure of analytic renormalization.

Let $A.R.\mathscr{F}_G(q)$ is an operation of analytic renormalization for a small parameter $\varepsilon$ of Feynman amplitude $\mathscr{F}_G(q)$ of graph $G$:

$$A.R.\mathscr{F}_G(q) = \sum_{r=1}^{n} \sum_{V_1,\ldots,V_r} \sum_{\substack{\{G_1,\ldots,G_r\} \\ V(G_j)=V_j}} \mathscr{F}_{G|G_1,\ldots,G_r}(q) \prod_{j=1}^{r} O(G_j) \tag{8}$$

where $n$ is a number of vertices of graph $G$, the second summation in (8) goes over unordered partitions of vertices of graph $G$ to $r$ nonempty sets $V_1,\ldots,V_r$. The third summation in (8) goes over all partitions of the graph $G$ to subgraphs $G_1,\ldots,G_r$ with corresponding sets of vertices $V_1,\ldots,V_r$. $G|G_1,\ldots,G_r$ is a graph that is obtained from $G$ by a contraction of the subgraphs $G_1,\ldots,G_r$. $O(G_j)$ is a counterterm of $G_j$.

The counterterms $O(G)$ are defined in such a way that the value of $A.R.\mathscr{F}_G(q)$ does not have a pole at $\varepsilon = 0$. This requirement and (8) specify a recursive algorithm of calculating the counterterms.

Consider the *renormalized projection Hamiltonian*:

$$H_{eff}(u_1, u_2) = \frac{1}{2} H_{0,\alpha,R} - A.R. < \exp(-u_1 H_1 - u_2 H_2) >_{\theta_1}^c . \tag{9}$$

Here the operation $A.R.$ applies to all Feynman amplitudes $\mathscr{F}_G(q)$ appearing in (6), and ensures that the Hamiltonian does not have ultraviolet divergences when $\varepsilon \to 0$.

It can be shown that for the Hamiltonian (9) the following representation is true:

$$H_{eff}(u_1, u_2) = \frac{1}{2} H_{0,\alpha,R} - < \exp(-u_1 w_1(u_2) H_1 - w_2(u_2) H_2) >_{\theta_1}^c$$

133

$$w_1(u) = 1 + \sum_{n \geq 1} O_1(n) \frac{(-u)^n}{n!}, \quad O_1(n) = \sum_{G_1} O(G_1) S_N(G_1),$$
$$w_2(u) = -\sum_{n \geq 1} O_2(n) \frac{(-u)^n}{n!}, \quad O_2(n) = \sum_{G_2} O(G_2) S_N(G_2). \tag{10}$$

The summation $\sum_{G_1}$ goes over one-particle irreducible (1PI) Feynman graphs with $n$ vertices of degree 4 and one vertex of degree 2 and having 2 external lines. The summation $\sum_{G_2}$ goes over 1PI Feynman graphs with $n$ vertices of degree 4 and having 4 external lines.

Action of RG transformation on a renormalized projection Hamiltonian (9) is given by the differential operator

$$R_{\tau,\alpha} H_{eff}(u_1, u_2) = \exp\left(\tau \ln p \sum_{j=1}^{2} \beta_j \frac{\partial}{\partial u_j}\right) H_{eff}(u_1, u_2), \tag{11}$$

$$\beta_2(u_2) = (2\alpha - 3d) \frac{w_2(u_2)}{w_2'(u_2)}, \tag{12}$$

$$\beta_1(u_1, u_2) = u_1 \left(\alpha - d - (2\alpha - 3d) \frac{w_1'(u_2)}{w_1(u_2)} \frac{w_2(u_2)}{w_2'(u_2)}\right). \tag{13}$$

To find a fixed point of the transformation $R_{\tau,\alpha}$ in a form of renormalized projection Hamiltonian $H_{eff}(u_1, u_2)$ we have to solve a system of equations in $u_1, u_2$:

$$\beta_1(u_1, u_2) = 0, \quad \beta_2(u_2) = 0.$$

It turns out that the system has two solutions as a formal series in $\varepsilon$: a trivial solution $u_1 = 0$, $u_2 = 0$ corresponding to the Gaussian fixed Hamiltonian

$$H_{eff}(0,0) = \frac{1}{2} H_{0,\alpha,R},$$

and a solution

$$u_1 = 0, \quad u_2 = u^* = c_1 \varepsilon + c_2 \varepsilon^2 + \ldots, \quad c_1 \neq 0,$$

which corresponds to a non-Gaussian fixed Hamiltonian $H_{eff}(0, u^*)$.

Critical exponent $\eta$ (*anomalous dimension*) is defined by the equation $\eta = 2 + d - \alpha = (4-d)/2 - \varepsilon$.

Critical exponent $\nu$ is calculated using the formula

$$\nu = \frac{\tau \ln p}{\ln \lambda_1},$$

where $\lambda_1$ is the largest eigenvalue of the differential of $R_{\tau,\alpha}$ at the non-Gaussian fixed point.

It can be shown that the Hamiltonian

$$\left.\frac{\partial}{\partial u_1} H_{eff}(u_1, u_2)\right|_{u_1=0, u_2=u^*}$$

is an eigen-Hamiltonian of the differential of $R_{\tau,\alpha}$ at the fixed point $H_{eff}(0, u^*)$ with corresponding eigenvalue $\lambda_1 = p^{\tau \frac{\partial \beta_1}{\partial u_1}}\Big|_{u_1=0, u_2=u^*}$.

Hence the critical exponent $\nu$ is equal to

$$\nu = \left(d/2 + \varepsilon - 2\varepsilon \frac{w_1'(u^*) w_2(u^*)}{w_1(u^*) w_2'(u^*)}\right)^{-1}. \tag{14}$$

Up to a second order in $\varepsilon$ we have the following expression for a model in $p$-adic space:

$$\nu^{-1} = d/2 + \varepsilon - 2\frac{N+2}{N+8}\varepsilon + 8\varepsilon^2 \frac{(N+2)(7N+20)}{(N+8)^3} A_p(d) + O(\varepsilon^3), \tag{15}$$

$$A_p(d) = -\frac{1 + 4p^{d/2} + p^d}{2(p^d - 1)} \ln p.$$

$A_p(d)$ may be represented also in the following way:

$$A_p(d) = -\frac{1}{2}\left(\ln p - 2\psi_p(d/4) + \psi_p(d/2)\right),$$

$$\psi_p(x) = f_p'(x)/f_p(x), \quad f_p(x) = \left(1 - p^{-2x}\right)^{-1}.$$

It is known, that for a model in the Euclidean space we have

$$\nu^{-1} = d/2 + \varepsilon - 2\frac{N+2}{N+8}\varepsilon + 8\varepsilon^2 \frac{(N+2)(7N+20)}{(N+8)^3} A(d) + O(\varepsilon^3), \tag{16}$$

$$A(d) = -\frac{1}{2}(-\gamma - 2\psi(d/4) + \psi(d/2)), \quad \psi(x) = \Gamma'(x)/\Gamma(x),$$

where $\gamma = 0.577216...$ is Euler constant.

Comparison of (16) and (15) shows an interesting similarity of expressions up to a second order in $\varepsilon$. Using the equations

$$-\gamma = \lim_{x \to 0}\left(\frac{\Gamma(x)}{\operatorname{Res}\Gamma(0)} - \frac{1}{x}\right), \quad \ln p = \lim_{x \to 0}\left(\frac{f_p(x)}{\operatorname{Res} f_p(0)} - \frac{1}{x}\right).$$

we conclude the equation (16) coincide with (15) if we replace $\Gamma(x)$ by $f_p(x)$. Relation between $\Gamma(x)$ and $f_p(x)$ is well known in the theory of Riemann Zeta-function.

Up to a third order in $\varepsilon$ we have

$$\begin{aligned}
v^{-1} &= d/2 + \varepsilon - 2\frac{N+2}{N+8}\varepsilon + 8\varepsilon^2\frac{(N+2)(7N+20)}{(N+8)^3}A_p(d) + 8\varepsilon^3\frac{N+2}{(N+8)^5(p^d-1)^2} \\
&\times (N^3(5 + 38p^{d/2} + 88p^d + 38p^{3d/2} + 5p^{2d}) \\
&+ 8N^2(3 + 35p^{d/2} + 65p^d + 35p^{3d/2} + 3p^{2d}) \\
&+ 8N(45 + 512p^{d/2} + 962p^d + 512p^{3d/2} + 45p^{2d}) \\
&+ 32(41 + 424p^{d/2} + 834p^d + 424p^{3d/2} + 41p^{2d}))(\ln p)^2 + O(\varepsilon^4).
\end{aligned} \tag{17}$$

## 3. $(4-d)$ – EXPANSION

In the present section we will use the expansion of critical exponents in powers of the parameter $\delta = 4 - d$.

Let' denote

$$H_{0,R}(\sigma) = \sum_{i=1}^{N}\int_{q_1,q_2\in\Omega_R}\delta(q_1+q_2)|q_1|^2\sigma_i(q_1)\sigma_i(q_2)\,dq.$$

The Hamiltonian $\frac{1}{2}H_{0,R}$ corresponds to a Gaussian field with the propagator

$$<\sigma_i(q_1)\sigma_j(q_2)> = \delta_{i,j}\delta(q_1+q_2)\theta_2(q_1),$$

where $\theta_2(q) = |q|^{-2}(1-\chi(q))$. Denote by $<\cdot>^c_{\theta_2}$ the operation of connected Wick ordering for this Gaussian field.

Let $D.R.\mathscr{F}_G(q)$ is an operation of dimensional renormalization in $\delta$ of Feynman amplitude $\mathscr{F}_G(q)$ corresponding to a graph $G$.

Consider the renormalized projection Hamiltonian

$$H_{eff}(u_0,u_1,u_2) = \frac{1}{2}H_{0,R} - D.R. <\exp(-u_0H_{0,\infty} - u_1H_1 - u_2H_2)>^c_{\theta_2}. \tag{18}$$

For the Hamiltonian (18) the following representation holds:

$$H_{eff}(u_0,u_1,u_2) = \frac{1}{2}H_{0,R} - <\exp(-v_0H_{0,\infty} - v_1H_1 - v_2H_2)>^c_{\theta_2},$$

$$v_0 = u_0 + (1+2u_0)w_0\left(\frac{u_2}{(1+2u_0)^2}\right),$$

$$v_1 = u_1w_1\left(\frac{u_2}{(1+2u_0)^2}\right), \quad v_2 = (1+2u_0)^2w_2\left(\frac{u_2}{(1+2u_0)^2}\right),$$

$$w_0(u) = -\sum_{n\geq 1}O_0(n)\frac{(-u)^n}{n!}, \quad O_0(n) = \sum_{G_0}O(G_0)S_N(G_0),$$

$$w_1(u) = 1 + \sum_{n \geq 1} O_1(n) \frac{(-u)^n}{n!}, \quad O_1(n) = \sum_{G_1} O(G_1) S_N(G_1),$$

$$w_2(u) = -\sum_{n \geq 1} O_2(n) \frac{(-u)^n}{n!}, \quad O_2(n) = \sum_{G_2} O(G_2) S_N(G_2).$$

Here the summation $\sum_{G_0}$ runs over graphs with 2 external lines and with $n$ vertices of degree 4. The summation $\sum_{G_1}$ runs over graphs with 2 external lines having $n$ vertices of degree 4 and a single vertex of degree 2. The summation $\sum_{G_2}$ runs over graphs with 4 external lines and generated by $n$ vertices of degree 4. $O(G)$ is a counterterm of $G$.

The action of RG transformation on the renormalized projection Hamiltonian (18) is given by the differential operator

$$R_{\tau,\alpha} H_{eff}(u_0, u_1, u_2) = \exp\left(\tau \ln p \sum_{j=0}^{2} \beta_j \frac{\partial}{\partial u_j}\right) H_{eff}(u_0, u_1, u_2), \tag{19}$$

where

$$\beta_0 = \frac{1 + 2u_0}{2}(\alpha - 3/2d - \zeta(\tilde{u})/2), \quad \beta_1 = u_1 (\alpha - d - \rho_1(\tilde{u}) \rho_2(\tilde{u}) \zeta(\tilde{u})),$$

$$\beta_2 = (1 + 2u_0)^2 (\zeta(\tilde{u}) \rho_2(\tilde{u}) + \tilde{u}(2\alpha - 3d - \zeta(\tilde{u}))),$$

$$\rho_0(u) = \frac{w_0'(u)}{\frac{1}{2} + w_0(u)}, \quad \rho_1(u) = \frac{w_1'(u)}{w_1(u)}, \quad \rho_2(u) = \frac{w_2(u)}{w_2'(u)},$$

$$\zeta(u) = \frac{4 - d}{1 - 2\rho_0(u) \rho_2(u)}, \quad \tilde{u} = u_2 (1 + 2u_0)^{-2}.$$

To find a fixed point of the transformation $R_{\tau,\alpha}$ in a form of renormalized projection Hamiltonian $H_{eff}(u_0, u_1, u_2)$ we must solve a system of equations in $\alpha, u_1, \tilde{u}$:

$$\beta_0 = 0, \quad \beta_1 = 0, \quad \beta_2 = 0.$$

It has two solutions as a formal series in $\delta$: the solution $u_1 = 0$, $\tilde{u} = 0$, $\alpha = 2 + d$ corresponding to the Gaussian fixed Hamiltonian, and the solution

$$u_1 = 0, \quad \tilde{u} = u^* = c_1 \delta + c_2 \delta^2 + \ldots, \quad c_1 \neq 0,$$
$$\alpha = 3/2d + \zeta(u^*)/2, \tag{20}$$

corresponding to the non-Gaussian fixed Hamiltonian $H_{eff}(u_0, 0, u^*(1 + 2u_0)^2)$. A value of $u^*$ may be obtained from the equation

$$\zeta(u^*) \rho_2(u^*) = 0.$$

The value of $u_0$ may be chosen arbitrarily and can be removed by a simple rescaling of the field $\sigma(q)$.

The fundamental difference between the p-adic and Euclidean cases is that in p-adic case we have $w_0(u) = 0$. The reason is that the Feynman diagrams with two external lines containing only vertices of degree 4 do not diverge at $\delta \to 0$ and have zero counterterms. As a consequence we have $\rho_0 = 0, \zeta(u) = 4 - d, \alpha = 2 + d$ and $\eta = 2 + d - \alpha = 0$. In the Euclidean case $w_0 \neq 0$, i.e. there exist the Feynman diagrams with non-zero counterterms containing only vertices of degree 4 with two external lines.

Consider for example the diagram

 which we denote by $I(q)$. In the p-adic case the expression for $I(q)$ has a form

$$I(q) = \frac{f_p(1 - \delta/2)^3}{f_p(1)^3} \frac{f_p(-1 + \delta)}{f_p(3 - 3\delta/2)} |q|^{2-2\delta},$$

and does not have a singularity at $\delta = 0$. In the Euclidean case we have

$$I(q) = \pi^{4-\delta} \frac{\Gamma(1 - \delta/2)^3}{\Gamma(1)^3} \frac{\Gamma(-1 + \delta)}{\Gamma(3 - 3\delta/2)} |q|^{2-2\delta}.$$

The term $\Gamma(-1 + \delta)$ has a pole at $\delta = 0$. Therefore in the Euclidean case the above diagram has a non-zero counterterm.vsj
Hence, in contrast to the p-adic case in the Euclidean case we have $\eta \neq 0$.vsj
It can be shown that the Hamiltonian $\frac{\partial}{\partial u_1} H_{eff}(u_0, u_1, u_2)\Big|_{u_1 = 0, u_2 = u^*(1 + 2u_0)^2}$ is an eigen-Hamiltonian of the differential of $R_{\tau,\alpha}$ at the fixed point $H_{eff}(u_0, 0, u^*(1 + 2u_0)^2)$ with corresponding eigenvalue

$$\lambda_1 = p^{\tau \frac{\partial \beta_1}{\partial u_1}}\Big|_{u_1 = 0, u_2 = u^*(1 + 2u_0)^2}.$$

It allows us to get the following expression for the critical exponent $\nu$:

$$\nu^{-1} = \frac{\partial \beta_1}{\partial u_1}\Big|_{u_1 = 0, u_2 = u^*(1 + 2u_0)^2} = d/2 + \zeta(u^*)/2 - \zeta(u^*)\rho_1(u^*)\rho_2(u^*) \qquad (21)$$

Up to a third order in $\delta$ we have

$$\begin{aligned}\nu^{-1} =\ & 2 - \frac{N+2}{N+8}\delta + 2\delta^2 \frac{(N+2)(7N+20)}{(N+8)^3} A_p(4) + \delta^3 \frac{N+2}{(N+8)^5(p^4 - 1)^2} \\ & \times (N^3(5 + 24p^2 + 74p^4 + 24p^6 + 5p^8) + 8N^2(3 + 2p^2 + 32p^4 + 2p^6 + 3p^8) \\ & + 40N(9 + 64p^2 + 154p^4 + 64p^6 + 9p^8) \\ & + 32(41 + 344p^2 + 754p^4 + 344p^6 + 41p^8))\ln^2 p + O(\delta^4). \end{aligned} \qquad (22)$$

By direct computations it can be shown that the expression (17) becomes equal to (22) if in (17) we substitute

$$d = 4 - \delta, \quad \varepsilon = \alpha - 3d/2 = 2 + d - 3d/2 = \delta/2,$$

and expand (17) in $\delta$.

## ACKNOWLEDGMENTS

We are grateful to Prof. Branko Dragovich for the hospitality and excellent organization of the 2nd International Conference on $p$-Adic Mathematical Physics, Belgrade (2005).

## REFERENCES

1. Aref'eva I.Ya., Dragovich B., Volovich I.V., *P*-adic superstrings, *Phys. Lett.* **B 214**, 339–346 (1988).
2. Bleher P.M. and Missarov M.D., The Equations of Wilson's Renormalization Group and Analytic Renormalization, *Commun. Math. Phys.* **74**, 235–272 (1980).
3. Bleher P.M., Sinai Ya.G., Investigation of the critical point in models of the type of Dyson's hierarchical models, *Commun. Math. Phys.* **33**, 23 (1973).
4. Brekke L., Freund P.G.O., *p*-Adic numbers in Physics, *Phys. Rep.* **233**, no. 1, 2–66 (1993).
5. Dragovich B., On *p*-adic and adelic generalization of quantum field theory, *Nucl. Phys. B (Proc. Suppl.)* **102/103**, 150–155 (2001).
6. Kochubei A.N., *Pseudo-differential equations and stochastics over non-Archimedean fields*, New York: M. Dekker, 2001.
7. Khrennikov A., *p-Adic valued distributions in mathematical physics*, Dordrecht: Kluwer, 1994.
8. Khrennikov A., *Non-Archimedean analysis: Quantum paradoxes, dynamical systems and biological models*, Dordrecht: Kluwer, 1997.
9. Lerner E.Yu., Missarov M.D., *p*-adic Feynman and String Amplitudes, *Commun. Math. Phys.* **121**, no. 1, 35–48 (1989).
10. Lerner E.Yu., Missarov M.D., Fixed points of renormalization group in the hierarchical fermionic model, *J. Stat. Phys.* **76**, no. 3/4. 805–817 (1994).
11. Missarov M.D., *Renormalization group and renormalization theory in p-adic and adelic scalar models*, Dynamical systems and statistical mechanics, ed. Ya.G. Sinai (*Adv. Sov. Math.* **3**, Amer. Math. Soc.), pp. 143–161 (1991).
12. Missarov M.D., RG-invariant curves in the fermionic hierarchical model, *Theor. Math. Phys.* **114**, no. 3, 255–265 (1998).
13. Missarov M.D., Critical phenomena in the fermionic hierarchical model, *Theor. Math. Phys.* **117**, no. 3, 1483–1498 (1999).
14. Missarov M.D., Continuum limit in the fermionic hierarchical model, *Theor. Math. Phys.* **118**, no. 1, 32–40 (1999).
15. Missarov M.D., **Renormalization group solution of fermionic Dyson model**, In: Asymptotic Combinatorics with Application to Mathematical Physics, V.A. Malyshev and A.M. Vershik (eds.), Kluwer Academic Publishers, Printed in Netherlands, pp. 151–166 (2002).
16. Vladimirov V.S., Volovich I.V., Zelenov E.I., *p-Adic Analysis and Mathematical Physics*, World Scientific Publ., Singapoure, 1994.
17. Volovich I.V., *p*-adic string, *Class. Quantum Grav.* **4**, L83–L87 (1987).

# On Phase Transitions for $p$-Adic Potts Model with Competing Interactions on a Cayley Tree

F. M. Mukhamedov*,†, U. A. Rozikov** and J. F. F. Mendes*

*Departamento de Fisica, Universidade de Aveiro, Campus Universitário de Santiago, 3810-193 Aveiro, PORTUGAL
emails: farruh@fis.ua.pt, jfmendes@fis.ua.pt
†Institute of Mathematics, 29, F. Hodjaev str., Tashkent, 700125, UZBEKISTAN
**Institute of Mathematics, 29, F. Hodjaev str., Tashkent, 700143, UZBEKISTAN
email: rozikovu@yandex.ru

**Abstract.** In the paper we consider three state $p$-adic Potts model with competing interactions on a Cayley tree of order two. We reduce a problem of describing of the $p$-adic Gibbs measures to the solution of certain recursive equation, and using it we will prove that a phase transition occurs if and only if $p = 3$ for any value (non zero) of interactions. As well, we completely solve the uniqueness problem for the considered model in a $p$-adic context. Namely, if $p \neq 3$ then there is only a unique Gibbs measure the model.

**Keywords:** $p$-adic field, Potts model, Cayley tree, Gibbs measure, phase transition, uniqueness.
**PACS:** 02.50.Ga, 05.20.-y, 05.70.Fh, 64.60.Cn.

## 1. INTRODUCTION

A non-Kolmogorovian probability models [7], [8] in that probabilities belong to the filed of $p$-adic numbers $\mathbf{Q}_p$ were developed in connection with $p$-adic quantum models (see for example, [1], [6], [14], [22]). In [7], [9] a measure-theoretical axiomatics of the $p$-adic probability theory were proceeded. In [10], [12] certain, various limit theorems for $p$-adic valued probabilities have been proved. In [11] the theory of stochastic processes with values in $p$-adic and more general non-Archimedean fields having probability distributions with non-Archimedean values has been developed. There a non-Archimedean analogue of the Kolmogorov theorem was proved, that gives the opportunity to construct wide classes of stochastic processes by using finite dimensional probability distributions. This allowed us to begin the study and the development of certain problems of statistical mechanics in a context of the $p$-adic probability theory. In [16], [17] we have developed the $p$-adic probability theory approaches to study some models with nearest neighbor interactions on the Cayley tree, such as Ising and Potts models[1]. In those papers we investigated the set of $p$-adic Gibbs measures and a problem of phase transitions. Note that the $p$-adic Gibbs measures, associated with those models, enable Markov property.

---
[1] The classical (real value) contra parts of such models were considered in [3], [5].

In the present paper we consider three state $p$-adic Potts models with competing interactions on a Cayley tree order two. We note that the models on a Cayley tree with competing interactions have been studied extensively (see Refs. [15], [20]) since the appearance of the Vannimenus model (see Ref. [21]), in which the physical motivations for the urgency of study such models was presented. In all of these works no exact solutions of the phase transition problem were found, but some solutions for specific parameter values were presented. In this paper we can completely solve the uniqueness problem for the considered model in a $p$-adic context. Basically, in the real case there are only sufficient conditions (like Dobrushin condition [4]) for the uniqueness of a Gibbs measure of certain models. We show that for the model under consideration there is a phase transition if and only if $p = 3$ as well.

## 2. PRELIMINARIES

### 2.1. $p$-adic numbers and measures

Let $\mathbf{Q}$ be the field of rational numbers. Throughout the paper $p$ will be a fixed prime number. Every rational number $x \neq 0$ can be represented in the form $x = p^r \frac{n}{m}$, where $r, n \in \mathbf{Z}$, $m$ is a positive integer and $p, n, m$ are relatively prime. The $p$-adic norm of $x$ is given by $|x|_p = p^{-r}$ and $|0|_p = 0$. This norm satisfies so called the strong triangle inequality

$$|x+y|_p \leq \max\{|x|_p, |y|_p\}.$$

This is an ultrametricity of the norm. The completion of $\mathbf{Q}$ with respect to $p$-adic norm defines the $p$-adic field which is denoted by $\mathbf{Q}_p$. Let $B(a,r) = \{x \in \mathbf{Q}_p |\ |x-a|_p < r\}$, where $a \in \mathbf{Q}_p$, $r > 0$. By $\log_p$ and $\exp_p$ we mean $p$-adic logarithm and exponential which are defined as series with the usual way (see, for more details [13]). The domain of converge for them are $B(1,1)$ and $B(0, p^{-1/(p-1)})$, respectively. These two functions have the following properties (see [13, 22]):

$$|\exp_p(x)|_p = 1, \quad |\exp_p(x) - 1|_p = |x|_p < 1, \quad |\log_p(1+x)|_p = |x|_p < p^{-1/(p-1)} \quad (1)$$

and

$$\log_p(\exp_p(x)) = x, \quad \exp_p(\log_p(1+x)) = 1+x.$$

Let $(X, \mathscr{B})$ be a space, where $\mathscr{B}$ is an algebra of subsets $X$. A function $\mu : \mathscr{B} \to \mathbf{Q}_p$ is said to be a *p-adic measure* if for any $A_1, ..., A_n \subset \mathscr{B}$ such that $A_i \cap A_j = \emptyset$ ($i \neq j$)

$$\mu\left(\bigcup_{j=1}^n A_j\right) = \sum_{j=1}^n \mu(A_j).$$

A $p$-adic measure is called *a probability measure* if $\mu(X) = 1$. A $p$-adic probability measure $\mu$ is called *bounded* if $\sup\{|\mu(A)|_p\ |\ A \in \mathscr{B}\} < \infty$.

For more detail information about $p$-adic measures we refer to [7, 9].

## 2.2. The Cayley tree

The Cayley tree $\Gamma^k$ of order $k \geq 1$ is an infinite tree, i.e., a graph without cycles, such that each vertex of which lies on $k+1$ edges. Let $\Gamma^k = (V, \Lambda)$, where $V$ is the set of vertices of $\Gamma^k$, $\Lambda$ is the set of edges of $\Gamma^k$. The vertices $x$ and $y$ are called *nearest neighbor*, which is denoted by $l = <x,y>$ if there exists an edge connecting them. A collection of the pairs $<x,x_1>, ..., <x_{d-1},y>$ is called *a path* from $x$ to $y$. The distance $d(x,y), x,y \in V$ is the length of the shortest path from $x$ to $y$ in $V$.

For the fixed $x^0 \in V$ we set

$$W_n = \{x \in V | d(x, x^0) = n\}, \quad V_n = \bigcup_{m=1}^{n} W_m,$$

$$L_n = \{l = <x,y> \in L | x, y \in V_n\},$$

for a fixed point $x^0 \in V$.

Denote

$$S(x) = \{y \in W_{n+1} | d(x,y) = 1\}, \quad x \in W_n.$$

The defined set is called the set of *direct successors*. Observe that any vertex $x \neq x^0$ has $k$ direct successors and $x^0$ has $k+1$.

Two vertices $x, y \in V$ is called *one level next-nearest-neighboring vertices* if there is a vertex $z \in V$ such that $x, y \in S(z)$ and they are denoted by $> x, y <$. In this case the vertices $x, z, y$ are called *ternary* and denoted by $< x, z, y >$.

In the sequel we will consider semi-infinite Cayley tree $J^2$ of order 2, i.e. an infinite graph without cycles with 3 edges issuing from each vertex except for $x^0$ and with 2 edges issuing from the vertex $x^0$.

## 2.3. The model

Let $\mathbf{Q}_p$ be the field of $p$-adic numbers. By $\mathbf{Q}_p^{q-1}$ we denote $\underbrace{\mathbf{Q}_p \times ... \times \mathbf{Q}_p}_{q-1}$. The norm $\|x\|_p$ of an element $x \in \mathbf{Q}_p^{q-1}$ is defined by $\|x\|_p = \max_{1 \leq i \leq q-1} \{|x_i|_p\}$, here $x = (x_1, ..., x_{q-1})$. By $xy$ we mean the bilinear form on $\mathbf{Q}_p^{q-1}$ defined by

$$xy = \sum_{i=1}^{q-1} x_i y_i, \quad x = (x_1, \cdots, x_{q-1}), y = (y_1, \cdots, y_{q-1}).$$

Let $\Psi = \{\eta_1, \eta_2, ..., \eta_q\}$, where $\eta_1, \eta_2, ..., \eta_q$ are elements of $\mathbf{Q}_p^{q-1}$ such that $\|\eta_i\|_p = 1$, $i = 1, 2, ..., q$ and

$$\eta_i \eta_j = \begin{cases} 1, & \text{for } i = j, \\ 0, & \text{for } i \neq j \end{cases} (i,j = 1,2,...,q-1), \quad \eta_q = \sum_{i=1}^{q-1} \eta_i. \tag{2}$$

Let $h \in \mathbf{Q}_p^{q-1}$, then we have $h = \sum_{i=1}^{q-1} h_i \eta_i$ and

$$h\eta_i = \begin{cases} h_i, & \text{for } i = 1, 2, ..., q-1, \\ \sum_{i=1}^{q-1} h_i, & \text{for } i = q \end{cases} \quad (3)$$

We consider the *p*-adic Potts model where spin takes values in the set $\Psi$ and is assigned to the vertices of the tree $J^2 = (V, \Lambda)$. A configuration $\sigma$ on $V$ is then defined as a function $x \in V \to \sigma(x) \in \Psi$; in a similar fashion one defines a configuration $\sigma_n$ and $\sigma^{(n)}$ on $V_n$ and $W_n$ respectively. The set of all configurations on $V$ (resp. $V_n$, $W_n$) coincides with $\Omega = \Psi^V$ (resp. $\Omega_{V_n} = \Psi^{V_n}$, $\Omega_{W_n} = \Psi^{W_n}$). One can see that $\Omega_{V_n} = \Omega_{V_{n-1}} \times \Omega_{W_n}$. Using this, for given configurations $\sigma_{n-1} \in \Omega_{V_{n-1}}$ and $\sigma^{(n)} \in \Omega_{W_n}$ we define their concatenations by

$$\sigma_{n-1} \vee \sigma^{(n)} = \left\{ \{\sigma_n(x), x \in V_{n-1}\}, \{\sigma^{(n)}(y), y \in W_n\} \right\}.$$

It is clear that $\sigma_{n-1} \vee \sigma^{(n)} \in \Omega_{V_n}$.

The Hamiltonian $H_n : \Omega_{V_n} \to \mathbf{Q}_p$ of the *p*-adic Potts model with competing interactions has the form

$$\begin{aligned} H_n(\sigma) &= -\sum_{<x,y> \in L_n} J_{x,y} \delta_{\sigma(x), \sigma(y)} - \sum_{>x,y<: x,y \in V_n} K_{x,y} \delta_{\sigma(x), \sigma(y)} \\ &\quad - H \sum_{x \in V_n} \delta_{\eta_3, \sigma(x)}, \quad n \in \mathbf{N}, \end{aligned} \quad (4)$$

here $\sigma \in \Omega_{V_n}$, $\delta$ is the Kronecker symbol and

$$\begin{cases} |J_{x,y}|_p < p^{-1/(p-1)}, & \forall <x,y>, \\ |K_{u,v}|_p < p^{-1/(p-1)}, & \forall >u,v<, \\ |H|_p < p^{-1/(p-1)}. \end{cases} \quad (5)$$

## 3. EXISTENCE OF PHASE TRANSITION

In this section we give a construction of Gibbs measures for the three state ($q = 3$) *p*-adic Potts model with competing interactions on a semi-infinite Cayley tree $J^2$ of order 2, and establish a phase transition for it.

In the sequel we will assume that the condition (5) is satisfied. Let $\mathbf{h} : x \in V \to h_x \in \mathbf{Q}_p^{q-1}$ be a function of $x \in V$ such that $\|h_x\|_p < p^{-1/(p-1)}$ for all $x \in V$. Given $n = 1, 2, ...$ consider a *p*-adic probability measure $\mu_{\mathbf{h}}^{(n)}$ on $\Omega_{V_n}$ defined by

$$\mu_{\mathbf{h}}^{(n)}(\sigma) = Z_n^{-1} \exp_p\{-H_n(\sigma) + \sum_{x \in W_n} h_x \sigma(x)\}, \quad (6)$$

Here, as before, $\sigma \in \Omega_{V_n}$ and $Z_n$ is the corresponding partition function:

$$Z_n = \sum_{\tilde{\sigma} \in \Omega_{V_n}} \exp_p\{-H(\tilde{\sigma}) + \sum_{x \in W_n} h_x \tilde{\sigma}(x)\}.$$

Note that the measures $\mu_{\mathbf{h}}^{(n)}$ are well defined, since from (5), $\|h_x\|_p < p^{-1/(p-1)}$ and the strong triangle inequality one gets

$$\left| H_n(\sigma) + \sum_{x \in W_n} h_x \sigma(x) \right|_p < p^{-1/(p-1)}$$

for any $n \in \mathbf{N}$, which enables the existence of the measures (6). The compatibility condition for $\mu_{\mathbf{h}}^{(n)}, n \geq 1$ is given by the equality

$$\sum_{\sigma^{(n)} \in \Omega_{W_n}} \mu_{\mathbf{h}}^{(n)}(\sigma_{n-1} \vee \sigma^{(n)}) = \mu_{\mathbf{h}}^{(n-1)}(\sigma_{n-1}). \qquad (7)$$

We note that an analog of the Kolmogorov extension theorem for distributions can be proved for the $p$-adic measures given by (6) (see [11]). Then according to the Kolmogorov theorem there exists a unique $p$-adic measure $\mu_{\mathbf{h}}$ on $\Omega$ such that for every $n = 1, 2, \ldots$ and $\sigma_n \in \Omega_n$ the equality holds

$$\mu_{\mathbf{h}}\left(\{\sigma|_{V_n} = \sigma_n\}\right) = \mu_{\mathbf{h}}^{(n)}(\sigma_n),$$

which will be called a *p-adic Gibbs measure* for the considered model. It is clear that the measure $\mu_{\mathbf{h}}$ depends on the function $\mathbf{h}$. By $\mathscr{S}$ we denote the set of all $p$-adic Gibbs measures associated with functions $\mathbf{h} = (h_x, x \in V)$. If $|\mathscr{S}| \geq 2$, then we say that for this model there exists *a phase transition*, otherwise, we say there is *no phase transition* ( here $|A|$ means the cardinality of a set $A$). In other words, the phase transition means that there are two different functions $\mathbf{h} = (h_x, x \in V)$ and $\mathbf{s} = (s_x, x \in V)$ for which there exist two $\mu_{\mathbf{h}}$ and $\mu_{\mathbf{s}}$ $p$-adic Gibbs measures on $\Omega$, respectively.

Using (6) and the argument of the proof of Theorem 3.2 [16] we may obtain that the measures $\mu_{\mathbf{h}}^{(n)}$, $n = 1, 2, \ldots$ satisfy the compatibility condition (7) if and only if for any $x \in V$ the following recursive equation holds:

$$\begin{cases} h_{x,1} = \log \dfrac{\theta_1 F_1\left(\theta_{xy}, \theta_{xz}, \kappa_{yz}; \exp_p(h_y), \exp_p(h_z)\right)}{F_2\left(\theta_{xy}, \theta_{xz}, \kappa_{yz}; \exp_p(h_y), \exp_p(h_z)\right)} \\[2mm] h_{x,2} = \log \dfrac{\theta_1 F_1\left(\theta_{xy}, \theta_{xz}, \kappa_{yz}; \exp_p(h_y)^t, \exp_p((h_z)^t)\right)}{F_2\left(\theta_{xy}, \theta_{xz}, \kappa_{yz}; \exp_p((h_y)^t), \exp_p((h_z)^t)\right)} \end{cases} \qquad (8)$$

here $<y,x,z>$ ternary vertices, $\theta_{xy} = \exp_p\{J_{xy}\}$, $\kappa_{xy} = \exp_p\{K_{xy}\}$, $\theta_1 = \exp_p(H)$ and for given vector $h = (h_1,h_2)$ by $\exp_p(h)$ and $h'$ we have denoted the vectors $(\exp_p(h_1), \exp_p(h_2))$ and $(h_2, h_1)$ respectively, and $F_i : \mathbf{Q}_p^3 \times \mathbf{Q}_p^4 \to \mathbf{Q}_p$, $(i=1,2)$ functions are defined by

$$\begin{cases} F_1(\alpha_1,\alpha_2,\beta;h,r) = \alpha_1\alpha_2\beta h_1h_2r_1r_2 + \alpha_1 h_1h_2(r_1+r_2) + \alpha_2 r_1 r_2(h_1+h_2) \\ \qquad\qquad\qquad\quad + \beta(h_1r_1 + h_2r_2) + h_1r_2 + h_2r_1 \\ F_2(\alpha_1,\alpha_2,\beta;h,r) = \beta h_1h_2r_1r_2 + \alpha_1 h_1 r_1 r_2 + \alpha_2 h_1 h_2 r_1 + h_2 r_1 r_2 + h_1 h_2 r_2 \\ \qquad\qquad\qquad\quad + \alpha_1 h_1 r_2 + \alpha_2 h_2 r_1 + \beta h_2 r_2 + \alpha_1 \alpha_2 \beta h_1 r_1 \end{cases} \quad (9)$$

where $h = (h_1, h_2), r = (r_1, r_2)$.

Consequently, the problem of describing of $\mathscr{S}$ is reduced to the finding of solutions of the functional equation (8).

Write
$$\Sigma = \{\mathbf{h} = (h_x \in \mathbf{Q}_p^2, x \in V) : h_x \text{ satisfies the equation (8)}\}.$$

To prove the existence of phase transition it suffices to show that there are two different functions in $\Sigma$. The description of arbitrary elements of the set $\Sigma$ is a complicated problem.

Therefore, assume that $J_{xy} = J$, $K_{xy} = K$ and $H = 0$. In this paper we restrict ourselves to the description of translation - invariant elements of $\Sigma$, i.e. in which $h_x = h$ is independent on $x$.

Let $h_x = h = (h_1, h_2)$ for all $x \in V$. Then using (9) we can reduce (8) to the following form

$$\begin{cases} u_1 = \frac{\theta^2 \kappa u_1^2 u_2^2 + 2\theta(u_1^2 u_2 + u_1 u_2^2) + \kappa(u_1^2 + u_2^2) + 2u_1 u_2}{\kappa u_1^2 u_2^2 + 2\theta u_1^2 u_2 + 2u_1 u_2^2 + 2\theta u_1 u_2 + \kappa u_2^2 + \theta^2 \kappa u_1^2}, \\ u_2 = \frac{\theta^2 \kappa u_1^2 u_2^2 + 2\theta(u_1^2 u_2 + u_1 u_2^2) + \kappa(u_1^2 + u_2^2) + 2u_1 u_2}{\kappa u_1^2 u_2^2 + 2\theta u_1 u_2^2 + 2u_1^2 u_2 + 2\theta u_1 u_2 + \kappa u_1^2 + \theta^2 \kappa u_2^2}, \end{cases} \quad (10)$$

here $u_1 = \exp_p(h_1)$, $u_2 = \exp_p(h_2)$ and $\theta = \exp_p(J)$, $\kappa = \exp_p(K)$.

From (10) it is easily seen that the lines $u_1 = u_2$, $u_1 = 1$ and $u_2 = 1$, are invariant for the equation. Therefore, it is enough to consider the equation on the line $u_2 = 1$, since other cases can be reduced to this case. So, we rewrite (10) as follows

$$u = \frac{(\theta^2 \kappa + 2\theta + \kappa)u^2 + 2(\theta+1)u + \kappa}{2(\kappa+1)u^2 + 4\theta u + \theta^2 \kappa} \quad (11)$$

It is evident that $u = 1$ is a solution of (11), but to exist a phase transition we are interested for other solutions one. After some simple algebra we find the following equation

$$2(\kappa+1)u^2 + (2+2\theta - \kappa - \theta^2 \kappa)u - \kappa = 0. \quad (12)$$

We have to find a solution of (12) such that $|u-1|_p < p^{-1/(p-1)}$.

Rewriting (12) as follows

$$2(\kappa+1)(u^2-1)+(2+2\theta-\kappa-\theta^2\kappa)(u-1)+2\theta-\theta^2\kappa+4=0$$

we infer that if $|2\theta-\theta^2\kappa+4|_p = 1$ then (12) does not have a needed solution. Therefore, we should require that $|2\theta-\theta^2\kappa+4| \leq \frac{1}{p}$. We know (see (1)) that from the properties of the exponential function the parameters $\theta$ and $\kappa$ satisfy the following inequalities

$$|\theta-1|_p \leq \frac{1}{p}, \quad |\kappa-1|_p \leq \frac{1}{p}. \tag{13}$$

Using these inequalities we derive that the inequality $|2\theta-\theta^2\kappa+4| \leq \frac{1}{p}$ is valid if and only if $p = 3$. Therefore, let us assume that $p = 3$. Denote

$$P(x) = 2(\kappa+1)x^2 + (2+2\theta-\kappa-\theta^2\kappa)x - \kappa.$$

For $P(x)$ we have $P(1) = 2\theta-\theta^2\kappa+4$, $P'(1) = 6+2\theta-\theta^2\kappa+3\kappa$, hence by means of (13) one gets

$$|P(1)|_3 \leq \frac{1}{3}, \quad |P'(1)|_3 = 1.$$

So according to the Hensel's lemma (see [13]) there is a solution $U \in \mathbf{Q}_3$ of the equation $P(x) = 0$ such that $|U-1|_3 \leq \frac{1}{3}$. Consequently, for the model under consideration there is a phase transition at $p = 3$ for every $J, K$ such that $0 < |J|_3 \leq \frac{1}{3}, 0 < |K|_3 \leq \frac{1}{3}$.

## 4. THE UNIQUENESS OF THE GIBBS MEASURE

From the previous section we infer that if $p \neq 3$ then the equation (11) has a unique solution. Note that this solution corresponds to the case $h_x = h$. Therefore, in general, does there exist a phase transition in this case or not? In this section we are going to answer this question. Here we will consider two cases.

### 4.1. Non-homogeneous case

Let us assume that $H = 0$ and (5) be satisfied. Then it is not hard to check that $\mathbf{h} = (h_x = 0, x \in V)$ is a solution for (8). We will to prove that any other solution of (8) coincides with this one. To show it we have to estimate the value $\|h_x\|_p$. Denote $u_{x,i} = \exp_p(h_{x,i})$, $i = 1, 2$. Then from (1) we find

$$|h_{x,i}|_p = |u_{x,i}-1|_p, \quad i = 1,2. \tag{14}$$

hence we have

$$|u_{x,1}-1|_p = \left|\frac{F_1(\theta_{xy},\theta_{xz},\kappa_{yz};u_y,u_z) - F_2(\theta_{xy},\theta_{xz},\kappa_{yz};u_y,u_z)}{F_2(\theta_{xy},\theta_{xz},\kappa_{yz};u_y,u_z)}\right|_p$$

$$= \left| \kappa_{yz} u_{y,1} u_{z,1} (\theta_{xy}\theta_{xz} - 1)(u_{y,2}u_{z,2} - 1) + u_{y,1} u_{z,2}(\theta_{xy} - 1)(u_{y,2} - 1) \right.$$

$$\left. + u_{y,2} u_{z,1}(\theta_{xz} - 1)(u_{z,1} - 1) + u_{y,1} u_{z,1}(\theta_{xy} - \theta_{xz})(u_{y,2} - u_{z,2}) \right|_p$$

$$\leq \frac{1}{p}\max\left\{|u_{y,2}u_{z,2} - 1|_p, |u_{y,2} - 1|_p, |u_{z,1} - 1|_p, |u_{y,2} - u_{z,2}|_p\right\}$$

$$= \frac{1}{p}\max\left\{|h_{y,2} + h_{z,2}|_p, |h_{y,2}|_p, |h_{z,1}|_p, |h_{y,2} - h_{z,2}|_p\right\}$$

$$\leq \frac{1}{p}\max\left\{\|h_y\|_p, \|h_z\|_p\right\} \tag{15}$$

here we again used (1) and

$$|F_2(\theta_{xy}, \theta_{xz}, \kappa_{yz}; u_y, u_z)|_p = 1$$

which is valid if $p \neq 3$.

Analogously reasoning we derive

$$|u_{x,2} - 1|_p \leq \frac{1}{p}\max\left\{\|h_y\|_p, \|h_z\|_p\right\}. \tag{16}$$

The equality (14) with (15),(16) implies that

$$\|h_x\|_p \leq \frac{1}{p}\max\left\{\|h_y\|_p, \|h_z\|_p\right\}. \tag{17}$$

Take an arbitrary $\varepsilon > 0$. Let $n_0 \in \mathbf{N}$ be such that $\frac{1}{p^{n_0}} < \varepsilon$. Now iterating (17) $n_0$ times one gets

$$\|h_x\|_p \leq \frac{1}{p^{n_0}} < \varepsilon$$

this means that $h_x = 0$ for all $x \in V$. Thus, the $p$-adic Gibbs measure is unique.

## 4.2. Homogeneous case

In this subsection we will assume that $J_{xy} = J$, $K_{xy} = K$ and $H \neq 0$. Let us suppose that $h_x = h = (h_1, h_2)$ for all $x \in V$. In this case one easily sees that (8) invariant with respect to $u_1 = u_2$. Therefore, we are looking for a solution of (8) of the form $(h_1, h_1)$. Then it can be rewritten as follows

$$u = \theta_1 \frac{\theta^2 \kappa u^2 + 4\theta u + 2(\kappa + 1)}{\kappa u^2 + 2(\theta + 1)u + \theta^2\kappa + 2\theta + \kappa}, \quad u = \exp_p(h_1). \tag{18}$$

Denote

$$f(x) = \theta_1 \frac{\theta^2 \kappa x^2 + 4\theta x + 2(\kappa + 1)}{\kappa x^2 + 2(\theta + 1)x + \theta^2\kappa + 2\theta + \kappa}. \tag{19}$$

Let us show that $f(B(1,p^{-1/(p-1)})) \subset B(1,p^{-1/(p-1)})$. Indeed, let $|x-1|_p < p^{-1/(p-1)}$, then

$$\begin{aligned}
|f(x)-1|_p &= \left|\frac{(\theta^2\theta_1\kappa-\kappa)x^2+(4\theta\theta_1-2\theta-2)x+2(\kappa+1)\theta_1-\theta^2\kappa-2\theta-\kappa}{\kappa x^2+2(\theta+1)x+\theta^2\kappa+2\theta+\kappa}\right|_p \\
&\leq \max\{|(\theta^2\kappa\theta_1-\kappa)(x^2-1)|_p, |(4\theta\theta_1-2\theta-2)(x-1)|_p, \\
&\qquad |(\theta^2\kappa+4\theta+2(\kappa+1))(\theta_1-1)|_p\} \\
&\leq \max\{|(x-1)|_p, (\theta_1-1)|_p\} < p^{-1/(p-1)}
\end{aligned}$$

here we have used the equality $|\kappa x^2+2(\theta+1)x+\theta^2\kappa+2\theta+\kappa|_p = 1$ which is valid if $p \neq 3$.

Now after some algebra we derive

$$\begin{aligned}
|f(x)-f(y)|_p &= \Big|2\theta\kappa xy(2-\theta-\theta^2) \\
&\quad +\kappa(x+y)(2(\kappa+1)-\theta^2(\theta^2\kappa+2\theta+\kappa)) \\
&\quad +4((\theta+1)(\kappa+1)-\theta^2\kappa-2\theta-\kappa)\Big|_p |x-y|_p \\
&\leq \max\{|2-\theta-\theta^2|, \\
&\qquad |2(\kappa+1)-\theta^2(\theta^2\kappa+2\theta+\kappa)|_p, \\
&\qquad |(\theta+1)(\kappa+1)-\theta^2\kappa-2\theta-\kappa|_p\} |x-y|_p. \qquad (20)
\end{aligned}$$

Using (13) one can be shown that

$$\begin{cases} |2-\theta-\theta^2| \leq \frac{1}{p}, \\ |2(\kappa+1)-\theta^2(\theta^2\kappa+2\theta+\kappa)|_p \leq \frac{1}{p}, \\ |(\theta+1)(\kappa+1)-\theta^2\kappa-2\theta-\kappa|_p \leq \frac{1}{p}, \end{cases}$$

The last inequalities with (20) imply that

$$|f(x)-f(y)|_p \leq \frac{1}{p}|x-y|_p. \qquad (21)$$

Thus the inequality (21) yields that $f$ is a contraction of $B(1,p^{-1/(p-1)})$, hence $f$ has a unique fixed point $\zeta \in B(1,p^{-1/(p-1)})$. Let $\xi = \log_p \zeta$ and $\bar{\xi} = (\xi,\xi)$. Then using the same argument as (15) we may obtain

$$\|h_x - \bar{\xi}\|_p \leq \frac{1}{p}\max\{\|h_y-\bar{\xi}\|_p, \|h_z-\bar{\xi}\|_p\}.$$

Now using the argument of the previous subsection we get that $h_x = \bar{\xi}$ for all $x \in V$. Thus, the $p$-adic Gibbs measure is unique.

## 5. CONCLUSIONS

In the paper we have considered three state $p$-adic Potts model with competing interactions on a Cayley tree order two. We reduced a problem of describing of the $p$-adic Gibbs measures to the solution of certain recursive equation, and using it we proved that a phase transition occurs if and only if $p = 3$ for any value (non zero) of interactions. If $p \neq 3$ we showed that there is only a unique Gibbs measure for the inhomogeneous Potts model with zero external filed, as well as we established that result for homogeneous model but with non-zero external filed. From these results we conclude that a phase transition depends only on a value of $p$. These results are totally different from the real case, since in this setting there is a phase transition on some constraints for the interaction parameters (see [3], [18]). When a $p$-adic Gibbs measure is unique, then by means of a method of paper [16] one can be shown that the measure is bounded as well. The results concerning the uniqueness of the Gibbs measures extend results obtained in [16, 17]. We hope our results will force to study certain limit theorems for such kind of measures, since they naturally appear from some Hamiltonian systems, and on the other hand, these measures enable a Markov property. We also hope that these investigations give some opportunity to study Hamiltonian systems over networks in a $p$-adic setting [2].

## ACKNOWLEDGMENTS

The authors would like to express their gratitude to the organizers of the 2nd International Conference on $p$-Adic Mathematical Physics, Belgrade, for an invitation. F. M. thanks the FCT (Portugal) grant SFRH/BPD/17419/2004. F. M. and U. R thank also for grant Φ-2.1.56 of CST of Uzbekistan. U. R. acknowledges prof. M. Cassandro for an invitation to "La Sapienza" University and NATO Reintegration Grant FEL. RIG.980771.

## REFERENCES

1. I.Ya. Aref'eva, B. Dragovich, P.H. Frampton and I.V. Volovich, *Mod.Phys. Lett.* **A6**, 4341–4358 (1991).
2. S.N. Dorogovtsev, A.V. Goltsev and J.F.F. Mendes, *Phys. Rev. E* **67**, 026123 (2003).
3. N.N. Ganikhodjaev and U.A. Rozikov, *Osaka Jour. Math.* **37**, 373–383 (2000).

4. H.O. Georgii, *Gibbs measures and phase transitions*, Walter de Gruyter, Berlin, 1988.
5. S. Katsura and M. Takizawa, *Progr. Theor. Phys.* **51**, 82–98 (1974).
6. A.Yu. Khrennikov, *J. Math. Phys.* **32**, 932–936 (1991).
7. A.Yu. Khrennikov, *p-adic Valued Distributions in Mathematical Physics*. Kluwer Academic Publisher, Dordrecht, 1994.
8. A.Yu. Khrennikov, *p*-adic probability distribution of hidden variables, *Physica A* **215** (1995), 577–587.
9. A.Yu. Khrennikov, *Indag. Mathem. N.S.* **7**, 311–330 (1996).
10. A.Yu. Khrennikov, *Statis. Probab. Lett.* **51**, 269–276 (2001).
11. A.Yu. Khrennikov and S. Ludkovsky, *Markov Process. Related Fields* **9**, 131–162 (2003).
12. A.Yu. Khrennikov and S. Yamada, *Theor. Probab. Appl.* **49**, 65–76 (2005).
13. N. Koblitz, *p-adic numbers, p-adic analysis and zeta-function*, Berlin, Springer, 1977.
14. E. Marinary and G. Parisi, *Phys. Lett. B* **203**, 52–56 (1988).
15. M. Mariz, C. Tsalis and A.L. Albuquerque, *Jour. of Stat. Phys.* **40**, 577–592 (1985).
16. F.M. Mukhamedov and U.A. Rozikov, *Indag. Mathem. N.S.* **15**, 85–100 (2004).
17. F.M. Mukhamedov and U.A. Rozikov, *Infin. Dimens. Anal. Quantum Probab. Relat. Top.* **8**, 277–290 (2004).
18. F.M. Mukhamedov and U.A. Rozikov, *J. Stat. Phys.* **114** 825–848 (2004); **119**, 427–446 (2005).
19. D. Ruelle, *Statistical Mechanics: Rigorus Results*, Benjamin, 1969.
20. C.R. da Silca and S. Coutinho, *Phys. Review B* **34**, 7975–7985 (1986).
21. J. Vannimenus, *Z. Phys. B* **43**, 141 (1981).
22. V.S. Vladimirov, I.V. Volovich and E.I. Zelenov, *p-adic Analysis and Mathematical Physics*, Singapore, World Scientific, 1994.

# From Data to the Physics Using Ultrametrics: New Results in High Dimensional Data Analysis

Fionn Murtagh

*Department of Computer Science, Royal Holloway,*
*University of London Egham TW20 0EX, ENGLAND*
*email:* fmurtagh@acm.org

**Abstract.** We begin with pervasive ultrametricity due to high dimensionality and/or spatial sparsity. How extent or degree of ultrametricity can be quantified leads us to the discussion of varied practical cases when ultrametricity can be partially or locally present in data. We show how the ultrametricity can be assessed in text or document collections, in time series signals, and in other areas. We conclude with a discussion of ultrametricity in astrophysics, relating to observational cosmology.

**Keywords:** Multivariate data analysis, cluster analysis, hierarchy, factor analysis, correspondence analysis, ultrametric, p-adic, phylogeny.
**PACS:** 02.50.Sk, 02.10.De, 89.75.Hc, 89.75.Fb, 95.80.+p, 98.52.Cf.

## 1. INTRODUCTION

P-adic algebraic or ultrametric topological representation is important for observational data analysis because of the insights and benefits that such an approach provides.

Firstly ultrametricity is a pervasive property of observational data. It arises as a limit case when data dimensionality or sparsity grows. More strictly such a limit case can be a regular lattice structure. Ultrametricity is one possible representation for it. Notwithstanding alternative representations, ultrametricity offers computational efficiency (related to tree depth/height being logarithmic in number of terminal nodes), linkage with dynamical or related functional properties (phylogenetic interpretation), and well understood processing tools based on well known p-adic or ultrametric theory (example: ultrametric wavelet transform).

Secondly, practical data sets (derived from, or observed in, databases and data spaces) present some but not exclusively ultrametric characteristics. This can be used for forensic data exploration (fingerprinting data sets, as we discuss below). Or, it can be used to expedite search and discovery in information spaces. Indeed we would like to go a lot further, and gain new insights into data (and observed phenomena and events) through ultrametric or p-adic representations. We see this as a program of work for the near future.

# 2. ULTRAMETRICITY AND DIMENSIONALITY

## 2.1. Very High Dimensions are Naturally Ultrametric

Bellman's [2] "curse of dimensionality" relates to exponential growth of hypervolume as a function of dimensionality. Problems become tougher as dimensionality increases. In particular problems related to proximity search in high-dimensional spaces tend to become intractable.

But in very high dimensions the situation is quite different. In very high dimensions there is no longer a "curse of dimensionality" as we will see in sections 2.2 and 2.3 below. High dimensionality implies relative sparsity. So our results are equally valid (i) for high spatial dimensionality, and (ii) for spatially sparse clouds of data points.

In a way, a "trivial limit" (Treves [27]) case is reached as dimensionality increases. This makes high dimensional proximity search very different, and given an appropriate data structure – such as a binary hierarchical clustering tree – we can find nearest neighbors in worst case $O(1)$ or constant computational time [19]. The proof is simple: the tree data structure affords a constant number of edge traversals.

The fact that limit properties are "trivial" makes them no less interesting to study. Let us refer to such "trivial" properties as (structural or geometrical) regularity properties (e.g. all points lie on a regular lattice). First of all, the symmetries of regular structures in our data may be of importance. Secondly, "islands" or clusters in our data, where each "island" is of regular structure, may be exploitable. Thirdly, the mention of exploitability points to the application areas targeted: in this paper, we focus on search and matching and show some ways in which ultrametric regularity can be exploited in practice. Fourthly, and finally, regularity by no means implies complete coverage (e.g., existence of all pairwise linkages) so that interesting or revealing structure will be present in real data sets.

## 2.2. Distance Properties in Very Sparse Spaces

In Murtagh [19] we discussed the finding that ultrametricity is a natural property of spaces in the (spatially) sparse limit. By this we mean that as spatial dimensionality and spatial sparseness increase, so too do ultrametric properties.

The triangular inequality holds for a metric space: $d(x,z) \leq d(x,y) + d(y,z)$ for any triplet of points $x,y,z$. In addition the properties of symmetry and positive definiteness are respected. The "strong triangular inequality" or ultrametric inequality is: $d(x,z) \leq \max\{d(x,y), d(y,z)\}$ for any triplet $x,y,z$. An ultrametric space implies respect for a range of stringent properties. For example, the triangle formed by any triplet is necessarily isosceles, with the two large sides equal; or is equilateral.

Murtagh [19], and earlier work by Rammal et al. [24, 25], has demonstrated the pervasiveness of ultrametricity, by focusing on the fact that sparse high-dimensional data tend to be ultrametric. One reason for this is as follows.

As dimensionality grows, so too do distances (or indeed dissimilarities, if they do not satisfy the triangular inequality). The least change possible for dissimilarities to become distances has been formulated in terms of the smallest additive constant needed, to be added to all dissimilarities [26, 5, 6, 23]. Adding a sufficiently large constant to all dissimilarities transforms them into a set of distances. Through addition of a larger constant, it follows that distances become approximately equal, thus verifying a trivial case of the ultrametric or "strong triangular" inequality. Adding to dissimilarities or distances may be a direct consequence of increased dimensionality.

For a close fit or good approximation, the situation is not as simple for taking dissimilarities, or distances, into ultrametric distances. If we want a close fit to the given dissimilarities then a good choice would avail either of the maximal inferior, or subdominant, ultrametric; or the minimal superior ultrametric. Stepwise algorithms for these are commonly known as, respectively, single linkage hierarchical clustering; and complete link hierarchical clustering. (See [3, 16, 17] and other texts on hierarchical clustering.)

## 2.3. Quantifying Degree of Ultrametricity

Summarizing a full description in Murtagh [19] we explored two measures quantifying how ultrametric a data set is.

Firstly, Rammal et al. [25] used discrepancy between each pairwise distance and the corresponding subdominant ultrametric. Now, the subdominant ultrametric is also known as the ultrametric distance resulting from the single linkage agglomerative hierarchical clustering method. Closely related graph structures include the minimal spanning tree, and graph (connected) components. While the subdominant provides a good fit to the given distance (or indeed dissimilarity), it suffers from the "friends of friends" or chaining effect.

Secondly, Lerman [16] developed a measure of ultrametricity, termed H-classifiability, using ranks of all pairwise given distances (or dissimilarities). The isosceles (with small base) or equilateral requirements of the ultrametric inequality impose constraints on the ranks. The interval between median and maximum rank of every set of triplets must be empty for ultrametricity. We have used extensively Lerman's measure of degree of ultrametricity in a data set. Taking ranks provides scale invariance. But the limitation of Lerman's approach, we find, is that it is not reasonable to study ranks of real-valued distances defined on a large set of points.

Thirdly, our own measure of extent of ultrametricity [19] can be described algorithmically. We assume a Euclidean metric. We examine triplets of points (exhaustively if possible, or otherwise through sampling), and determine the three angles formed by the associated triangle. We select the smallest angle formed by the triplet points. Then we check if the other two remaining angles are approximately equal. If they are equal then our triangle is isosceles with small base, or equilateral (when all triangles are equal). The approximation to equality is given by 2 degrees (0.0349 radians). Our motivation for the approximate ("fuzzy") equality was that it makes our approach robust and independent of measurement precision.

Studies are discussed in Murtagh [19] showing how numbers of points in our clouds of data points are irrelevant; but what counts is the ambient spatial dimensionality. Among cases looked at are statistically uniformly (hence "unclustered", or without structure in a certain sense) distributed points, and statistically uniformly distributed hypercube points (so the latter are random 0/1 valued vectors). Using our ultrametricity measure, there is a clear tendency to ultrametricity as the spatial dimensionality (hence spatial sparseness) increases.

## 3. LOCAL OR PARTIAL ULTRAMETRICITY

### 3.1. Approximating Local Ultrametricity

Now we look at data where some triangles are consistent with ultrametric properties, while others are not.

It has long been known [8, 28] that forms of data structuring, and more particularly data clustering, can be used to expedite search problems in high dimensions. Some of the work of Chávez and Navarro and their colleagues provides an explanation as to why and how clustering can be exploited for high dimensional proximity search.

In large data sets, i.e. large $n$ or number of observations, a clever way to expedite proximity searching (in particular nearest neighbor finding) in metric spaces is as follows. The metric property implies that the triangular inequality holds. We have a given point and we are looking for its nearest neighbor. We use a third point, called a pivot point. Such a pivot point is carefully selected at the start of the processing, and all necessary distances to it are stored. Through the triangular inequality, we then form a bound on the best potential nearest neighbor distance. Thereby we limit the region within which the search is carried out. See [7, 8, 9, 4]. As pointed out in Murtagh [19], the bounding rule, or rejection rule, that ensues, is forcing retained triangles to be isosceles. This is interesting because it can be viewed as finding locally ultrametric relationships.

In [7, 8], the ambient spatial dimension is termed the "representational dimension", or embedding dimension, $m$. (This is dimensionality, $m$: we have for example $x \in \mathbb{R}^m$.) Search is subject to the curse of dimensionality when addressed in all generality in $\mathbb{R}^m$. However there is often a smaller "intrinsic dimensionality", or average local dimensionality (e.g. when the data are clustered, or lie on a surface of dimension $< m$). This can be exploited to provide fast proximity searching opportunities. However it is difficult in general to define the intrinsic dimensionality.

These authors define intrinsic dimensionality of a metric space as: $\rho = \frac{\mu^2}{2\sigma^2}$ where $\mu$ and $\sigma^2$ are, respectively, the mean and variance of the distances.

So, firstly, the intrinsic dimensionality grows with the mean distance. We have observed that ultrametricity increases with average distance both by simulations in Murtagh [19], and also through the argument of a simple additive transformation (in section 2.2 above). Secondly, the intrinsic dimensionality grows with inverse variance. Small variance of distances implies equilateral triangles between point triplets, and therefore implies ultrametricity.

We see therefore that the intrinsic dimensionality of [7, 9] affords another definition of ultrametricity. We have already observed how their fast, pivot-based proximity rule can be interpreted as local enforcement of the ultrametric inequality. We conclude from these observations that local or global ultrametricity (i.e., high values of Chávez and Navarro's $\rho$, or high local contributions to $\rho$) permit fast proximity search.

## 3.2. The Case of High Dimensional Gaussian Clouds

We have already seen that high dimensions lead to "natural" ultrametricity. In this section we discuss the insightful work of [15], viz.

- algorithms for registering (finding an exact match between) two high dimensional Gaussian point sets; and
- an algorithm for partitioning a high dimensional Gaussian point set into two subsets.

What we have discussed in sections 2.1 and 2.2 above allows us to accept that a "trivial limit" of similar distances may be arrived at in high dimensional spaces. But this opens up the intriguing perspective on more than one high dimensional point cloud, each of which has simple simplex structure, but different clouds have different such structures.

Consider two clouds of points in the same space, of typical values $x$ and $y$, respectively. We consider the following:

- The average variances of $x$ and $y$ points, over all dimensions, are respectively constant $\sigma^2$ and $\tau^2$.
- The average over dimensions of the distance squared between the means of $x$ points, and of $y$ points, on each dimension is constant $\mu^2$.

From the first bullet point here, when rescaled by $m^{-1/2}$, points in the $x$ and $y$ clouds are asymptotically located at the vertices of regular simplexes with edge lengths, respectively, $2\sigma^{1/2}$ and $2\tau^{1/2}$.

From the above two points, the independence of the two clouds, and the Law of Large Numbers, the following result holds.

$$\left[\frac{1}{m}\sum_{k=1}^{m}(x_{ik}-y_{jk})^2\right]^{1/2} \xrightarrow{p} (\sigma^2+\tau^2+\mu^2)^{1/2} \qquad (1)$$

Interpretation:

All points $x$ are equidistant with distance $2^{1/2}\sigma$.

All points $y$ are equidistant with distance $2^{1/2}\tau$.

Points $x$ tend in probability to be all the same distance from *all* points in $y$ with distance $(\sigma^2+\tau^2+\mu^2)^{1/2}$, from result (1) above.

For the matching or registering of two Gaussian point sets, we require that: $\sigma = \tau$, and $\mu = 0$, so that we have as a result just one point set.

For the partitioning of a high dimensional Gaussian point cloud into two clouds, we require the following:

- $\frac{1}{2} n_1(n_1 - 1)$ distances equal to $2^{1/2}\sigma$
- $\frac{1}{2} n_2(n_2 - 1)$ distances equal to $2^{1/2}\tau$
- $\frac{1}{2}(n_1 + n_2)(n_1 + n_2 - 1) - \frac{1}{2} n_1(n_1 - 1) - \frac{1}{2} n_2(n_2 - 1) = n_1 n_2$ distances equal to $(\sigma^2 + \tau^2 + \mu^2)^{1/2}$.

An algorithm based on Gaussian mixture modeling is then as follows. We seek clusters of distances of (possibly very) small variance to define the $x, y, \ldots$ clouds. All other distances can be considered as (by assumption, Poisson) background noise. Model-based clustering [14] can accommodate such a model (although the computational aspect in high dimensions may be troublesome), and provide Bayes factor support for the inherent number of clusters present.

Full exploitation of high dimensional Gaussian cloud merging or subdivision has yet to be undertaken.

# 4. INCREASING ULTRAMETRICITY THROUGH DATA RECODING

## 4.1. Ultrametricity of Time Series

In Murtagh [20] we use the following coding to show that chaotic time series are less ultrametric than, say, financial, biomedical or meteorological time series; random generated (uniformly distributed) time series data are remarkably similar in their ultrametric properties; and ultrametricity can be used to distinguish various types of biomedical (EEG) signals.

A time series can be easily embedded in a space of dimensionality $m$, by taking successive intervals of length $m$, or a delay embedding of order $m$. Thus we define points

$$\mathbf{x}_r = (x_{r-m+1}, x_{r-m+2}, \ldots, x_{r-1}, x_r)^t \in \mathbb{R}^m$$

where $t$ denotes vector transpose. Based on previous results we expect that as the dimension $m$ grows, then the point set in $\mathbb{R}^m$ becomes more ultrametric. This finding is borne out below.

Given any $\mathbf{x}_r = (x_{r-m+1}, x_{r-m+2}, \ldots, x_{r-1}, x_r)^t \in \mathbb{R}^m$, let us consider the set of $s$ such contiguous intervals determined from the time series of overall size $n$. For convenience we will take $s = \lfloor n/m \rfloor$ where $\lfloor . \rfloor$ is integer truncation. The contiguous intervals could be overlapping but for exhaustive or near-exhaustive coverage it is acceptable that they be non-overlapping. In our work, the intervals were non-overlapping. The quantification of the ultrametricity of the overall time series is provided by the aggregate over $s$ time intervals of the ultrametricity of each $\mathbf{x}_r$, $1 \leq r \leq s$.

We seek to directly quantify the extent of ultrametricity in time series data. In [25, 19] it was shown how increase in ambient spatial dimensionality leads to greater ultrametricity. However it is not satisfactory from a practical point of view to simply increase the embedding dimensionality $m$ insofar as short memory relationships are of greater practical relevance (especially for prediction). The greatest possible value of $m > 1$ is the total length of the time series, $n$. Instead we will look for an ultrametricity measurement approach for given and limited sized dimensionalities $m$. Our experimental results for real and for random data sets are for "window" lengths $m = 5, 10, \ldots, 105, 110$.

We seek local ultrametricity, i.e. hierarchical structure, by studying the following: Euclidean distance squared, $d_{jj'} = (x_{rj} - x_{rj'})^2$ for all $1 \le j, j' \le m$ in each time window, $\mathbf{x}_r$.

We enforce sparseness [24, 25, 19] on our given distance values, $\{d_{jj'}\}$. We do this by linearly approximating each value $d_{jj'}$, in the range $\max_{jj'} d_{jj'} - \min_{jj'} d_{jj'}$, by an integer in $1, 2, \ldots p$. Note that the range is chosen with reference to the currently considered time series window, $1 \le j, j' \le m$. Note too that the value of $p$ must be specified. In our work we set $p = 2$. Thus far, the recoded value, $d'_{jj'}$ is not necessarily a distance. With the extra requirement that $d'_{jj'} \longrightarrow 0$ whenever $j = j'$ it can be shown that $d'_{jj'}$ is a metric.

To summarize, in our coding, a small pairwise transition is mapped onto a value of 1; and an exceptionally large pairwise transition is mapped onto a value of 2. A pairwise transition is defined not just for data values that are successive in time but for any pair of data values in the window considered.

## 4.2. Ultrametricity of Text

In [21], (in principle all, but in practice a set of the few hundred most frequent) words appearing in a text are used to "fingerprint" it. Rare words in a text corpus may be appropriate for querying the corpus for relevant texts, but such words are of little help for inter-text characterization and comparison. We also use entire words, with no stemming or other preprocessing. A full justification for such an approach to textual data analysis can be found in Murtagh [22].

So our methodology for studying a set of texts is to characterize each text with numbers of terms appearing in the text, for a set of terms. The $\chi^2$ distance is an appropriate weighted Euclidean distance for use with such data [3, 21]. Consider texts $i$ and $i'$ crossed by words $j$. Let $k_{ij}$ be the number of occurrences of word $j$ in text $i$. Then, omitting a constant, the $\chi^2$ distance between texts $i$ and $i'$ is given by $\sum_j 1/k_j (k_{ij}/k_i - k_{i'j}/k_{i'})^2$. The weighting term is $1/k_j$. The weighted Euclidean distance is between the *profile* of text $i$, viz. $k_{ij}/k_i$ for all $j$, and the analogous *profile* of text $i'$.

Correspondence analysis allows us to project the space of documents (we could equally well explore the terms in the *same* projected space) into a Euclidean space. It maps the all pairs $\chi^2$ distance into the corresponding Euclidean distance. In the resulting factor space, we use our triangle-based approach for quantifying how ultrametric the data are.

We did this for a large number of texts (novels – Jane Austen, James Joyce, technical reports – airline accident reports, fairy tales – Brothers Grimm, dream reports, Aristotle's *Categories*, etc.), finding consistent degree of ultrametricity results over texts of the same sort.

Some very intriguing ultrametricity characterizations were found in our work. For example, we found that the technical vocabulary of air accidents did not differ greatly in terms of inherent ultrametricity compared to the Brothers Grimm fairy tales. Secondly we found that novelist Austen's works were distinguishable from the Grimm fairy tales. Thirdly we found dream reports to be have higher ultrametricity level than the other text collections.

## 4.3. Data Recoding in the Correspondence Analysis Tradition

If the $\chi^2$ distance (see above, section 4.2) is used on data tables with constant marginal sums then it becomes a weighted Euclidean distance. This is important for us, because it means that we can directly influence the analysis by equi-weighting, say, the table rows in the following way: we double the row vector values by including an absence (0 value) whenever there is a presence (1 value) and vice versa. Or for a table of percentages, we take both the original value $x$ and $100 - x$. In the correspondence analysis tradition [3, 20] this is known as *doubling* (*dédoublement*).

More generally, booleanizing, or making qualitative, data in this way, for a varying (value-dependent) number of target value categories (or modalities) leads to the form of coding known as *complete disjunctive form*.

Such coding increases the embedding dimension, and data sparseness, and thus may encourage degree of ultrametricity. That it can do more we will now show.

The iris data has been very widely used as a toy data set since Fisher used it in 1936 ([10], taking from a 1935 article by Anderson) to exemplify discriminant analysis. It consists of 150 iris flowers, each characterized by 4 petal and sepal, width and breadth, measurements. On the one hand, therefore, we have the 150 irises in $\mathbb{R}^4$. Next, each variable value was recoded to be a rank (all ranks of a given variable considered) and the rank was boolean-coded (viz., for the top rank variable value, 1000..., for the second rank variable value, 0100..., etc.). Following removal of zero total columns, the second data sets defined the 150 irises in $\mathbb{R}^{123}$. Actually, this definition of the 150 irises is in fact in $\{0,1\}^{123}$.

Our triangle-based measure of the degree of ultrametricity in a data set (here the set of irises), with 0 = no ultrametricity, and 1 = every triangle an ultrametric-respecting one, gave the following: for irises in $\mathbb{R}^4$, 0.017; and for irises in $\{0,1\}^{123}$: 0.948.

This provides a nice illustration of how recoding can dramatically change the picture provided by one's data. Furthermore it provides justification for data recoding if the ultrametricity can be instrumentalized by us in some way.

It is with such instrumentalization of local ultrametricity that we are now working, applied to chemical structure databases. The clustering of chemical compounds, based on chemical descriptors or representations, is important in the pharmaceutical and chemical sectors. It is used for screening and knowledge discovery in large databases of chemical compounds. A chemical compound is encoded as a bit string (i.e. a set of boolean or 0/1 values). We have started to look at a set of 1.2 million chemical compounds, each characterized (in a given descriptor or coding system) by 1052 variables. Through (i) normalizing to take variable colum sums into account, and then (ii) using the Euclidean distance, we find the following: (i) many resultant distances are zero, pointing to membership in the same cluster; and (ii) of non-zero distances, triangles that are isosceles with small base predominate. In other words, we have a simple but most effective way to embed our data in an ultrametric space.

## 5. ULTRAMETRICITY IN OBSERVATIONAL COSMOLOGY

In [1], the authors suggest that for large scale galactic structures the ancestor distance – tree distance and hence ultrametric – is important in addition to the distance measured by traveling light waves. In this section we will discuss how this can be instrumentalized.

Clearly one way to construct a hierarchical classification of galaxies is to have observed physical descriptors for each galaxy (related to morphology, motion, photometry, etc.). In such a parameter space it is straightforward to carry out a hierarchical clustering of the galaxies (see some references in [18]). This gives us a hierarchy from which we can read off an ultrametric topology on our galaxy set. But to infer ancestor relations is quite another matter, and such an approach is not suitable for such an objective.

In order to more directly derive hierarchical relationships from positional data, we took data on just over 118,000 galaxies from the Sloan Digital Sky Survey (SDSS), using right ascension (RA), declination (dec) and photometric redshift. We have an immediate issue to address in regard to data encoding: the redshift is far less precisely known, it is substantively different, and is error-prone. Our processing procedure started with binning the data into a regular 3-dimensional grid of dimensions $50 \times 50 \times 50$ (for separate but in each case contiguous parts of the sky that we analyzed). Then we used Lerman's H-classifiability ultrametricity measure, rather than our own, in view of the possible benefit of rank order information in this case. For three parts of the sky, using respectively 589, 554 and 715 galaxies, we found H-classifiability values of 0.15, 0.12, 0.15. For comparison, corresponding (statistically) uniformly distributed positions and redshifts were also used, and processed in the same way. The H-classifiability values (0 = fully ultrametric, 1 = non-ultrametric) were, respectively: 0.19, 0.18 and 0.17. At least we see that we find greater ultrametricity than random values.

Our methodology has not done justice to the data although it must be admitted that data analysis that involves redshifts (in particular) is problematic. Perhaps a better way to explore observational data on galaxies is being pursed by Fraix-Burnet [11, 12, 13]. This work takes observational data (or simulated data, to validate the methodology) on galaxy morphologies. Then a taxonomical approach is pursued where presence or absence of features are processed in order to build a graph of relationships. This procedure yields a robust hierarchy. The stated aim of the work, "to evaluate astrocladistics in reconstructing phylogenies of galaxies", in order "to formalize the concept of galaxy formation and to identify the processes of diversification", has led (in this early stage of the work) to strong evidence for hierarchical branching processes.

## 6. CONCLUSIONS

It has been our aim in this work to link observed data with an ultrametric topology for such data. The traditional approach in data analysis, of course, is to impose structure on the data. This is done, for example, by using some agglomerative hierarchical clustering algorithm. We can always do this (modulo distance or other ties in the data). Then we can assess the degree of fit of such a (tree or other) structure to our data.

For our purposes, here, this is unsatisfactory.

Firstly, our aim was to show that ultrametricity can be naturally present in our data, globally or locally. We did not want any "measuring tool" such as an agglomerative hierarchical clustering algorithm to overly influence this finding. (Unfortunately [25] suffers from precisely this unhelpful influence of the "measuring tool" of the subdominant ultrametric.)

Secondly, let us assume that we did use hierarchical clustering, and then based our discussion around the goodness of fit. This again is a traditional approach used in data analysis, and in statistical data modeling. But such a discussion would have been unnecessary and futile. For, after all, if we have ultrametric properties in our data then many of the widely used hierarchical clustering algorithms will give precisely the same outcome, and furthermore the fit is by definition exact.

In linking data with an ultrametric view of it we have, in this article, proceeded a little in the direction of exploiting this achievement. Processing tools available to us include, for example, the ultrametric wavelet transform studied by S.V. Kozyrev, A.Yu. Khrennikov and others (including implementation and case studies for filtering of tabular data by ourselves). The full unleashing of the potential of such data handling and processing tools still remains to be fully investigated.

While some applications, like discrimination between time series signals, or texts, have been covered here, other applications like chemical compound database search and discovery, and analysis of large scale cosmological structures, have just been opened up. In the distance there looms the challenge of analysis of networks of enormous size (internet, or biological). There is a great deal of work to be accomplished.

## REFERENCES

1. M.V. Altaisky and B.G. Sidharth, p-Adic physics below and above Planck scales, *Chaos, Solitons and Fractals* **10**, 167–176 (1999).
2. R. Bellman, *Adaptive Control Processes: A Guided Tour*, Princeton University Press, 1961.
3. J.P. Benzécri, *La Taxinomie*, 2nd ed., Paris, Dunod, 1979.

4. D. Bustos, G. Navarro and E. Chávez, Pivot selection techniques for proximity searching in metric spaces, *Pattern Recognition Letters* **24**, 2357–2366 (2003).
5. F. Cailliez and J.P. Pagès, *Introduction à l'Analyse de Données*, SMASH (Société de Mathématiques Appliquées et de Sciences Humaines), Paris, 1976.
6. F. Cailliez, The analytical solution of the additive constant problem, *Psychometrika* **48**, 305–308 (1983).
7. E. Chávez and G. Navarro, Measuring the dimensionality of general metric spaces, Technical Report TR/DCC-00-1, Department of Computer Science, University of Chile, 2000.
8. E. Chávez, G. Navarro, R. Baeza-Yates and J.L. Marroquín, Proximity searching in metric spaces, *ACM Computing Surveys* **33**, 273–321 (2001).
9. E. Chávez and G. Navarro, Probabilistic proximity search: fighting the curse of dimensionality in metric spaces, *Information Processing Letters* **85**, 39–56 (2003).
10. R.A. Fisher, The use of multiple measurements in taxonomic problems, *The Annals of Eugenics* **7**, 179–188 (1936).
11. D. Fraix-Burnet, Astrocladistics web page, `www-laog.obs.ujf-grenoble.fr/~fraix/astroclad.htm`
12. D. Fraix-Burnet, P. Choler, E. Douzery and A. Verhamme, Astrocladistics: a phylogenetic analysis of galaxy evolution. I. Character evolutions and galaxy histories, *Journal of Classification*, submitted, 2005.
13. D. Fraix-Burnet, E. Douzery, P. Choler and A. Verhamme, Astrocladistics: a phylogenetic analysis of galaxy evolution. II. Formation and diversification of galaxies, *Journal of Classification*, submitted, 2005.
14. C. Fraley and A.E. Raftery, How many clusters? Which clustering method? Answers via model-based cluster analysis, *The Computer Journal* **41**, 578–588 (1998).
15. P. Hall, J.S. Marron and A. Neeman, Geometric representation of high dimension low sample size data, *Journal of the Royal Statistical Society B* **67**, 427–444 (2005).
16. I.C. Lerman, *Classification et Analyse Ordinale des Données*, Paris, Dunod, 1981.
17. F. Murtagh, *Multidimensional Clustering Algorithms*, Physica-Verlag, 1985.
18. F. Murtagh and A. Heck, *Multivariate Data Analysis*, Kluwer, 1987.
19. F. Murtagh, On ultrametricity, data coding, and computation, *Journal of Classification* **21**, 167–184 (2004).
20. F. Murtagh, Identifying the ultrametricity of time series, *European Physical Journal B* **43**, 573–579 (2005).
21. F. Murtagh, A note on local ultrametricity in text, *Literary and Linguistic Computing*, submitted, 2005.
22. F. Murtagh, *Correspondence Analysis and Data Coding with R and Java*, Chapman & Hall/CRC, 2005.
23. E. Neuwirth and L. Reisinger, Dissimilarity and distance coefficients in automation-supported thesauri, *Information Systems* **7**, 47–52 (1982).
24. R. Rammal, J.C. Angles d'Auriac, and B. Doucot, On the degree of ultrametricity, *Le Journal de Physique – Lettres* **46**, L945–L952 (1985).
25. R. Rammal, G. Toulouse, and M.A. Virasoro, Ultrametricity for physicists, *Reviews of Modern Physics* **58**, 765–788 (1986).
26. W.S. Torgerson, *Theory and Methods of Scaling*, Wiley, 1958.
27. A. Treves, On the perceptual structure of face space, *BioSystems* **40**, 189–196 (1997).
28. C.J. van Rijsbergen, *Information Retrieval*, 2nd ed. Butterworths, 1979.

# The Arithmetic of Discretized Rotations

Franco Vivaldi

*School of Mathematical Sciences, Queen Mary, University of London,
London E1 4NS, UK
email:* f.vivaldi@qmul.ac.uk

**Abstract.** We consider the problem of planar rotation by an irrational angle, where the space is discretized to a lattice by means of a round-off procedure which preserves invertibility. For a dense set of values of the rotational angle, this mapping admits an embedding into a dynamical system which is expanding with respect to a non-archimedean metric, and which has a complete symbolic dynamics. We consider the arithmetical phenomena that arise in such systems, and their relation to the question of pseudo-randomness in discrete dynamics. The exposition is organized around the concept of *minimal modules,* the lattices of minimal complexity which support periodic orbits.

**Keywords:** Round-off errors, discretized rotations, *p*-adic numbers.
**MSC 2000:** 65P20, 37D20, 11S99.

## 1. INTRODUCTION

Dynamical systems with discrete phase space present a specific set of challenges. How does one choose a topology? How does one characterize predictable vs. unpredictable motions? Approaching these questions involves looking at asymptotics, whereby some natural system parameter (the size, typically) becomes large. However, on a discrete phase space, the study of orbit asymptotics is often difficult, sometimes intractable.

In this paper we explore some of these issues in the analysis of the lattice map

$$\Phi : \mathbb{Z}^2 \mapsto \mathbb{Z}^2 \qquad (x,y) \mapsto (\lfloor \alpha x \rfloor - y, x) \qquad \alpha = 2\cos(2\pi\theta) \qquad (1)$$

where $\lfloor \cdot \rfloor$ is the floor function —the largest integer not exceeding its argument. One verifies that $\Phi$ is invertible. Without the floor function, equation (1) represents a one-parameter family of linear maps of the plane, conjugate to rotation by the angle $\theta$. The floor function models the effect of round-off, pushing the point $(\alpha x - y, x)$ to the nearest lattice point on the left.[1] The lattice map should be regarded as a discrete approximation of the planar map: the discretization length is fixed, and the limit of vanishing discretization corresponds to motions at infinity.

The family of maps (1) displays a vast range of mathematical phenomena, which have been object of intense investigations [1, 2, 3, 4, 5, 6, 7, 8]. Here we are interested in a

---
[1] This procedure is arithmetically nicer than rounding to the nearest lattice point, while producing analogous dynamical phenomena.

special set of *rational* values of the parameter $\alpha$, namely

$$\alpha = \frac{q}{p} \qquad p \text{ prime} \qquad |q| < 2p. \tag{2}$$

It turns out that, for this choice of parameters, the map $\Phi$ admits an embedding into an expanding map of the $p$-adic integers $\mathbb{Z}_p$, which features a complete symbolic dynamics on $p$ symbols. The set $\mathbb{Z}_p$ consists of all expressions of the type

$$\chi = \sum_{k=m}^{\infty} c_k p^k \qquad c_k \in \{0, \ldots, p-1\}, \quad m \geq 0, \quad c_m \neq 0 \tag{3}$$

which converge with respect to the $p$-adic absolute value

$$|\chi|_p = \frac{1}{p^m}.$$

The function $|\cdot|_p$ assumes discrete values, and satisfies the ultrametric inequality

$$|\chi + \chi'|_p \leq \max(|\chi|_p, |\chi'|_p). \tag{4}$$

With the topology induced by $|\cdot|_p$, the set $\mathbb{Z}_p$ is Cantor set.

An expanding map with a complete symbolic dynamics is a very favorable dynamical scenario; it calls to mind the dynamical system given by multiplication by a prime $p$ on the circle

$$\Psi : \quad x \mapsto px \,(\mathrm{mod}\, 1), \tag{5}$$

also an expanding map with a complete symbolic dynamics on $p$ symbols [9, section 1.7]. This is a much-studied toy model of ergodic theory, which has direct connections with deep arithmetical phenomena. See [10], for a friendly introduction to the question of algorithmic complexity in this dynamical system.

By comparing and contrasting the behaviour of the maps $\Phi$ and $\Psi$, we will put into perspective the dynamical and arithmetical nature of round-off fluctuations. The exposition will be organized around the concept of *minimal modules,* which are lattices of minimal complexity that support periodic orbits. These are relevant to the study of pseudo-randomness and complexity in discrete dynamical systems. (For a parallel with quantum dynamics, see [11].) In both maps, the positive metric entropy of the embedding dynamical system provides an essential substrate for the existence of irregular motion on a lattice. At the same time, the presence of minimal modules guarantees —in ways which are still not fully understood— good properties of pseudo-randomness. The combined effect of these two ingredients is of great theoretical interest, and also of value in applications.

This paper is a new synthesis of research published elsewhere [12, 5, 6, 7, 13]. I assume some basic knowledge of integer and $p$-adic arithmetic: for background bibliography, see [14] and [15].

## 2. PERIODIC ORBITS AND MINIMAL MODULES

If we represent the numbers in the unit interval in base $p$

$$x = c_0 p^{-1} + c_1 p^{-2} + c_2 p^{-3} + \cdots \qquad c_k \in \{0, \ldots, p-1\} \qquad (6)$$

we find that the action of $\Psi$ corresponds to the left digit shift

$$\Psi(x) = c_1 p^{-1} + c_2 p^{-2} + c_3 p^{-2} + \cdots.$$

The above dynamics is conjugate almost everywhere to the Bernoulli shift $\sigma$ on the space of semi-infinite $p$-symbol sequences. This space, equipped with the usual topology, is a Cantor set. The conjugacy between $\Psi$ and $\sigma$ fails on the rationals whose denominator is a power of $p$, because these have two distinct representations as digit sequences. Disconnecting the interval at this dense set of points, we obtain a Cantor set.

The periodic orbits of $\Psi$ comprise the set of rationals with denominator coprime to $p$, which is dense on the circle. The restriction of the map to such set is invertible. Because in equation (6) we may choose the coefficients $c_k$ arbitrarily, the symbolic dynamics is complete. From this one infers that orbits of all periods exist, and that their number grows exponentially with the period. Specifically, given an arbitrary periodic code $(\overline{c_0, c_1, \ldots, c_{t-1}})$ with minimal period $t$, which represents a $t$-cycle, one finds

$$x = \sum_{k=0}^{\infty} c_k p^{-(k+1)} = \frac{1}{p^t - 1} \sum_{k=0}^{t-1} c_k p^{t-k-1}. \qquad (7)$$

As an indicator of complexity, we define the *height* $h$ of a reduced rational number $a/b$, as $h(a/b) = \max(|a|, |b|)$. In general, the height of a periodic point grows exponentially with the period; however, for some codes, the numerator and denominators may have a large common factor, resulting in periodic points of small height. Thus, for every positive integer $t$, we consider the smallest positive integer $m$ such that the map $\Psi$ has a $t$-periodic point with denominator $m$. We denote such $m$ by $M(t)$.

The $t$-periodic points with denominator $M(t)$ belong to the $\mathbb{Z}$-module[2]

$$\Lambda(t) = M(t)^{-1} \mathbb{Z}/\mathbb{Z}$$

which consists of $M(t)$ equally spaced points on the unit circle. We call such module the *minimal module* for the period $t$; the $t$-cycles on $\Lambda(t)$ are those of minimal height.

**Proposition 1.** *The function $t \mapsto \Lambda(t)$ is injective.*

This is true, because all points on $\Lambda(t)$ whose numerator is coprime to $M(t)$ have the same period $t$, from elementary properties of congruences. For all other points, cancelation takes place, and hence $\Lambda(t)$ cannot be a minimal module for their period.

---

[2] An additive group, closed under multiplication by integers.

The quantity $M(t)$ features wild fluctuations. For instance, for $p = 2$

$$\begin{array}{c|cccc} t & 17 & 18 & 19 & 20 \\ \hline M(t) & 131071 & 19 & 524287 & 25 \end{array} \qquad (8)$$

For general $p$, we have the bounds

$$M(1) = 1; \qquad t+1 \leq M(t) \leq p^t - 1 \qquad t > 1. \qquad (9)$$

The upper bound follows from (7), while the lower bound derives from the fact that the minimal module $\Lambda(t)$ always includes the fixed point at zero, and therefore cannot have fewer than $t + 1$ points. The data in (8) show that these bounds are sharp.

The $t$-periodic points with denominator $M(t)$ are the rationals $0 < a/M(t) < 1$ with $a$ coprime to $M(t)$. There are $\phi(M(t))$ of them, where $\phi$ is Euler's function [14, chapter XVI]; hence the minimal module contains $\phi(M(t))/t$ cycles of minimal period $t$. Roughly speaking, the smaller the number of cycles on the module, the smaller their height, that is, the amount of information needed to specify them. The extreme situation corresponds to the lower bound in (9): this is attained precisely when $M = t + 1$ is a prime number, and $p$ is a primitive root modulo $M$, namely a generator of the multiplicative group of the ring $\mathbb{Z}/M\mathbb{Z}$. At these values of $t$, the $t$-cycle consists of $t$ equally spaced points, namely the whole lattice $\Lambda(t)$, excluding the origin. This is an extreme form of spatial uniformity, exemplified by the value $t = 18$ in (8).

To analyze uniform distribution we proceed as follows. We choose, for definiteness, the point $M(t)^{-1}$ as the initial condition for a representative $t$-cycle on the minimal module; then we consider the Weil sum

$$W(t) = \frac{1}{t} \sum_{k=0}^{t-1} e^{2\pi i x_k} \qquad x_k = \Psi^k(M(t)^{-1}) \qquad (10)$$

which represents the barycenter of the orbit. The value of $W(t)$ lies within the closed unit circle. Let now $T$ be an infinite set of positive integers. Then, according to Weil's criterion [16], the sequence of $t$-cycles with $t \in T$ is uniformly distributed iff

$$\lim_{\substack{t \to \infty \\ t \in T}} W(t) = 0. \qquad (11)$$

If on the minimal module there is a single $t$-cycle, then the Weil's sum becomes a Ramanujan's sum [14, section 16.6], which can be evaluated explicitly

$$W(t) = \frac{1}{t} \sum_{\substack{0 \leq k < M \\ \gcd(k,M)=1}} \exp\left(2\pi i \frac{k}{M}\right) = \frac{1}{t} \mu(M) \qquad t = \phi(M(t)) \qquad (12)$$

where $\mu$ is the Möbius function [14, chapter XVI]. Because $|\mu(x)| \leq 1$, we get a strong form of uniform distribution (e.g., the values $t = 18, 20$ in (8)). Uniform distribution in the sense (11) holds for a much larger set of orbits than the one described above, namely orbits for which $M(t)/t$ is small enough —see [17, 18] for details and generalizations.

So, in equation (11), we are led to consider the set $T = T^*$ of periods $t$ at which the function $M(t)$ attains the lower bound in (9). We have seen that these are precisely the integers $t$ for which $r = t+1$ is prime, and $p$ is a primitive root modulo $r$. The study of such primes $r$ is a classic arithmetical problem, known under the heading of Artin's conjecture [19]. If one assumes the generalized Riemann hypothesis[3] (GRH), then it is possible to show that the set $T^*$ is not only infinite, but also has positive density among the primes (see below); this density is given by the so-called *Artin's constant* [20, page 304]

$$A = 0.373955813619202\ldots.$$

Combining the above with the prime number theorem [21], we have that the number of periods $t < N$ for which the minimal module $\Lambda(t)$ achieves the lower bound in (9) admits the asymptotic estimate

$$\#\{t : t \in T^*, t < N\} \sim A \frac{N}{\log(N)}. \tag{13}$$

Without GRH, we do not even know if the set $T^*$ is infinite. Defining the *density* $D(X)$ of a set $X \subset \mathbb{N}$ as

$$D(X) = \lim_{N \to \infty} \frac{1}{N} \#\{n < N : n \in X\} \tag{14}$$

we see that $D(T^*) = 0$. The notion of density can be defined in an obvious way on sets other than $\mathbb{N}$, such as the primes, $\mathbb{Z}, \mathbb{Z}^2$, etc.

With a bit of extra work, one could also arrive to a conjectured asymptotic form for the set of periods $t$ satisfying (12), which involves considering the moduli $m$ having cyclic multiplicative group. We shall not pursue this matter here.

Minimal modules are difficult to compute, in the sense that they lead to non-polynomial time algorithms. A direct approach consists of constructing the ascending sequence $d_i(t)$ of the divisor of $p^t - 1$ greater than $t$, and then determining the multiplicative order of $p$ modulo each of them. The smallest $d_i(t)$ for which $p$ has order $t$ is $M(t)$. This procedure requires factoring $p^t - 1$, which for large $t$ is plainly unfeasible. Alternatively, one may construct the elements of the ascending sequence of the integers $m_i(t)$ such that $t | \phi(m_i(t))$, which lie in the range $t + 1 \leq m_i \leq p^t - 1$. For each $m_i$, we check whether or not the order of $p$ modulo $m_i$ is $t$: the smallest $m_i$ for which this is true is $M(t)$. Constructing the $m$-sequence requires evaluating $\phi(kt)^{-1}$ for $k = 1, 2, \ldots$ (or $k = 2, 4, \ldots$, if $t$ is odd); the difficulty originates from the need of knowing the prime factorization of $t$, which is a non-polynomial time problem [22].

We conclude this section with some brief remarks on related problems.

The constructions described above can be reformulated in higher dimensions, for hyperbolic toral automorphisms [23, 24]. The rationals $\mathbb{Q}$ on the circle are replaced by algebraic numbers $\mathbb{Q}(\lambda)$ on the torus, where $\lambda$ is an eigenvalue of the automorphism.

---

[3] The analogue of the celebrated conjecture of Riemann, for the zeta function of an algebraic number field.

The multiplicative constant $p$ in (5) is replaced by $\lambda$; the periodic orbits with prime denominator —out of which the set $T^*$ was constructed— are replaced by the so-called *ideal orbits,* which belong to prime ideals in the ring $\mathbb{Z}[\lambda]$ (see below, for explanation of the terminology). The arithmetical properties of $T^*$ are similar, involving again Artin's constant.

The upper bound in (9) is also of interest. It is attained precisely when $p = 2$, $t$ is prime, and $2^t - 1$ is also prime —a so-called *Mersenne prime.* The orbit with initial condition $M(t)^{-1}$ now has the same symbolic sequence as a rotation: these are the *sturmian sequences,* of minimal complexity [25]. The spatial distribution of these orbits is, in a sense, as far as possible from being uniform. The problem of the infinitude of Mersenne primes —the largest know primes— is also unsolved [20].

Finally, the map $\Psi$, which is defined over the circle, may also be represented over the $p$-adics. It is a contraction on $\mathbb{Q}_p$ (the field of fractions of $\mathbb{Z}_p$, obtained by removing the constraint $m \geq 0$ in (3)), and an isometry on $\mathbb{Q}_r$, for all primes $r \neq p$. In the former case, the dynamics is trivial, since every point is attracted to the origin at a constant rate. In the latter case, the dynamics is an 'irrational rotation'. This terminology is justified as follows. For $r \neq p$, we have $|p|_r = 1$, namely $p$ lies on the unit circle in $\mathbb{Q}_r$. Equivalently, $p$ is a *unit* in $\mathbb{Z}_r$.[4] Furthermore, $p$ is not a root of unity, because there is no positive integer $k$ for which $p^k = 1$; as a result, the only periodic point in $\mathbb{Q}_r$ is the fixed point at the origin. Thus the space foliates into the union of invariant circles, and each circle in turn decomposes into a finite number of uniquely ergodic components, the same number of components for each circle [12]. This dynamical system has zero metric entropy; some aspects of computational complexity emerging in this context are discussed in [13]. Here we just mention that, in the limit of large $k$, the period of orbits of $\Psi$ with denominator $r^k$ is computable in polynomial time, thanks to the analytical properties of the $p$-adic logarithmic function. For this reason, these discrete motions should be regarded as being regular, predictable.

## 3. DISCRETIZED ROTATIONS

The current state of affairs regarding the round-off map $\Phi$ defined in (1) is summarized by the following

**Conjecture 1.** *All orbits of $\Phi$ are periodic.*

This conjecture has been proved only for finitely many parameter values, corresponding to the rotation number $\theta$ being rational with denominator $5, 8, 10, 12$ (eight cases in all) [8]. For these values, $\alpha$ is a quadratic irrational, and one can exploit the presence of exact scaling to develop computer-assisted proofs. Such proofs apply to a set of full density of initial conditions; the more tedious (albeit conceptually similar) proof for all initial conditions was carried out only in one case: $\theta = 3/10$. Although unbounded orbits have

---

[4] A unit $\eta$ in the ring $\mathbb{Z}_r$ is a number such that $\eta^{-1} \in \mathbb{Z}_r$.

not been found, the possibility that boundedness could fail for a zero-density set should not be discounted. In [8], an unbounded orbit was constructed for a map obtained from (1) via a simple modification of the rounding procedure.

The periodicity conjecture was recently formulated for a system closely related to (1), with the ceiling instead of the floor function [26]. This dynamical system originated in number theory, in connection with the so-called shift radix systems, and the theory of Pisot and Salem numbers. A boundedness proof for the parameter $\theta = 1/5$ is given in [27].

In the present case —cf. equation (2)— the parameter $\alpha$ is rational; from (1) it then follows that $\theta$ is irrational, apart from finitely many exceptions [28, chapter 2]. Proving the periodicity conjecture for irrational rotational angles seems quite difficult.

When $\alpha = q/p$ in (1), we define a symbolic dynamics on $p$ symbols as follows

$$c : \mathbb{Z}^2 \to \{0, \ldots, p-1\} \qquad (x,y) \mapsto x \,(\mathrm{mod}\, p). \tag{15}$$

The next three theorems were proved in [5] (in a slightly more general setting). The first result characterizes the cylinder sets of the symbolic dynamics, namely the set of lattice points whose symbol sequence begins with a specified code.

**Theorem 1.** *There exists a nested sequence of lattices*

$$L_1 \supset L_2 \supset L_3 \supset \cdots$$

*with $|\mathbb{Z}^2/L_k| = p^k$, with the property that two points in $\mathbb{Z}^2$ have the same $k$-code if and only if they are congruent modulo $L_k$.*

It follows that *all* finite codes of the symbolic dynamics (15) are represented by orbits in $\mathbb{Z}^2$; furthermore, any $k$-cylinder set in $\mathbb{Z}^2$ has density $p^{-k}$ —see equation (14), and the following remark.

Consider now the identity

$$f(x) = x^2 - qx + p^2 \equiv x(x-q) \,(\mathrm{mod}\, p). \tag{16}$$

The polynomial $f(x)$ is irreducible in $\mathbb{Q}$, having negative discriminant. Moreover, the factors of $f(x)$ modulo $p$ are *distinct,* because $p$ and $q$ are coprime. Let $\lambda$ be a root of $f(x)$, and consider the ring

$$\mathbb{Z}[\lambda] = \{m + n\lambda \,:\, m, n \in \mathbb{Z}\}.$$

The factorization (16) implies that the prime $p$ splits in $\mathbb{Z}[\lambda]$ into the product of two distinct prime ideals: $p\mathbb{Z}[\lambda] = P\overline{P}$ (see, e.g., [28, chapter 3]). Recall that an ideal in a ring is an additive group closed under multiplication by ring elements. Geometrically, these ideals are two-dimensional lattices, and the product $P\overline{P}$ alluded above is defined as the set of all finite sums $\pi_1 \overline{\pi}_1 + \cdots + \pi_s \overline{\pi}_s$ with $\pi_i \in P$ and $\overline{\pi}_i \in \overline{P}$.

The next result shows that the discrete phase space $\mathbb{Z}^2$, as well as the lattices $L_k$, are in fact more than two-dimensional lattices

**Theorem 2.** *The embedding*

$$\mathscr{L}_1 : \mathbb{Z}^2 \to \mathbb{Z}[\lambda] \qquad (x,y) \mapsto px - \lambda y$$

*defines an isomorphism $\mathbb{Z}^2 \sim P$ of $\mathbb{Z}$-modules such that, for all $k \geq 1$, we have $L_k \sim P^{k+1}$.*

This result provides the phase space with a multiplicative structure. Now, for all positive $k$, the finite ring $\mathbb{Z}[\lambda]/P^k$ is isomorphic to $\mathbb{Z}/p^k\mathbb{Z}$. Under this isomorphism, an element $\zeta$ in $\mathbb{Z}[\lambda]$ maps to a unique residue class modulo $p^k$, and the sequence of these residue classes converges to a $p$-adic number. In particular, the roots of $f(x)$ in $\mathbb{C}$ correspond to the roots of $f(x)$ in $\mathbb{Q}_p$. We denote the latter by $\varphi$ and $\overline{\varphi}$; they are identified as follows (cf. (16)):

$$\varphi \equiv 0 \,(\mathrm{mod}\, p) \qquad \overline{\varphi} \equiv q \,(\mathrm{mod}\, p). \qquad (17)$$

Because $q$ is coprime to $p$, we have that $|\overline{\varphi}|_p = 1$, that is, $\overline{\varphi}$ is a $p$-adic unit, while $|\varphi|_p < 1$. The computation of $\varphi$ and $\overline{\varphi}$ is performed efficiently with the $p$-adic Newton's method, which is superconvergent [29, chapter 2]. From equations (17) we see that the initial condition for the Newton recursive sequence $\varphi_k \to \varphi$ is $\varphi_0 = 0$, whereas $\overline{\varphi}_0 = q$.

Identifying $\lambda$ with $\varphi$ defines an embedding $\mathscr{L}_2$ of $\mathbb{Z}[\lambda]$ into the ring $\mathbb{Z}_p$ of $p$-adic integers

$$\mathscr{L}_2 : \mathbb{Z}[\lambda] \to \mathbb{Z}_p \qquad x + \lambda y \mapsto x + \varphi y.$$

Composing $\mathscr{L}_1$ and $\mathscr{L}_2$, and scaling gives us the following

**Theorem 3.** *The dense embedding*

$$\mathscr{L} : \mathbb{Z}^2 \mapsto \mathbb{Z}_p \qquad (x,y) \mapsto \frac{1}{p}\mathscr{L}_2(\mathscr{L}_1(x,y)) = x - \frac{\varphi}{p} y \qquad (18)$$

*has the property that the mapping $\Phi^* = \mathscr{L} \circ \Phi \circ \mathscr{L}^{-1}$ can be extended continuously from $\mathscr{L}(\mathbb{Z}^2)$ to the whole of $\mathbb{Z}_p$, giving*

$$\chi_{t+1} = \Phi^*(\chi_t) = \sigma(\overline{\varphi} \chi_t) \qquad (19)$$

*where $\overline{\varphi} = \mathscr{L}(n,p)$ and $\sigma$ is the shift mapping.*

The shift $\sigma$ is defined like its archimedean counterpart; if $\chi = c_0 + c_1 p + c_2 p^2 + \cdots$, then

$$\sigma(\chi) = c_1 + c_2 p + c_3 p^2 + \cdots. \qquad (20)$$

The $p$-adic shift mapping has been studied in [30, 31, 32].

The above result represents the round-off problem on $\mathbb{Z}^2$ as a sub-system of an expanding map over the $p$-adics, featuring a complete symbolic dynamics over $p$ symbols, and a dense set of unstable periodic orbits. It preserves the standard probability measure on $\mathbb{Z}_p$ (the additive Haar measure), obtained by assigning to each residue class modulo $p^k$ the measure $p^{-k}$. This is just the natural measure on $\mathbb{Z}_p$ as a Cantor set.

This system has a lot in common with the Bernoulli shift on $p$-symbols. For instance, the metric entropy is the same, namely $\log(p)$; the periodic points also have a similar structure, as we shall see below.

# 4. CODING FUNCTION AND PERIODIC ORBITS

We consider the function $\mathscr{C}$ that maps the initial point $\chi \in \mathbb{Z}_p$ of an orbit of $\Phi^*$ into the corresponding symbolic code $(c_0, c_1, \ldots)$ —cf. equation (15). The latter is easily defined, for instance, as the limit of the codes through the points $(x^{(k)}, 0) \in \mathbb{Z}^2$, for a rational sequence $x^{(k)} \to \chi$. If we represent the code as a $p$-adic integer

$$\mathscr{C}(\chi) = \sum_{k=0}^{\infty} c_k p^k,$$

then we have

**Theorem 4.** *The function $\mathscr{C}$ defines an isometric bijection of $\mathbb{Z}_p$, which is nowhere differentiable.*

Of interest is the fact that a similar property holds for the coding function of the $p$-adic version of the so-called $3x+1$ problem [33, theorem 10.4].

This result establishes a one-to-one correspondence between periodic codes and periodic orbits, while the metric-preserving property of $\mathscr{C}$ show that the periodic orbits are dense and uniformly distributed with respect to the Haar measure. Just like for the $t$-cycle of $\Phi$, those of $\Phi^*$ are all unstable, with multiplier $p^{-t}$, where $t$ is the period.

Next we characterize the periodic points arithmetically. We define

$$a_1 = 1; \quad b_1 = 0; \quad \begin{cases} a_{k+1} = qa_k + pb_k \\ b_{k+1} = -pa_k \end{cases} \quad k \geq 1, \tag{21}$$

and

$$U_{t,r} = p^r a_{t-r} + p^{t-r} a_r \quad r \geq 0, \quad t \geq 1. \tag{22}$$

**Theorem 5.** *The periodic point $\chi$ corresponding to the $t$-periodic code $(\overline{c_0, \ldots, c_{t-1}})$ takes the form*

$$\chi = \frac{1}{B(t)} \left( x - \frac{\varphi}{p} y \right) \tag{23}$$

*where $x$ and $y$ are integers given by*

$$x = \sum_{r=0}^{t-1} c_r U_{t,r} \quad y = \sum_{r=0}^{t-1} c_r U_{t,r+1} \tag{24}$$

*while*

$$B(t) = a_t q + 2pb_t - 2p^t \tag{25}$$

*is an integer coprime to $p$.*

We verify directly that the periodic points $\chi$ in equation (23) are $p$-adic integers. Firstly, we note that $|\varphi|_p < 1$, and hence $|\varphi/p|_p \leq 1$, while $|B(t)|_p = 1$. Using the ultrametric inequality (4), we then get $|\chi|_p \leq 1$. Comparison with (3) confirms that $\mathbb{Z}_p$ is precisely the closed $p$-adic unit disc.

Comparing expressions (7) and (23) we see that the structure of the cycles of the maps $\Phi^*$ and $\Psi$ is very similar. The role or the exponential sequence $p^t - 1$ at denominator is here played by the sequence $B(t)$, which grows exponentially at the same rate. The numerators of the periodic points store code information in a similar manner.

The domain of definition of the embedding map $\mathscr{L}$, defined in equation (18), can be extended from $\mathbb{Z}^2$ to the set $\mathbb{U}_p^2$, where $\mathbb{U}_p$ is the set of rationals with denominator coprime to $p$, that is, $\mathbb{U}_p = \mathbb{Q} \cap \mathbb{Z}_p$. The map $\mathscr{L}$ remains invertible on the extended domain, and since $B(t)$ is coprime to $p$, using $\mathscr{L}^{-1}$, all the $p$-adic periodic orbits of $\Phi^*$ can be lifted to the plane.[5] Specifically, the $t$-cycles belong to the $\mathbb{Z}$-module $B(t)^{-1}\mathbb{Z}^2$. Those corresponding to orbits of the round-off mapping $\Phi$ must lie in the sub-module $\mathbb{Z}^2$, which is the case when both $x$ and $y$ are divisible by $B(t)$. Thus, for any positive integer $t$, we consider the smallest integer $m$ such that there exists a $t$-cycle in $\frac{1}{m}\mathbb{Z}^2$. We denote such $m$ by $M(t)$, and —as we did before— we call $\Lambda(t) = M(t)^{-1}\mathbb{Z}^2$ the *minimal module* of the map $\Phi^*$ for the period $t$.

In place of (9) now we have the bounds

$$1 \leq M(t) \leq B(t).$$

As noted above, the sequence $B(t)$ plays the role of the sequence $p^t - 1$ for the shift map. Comparing the respective lower bounds is rather more interesting. The set $T^*$ for the shift map —the set of integers $t$ for which $p$ is a primitive root modulo $t+1$— is here represented by the set of periods $t$ for which the minimal module of $\Phi^*$ is equal to $\mathbb{Z}^2$. In other words, $T^*$ are the periods the orbits of the round-off mapping!

Plainly, the function $t \to \Lambda(t)$ is not injective —cf. proposition 1. The permitted periods $T^*$ are characterized via the multiplicity function $\kappa$, which counts the number of distinct orbits on $\mathbb{Z}^2$ having period $t$. Thus, letting $\tau : \mathbb{Z}^2 \to \mathbb{N} \cup \infty$ be the (possibly infinite) period of the orbit through $z$, we define

$$\kappa : \mathbb{N} \to \mathbb{N} \qquad t \mapsto \frac{\#\{z \in \mathbb{Z}^2 : \tau(z) = t\}}{t}. \qquad (26)$$

Using diophantine approximations, it is not difficult to show that the function $\kappa$ is well-defined —the numerator in (26) is finite— for a set of (irrational) rotation angles $\theta$ having full Lebesgue measure [1]. The periodicity of the round-off map was briefly investigated in [6]; roughly speaking, $T^*$, which is the support of $\kappa$, consists of denominators of good rational approximants of $\theta$. The best approximants —the denominators of the convergents of $\theta$— were found to correspond to the local maxima of the multiplicity function. As for the map $\Psi$, the density of $T^*$ in $\mathbb{N}$ appears to be zero, but with a faster decay rate than (13). However, the deep structure of the function $\kappa$ remains unexplored; this investigation seems very worthwhile, involving probabilistic aspects of diophantine approximations.

---

[5] One could also extend the round-off mapping $\Phi$ to $\mathbb{U}_p^2$, as the conjugate of $\Phi^*$ under $\mathscr{L}^{-1}$ —see [5].

The symbol sequences generated by the round-off orbits are good pseudo-random sequences (a comparison with [32] is instructive). Some rigorous results will be found in [7], most notably a central limit theorem governing the departure of round-off orbits from exact orbits. However, as is often the case in this type of problems, the most interesting properties are difficult to analyze. Experimentally, the period of the sequences is found to grow —on average— proportionally to the height of the initial condition (the 'seed' of the pseudo-random sequence), while displaying at the same time huge fluctuations. The latter are symptoms of difficulties in computing the period function $\tau$, which is a good candidate for a non-polynomial time problem. The symbol sequences corresponding to $\alpha = 1/2$ were tested at Hewlett Packard [34]. They show an optimal degree of pseudo-randomness for short times, and some faint correlations at certain larger times, related to the local maxima of the multiplicity function $\kappa$.

Finally, we examine the question of uniform distribution. Let $z \in \mathbb{Z}^2$, and $z_k = \Phi^k(z)$. Using the code (15), we construct the symbol sequence $(c_0, c_1, \ldots)$, where $c_k = c(z_k)$. Using this sequence in the representation (6), we define a real number $x = x(z)$ in the unit interval. By analogy with (10), we then let

$$W(z) = \frac{1}{\tau} \sum_{k=0}^{\tau-1} e^{2\pi i x_k} \qquad x_k = x(z_k) \tag{27}$$

where $\tau = \tau(z)$ is the period of the orbit through $z$. In accordance with conjecture 1, we assume that $\tau$ is finite. It is clear that the value of the sum (27) depends only on the orbit $\mathcal{O}$ of $z$ and not on the choice of the point $z$ within $\mathcal{O}$; accordingly, we write $W = W(\mathcal{O})$. Let now $(\mathcal{O}_0, \mathcal{O}_1, \ldots)$ be an infinite sequence of distinct $\Phi$-orbits. We say that this sequence is *uniformly distributed* if

$$\lim_{k \to \infty} W(\mathcal{O}_k) = 0$$

which should be compared with (11). We conclude this paper with the following

**Conjecture 2.** *Any sequence of distinct $\Phi$-orbits is uniformly distributed.*

## ACKNOWLEDGMENTS

This work was partially supported by EPSRC grant No GR/S62802/01.

## REFERENCES

1. F. Vivaldi, *Exp. Math.* **3**, 303–315 (1994).
2. J.H. Lowenstein, S. Hatjispyros, and F. Vivaldi, *Chaos* **7**, 49–66 (1997).

3. J.H. Lowenstein, and F. Vivaldi, *Nonlinearity* **11**, 1321–1350 (1998).
4. J.H. Lowenstein, and F. Vivaldi, *Chaos* **10**, 747–755 (2000).
5. D. Bosio, and F. Vivaldi, *Nonlinearity* **13**, 309–322 (2000).
6. D. Bosio, *Round-off errors and p-adic numbers*, Ph.D. thesis, Queen Mary, University of London, 2000.
7. F. Vivaldi, and I. Vladimirov, *Int. J. of Bifurcations and Chaos* **13**, 3373–3393 (2003).
8. K.L. Kouptsov, J.H. Lowenstein, and F. Vivaldi, *Nonlinearity* **15**, 1795–1482 (2002).
9. A. Katok, and B. Hasselblat, *Introduction to the modern theory of dynamical systems*, Cambridge University Press, Cambridge, 1997.
10. J. Ford, *Physics Today* **36**, 40–47 (1983).
11. B. Chirikov, and F.Vivaldi, *Physica D* **129**, 223–235 (1999).
12. D.K. Arrowsmith, and F. Vivaldi, *Physica D* **71**, 222–236 (1994).
13. J. Pettigrew, J.A.G. Roberts, and F. Vivaldi, *Chaos* **11**, 849–857 (2001).
14. G.H. Hardy, and E.M. Wright, *An introduction to the theory of numbers*, Oxford University Press, Oxford, 1979.
15. F.Q. Gouvêa, *p-adic numbers: an introduction*, Springer-Verlag, Berlin, 1993.
16. L. Kuipers, and Niederreiter, H. *Uniform Distribution of Sequences*, Wiley, New York, 1974.
17. J.B. Friedlander, C. Pomerance, and I.E. Shparlinski, *Math. Comp.* **70**, 1591–1605 (2001).
18. J.B. Friedlander, and I.E. Shparlinski, *Math. Comp.* **70**, 1575–1589 (2001).
19. M. RamMurty, *Math. Intelligencer* **10**, 59–67 (1988).
20. P. Ribenboim, *The book of prime number records*, Springer-Verlag, New York, 1988.
21. G. Tenenbaum, and M.M. France, *The prime numbers and their distribution*, AMS, Providence, Rhode Island, 2000.
22. H. Cohen, *A course in computational algebraic number theory*, Springer-Verlag, New York, 1996.
23. M. Bartuccelli, and F. Vivaldi, *Physica D* **39**, 194–204 (1989).
24. M.D. Esposti, and S. Isola, *Nonlinearity* **8**, 827–842 (1995).
25. N.P. Fogg, *Substitutions in Dynamics, Arithmetics and Combinatorics*, Springer-Verlag, Berlin, 2002.
26. S. Akiyama, H. Brunotte, A. Pethö, and J.M. Thuswaldner, Generalized radix representations and dynamical systems ii (2005), to appear in *Acta Arith*.
27. V. Akiyama, H. Brunotte, A. Pethö, and W. Steiner, Remarks on a conjecture on certain integer sequences (2005), preprint, Niigata University.
28. D.A. Marcus, *Number fields*, Springer-Verlag, New York, 1977.
29. D.A. Serre, *A course in arithmetic*, Springer-Verlag, New York, 1973.
30. E. Thiran, D. Verstegen, and J. Weyers, *J. Stat. Phys.* **54**, 893–913 (1989).
31. D.K. Arrowsmith, and F. Vivaldi, *Phys. Lett. A* **176**, 292–294 (1993).
32. F. Woodcock, and N.P. Smart, *Exp. Math.* **7**, 334–342 (1998).
33. G.J. Wirsching, *The dynamical system generated by the $3n + 1$ function*, Springer-Verlag, Berlin, 1998.
34. J. Castejon (2001), private communication.

# $p$-Adic Models of Turbulence

S. Fischenko* and E. Zelenov[†]

*TRIUMPH Ltd. Scientific Enterprise, Pionerskaya St., 8A, 141070, Moscow Region, Kaliningrad, RUSSIA

[†]Steklov Mathematical Institute, Vavilov St., 42, GSP-1, 117966, Moscow, RUSSIA[1]
email: zelenov@mph.mian.su

**Abstract.** It is shown that $p$-adic numbers and ultrametric topology are very useful for describing turbulence. $p$-Adic numbers are proved to give us a natural and systematic approach to cascade models. $p$-Adic scalar model of turbulence is suggested.

## 1. INTRODUCTION

Universal power-law behavior of two-point correlation function of velocity in so-called inertial range is one of the most interesting property of turbulent flow. Such a behavior leads us to conclusion about existence of universal mechanism for the flow self order at high Reynolds numbers [1].

In spite of variety approaches the conclusion concerning multifractal behavior of rate of turbulent energy dissipation seems to be most promising. The multifractal behavior appears explicitly in so-called cascade models. Advanced versions of such models give rather realistic singularity spectrum for multifractal measure connected with rate of turbulent energy dissipation (see, for example [2]), but these models are not dynamical in fact.

Note that if a measure (e.g. connected with turbulent energy dissipation) has non-trivial singularity spectrum, then corresponding characteristics of turbulence (e.g. velocity field) are singular and it hardly probable that they can be given as a solution of a differential or integral equation.

We suggest here one of the possible approaches to this problem using $p$-adic numbers. Note that the idea that $p$-adic numbers and ultrametric topology are rather natural to use in turbulence theory was suggested by several authors [3].

The paper is organized as follows. Sections 2 and 3 are devoted to the notion of eddies space and its connection with real coordinate space. In Section 4 we suggest a $p$-adic approach to multifractal cascade models. In Section 5 we consider properties of $p$-adic dynamical model.

About $p$-adic numbers and $p$-adic analysis see, for example [4].

---

[1] The work was supported by TRIUMPH Ltd. Sci. Ent. Preprint SMI-MPH-01/93.

## 2. EDDIES SPACE

*p*-Adic numbers are contained in fact in Richardson picture for development of turbulence [5]. A developed turbulence (within inertial range) appears to be considered as a collection of pieces of liquid referred to as eddies of different sizes or scales. Usual approach for identifying an eddy is to indicate its scale and coordinate of its arbitrary point (because eddies of the same scale are not intersected).

We suggest here another natural and one-to-one way to identify eddies. Let us consider the cascade process of division of eddies as a graph. Eddies are vertices of the graph. Two vertices are connected by edge if one of the corresponding eddy gives birth to another. With ordinary assumptions the graph in question will be a tree. We slightly simplify the picture assuming that there is only one macroscale eddy (the ancestor of all others eddies). Moreover we assume that each eddy is divided into $p$ eddies, where $p$ is a prime number. The last assumption is not so crucial for us but it is more pleasant to operate with the field of $p$-adic numbers rather than the ring of $m$-adic numbers.

Thus we get a homogeneous tree, every of its vertices excluding one (starting vertex) connected by edge with $p+1$ another vertices. This tree equipped with standard metric $d$ is called *eddies space*. Generally speaking, the process of division of eddies is finite because of viscosity (there exists the so-called Kolmogorov's scale). So in the case of nonzero viscosity the eddies space forms a finite tree, but nevertheless we consider finite trees as a subtree of an infinite tree.

Now we describe the boundary of the eddies space. An infinite path in the tree means an infinite sequence of adjacent vertices $q_0, q_1, \ldots$ without repetitions i.e. $q_{i-1} \neq q_{i+1}$ for all $i$. If $q_0$ denotes the starting vertex of the tree then the set of all these infinite paths forms boundary of the tree by definition.

**Lemma 1.** *There exists one-to-one correspondence between boundary points of the eddies space and the set $Z_p$ of p-adic integers.*

For the *Proof* of the Lemma 1 see for example [6].

Lemma 1 gives us another realization of the eddies space. In fact let us consider the set of all infinite paths in the eddies tree with starting point $q$. It is known that they form a disk in $Z_p$. Moreover, there is one-to-one correspondence between the set of disks in $Z_p$ and vertices of the eddies space, see [7]. Note that the set of $p$-adic integers forms a compact Abelian group and thus there is an invariant Haar measure $\mu$ normalized by the condition $\mu(Z_p) = 1$. By the definition size (or scale) of an eddy means the measure of corresponding disk in $Z_p$.

Thus we get the following: eddies are considered as disks in the set of $p$-adic integers (note that $Z_p$ forms a disk too). Measure of a disk gives the scale of the eddy (we have therefore the discrete set of scales $p^{-n}$, $n = 0, 1, 2, \ldots$). The set of $p$-adic integers itself as a disk forms a unique macro-eddy, but as a set of points $Z_p$ is to be considered as a region of energy dissipation in eddies space at zero viscosity limit.

# 3. EDDIES SPACE AND REAL SPACE

As noted above eddies tree (or eddies space) gives another but one-to-one identification of eddies. As we restrict our consideration to scalar models we suppose that in real coordinate space an eddy occupies a segment included in the segment $[0,1]$ on real numbers line. Consequently, the following two mappings will exist:

$$\mathscr{F} : \text{set of disks in } Z_p \longrightarrow \text{set of segments in } [0,1].$$

$$\mathscr{G} : Z_p \longrightarrow [0,1].$$

Therewith both of the mappings $\mathscr{F}$ and $\mathscr{G}$ are injections. Moreover the last mapping will be continuous, because if two eddies descend from common eddy close to them (i.e. these eddies are close to each other in $p$-adic topology), then they are close in real coordinate space. The mappings in question are not uniquely defined by these properties. We define below a one-parameter class of such mappings.

**Lemma 2.** *The mapping $\mathscr{G}_\alpha : Z_p \longrightarrow [0,1]$ defined by the formula*

$$\mathscr{G}_\alpha\left(\sum_{i=0}^{\infty} x_i p^i\right) = 1/p \sum_{i=0}^{\infty} x_i p^{-\alpha i}$$

*is injection and continuous for all $\alpha > 1$.*

Proof. Let $x, y \in Z_p$, $x \neq y$ have the following canonical representations $x = \sum_{i=0}^{\infty} x_i p^i$ and $y = \sum_{i=0}^{\infty} y_i p^i$ with $x_0 = y_0, \ldots, x_{n-1} = y_{n-1}, x_n \neq y_n$ respectively and let $x_n > y_n$ for definiteness. Then we have

$$\mathscr{G}_\alpha(x) - \mathscr{G}_\alpha(y) = 1/p\left(p^{-\alpha n}(x_n - y_n) + \sum_{i=n+1}^{\infty} p^{-\alpha i}(x_i - y_i)\right) \geq 1/p\left(p^{-\alpha n}(x_n - y_n)\right.$$

$$\left. + \sum_{i=n+1}^{\infty} p^{-\alpha i}(p-1)\right) = 1/p\left(p^{-\alpha n}\left(x_n - y_n - \frac{p-1}{p^\alpha - 1}\right)\right) > 0,$$

therefore the mapping in question is injection.
By analogy to the above calculations we get

$$|\mathscr{G}_\alpha(x) - \mathscr{G}_\alpha(y)| \leq \frac{1}{p} p^{-\alpha n}\left|x_n - y_n + \frac{p-1}{p^\alpha - 1}\right| \leq |x-y|_p^\alpha,$$

that means that the mapping in question is continuous. □

Let $B_{-n}(a)$, $n \geq 0$ denotes $p$-adic disk of radius $p^{-n}$ with the center in $a$, i.e. $B_{-n}(a) = \{x \in Z_p : |x-a|_p \leq p^{-n}\}$. As it is known, each point of a $p$-adic disk is its center, but it is convenient to choose some canonical form for the center. In our notations we put $a = a_0 + a_1 p + \cdots + a_{n-1} p^{n-1}$.
It is easy to see that

$$\min_{x \in B_{-n}(a)} \mathscr{G}_\alpha(x) = \mathscr{G}_\alpha(a)$$

and
$$\max_{x \in B_{-n}(a)} \mathscr{G}_\alpha(x) = \mathscr{G}_\alpha(a) + \frac{1-p^{-1}}{1-p^{-\alpha}} p^{-\alpha n}.$$

The last formulas make natural the following definition of the mapping $\mathscr{F}_\alpha$: disks in $Z_p \longrightarrow$ segments in $[0,1]$:

$$\mathscr{F}_\alpha(B_{-n}(a))[\mathscr{G}_\alpha(a), \mathscr{G}_\alpha(a) + \frac{1-p^{-1}}{1-p^{-\alpha}} p^{-\alpha n}] I_{-n}(a) \subset [0,1].$$

Note the following interesting formula

$$\mu(B_{-n}(a))\left(\frac{1-p^{-\alpha}}{1-p^{-1}}\right)^{1/\alpha} |I_{-n}(a)|^{1/\alpha}, \tag{1}$$

here $|\,.\,|$ denotes the length of a segment on real numbers line. The last formula shows that $\mathscr{F}_\alpha$ maps the set of disks in $Z_p$ into a set of segments in $[0,1]$ with "holes".

The following lemma makes clear the meaning of the parameter $\alpha$.

**Lemma 3.** *The Hausdorff dimension of the image of the mapping $\mathscr{G}_\alpha$ equals $1/\alpha$.*

*Proof.* In fact the statement of the lemma directly follows from (1), nevertheless we present here the explicit calculations of this dimension using some sequence of iterations for constructing the image of the mapping in question. This procedure is the same as one uses to construct the standard Cantor set.

At zeroth step we consider the covering of $Z_p$ by set of disks of radii $1/p$ and by means of the mapping $\mathscr{F}_\alpha$ get the set of segments $I_{-1}(a_0)$, $a_0 = 0, 1, \ldots, p-1$. This set of segments doesn't cover all segment $[0,1]$, there are holes of total measure $\Delta_0 = \frac{p^{\alpha-1}-1}{p^{\alpha}-1}$. Thus the zeroth step of iteration can be considered as follows: the segment $[0,1]$ is "divided" into $\Gamma_0 = 1/\Delta_0$ parts and then we reject a part of them, leaving $N_0 = \frac{1-\Delta_0}{\Delta_0} = \frac{p-1}{p^{\alpha-1}-1}$.

By analogy to zeroth step at $n$-th step of iteration we get using the same notations: the total measure of holes $\Delta_n = p^{-\alpha n}\Delta_0$; the total measure of holes through all iterations up to $n$-th iteration

$$S_n = \Delta_0\left(1 + \frac{p-1}{p^\alpha} + \frac{p(p-1)}{p^{2\alpha}} + \cdots + \frac{p^{n-1}(p-1)}{p^{n\alpha}}\right),$$

and
$$N_n = \frac{1-S_n}{\Delta_n} p^n \frac{p^{\alpha-1}(p-1)}{p^{\alpha-1}-1}, \quad \Gamma_n = p^{n\alpha}\Gamma_0.$$

The iteration procedure explained above shows that the image of the mapping in question is selfsimilar fractal set with the dimension $D = \lim_{n\to\infty} \frac{\log_p N_n}{\log_p \Gamma_n} = 1/\alpha\ldots\ldots$ □

The proved Lemma makes clear the meaning of the parameter $\alpha - 1/\alpha$ is fractal dimension of energy dissipation region at zero viscosity.

# 4. ENERGY DISSIPATION

In previous section we discussed the relation between eddies space and real space. Here we consider functions on the eddies space, which correspond to rate of energy dissipation. Measures on the set of $p$-adic integers naturally appear in this way.

In fact, let $\varepsilon_q$ denotes the dissipation per unit time integrated over the volume of the eddy $q$ for all $q$ from eddies space. We have therefore a function $\varepsilon$ on the set of disks in $Z_p$. Note that $\varepsilon$ has definite sign. Moreover, this function is additive by its nature, because we consider conservative cascade process (i.e. energy of an eddy conserves through the division process [2]). Thus we get a measure $\varepsilon$ on the Borel algebra generated by set of disks. The last note restricts our consideration to the case of measures (if we are interested in behavior of energy dissipation). We call them *p-adic dissipation measures* for brevity.

As noted by several authors (see, for example [8]), rate energy dissipation per eddy need not follow the self-similar behavior expected in the inertial range because of it is not an inertial range quantity. Moreover there are a lot of results (both theoretical and experimental) relating to the multifractal nature of $\varepsilon$ and dissipation field (see [2] and references therein). So the dissipation field is described by a very singular function.

In our $p$-adic approach there exists another characteristic of energy dissipation – *p-adic dissipation field*. If measure $\varepsilon$ on $Z_p$ connected with dissipation has the form $\varepsilon(x)dx$, where $\varepsilon(x)$ denotes some real valued function on $Z_p$ and $dx$ is the Haar measure, then we call $\varepsilon(x)$ the $p$-adic dissipation field. The meaning of this notation is clear: rate of dissipation $\varepsilon_q$ in the volume of the eddy $q$ is defined by the formula

$$\varepsilon_q = \int_B \varepsilon(x)dx,$$

where $B$ denotes the disk corresponding to eddy $q$. $p$-adic dissipation field is in fact rate of dissipation per unit volume, but in eddies space.

The natural question concerning relations between real dissipation field and $p$-adic one arises. By means of mapping $\mathscr{F}_\alpha$ and (1) from previous section the measure on $Z_p$ defines the measure on Borel algebra generated by set of segments $I_{-n}(a)$, $n = 1, 2, \ldots$, $a = a_0 + a_1 p + \ldots + a_{n-1} p^{n-1}$. Therefore if $\varepsilon_R(x)$, $x \in [0,1]$ denotes the dissipation field in real space, then we have the following formula

$$\int_{B_n(a)} \varepsilon(x)dx = \int_{I_{-n}(a)} \varepsilon_R(y)dy,$$

which gives us the relation between $p$-adic dissipation field and real one.

It is our interest to study spectrum of generalized dimensions $D_q$ (see, for example [9,10]) for $p$-adic dissipation measure. For calculation of the spectrum in question we use the following standard method (see [11]). Let $\varepsilon(x)$, $x \in Z_p$ be a $p$-adic dissipation field and let

$$E_B = \int_B \varepsilon(x)dx,$$

denotes the total dissipation in the eddy, corresponding to disk $B$. Then we consider the asymptotic expansion of the sum

$$\Sigma_n = \sum_{B_{-n}} E^q_{B_{-n}} \sim E^q_{Z_p} |\mathscr{F}_\alpha(B_{-n})|^{D_q(q-1)},$$

where $n \to \infty$ and summation is carried out over all disks of radius $p^{-n}$. Using (1) we rewrite the last formula in the following way

$$\Sigma_n = E^q_{Z_p}(p^{-n})^{\frac{D_q}{D}(q-1)}, \qquad (2)$$

where $D = 1/\alpha$ is fractal dimension of dissipation region in eddies space at zero viscosity limit, see Section 3. $D$ is also in fact the fractal dimension of the support of the measure connected with dissipation in real space.

It is important to underline, that in spite of $p$-adic machinery we obtain the spectrum of generalized dimensions for the cascade process in real space.

Let us give an example of such calculations.

**Example 1.** *Let $p$-adic dissipation field is constant, then $D_q = D$ for all $q$.*

*Proof.* If $\varepsilon$ is constant $p$-adic dissipation field, then $E_{B_{-n}} = p^{-n}$. As the total number of disks of radius $p^{-n}$ in $Z_p$ equals $p^n$, then we have

$$\Sigma_n \sim \varepsilon^q (p^{-n})^{(q-1)}.$$

□

Thus in the case of constant $p$-adic dissipation field we see that the singularity spectrum consists of single point $D < 1$. The same result appeared in $\beta$-model (see [12]), relation $D < 1$ reflects the fractal behavior of support of the measure. More detailed analysis shows that the case of constant dissipation field is in fact $\beta$-model. Although fractal support of the measure appears not because of existence of 'active' and 'passive' eddies, but because of additional decreasing of their volumes.

Let us now consider the case when the spectrum is not trivial. It is interesting, that very simple $p$-adic dissipative measure have nontrivial spectrum.

**Example 2.** *Let $p$-adic dissipation field has the form $\varepsilon(x) = |x|_p^{\Delta-1}$, $0 < \Delta < 1$. Then we get the following singularity spectrum:*

$$D_q = \begin{cases} D & \text{if } q < 1/(1-\Delta), \\ \Delta D \frac{q}{q-1} & \text{if } q \geq 1/(1-\Delta). \end{cases}$$

*Proof.* Let $B_{-n}(a) = a_0 + a_1 p + \ldots + a_{n-1} p^{n-1}$ denotes disk of radius $p^{-n}$ with the center in $a$. Then we have

$$E_{B_{-n}(a)} = \begin{cases} |a|_p^{\Delta-1} p^{-n} & \text{if } a \neq 0, \\ (1-1/p)\frac{p^{-n\Delta}}{1-p^{-\Delta}} & \text{if } a = 0. \end{cases} \qquad (3)$$

In fact, the first row directly follows from the relation $|x+a|_p = |a|_p$ valid for all $|x|_p < |a|_p$.

The second row in (3) is the result of following simple calculations ($B_{-n}$ denotes here disk with the center in zero):

$$\int_{B_{-n}} |x|_p^{\Delta-1} dx = \sum_{j=-\infty}^{-n} \int_{|x|_p=p^j} |x|_p^{\Delta-1} dx = \sum_{j=-\infty}^{-n} p^{j(\Delta-1)} \left( \int_{B_j} dx - \int_{B_{j-1}} dx \right)$$

$$= \sum_{j=-\infty}^{-n} p^{j(\Delta-1)}(p^j - p^{j-1})(1-1/p) \sum_{j=n}^{\infty} p^{-\Delta j} = (1-1/p)\frac{p^{-n\Delta}}{1-p^{-\Delta}}.$$

In order to calculate $\Sigma_n$ we note, that the number of disks $B_{-n}(a)$ with $|a|_p = p^{-k}$, $k = 0, 1, 2, \ldots, n-1$ equals $(p-1)p^{n-k-1}$. Therefore

$$\Sigma_n = \sum_a E_{B_{-n}(a)}^q = \left(\frac{1-p^{-1}}{1-p^{-\Delta}}\right)^q p^{-nq\Delta} + p^{-nq}((p-1)p^{n-1} + p^{-q(\Delta-1)}(p-1)p^{(n-2)}$$

$$+ p^{-2q(\Delta-1)}(p-1)p^{n-3} + \ldots + p^{(n-1)q(\Delta-1)}(p-1))$$

$$= \left(\frac{1-p^{-1}}{1-p^{-\Delta}}\right)^q p^{-nq\Delta} + (1-p^{-1})p^{-n(q-1)}\frac{1-p^{-nq\Delta+n(q-1)}}{1-p^{-q\Delta+q-1}}.$$

From the last formula we have that in the case of $q < 1/(1-\Delta)$ the needed asymptotic expansion is given by the second term and has the form

$$\Sigma_n \sim p^{-n(q-1)},$$

and $D_q = D$. In the case of $q > 1/(1-\Delta)$ we see that both terms in the formula in question have the same order and

$$\Sigma_n \sim p^{-n(q-1)\Delta \frac{q}{q-1}},$$

hence $D_q = \Delta \frac{q}{q-1}$. □

From the Example 2 we see that a very simple *p*-adic dissipative measure has nontrivial spectrum of generalized dimensions. The spectrum (3) is not realistic (on the contrary to such models as random $\beta$-model [13] or binomial model [14]), but there are no doubts that it is not so difficult to find a *p*-adic dissipative measure with more realistic spectrum [15]. Thus *p*-adic measures seem to give a systematic approach to multifractal cascade models. As such measures are regular then it seems to be probable that they can be derived from some dynamical equations. This is the way to obtain a multifractal dynamical model. Such an attempt is done in the next section.

# 5. SCALAR DYNAMICAL MODEL

## 5.1. Model equation

Here we present a $p$-adic scalar dynamical model for describing developed turbulence.

The idea is similar to that of concerning energy dissipation considered in previous sections. Remind that conservativeness of cascade energy transfer through eddies space allows us to consider measures (or functions) on the boundary of the eddies tree (i.e. $Z_p$). Our construction of dynamical model is based on momentum conservation law for eddies. As it will be shown later that also leads us to consideration of functions (or distributions) on the boundary of eddies tree. These functions are to be interpreted as $p$-adic velocity field. The velocity of an eddy is obtained by averaging of the $p$-adic velocity field over the disk corresponding to this eddy.

Let $B$ be the disk in $Z_p$ corresponding to the eddy $q$ (the vertex in eddies space) and let it be divided into $p$ disks $B_1, \ldots, B_p$ which correspond to eddies $q_1, \ldots, q_p$ respectively. If we also denote by $P, P_1, \ldots, P_p$ momenta of these eddies and suppose the conservation law for these momenta we get

$$P = P_1 + \ldots + P_p.$$

Hence as in the case of energy dissipation (see Section 4) we obtain a measure on $Z_p$. If we use notation $\pi$ for the measure in question the velocity $v_q$ of an eddy $q$ corresponding to the disk $B$ is given by the formula

$$v_q = \frac{\pi(B)}{\mu(B)},$$

where $\mu$ denotes the Haar measure on $Z_p$. In the last formula we suppose for simplicity that the eddies mass distribution is given by the Haar measure. In other cases $\mu$ is to be changed for another mass distribution measure.

As in the case of energy dissipation we restrict our consideration to the special cases of measure $\pi$, when it has the form

$$\pi = v(x)dx,$$

where $dx$ denotes the Haar measure on $Z_p$ and $v$ is a real valued function on $Z_p$. As we try to construct dynamical model we are to consider a one-parameter family of the measures $\pi$ of the form

$$\pi_t = v(x,t)dx,$$

where $t$ is real-valued time variable. The velocity $v_q(t)$ of the eddy $q$ is given by averaging of the function $v(x,t) : Z_p \times R \longrightarrow R$ over the corresponding disk $B$:

$$v_q(t) = \frac{\int_B v(x,t)dx}{\int_B dx}. \tag{4}$$

We call the function $v(x,t)$ the *p-adic velocity field*.

The following step is to consider the dynamics of the p-adic velocity field or in other words to suggest an equation describing this function.

By analogy to standard scalar models of turbulence (see [16–18]) the equation must include the first derivative with respect to time $t$ and square nonlinearity. The general type of such equation is

$$\frac{d}{dt}v(x,t) = \int_{Z_p}\int_{Z_p} F(x,t;\xi,\eta)v(\xi,t)v(\eta,t)d\xi d\eta + \int_{Z_p} \Lambda(x,t;\zeta)v(\zeta,t)d\zeta, \quad (5)$$

where $F(x,t;\xi,\eta)$ describes interaction and $\Lambda(x,t;\zeta)$ is responsible for both dissipation and external force.

In order to simplify our model we suppose that $F$ and $\Lambda$ do not depend on time variable $t$. We also suppose that $F$ is homogeneous with respect to the last two arguments, i.e. $F$ depends on only difference between them. And the last suggestion concerning the kernel $F$ is that $F$ is a superposition of two functions: the p-adic valued function $\phi$ and real valued function $\mathscr{F}$, the first function describes the universal mechanism of eddies interaction and the last one depends on the type of flow. We also suppose for simplicity that $\phi$ has the simplest linear form. Thus the kernel $F$ takes the form

$$F(x,t;\xi,\eta) = \mathscr{F}(x + \kappa(\xi - \eta)).$$

Substituting the last simplified expression for the kernel into equation (5) we obtain the following

$$\frac{d}{dt}v(x,t) = \int_{Z_p}\int_{Z_p} \mathscr{F}(x + \kappa(\xi - \eta))v(\xi,t)v(\eta,t)d\xi d\eta + \int_{Z_p} \Lambda(x,\zeta)v(\zeta,t)d\zeta. \quad (6)$$

The meaning of the p-adic parameter $\kappa$ will be clear later.

Before further analysis of the equation (6) we note that the equation (5) with another type of the kernel $F$ includes standard scalar models. In fact, let the kernel $F$ is independent on time variable and depends on the norm of the last two variables, that is

$$F(x,t;\xi,\eta) = \Phi(x,|\xi|_p,|\eta|_p).$$

Using the notations

$$v_\alpha(t) = \int_{|\xi|_p = p^{-\alpha}} v(\xi,t)d\xi, \quad \alpha = 1,2,\ldots$$

and

$$\Gamma^\gamma_{\alpha\beta} = \int_{|x|_p = p^\gamma} \Phi(x, p^\alpha, p^\beta)dx$$

it is easy to obtain from (5) the following equation (under the condition $\Lambda = 0$):

$$\frac{d}{dt}v_\gamma(t) = \sum_{\alpha,\beta} \Gamma^\gamma_{\alpha\beta} v_\alpha(t) v_\beta(t).$$

The last equation has the same form as that of for standard scalar models at zero viscosity and zero external force.

The equation (6) describes the $p$-adic velocity field, but we are most interesting in velocities of eddies in real coordinate space. The answer is: we have to consider averages of the solution of the equation (6) in the sense of (4) in order to obtain the needed velocities. Nevertheless, it is useful to obtain a direct equation for velocities of eddies. Before doing that we take the notation:

$$[v]_n(k,t) = p^n \int_{|x|_p \leq p^{-n}} v(k+x,t) dx.$$

It is clear, that the variable $k$ is in fact discrete and runs through the set $\Omega_n = Z_p/p^n Z_p$ of all eddies of the scale $p^{-n}$, because the function $[v]_n(k,t)$ is constant on each disk of the radius $p^{-n}$. Comparing the definition of $[v]_n$ with (4) we see that $[v]_n$ describes dynamics of all eddies of the scale $p^{-n}$.

**Lemma 4.** *Let $v(x,t)$ be a solution of the equation (6) with $\Lambda = 0$. Then the functions $[v]_n(k,t)$, $n = 1,2,\ldots$ satisfy the following infinite system of equations:*

$$\frac{d}{dt}[v]_{n-\theta}(k,t) = \sum_{l,m \in Z_p/Z_p} [\mathscr{F}]_{n-\theta}(k + \kappa(l-m))[v]_n(l,t)[v]_n(m,t), \qquad (7)$$

*where $|\kappa|_p = p^\theta$ and summation over cosets means that functions under summation do not depend on choice of a representative in a coset.*

*Proof* of the Lemma is simple exercise in $p$-adic integration. In fact

$$\int_{Z_p}\int_{Z_p} d\xi d\eta \mathscr{F}(k + \kappa(\xi - \eta)) p^n \int_{|\xi'|_p \leq p^{-n}} d\xi' v(\xi + \xi',t) p^n \int_{|\eta'|_p \leq p^{-n}} d\eta' v(\eta + \eta',t)$$

$$= p^{2n} \int_{|\xi'|_p \leq p^{-n}} d\xi' \int_{|\eta'|_p \leq p^n} d\eta' \int_{Z_p}\int_{Z_p} d\xi d\eta \mathscr{F}(k + \kappa(\xi' - \eta'))$$

$$+ \kappa(\xi - \eta))v(\xi,t)v(\eta,t) = p^n \int_{|\xi|_p \leq p^{-n}} \frac{d}{dt} v(k + \kappa \zeta) = \frac{d}{dt}[v]_{n-\theta}(k,t).$$

The above calculations give us the integral form of the (7). The discrete form can be easily obtained as follows

$$\int_{Z_p}\int_{Z_p} d\xi d\eta \mathscr{F}(k + \kappa(\xi - \eta))[v]_n(\xi,t) v_n(\eta,t) =$$

$$= \sum_{l,m \in Z_p/p^n Z_p} [v]_n(l,t)[v]_n(m,t) \int_{|\xi|_p \leq p^{-n}} d\xi \int_{|\eta|_p \leq p^{-n}} d\eta \mathscr{F}(k + \kappa(l-m)) + \kappa(\xi - \eta)$$

$$= \sum_{l,m \in Z_p/p^n Z_p} [v]_n(l,t)[v]_n(m,t) [\mathscr{F}]_{n-\theta}(k + \kappa(l-m)). \qquad \square$$

The interpretation of the system (7) is quite obvious: the dynamics of disks of scale $p^{-n}$ defines the dynamics of disks of scale $p^{-n+\theta}$. From this point we choose $\theta = -1$, that means that dynamics on the '$n$-th level' defines that on the '$n+1$-th level'. We see that the system (7) reflects the cascade process of eddies division and in fact can be considered as a kinetic equation for eddies. But up to now this system is infinite which corresponds to zero Kolmogorov's scale (i.e. the case of zero viscosity). In order to make the system finite we put viscosity into it. The following Lemma gives one of the possible ways to 'switch on' the viscosity.

**Lemma 5.** *In the case of nonzero viscosity but if the viscosity kernel $\Lambda$ satisfies the relation*

$$[\Lambda]_N = 0$$

*(the average means with respect to the second argument of $\Lambda(x,\xi)$ for sufficiently large $N$, the velocities $[v]_n(k,t)$ of eddies still satisfy the system (7) but in the case of $n < N$.*

*Proof.* It is sufficient to prove that the following relation

$$\int_{Z_p} \Lambda(x,\zeta)[v]_n(\zeta,t)d\zeta = 0$$

is valid for all $0 < n < N$. The last formula is the direct consequence of the following one for all $n \leq N$:

$$\int_{Z_p} \Lambda(x,\zeta)[v]_n(\zeta,t)d\zeta = p^N \int_{|\xi|_p \leq p^{-N}} \int_{Z_p} \Lambda(x,\zeta)[v]_n(\zeta-\xi,t)d\zeta d\xi$$

$$= \int_{Z_p} [\Lambda(x,\zeta)]_N [v]_n(\zeta,t)d\zeta = 0. \qquad \square$$

The meaning of the parameter $N$ is clear – it corresponds to the boundary of the inertial range, because if $p^{-n} \geq p^{-N}$ then viscosity doesn't influence the velocity field. Hence $p^{-N}$ is Kolmogorov's scale in our model.

## 5.2. Scale transformation

As it is known, the ideas of scale transformation and scale invariantness play an important role in turbulence. In this subsection we suggest the scale transformation for our dynamical model.

Let $f$ be a real valued function on $Z_p$. Its scale transformation $\mathscr{S}f$ is the following function by the definition:

$$\mathscr{S}f(x) = \frac{1}{p}f(\frac{1}{p}x), \quad x \in Z_p.$$

It is easy to see that the transformation in question acts as follows: it $p$-times contracts the support of the function and the amplitude of the function under transformation $p$-times decreases. The properties of equation (6) under the scale transformation is given by the following lemma.

**Lemma 6.** *The equation (6) with zero viscosity is invariant under scale transformation if and only if the kernel of interaction satisfies the property:*

$$\mathscr{F}(px) = p^3 \mathscr{F}(x). \tag{8}$$

*Proof.* By the definition of scale transformation we have

$$\int_{Z_p}\int_{Z_p} \mathscr{F}(x+\kappa(\xi-\eta))\mathscr{S}v(\xi,t)\mathscr{S}v(\eta,t)d\xi\,d\eta$$

$$= 1/p^2 \int_{Z_p}\int_{Z_p} \mathscr{F}(x+\kappa(\xi-\eta))v(1/p\xi,t)v(1/p\eta,t)d\xi d\eta$$

$$= 1/p^4 \int_{Z_p}\int_{Z_p} \mathscr{F}(p(1/px+\kappa(\xi-\eta)))v(\xi,t)v(\eta,t)d\xi d\eta = \frac{d}{dt}\mathscr{S}v(x,t)$$

if the equation (8) is valid.
If now the equation (6) is scale invariant, we similarly get

$$\int_{Z_p}\int_{Z_p} (1/p^3 \mathscr{F}(p(x+\kappa(\xi-\eta)))) - \mathscr{F}(x+\kappa(\xi-\eta))v(\xi,t)v(\eta,t)d\xi\,d\eta = 0$$

for an arbitrary function $v(x,t)$, hence we get (8). □

### 5.3. Eddy living time

This section presents the evaluation of an important parameter of our model – eddy living time. This time depends on the scale of an eddy and we denote it by $\tau_n$ for the scale $p^{-n}$.

In order to evaluate the needed time $\tau_n$ we at first represent our equation (6) in the equivalent integral form and then consider the sequence of approximations to the solution of the equation with a special initial condition at zero time

$$v(x,t) = v(0,t) + \int_0^t d\tau \int_{Z_p}\int_{Z_p} \mathscr{F}(x+\kappa(\xi-\eta))v(\xi,t)v(\eta,t)d\xi\,d\eta. \tag{9}$$

If we want to evaluate the living time for eddy of scale $\lambda_n = p^{-n}$ we choose the initial condition of the form $v(x,0) = \delta_n(v)h_n(x)$, where

$$h_n(x) = \begin{cases} 1 & \text{if } |x|_p \leq p^{-n}, \\ 0 & \text{if } |x|_p > p^{-n}, \end{cases}$$

and the constant $\delta_n(v)$ denotes the difference of the velocity at scale $\lambda_n$. The sequence of iterations satisfies the following system of equations:

$$v_0(x,t) = v(x,0),$$

$$v_n(x,t) = A(v_{n-1})(x,t),$$

where $A$ denotes the following integral operator on the space $C(B_n \times [0, \tau_n])$ of continuous functions in the cylinder:

$$A(v)(x,t) = v(x,0) + \int_0^t \int_{Z_p} \int_{Z_p} \mathscr{F}(x + \kappa(\xi - \eta)) v(\xi, \tau) v(\eta, \tau).$$

Let $\|v\|_n$ and $\|v\|_n^{(1)}$ denote the following standard norms:

$$\|v\|_n = \max_{x \in B_n, t \in [0, \tau_n]} |v(x,t)|,$$

$$\|v(x,t)\|_n^{(1)} = \int_{B_n} |v(x,t)| dx.$$

**Lemma 7.** *The following inequality is valid for the sequence of iterations:*

$$|Av_{i+1} - Av_i| \leq p \|\mathscr{F}\|_0^{(1)} \tau_n \delta_n(v) / \lambda_n \|v_{i+1} - v_i\|_n. \tag{10}$$

*Proof.* The proof is standard, so we present here only the sketch of the proof. By means of obvious transformations we get for all $u, v \in C(B_n \times [0, \tau_n])$

$$|Au - Av| =$$

$$\left| \int_0^t \int_{Z_p} \int_{Z_p} d\xi d\eta (u(\xi, \tau) - v(\xi, \tau))(\mathscr{F}(x + \kappa(\xi - \eta))u(\eta, \tau) + (\mathscr{F}(x + \kappa(\xi - \eta))v(\eta, \tau)) \right|$$

$$\leq \|u - v\| \int_0^t \int_{B_n} (|u(\eta, \tau)| + |v(\eta, \tau)|) d\eta \, p \int_{B_n} |\mathscr{F}(\zeta)| d\zeta$$

$$= p \|\mathscr{F}\|_n^{(1)} \int_0^t (\|u\|_n^{(1)} + \|v\|_n^{(1)}) d\tau \|u - v\|_n.$$

The needed equality (10) follows directly from the last formula and the following one

$$\|\mathscr{F}\|_n^{(1)} = \lambda_n^{-2} \|\mathscr{F}\|_0^{(1)},$$

where we take into account the condition of scale invariantness $\mathscr{F}(px) = p^3 \mathscr{F}(x)$. $\square$

The evaluation of eddy living time $\tau_n$ appears from the condition of convergence of the sequence of iterations on the interval $(0, \tau_n)$. Thus taking into account the fixed point theorem we get from the Lemma 7 the following value for $\tau_n$

$$\tau_n \approx 1 / \|\mathscr{F}\|_0^{(1)} \frac{\lambda_n}{\delta_n(v)}. \tag{11}$$

## 5.4. Velocity variation

An eddy in turbulence is characterized by not only its velocity or size but the difference (or variation) of velocity on the scale of the eddy. It is the velocity variation that we obtain in experiment. Let us try to understand the properties of the variation in an axiomatic way. Let $f$ be a real valued function on the set of $Z_p$ of $p$-adic integers. The function $f$ is locally constant of order $n$, $n = 0, 1, 2, \ldots$ if the relation $f(x+\xi) = f(x)$ is valid for all $\xi \in p^n Z_p$ and $x \in Z_p$. For example, $[f]_n(x)$ is locally constant of order $n$ (see Section 5.1).

**Definition 1.** *Let $f, g$ be functions from $C(Z_p)$. The variation $\delta_n$ of the velocity of eddy of scale n is functional on $C(Z_p)$ which satisfies the following properties:*

(i) $\delta_n(f) = 0$ if and only if $[f]_{n+1}(x)$ is locally constant of order n;
(ii) $\delta_n(f+g) \leq \delta_n(f) + \delta_n(g)$;
(iii) $\delta_n(\lambda f) = |\lambda| \delta_n(f)$;
(iv) $\delta_n(\mathscr{S}f) = 1/p \, \delta_{n+1}(f)$.

The property (i) in the definition means that eddy with scale n is really present in the $p$-adic velocity field if there exists nonzero variation of order n of the corresponding function on the boundary of the eddies tree. The last statement becomes more clear if we reformulate (i) in the following equivalent form:

(í) $\delta_n(f) = 0$ if and only if $[f]_{n+1}(x) \equiv [f]_n(x)$.

Properties (ii) and (iii) are rather natural in order to obtain the energy norm, we discuss them below. Property (iv) reflects the idea of scale invariantness.

Let $v(x,t)$ be a solution of the equation (6). We define the turbulent energy $E_v(t)$ of this solution by the formula

$$E_v(t) = 1/2 \sum_{n=0}^{\infty} \delta_n^2(v)(t).$$

Note that by statements (ii) and (iii) of the Definition 1 the functional $\sqrt{E_f} = \|f\|$ defines the seminorm on the space $C(Z_p)$ and $\|f\| = 0$ if and only if $f$ is constant on $Z_p$. It means that constant functions on the boundary of the eddies space really do not 'contain' eddies excluding that of macroscale order.

An example of the variation is given by the following lemma.

**Lemma 8.** *Let $\Omega_n$ denotes the set of all disks of radii $p^{-n}$ in $Z_p$. Then the following functional on $C(Z_p)$*

$$\delta_n(f) = \max_{A \in \Omega_n} \max_{B \in \Omega_{n+1}, B \subset A} |[f]_A - [f]_B| \qquad (12)$$

*defines the variation in the sense of Definition 1.*

*Proof.* (i) directly follows from the equation

$$[f]_A = 1/p \sum_{B \in \Omega_{n+1}, B \subset A} [f]_B.$$

(ii) and (iii) are obvious consequences of linearity property for average $[.]_A$

For the proof of (iv) we consider the isomorphism of sets $\Omega_n$ and $\Omega_{n+1}$ defined by the mapping $\phi$ on $Z_p$, $\phi : x \to 1/px$. This mapping satisfies the property: if $A \subset B$, then $\phi(A) \subset \phi(B)$. Moreover it is easy to obtain the formula

$$[\mathscr{S}f]_A = 1/p[f]_{\phi(A)}, A \in \Omega_n,$$

that gives us (iv). □

Let us note the following useful corollary of the Definition 1: variation $\delta_n$ has the property

$$\delta_n(f) = p^n \delta_0(\mathscr{S}^n f), \qquad (13)$$

which gives us a method to calculate variations of every order.

## 5.5. Kolmogorov's law

We discuss below the Kolmogorov–Obuchov law [19,20] for our scalar model. As usual the Kolmogorov–Obuchov law in scalar models means the existence of stationary solution for velocity field with power behavior of velocity variation with respect to scales.

In our case Kolmogorov's behavior corresponds to the stationary solution $v(x,t) = v(x)$ of the equation (6) with the following property:

$$\delta_n(v) \sim \lambda_n^\gamma,$$

where $\lambda_n = p^{-n}$.

Such a behavior in our model is given by homogeneous functions. In fact, let us assume that our equation (6) has the stationary solution $v(x) = |x|_p^{-\gamma}$. Then by means of (12) we get

$$\delta_n(v) = p^n \delta_0(p^{-n}|p^{-n}x|_p^{-\gamma}) = \lambda_n^\gamma \delta_0(v).$$

The following step is to understand when the equation (6) has the homogeneous stationary solution. At first we rewrite our equation in the convolution form.

**Lemma 9.** *The equation (6) in the case of $\Lambda = 0$ is equivalent to the following one:*

$$\frac{d}{dt} v(x,t) = \mathscr{F}_\kappa * v * v_-(-x/\kappa, t), \qquad (14)$$

*where $\mathscr{F}_\kappa(x) = \mathscr{F}(-\kappa x)$, $v_-(x,t) = v(-x,t)$ and convolutions are carried out with respect to the first variable.*

*Proof.* It is quite obvious. In fact by simple transformations we have:

$$\int_{Z_p}\int_{Z_p} \mathscr{F}(x+\kappa(\xi-\eta))v(\xi,t)v(\eta,t)d\xi d\eta = \int_{Z_p}\int_{Z_p} \mathscr{F}(x+\kappa\zeta)v(\xi,t)v_-(-\xi+\zeta,t)d\xi d\zeta$$

$$= \int_{Z_p} \mathscr{F}(x+\kappa\zeta)v*v_-(\zeta,t)d\xi d\zeta = \int_{Z_p} \mathscr{F}(-x/\kappa-\zeta)v*v_-(\zeta,t)d\zeta \mathscr{F}_\kappa * v * v_-(-x/\kappa,t). \square$$

The solution of problem in question is given by the following lemma.

**Lemma 10.** *The equation (6) has the homogeneous stationary solution*

$$\pi_\alpha(x) = |x|_p^\alpha / \Gamma_p(\alpha)$$

*($\Gamma_p$ denotes p-adic gamma function) if and only if the following relation is valid*

$$\mathscr{F}(x) = -\pi_{-2\alpha} * \int_{Z_p} \Lambda(x,\xi) \pi_\alpha(\xi) d\xi. \tag{15}$$

*Proof.* By means of the formula

$$\pi_\alpha * \pi_\beta = \pi_{\alpha+\beta},$$

see [21], it is easy to verify that if the formula (15) is valid, then the function $\pi_\alpha(x)$ gives the stationary solution of the equation (6).

Let now function $\pi_\alpha(x)$ is the stationary solution of (6). That is

$$\mathscr{F}_\kappa * \pi_\alpha * \pi_\alpha(-x/\kappa) + \int_{Z_p} \Lambda(x,\xi) \pi_\alpha(\xi) d\xi = 0.$$

Therefore, the function $\mathscr{F}$ satisfies the equation

$$\mathscr{F}_\kappa * \pi_{2\alpha}(-x/\kappa) + \int_{Z_p} \Lambda(x,\xi) \pi_\alpha(\xi) d\xi = 0.$$

It is known, (see [21]) that the last equation has unique up to a constant solution of the form (15). □

From the discussion in Section 4 we know that homogeneous solutions lead to nontrivial spectrum of generalized dimensions for energy dissipation. We see that such type solutions appear in our dynamical model and though it seems to be natural way to obtain multifractal behavior directly from dynamical models. Lemma 10 gives us conditions of existence of homogeneous solution of an arbitrary degree. If we want to obtain '1/3' behavior, we are to take additional assumption. Following Kolmogorov [19], we suppose that rate of energy dissipation

$$\varepsilon_v = 1/2 \frac{\delta_n^2(v)}{\tau_n}$$

is constant. Substituting the expression (11) for eddy living time $\tau_n$, we obtain from the last formula

$$\varepsilon_v = p \|\mathscr{F}\|_0^{(1)} \frac{\delta_n^3(v)}{\lambda_n}.$$

From the last formula we see that

$$\delta_n(v) \sim \lambda_n^{1/3}$$

and Kolmogorov's '1/3' behavior corresponds to the stationary solution

$$v(x) = v_0 |x|_p^{-1/3}$$

of the equation (6).

## ACKNOWLEDGMENTS

Authors wish to thank participants of the Prof. Vladimirov seminar on mathematical physics at the Steklov Mathematical Institute for fruitful discussions.

## REFERENCES

1. Nelkin M., What do we know about self-similarity in fluid turbulence, *Journ. Stat. Phys.* **54**, 1–15 (1989).
2. Meneveau Ch., Sreenivasan K.R., The multifractal nature of turbulent energy dissipation, *J. Fluid Mech.* **224**, 429–484 (1991).
3. Volovich I.V., Private communication, Such a possibility is discussed by several authors in connection with new ideas in $p$-adic mathematical physics, (see Vladimirov V.S., Volovich I.V., Applications of $p$-adic numbers in mathematical physics, *Trudi MIAN* **200**, 88-99 (1991) (in Russian)) but we don't know published papers concerning this question.
4. Schikhof W.H., *Ultrametric calculus, An introduction to p-adic analysis*, Cambridge Univ. Press, 1984.
5. Big whorls have little whorls.
   Which feed on their velocity.
   Little whorls have lesser whorls
   And so on to viscosity
   (in the molecular sense).
   L. F. Richardson, 1922.
6. Serre J.P., Arbres, amalgames, $SL_2$, *Soc. Math. de France. Asterisque* **46**, 1977.
7. Cartier P., Geometrie et analyse sur les arbres. *Lect. Notes in Math.* **317**, 1973.
8. Kraichnan R.H., On Kolmogorov's inertial-range theories, *J. Fluid Mech.* **62** 305–330 (1974).
9. Hentschel H.G., Procaccia I., The infinite numbers of generalized dimensions of fractals and strange attractors, *Phisica* **8D** 435–444 (1983).
10. Halsey T.C., Jensen M.H., Kadanoff L.P., Prococcia I., Shraiman B.I., Fractal measures and their singularities: the characterization of strange sets, *Phys. Rev.* A **33**, 1141–1151 (1986).
11. Paladin G., Vulpiani A., Anomalous scaling laws in multifractal objects *Phys. Rep.* **156**, 147–225 (1987).
12. Frish U., Sulem P.L., Nelkin M., A simple dynamical model of intermittent fully developed turbulence, *J. Fluid Mech.* **87**, 719–736 (1978).
13. Benzi R., Paladin G., Parisi G., Vulpiani A., On the multifractal nature of fully developed turbulence and chaotic systems, *J. Phys. A: Math. Gen.* **17**, 3521–3531 (1984).
14. Meneveau C, Sreenivasan K.R., Simple multifractal cascade model for fully developed turbulence, *Phys. Rev. Lett.* **59**, 1424–1427 (1987).
15. Fischenko S., Zelenov E., Generalized dimensions and $p$-adic cascade models of fully developed turbulence, in preparation.
16. Gledzer E.B., System of hydrodynamical type with two integrals, *Dokl. Akad. Nauk USSR* **209**, 1046–1048 (1973) (in Russian).

17. Yamada M., Ohkitani K., The inertial subrange and non-positive Lyapunov exponents in fully developed turbulence, *Prog. Theor. Phys.* 79, 1265–1268 (1988); Temporal intermittency in the energy cascade process and local Lyapunov analysis in fully developed turbulence. *Prog. Theor. Phys.* **81**, 329–341 (1989).
18. Jensen M.H., Paladin G., Vulpiani A., Intermittency in a cascade model for three-dimensional turbulence, *Phys. Rev.* A **43**, 798–805 (1991).
19. Kolmogorov A.N., The local structure of turbulence in incompressible viscous fluid for very large Reynolds numbers, *Dokl. Acad. Nauk USSR* **30**, 299–303 (1941) (in Russian).
20. Kolmogorov A.N., A refinement of previous hypotheses concerning the local structure of turbulence in a viscous incompressible fluid at high Reynolds numbers, *J. Fluid Mech.* **13**, 82–85 (1962).
21. Vladimirov V.S., Generalized functions over the field of $p$-adic numbers, *Usp. Mat. Nauk* **43**, 17–53 (1988).

# p-ADIC ANALYSIS

# Pseudo-Differential Operators in the $p$-Adic Lizorkin Space [1]

S. Albeverio*, A. Yu. Khrennikov† and V. M. Shelkovich**

*Institut für Angewandte Mathematik, Universität Bonn, Wegelerstraße 6,
D-53115 Bonn, GERMANY
email: albeverio@uni-bonn.de

†International Center for Mathematical Modeling in Physics, Engineering and Cognitive science
MSI, Växjö University, S-35195, SWEDEN
email: Andrei.Khrennikov@msi.vxu.se

**Department of Mathematics, St.-Petersburg State Architecture and Civil Engineering University,
2 Krasnoarmeiskaya 4, 190005, St. Petersburg, RUSSIA
email: shelkv@vs1567.spb.ed

**Abstract.** The $p$-adic Lizorkin type spaces of test functions and distributions are introduced and a class of pseudo-differential operators on this spaces are constructed. The $p$-adic Lizorkin spaces are invariant under the above-mentioned pseudo-differential operators. This class of pseudo-differential operators contains the Taibleson fractional operators. Solutions of pseudo-differential equations are also constructed.

**Keywords:** $p$-adic Lizorkin space, $p$-adic distributions (generalized functions), pseudo-differential operators, fractional operators, pseudo-differential equations.
**PACS:** 02.30.Nw, 02.30.Px, 02.30.Tb.

## 1. INTRODUCTION

It is well known that there are a lot of papers where different applications of $p$-adic analysis to physical problems, cognitive sciences and psychology are studied [3]–[5], [9]–[12], [19]–[21] (see also the references therein).

The field $\mathbb{Q}_p$ of $p$-adic numbers is defined as the completion of the field of rational numbers $\mathbb{Q}$ with respect to the non-Archimedean $p$-adic norm $|\cdot|_p$. This norm is defined as follows: $|0|_p = 0$; if an arbitrary rational number $x \neq 0$ is represented as $x = p^\gamma \frac{m}{n}$, where $\gamma = \gamma(x) \in \mathbb{Z}$, and $m$ and $n$ are not divisible by $p$, then $|x|_p = p^{-\gamma}$. This norm in $\mathbb{Q}_p$ satisfies the strong triangle inequality $|x + y|_p \leq \max(|x|_p, |y|_p)$. We recall that there exists a $p$-adic analysis connected with the mapping $\mathbb{Q}_p$ into $\mathbb{Q}_p$ and an analysis connected with the mapping $\mathbb{Q}_p$ into the field of complex numbers $\mathbb{C}$, there exist two

---

[1] This paper was supported in part by the grant of The Swedish Royal Academy of Sciences on collaboration with scientists of former Soviet Union and the EU-Network "Quantum Probability and Applications". The first and the third authors (S. A. and V. S.) were also supported in part by DFG Project 436 RUS 113/809/0-1.

types of $p$-adic physics models.

It is known that for the $p$-adic analysis related to the mapping $\mathbb{Q}_p \to \mathbb{C}$, the operation of partial differentiation is *not defined*, and the Vladimirov fractional operator $D^\alpha$ plays a corresponding role [19, IX]. Recall that large quantity of models connected with $p$-adic differential equations use the Vladimirov fractional operator and the theory of $p$-adic distributions (generalized functions) (see the above mentioned papers and books). However, in general, the spaces of test functions $\mathscr{D}(\mathbb{Q}_p)$ and distributions $\mathscr{D}'(\mathbb{Q}_p)$ *are not invariant* under the Vladimirov fractional operator, i.e., the operation $D^\alpha f$ is well defined only for some distributions $f \in \mathscr{D}'(\mathbb{Q}_p)$.

We recall that similar problems arise for the "$\mathbb{C}$-case" fractional operators (where all functions and distributions are complex or real valued defined on spaces with real or complex coordinates): in general, the Schwartzian test function space $\mathscr{S}(\mathbb{R}^n)$ *is not invariant* under fractional operators [15], [16]. To solve this problem in the excellent papers of P. I. Lizorkin [13], [14] a new type spaces *invariant* under fractional operators were introduced (see also [15], [16]).

**Contents of the paper.** In Sec. 2, we recall some facts from the $p$-adic theory of distributions. In Sec. 3, using ideas from [13], [14], [15], [16], the multidimensional $p$-adic Lizorkin type spaces of test functions $\Phi(\mathbb{Q}_p^n)$ and distributions $\Phi'(\mathbb{Q}_p^n)$, are constructed. In Sec. 4, we recall some facts on the multidimensional fractional operator $D^\alpha$ introduced by Taibleson [17, §2], [18, III.4.] defined in the space of distributions $\mathscr{D}'(\mathbb{Q}_p^n)$ for $\alpha \in \mathbb{C}$, $\alpha \neq -n$. In this section we also define the fractional operator $D_x^\alpha$ in the Lizorkin space of distributions for all $\alpha \in \mathbb{C}$. We prove that the family of fractional operators $D_x^\alpha$, $\alpha \in \mathbb{C}$, forms an Abelian group. By using the fractional operator, $p$-adic Laplacian and its powers are introduced. Note that such types of $p$-adic Laplacians were introduced in [8]. In Sec. 5, a class of pseudo-differential operators $A$ on the Lizorkin spaces are introduced. The Lizorkin spaces are *invariant* under our pseudo-differential operators. The fractional operator $D_x^\alpha$, $\alpha \in \mathbb{C}$ belongs to the class of the above mentioned pseudo-differential operators. The family of pseudo-differential operators $A$ with symbols $\mathscr{A}(\xi) \neq 0$, $\xi \in \mathbb{Q}_p^n \setminus \{0\}$, and, consequently, the family of fractional operators $D_x^\alpha$, $\alpha \in \mathbb{C}$, forms an Abelian group. In this paper solutions of pseudo-differential equations $Af = g$, $g \in \Phi'(\mathbb{Q}_p^n)$ are also constructed.

In fact, in order to define the one-dimensional fractional Vladimirov operators $D^{-1}$, the one-dimensional Lizorkin space of test functions $\Phi(\mathbb{Q}_p)$ was introduced in the book [19, IX.2] (compare with (8)). Moreover, according to [19, IX,(5.7),(5.8)], the eigenfunctions of Vladimirov's operator $D^\alpha$, satisfy condition (8), and, consequently, belong to the Lizorkin space $\Phi(\mathbb{Q}_p)$.

One can introduce the Lizorkin type space which is invariant under the Vladimirov multidimensional operator $D_\times^\alpha = D_{x_1}^{\alpha_1} \times \cdots \times D_{x_n}^{\alpha_n}$, where $D_{x_j}^{\alpha_j} = f_{-\alpha_j}(x_j)*$ is one-dimensional fractional Vladimirov's operator, $j = 1, 2, \ldots, n$. However, in this paper this problem is not considered.

The Lizorkin spaces are "natural" definition domains of the $p$-adic fractional operators and can play a key role in considerations related to the fractional operators problems.

## 2. $p$-ADIC DISTRIBUTIONS

Now we recall some facts from the theory of $p$-adic distributions (generalized functions). We shall systematically use the notations and results from [19]. Let $\mathbb{N}, \mathbb{Z}, \mathbb{C}$ be the sets of positive integers, integers, complex numbers, respectively. Denote by $\mathbb{Q}_p^* = \mathbb{Q}_p \setminus \{0\}$ the multiplicative group of the field $\mathbb{Q}_p$. The space $\mathbb{Q}_p^n = \mathbb{Q}_p \times \cdots \times \mathbb{Q}_p$ consists of points $x = (x_1, \ldots, x_n)$, where $x_j \in \mathbb{Q}_p$, $j = 1, 2 \ldots, n$, $n \geq 2$. The $p$-adic norm on $\mathbb{Q}_p^n$ is

$$|x|_p = \max_{1 \leq j \leq n} |x_j|_p, \quad x \in \mathbb{Q}_p^n. \tag{1}$$

Denote by $B_\gamma^n(a) = \{x : |x - a|_p \leq p^\gamma\}$ the ball of radius $p^\gamma$ with the center at a point $a = (a_1, \ldots, a_n) \in \mathbb{Q}_p^n$ and $B_\gamma^n = B_\gamma^n(0)$, $\gamma \in \mathbb{Z}$.

A complex-valued function $f$ defined on $\mathbb{Q}_p^n$ is called *locally-constant* if for any $x \in \mathbb{Q}_p^n$ there exists an integer $l(x) \in \mathbb{Z}$ such that

$$f(x+y) = f(x), \quad y \in B_{l(x)}^n.$$

Let $\mathcal{E}(\mathbb{Q}_p^n)$ and $\mathcal{D}(\mathbb{Q}_p^n)$ be the linear spaces of locally-constant $\mathbb{C}$-valued functions on $\mathbb{Q}_p^n$ and locally-constant $\mathbb{C}$-valued functions with compact supports (so-called test functions), respectively; $\mathcal{D}(\mathbb{Q}_p)$, $\mathcal{E}(\mathbb{Q}_p)$ [19, VI.1.,2.]. If $\varphi \in \mathcal{D}(\mathbb{Q}_p^n)$, according to Lemma 1 from [19, VI.1.], there exists $l \in \mathbb{Z}$, such that

$$\varphi(x+y) = \varphi(x), \quad y \in B_l^n, \quad x \in \mathbb{Q}_p^n.$$

The largest of such numbers $l = l(\varphi)$ is called the *parameter of constancy* of the function $\varphi$. Let us denote by $\mathcal{D}_N^l(\mathbb{Q}_p^n)$ the finite-dimensional space of test functions from $\mathcal{D}(\mathbb{Q}_p^n)$ having supports in the ball $B_N^n$ and with parameters of constancy $\geq l$ [19, VI.2.]. Denote by $\mathcal{D}'(\mathbb{Q}_p^n)$ the set of all linear functionals on $\mathcal{D}(\mathbb{Q}_p^n)$ [19, VI.3.].

The Fourier transform of $\varphi \in \mathcal{D}(\mathbb{Q}_p^n)$ is defined by the formula $F[\varphi](\xi) = \int_{\mathbb{Q}_p^n} \chi_p(\xi \cdot x)\varphi(x) d^n x$, $\xi \in \mathbb{Q}_p^n$, where $\chi_p(\xi \cdot x) = e^{2\pi i \sum_{j=1}^n \{\xi_j x_j\}_p}$, $\xi \cdot x$ is the scalar product of vectors, and the function $\chi_p(\xi_j x_j) = e^{2\pi i \{\xi_j x_j\}_p}$ for every fixed $\xi_j \in \mathbb{Q}_p$ is an additive character of the field $\mathbb{Q}_p$, $\{\xi_j x_j\}_p$ is the fractional part of a number $\xi_j x_j$, $j = 1, \ldots, n$ [19, VII.2.,3.]. It is known that the Fourier transform is a linear isomorphism $\mathcal{D}(\mathbb{Q}_p^n)$ into $\mathcal{D}(\mathbb{Q}_p^n)$. Moreover, according to [17, Lemma A.], [18, III,(3.2)], [19, VII.2.],

$$\varphi(x) \in \mathcal{D}_N^l(\mathbb{Q}_p^n) \quad \text{iff} \quad F[\varphi(x)](\xi) \in \mathcal{D}_{-l}^{-N}(\mathbb{Q}_p^n). \tag{2}$$

**Definition 1.** Let $\pi_\alpha$ be a multiplicative character of the field $\mathbb{Q}_p$ (see [19, III.2.]).

(a) According to [1], [2], a distribution $f_m \in \mathcal{D}'(\mathbb{Q}_p)$ is said to be *associated homogeneous (in the wide sense)* of degree $\pi_\alpha$ and order $m$, $m \in \mathbb{N} \cup \{0\}$, if

$$\left\langle f_m, \varphi\left(\frac{x}{t}\right)\right\rangle = \pi_\alpha(t)|t|_p \langle f_m, \varphi \rangle + \sum_{j=1}^m \pi_\alpha(t)|t|_p \log_p^j |t|_p \langle f_{m-j}, \varphi \rangle$$

for all $\varphi \in \mathscr{D}(\mathbb{Q}_p)$ and $t \in \mathbb{Q}_p^*$, where $f_{m-j} \in \mathscr{D}'(\mathbb{Q}_p)$ is an associated homogeneous distribution of degree $\pi_\alpha$ and order $m-j$, $j=1,2,\ldots,m$, i.e.,

$$f_m(tx) = \pi_\alpha(t)f_m(x) + \sum_{j=1}^m \pi_\alpha(t)\log_p^j|t|_p f_{m-j}(x), \quad t \in \mathbb{Q}_p^*.$$

If $m = 0$ we set that the above sum is empty.

(b) We say that a distribution $f \in \mathscr{D}'(\mathbb{Q}_p^n)$ is *associated homogeneous* (*in the wide sense*) of degree $\pi_\alpha$ and order $m$, $m \in \mathbb{N} \cup \{0\}$, if for all $t \in \mathbb{Q}_p^*$ we have

$$f_m(tx) = f_m(tx_1,\ldots,tx_n) = \pi_\alpha(t)f_m(x) + \sum_{j=1}^m \pi_\alpha(t)\log_p^j|t|_p f_{m-j}(x),$$

where $x = (x_1,\ldots,x_n) \in \mathbb{Q}_p^n$, $f_{m-j} \in \mathscr{D}'(\mathbb{Q}_p^n)$ is an associated homogeneous distribution of degree $\pi_\alpha$ and order $m-j$, $j=1,2,\ldots,m$. An *associated homogeneous* (*in the wide sense*) distribution of degree $\pi_{\alpha}(t) = |t|_p^{\alpha-1}$ and order $m$ is called *associated homogeneous* of degree $\alpha - 1$ and order $m$.

(c) Associated homogeneous distribution (in the wide sense) of order $m=1$ is called *associated homogeneous* distribution (see [6] and [1], [2]).

(d) Associated homogeneous distribution of degree $\pi_\alpha$ and order $m = 0$ is called *homogeneous* distribution of degree $\pi_\alpha$, i.e.,

$$f_0(tx) = f_0(tx_1,\ldots,tx_n) = \pi_\alpha(t)f_0(x), \quad x = (x_1,\ldots,x_n) \in \mathbb{Q}_p^n.$$

(for one-dimensional case see in [7, Ch.II,§2.3.], [19, VIII.1.]).

The theorem describing all one-dimensional *homogeneous* distributions was proved in [7, Ch.II,§2.3.], [19, VIII.1.]. The theorem describing all one-dimensional *associated homogeneous* (*in the wide sense*) distributions was proved in [1], [2].

The multidimensional homogeneous distribution $|x|_p^{\alpha-n} \in \mathscr{D}'(\mathbb{Q}_p^n)$ of degree $\alpha - n$ is constructed as follows. If $\operatorname{Re}\alpha > 0$ then the function $|x|_p^{\alpha-n}$ generates a regular functional

$$\langle |x|_p^{\alpha-n}, \varphi \rangle = \int_{\mathbb{Q}_p^n} |x|_p^{\alpha-n}\varphi(x)\,d^n x, \quad \forall \varphi \in \mathscr{D}(\mathbb{Q}_p^n), \qquad (3)$$

where $|x|_p$, $x \in \mathbb{Q}_p^n$ is given by (1). If $\operatorname{Re}\alpha \leq 0$ this distribution is defined by means of analytic continuation [17, (*)], [18, III,(4.3)], [19, VIII,(4.2)]:

$$\langle |x|_p^{\alpha-n}, \varphi \rangle = \int_{B_0^n} |x|_p^{\alpha-n}\big(\varphi(x) - \varphi(0)\big)\,d^n x$$

$$+ \int_{\mathbb{Q}_p^n \setminus B_0^n} |x|_p^{\alpha-n}\varphi(x)\,d^n x + \varphi(0)\frac{1-p^{-n}}{1-p^{-\alpha}}, \qquad (4)$$

for all $\varphi \in \mathscr{D}(\mathbb{Q}_p^n)$, $\alpha \neq \mu_j = \frac{2\pi i}{\ln p}j$, $j \in \mathbb{Z}$. The distribution $|x|_p^{\alpha-n}$ is an entire function of the complex variable $\alpha$ everywhere except the points $\mu_j$, $j \in \mathbb{Z}$, where it has simple poles with residues $\frac{1-p^{-n}}{\log p}\delta(x)$.

Similarly to the one-dimensional case [1], [2], one can construct the distribution $P(\frac{1}{|x|_p^n})$ called the principal value of the function $\frac{1}{|x|_p^n}$, $x \in \mathbb{Q}_p^n$:

$$\left\langle P\left(\frac{1}{|x|_p^n}\right), \varphi \right\rangle = \int_{B_0^n} \frac{\varphi(x) - \varphi(0)}{|x|_p^n} d^n x + \int_{\mathbb{Q}_p^n \setminus B_0^n} \frac{\varphi(x)}{|x|_p^n} d^n x, \qquad (5)$$

for all $\varphi \in \mathscr{D}(\mathbb{Q}_p^n)$. It is easy to show that this distribution is *associated homogeneous* of degree $-n$ and order 1 (see [1], [2]).

The Fourier transform of $|x|_p^{\alpha-n}$ is given by the formula from [17, Theorem 2.], [18, III,Theorem (4.5)], [19, VIII,(4.3)]

$$F[|x|_p^{\alpha-n}] = \Gamma_p^{(n)}(\alpha) |\xi|_p^{-\alpha}, \qquad \alpha \neq 0, n \qquad (6)$$

where the n-dimensional $\Gamma$-*function* $\Gamma_p^{(n)}(\alpha)$ is given by the following formulas (see [17, Theorem 1.], [18, III,Theorem (4.2)], [19, VIII,(4.4)]):

$$\Gamma_p^{(n)}(\alpha) \stackrel{def}{=} \lim_{k \to \infty} \int_{p^{-k} \leq |x|_p \leq p^k} |x|_p^{\alpha-n} \chi_p(u \cdot x) d^n x$$

$$= \int_{\mathbb{Q}_p^n} |x|_p^{\alpha-n} \chi_p(x_1) d^n x = \frac{1 - p^{\alpha-n}}{1 - p^{-\alpha}} \qquad (7)$$

where $|u|_p = 1$. Here $\Gamma_p^{(1)}(\alpha) = \Gamma_p(\alpha) = \int_{\mathbb{Q}_p} |x|_p^{\alpha-1} \chi_p(x) dx = \frac{1-p^{\alpha-1}}{1-p^{-\alpha}}$.

## 3. THE $p$-ADIC LIZORKIN SPACES

Let us consider the spaces

$$\Psi = \Psi(\mathbb{Q}_p^n) = \{\psi(\xi) \in \mathscr{D}(\mathbb{Q}_p^n) : \psi(0) = 0\}$$

and

$$\Phi = \Phi(\mathbb{Q}_p^n) = \{\phi : \phi = F[\psi], \psi \in \Psi(\mathbb{Q}_p^n)\}.$$

Here $\Psi, \Phi \subset \mathscr{D}(\mathbb{Q}_p^n)$. The space $\Phi(\mathbb{Q}_p^n)$ is called the $p$-adic *Lizorkin space of test functions*. The space $\Phi$ can be equipped with the topology of the space $\mathscr{D}(\mathbb{Q}_p^n)$ which makes $\Phi$ a complete space.

In view of (2), the following lemma holds.

**Lemma 1.** (a) $\phi \in \Phi(\mathbb{Q}_p^n)$ iff $\phi \in \mathscr{D}(\mathbb{Q}_p^n)$ and

$$\int_{\mathbb{Q}_p^n} \phi(x) d^n x = 0. \qquad (8)$$

(b) $\phi \in \mathscr{D}_N^l(\mathbb{Q}_p^n) \cap \Phi(\mathbb{Q}_p^n)$, i.e., $\int_{B_N^n} \phi(x) d^n x = 0$, iff $\psi = F^{-1}[\phi] \in \mathscr{D}_{-l}^{-N}(\mathbb{Q}_p^n) \cap \Psi(\mathbb{Q}_p^n)$, i.e.,

$$\psi(\xi) = 0, \qquad \xi \in B_{-N}^n.$$

In fact, for $n = 1$, this lemma was proved in [19, IX.2.]. Unlike the classical Lizorkin space, any function $\psi(\xi) \in \Phi$ is equal to zero not only at $\xi = 0$ but in a ball $B^n \ni 0$, as well. It follows from (8) that the space $\Phi(\mathbb{Q}_p^n)$ does not contain real-valued functions everywhere different from zero.

Let $\Phi' = \Phi'(\mathbb{Q}_p^n)$ denote the topological dual of the space $\Phi(\mathbb{Q}_p^n)$. We call it the *p-adic Lizorkin space of distributions*.

By $\Psi^\perp$ and $\Phi^\perp$ we denote the subspaces of functionals in $\mathscr{D}'$ orthogonal to $\Psi$ and $\Phi$, respectively. Thus

$$\Psi^\perp = \{f \in \mathscr{D}'(\mathbb{Q}_p^n) : f = C\delta, C \in \mathbb{C}\} \quad \text{and} \quad \Phi^\perp = \{f \in \mathscr{D}'(\mathbb{Q}_p^n) : f = C, C \in \mathbb{C}\}.$$

**Proposition 1.**
$$\Phi' = \mathscr{D}'/\Phi^\perp, \qquad \Psi' = \mathscr{D}'/\Psi^\perp.$$

*Proof.* This proposition can be proved in the same way as [15, Proposition 2.5.]. It follows from the well-known assertion: if $E$ is a topological vector space with a closed subspace $M$ then $E'$ can be identified with the quotient space $M' = E'/M^\perp$, where $M^\perp = \{f \in E' : \langle f, \varphi \rangle = 0, \forall \varphi \in M\}$. □

The space $\Phi'(\mathbb{Q}_p^n)$ can be obtained from $\mathscr{D}'(\mathbb{Q}_p^n)$ by "sifting out" constants. Thus two distributions in $\mathscr{D}'(\mathbb{Q}_p^n)$ differing by a constant are indistinguishable as elements of $\Phi'(\mathbb{Q}_p^n)$.

We define the Fourier transform of distributions $f \in \Phi'(\mathbb{Q}_p^n)$ and $g \in \Psi'(\mathbb{Q}_p^n)$ by the relations:

$$\begin{aligned} \langle F[f], \psi \rangle &= \langle f, F[\psi] \rangle, \quad \forall \psi \in \Psi(\mathbb{Q}_p^n), \\ \langle F[g], \phi \rangle &= \langle g, F[\phi] \rangle, \quad \forall \phi \in \Phi(\mathbb{Q}_p^n). \end{aligned} \quad (9)$$

By definition, $F[\Phi(\mathbb{Q}_p^n)] = \Psi(\mathbb{Q}_p^n)$ and $F[\Psi(\mathbb{Q}_p^n)] = \Phi(\mathbb{Q}_p^n)$, i.e., (9) give well defined objects.

Let $\Psi'_M(\mathbb{Q}_p^n)$ be a class of multipliers in $\Psi(\mathbb{Q}_p^n)$ and $\Phi'_*(\mathbb{Q}_p^n)$ a class of convolutes in $\Phi(\mathbb{Q}_p^n)$. Thus $\Phi'_*(\mathbb{Q}_p^n) = F[\Psi'_M(\mathbb{Q}_p^n)]$. It is clear that a distribution $f \in \Psi'(\mathbb{Q}_p^n)$ is a multiplier in $\Psi(\mathbb{Q}_p^n)$ if and only if $f \in \mathscr{E}(\mathbb{Q}_p^n \setminus \{0\})$. Since $\mathscr{E}(\mathbb{Q}_p^n) \subset \Psi'_M(\mathbb{Q}_p^n)$, according to the theorem from [19, VII.3.], the class of all compactly supported distributions from $f \in \mathscr{D}'(\mathbb{Q}_p^n)$ is a subset of $\Phi'_*(\mathbb{Q}_p^n)$.

**Lemma 2.** *The space $\Phi(\mathbb{Q}_p^n)$ is dense in $\mathscr{L}^\rho(\mathbb{Q}_p^n)$, $1 < \rho < \infty$.*

The proof of Lemma 2 is carried out practically word for word as the proof of the corresponding ("$\mathbb{C}$-case") Theorem 2.6 from [15, 2.2.]. For $n = 1$ and $\rho = 2$ the statement of Lemma 2 coincides with the lemma from [19, IX.4.]

## 4. MULTI-DIMENSIONAL FRACTIONAL OPERATOR

Let us introduce the distribution from $\mathscr{D}'(\mathbb{Q}_p^n)$

$$\kappa_\alpha(x) = \frac{|x|_p^{\alpha-n}}{\Gamma_p^{(n)}(\alpha)}, \quad \alpha \neq 0, n, \quad x \in \mathbb{Q}_p^n, \tag{10}$$

called the multidimensional *Riesz kernel* [17, §2], [18, III.4.], where the function $|x|_p$, $x \in \mathbb{Q}_p^n$ is given by (1). The Riesz kernel has a removable singularity at $\alpha = 0$ and according to [17, §2], [18, III.4.], [19, VIII.2], we have

$$\langle \kappa_\alpha(x), \varphi(x) \rangle = \frac{g_\alpha}{\Gamma_p^{(n)}(\alpha)} + \frac{1-p^{-n}}{(1-p^{-\alpha})\Gamma_p^{(n)}(\alpha)} \varphi(0)$$

$$= g_\alpha \frac{1-p^{-\alpha}}{1-p^{\alpha-n}} + \frac{1-p^{-n}}{1-p^{\alpha-n}} \varphi(0), \quad \forall \varphi \in \mathscr{D}(\mathbb{Q}_p^n),$$

where $g_\alpha$ is an entire function in $\alpha$. Passing to the limit, we obtain that

$$\langle \kappa_0(x), \varphi(x) \rangle \stackrel{def}{=} \lim_{\alpha \to 0} \langle \kappa_\alpha(x), \varphi(x) \rangle = \varphi(0),$$

for all $\varphi \in \mathscr{D}(\mathbb{Q}_p^n)$, i.e.,

$$\kappa_0(x) \stackrel{def}{=} \lim_{\alpha \to 0} \kappa_\alpha(x) = \delta(x). \tag{11}$$

Next, using (3), (7), (10), and taking into account (8), we define $\kappa_n(\cdot)$ as a distribution from the *Lizorkin space of distributions* $\Phi'(\mathbb{Q}_p^n)$:

$$\langle \kappa_n(x), \phi \rangle \stackrel{def}{=} \lim_{\alpha \to n} \langle \kappa_\alpha(x), \phi \rangle = \lim_{\alpha \to n} \int_{\mathbb{Q}_p^n} \frac{|x|_p^{\alpha-n}}{\Gamma_p^{(n)}(\alpha)} \phi(x) d^n x$$

$$= -\lim_{\beta \to 0} (1-p^{-n-\beta}) \int_{\mathbb{Q}_p^n} \frac{|x|_p^\beta - 1}{p^\beta - 1} \phi(x) d^n x$$

$$= -\frac{1-p^{-n}}{\log p} \int_{\mathbb{Q}_p^n} \log|x|_p \phi(x) d^n x, \quad \forall \phi \in \Phi(\mathbb{Q}_p^n),$$

where $|\alpha - n| \leq 1$. Similarly to the one-dimensional case [19, IX.2], the passage to the limit under the integral sign is justified by the Lebesgue dominated theorem [19, IV.4]. Thus,

$$\kappa_n(x) \stackrel{def}{=} \lim_{\alpha \to n} \kappa_\alpha(x) = -\frac{1-p^{-n}}{\log p} \log|x|_p. \tag{12}$$

If $\alpha \neq n$ then the Riesz kernel $\kappa_\alpha(x)$ is a *homogeneous* distribution of degree $\alpha - n$, and if $\alpha = n$ then the Riesz kernel is an *associated homogeneous* distribution of degree 0 and order 1 (see Definitions 1,(b),(d)).

With the help of (6), (11), we obtain the formula [17, (**)], [18, III,(4.6)], [19, VIII,(4.9),(4.10)]: $\kappa_\alpha(x) * \kappa_\beta(x) = \kappa_{\alpha+\beta}(x)$, $\alpha, \beta, \alpha + \beta \neq n$, which holds in the sense of the space $\mathscr{D}'(\mathbb{Q}_p^n)$. Taking into account formula (12), it is easy to see that

$$\kappa_\alpha(x) * \kappa_\beta(x) = \kappa_{\alpha+\beta}(x), \quad \alpha, \beta \in \mathbb{C}, \tag{13}$$

in the sense of the Lizorkin space $\Phi'(\mathbb{Q}_p^n)$.

The multi-dimensional Taibleson operator on the Lizorkin space is defined as the convolution:

$$(D_x^\alpha \phi)(x) \stackrel{def}{=} \kappa_{-\alpha}(x) * \phi(x) = \langle \kappa_{-\alpha}(x), \phi(x - \xi) \rangle, \quad x \in \mathbb{Q}_p^n, \tag{14}$$

where $\phi \in \Phi(\mathbb{Q}_p^n)$, $\alpha \in \mathbb{C}$.

**Lemma 3.** *The Lizorkin space $\Phi(\mathbb{Q}_p^n)$ is invariant under the Taibleson fractional operator $D_x^\alpha$ and $D_x^\alpha(\Phi(\mathbb{Q}_p^n)) = \Phi(\mathbb{Q}_p^n)$.*

*Proof.* In view of formula (6), $F[\kappa_\alpha(x)](\xi) = |\xi|_p^{-\alpha}$. Consequently, using the convolution formula [19, VII,(5.4)], we have $F[D_x^\alpha \phi](\xi) = |\xi|_p^{-\alpha} F[\phi](\xi)$, $\phi \in \Phi(\mathbb{Q}_p^n)$. Thus $F[\phi](\xi), |\xi_1|_p^{-\alpha} F[\phi](\xi) \in \Psi(\mathbb{Q}_p^n)$, $\alpha \in \mathbb{C}$ and $D_x^\alpha \phi \in \Phi(\mathbb{Q}_p^n)$. That is $D_x^\alpha(\Phi(\mathbb{Q}_p^n)) \subset \Phi(\mathbb{Q}_p^n)$. Since any function from $\Psi(\mathbb{Q}_p^n)$ can be represented as $\psi(\xi) = |\xi|_p^\alpha \psi_1(\xi)$, $\psi_1 \in \Psi(\mathbb{Q}_p^n)$, we have $D_x^\alpha(\Phi(\mathbb{Q}_p^n)) = \Phi(\mathbb{Q}_p^n)$. □

According to (14), (6), and [19, VII,(5.4)], we have

$$(D_x^\alpha \phi)(x) = F^{-1}\left[|\xi|_p^\alpha F[\phi](\xi)\right](x), \quad \phi \in \Phi(\mathbb{Q}_p^n). \tag{15}$$

Taking into account the convolution formula [19, VII.1.] and (14), we define $D^\alpha f$ of a distribution $f \in \Phi'(\mathbb{Q}_p^n)$ by the relation

$$\langle D_x^\alpha f, \phi \rangle \stackrel{def}{=} \langle f, D_x^\alpha \phi \rangle, \quad \forall \phi \in \Phi(\mathbb{Q}_p^n). \tag{16}$$

It is clear that $D_x^\alpha(\Phi'(\mathbb{Q}_p^n)) = \Phi'(\mathbb{Q}_p^n)$. Moreover, in view of (13), the family of operators $D_x^\alpha$, $\alpha \in \mathbb{C}$ on the Lizorkin space forms an Abelian group: if $f \in \Phi'(\mathbb{Q}_p^n)$ then $D_x^\alpha D_x^\beta f = D_x^\beta D_x^\alpha f = D_x^{\alpha+\beta} f$, $D_x^\alpha D_x^{-\alpha} f = f$, $\alpha, \beta \in \mathbb{C}$.

By analogy with [15], [16] (see also [8]), one can introduce the $p$-adic Laplacian

$$-\Delta f(x) \stackrel{def}{=} (D_x^2 f)(x), \quad f \in \Phi'(\mathbb{Q}_p^n).$$

and its powers

$$(-\Delta)^{\alpha/2} f(x) \stackrel{def}{=} (D_x^\alpha f)(x), \quad f \in \Phi'(\mathbb{Q}_p^n), \quad \alpha \in \mathbb{C}.$$

# 5. PSEUDO-DIFFERENTIAL OPERATORS AND EQUATIONS

Similarly to the representation (15), one can consider a class of pseudo-differential operators in the Lizorkin space of the test functions $\Phi(\mathbb{Q}_p^n)$

$$(A\phi)(x) = F^{-1}[\mathscr{A}(\xi)F[\phi](\xi)](x)$$
$$= \int_{\mathbb{Q}_p^n}\int_{\mathbb{Q}_p^n} \chi_p((y-x)\cdot\xi)\mathscr{A}(\xi)\phi(y)\,d^n\xi\,d^ny, \quad \phi \in \Phi(\mathbb{Q}_p^n) \tag{17}$$

with symbols $\mathscr{A}(\xi) \in \mathscr{E}(\mathbb{Q}_p^n \setminus \{0\})$.

In view of results of Sec. 3, functions $F[\phi](\xi)$ and $\mathscr{A}(\xi)F[\phi](\xi)$ belong to $\Psi(\mathbb{Q}_p^n)$, and, consequently, $(A\phi)(x) \in \Phi(\mathbb{Q}_p^n)$. Thus the pseudo-differential operators (17) are well defined and the Lizorkin space $\Phi(\mathbb{Q}_p^n)$ is invariant under them.

If we define a conjugate pseudo-differential operator $A^T$ as

$$(A^T\phi)(x) = F^{-1}[\mathscr{A}(-\xi)F[\phi](\xi)](x) = \int_{\mathbb{Q}_p^n} \chi_p(-x\cdot\xi)\mathscr{A}(-\xi)F[\phi](\xi)\,d^n\xi \tag{18}$$

then one can define operator $A$ in the Lizorkin space of distributions: for $f \in \Phi'(\mathbb{Q}_p^n)$ we have

$$\langle Af, \phi \rangle = \langle f, A^T\phi \rangle, \quad \forall \phi \in \Phi(\mathbb{Q}_p^n). \tag{19}$$

It is clear that

$$Af = F^{-1}[\mathscr{A}F[f]] \in \Phi'(\mathbb{Q}_p^n), \tag{20}$$

i.e., the Lizorkin space of distributions $\Phi'(\mathbb{Q}_p^n)$ is invariant under pseudo-differential operators $A$.

If $A, B$ are pseudo-differential operators with symbols $\mathscr{A}(\xi), \mathscr{B}(\xi) \in \mathscr{E}(\mathbb{Q}_p^n \setminus \{0\})$, respectively, then the operator $AB$ is well defined and represented by the formula

$$(AB)f = F^{-1}[\mathscr{A}\mathscr{B}F[f]] \in \Phi'(\mathbb{Q}_p^n).$$

If $\mathscr{A}(\xi) \neq 0$, $\xi \in \mathbb{Q}_p^n \setminus \{0\}$ then we define the inverse pseudo-differential by the formula

$$A^{-1}f = F^{-1}[\mathscr{A}^{-1}F[f]], \quad f \in \Phi'(\mathbb{Q}_p^n).$$

Thus the family of pseudo-differential operators $A$ with symbols $\mathscr{A}(\xi) \neq 0$, $\xi \in \mathbb{Q}_p^n \setminus \{0\}$ forms an Abelian group.

If the symbol $\mathscr{A}(\xi)$ of the operator $A$ is an *associated homogeneous* function then the operator $A$ is called an *associated homogeneous pseudo-differential operator*.

According to formulas (15), (10)–(12), and Definition 1, the operator $D_x^\alpha$, $\alpha \neq -n$ is a *homogeneous* pseudo-differential operator of degree $\alpha$ with the symbol $\mathscr{A}(\xi) = |\xi|_p^\alpha$ and $D_x^{-n}$ is a *homogeneous* pseudo-differential operator of degree $-n$ and order 1 with the symbol $\mathscr{A}(\xi) = P(|\xi|_p^{-n})$ (see (5))

Let us consider a pseudo-differential equation

$$Af = g, \quad g \in \Phi'(\mathbb{Q}_p^n), \tag{21}$$

where $A$ is a pseudo-differential operator (17), $f$ is the desired distribution.

**Theorem 1.** *If the symbol of a pseudo-differential operator $A$ is such that $\mathscr{A}(\xi) \neq 0$, $\xi \in \mathbb{Q}_p^n \setminus \{0\}$ then the equation (21) has the unique solution*

$$f(x) = F^{-1}\left[\frac{F[g](\xi)}{\mathscr{A}(\xi)}\right](x) = (A^{-1}g)(x) \in \Phi'(\mathbb{Q}_p^n).$$

*Proof.* Applying the Fourier transform to the left-hand and right-hand sides of equation $Af = g$, in view of representation (20), we obtain that $\mathscr{A}(\xi)F[f](\xi) = F[g](\xi)$. Since according to Sec. 3, $F[\Phi'(\mathbb{Q}_p^n)] = \Psi'(\mathbb{Q}_p^n)$, $F[\Psi'(\mathbb{Q}_p^n)] = \Phi'(\mathbb{Q}_p^n)$, and $\mathscr{A}(\xi)$ is a multiplier in $\Psi(\mathbb{Q}_p^n)$, we have $F[f](\xi) = \mathscr{A}^{-1}(\xi)F[g](\xi) \in \Psi'(\mathbb{Q}_p^n)$. Thus $f(x) = F^{-1}[\mathscr{A}^{-1}(\xi)F[g](\xi)](x) = (A^{-1}g)(x) \in \Phi'(\mathbb{Q}_p^n)$ is a solution of the problem (21).

Now we study solutions of the homogeneous problem (22). Let $f \in \mathscr{D}'(\mathbb{Q}_p^n)$ and $Af = 0$, i.e., according to (19), $\langle Af, \phi \rangle = \langle f, A^T \phi \rangle = 0$, for all $\phi \in \Phi(\mathbb{Q}_p^n)$. In view of (18), $A^T(\Phi(\mathbb{Q}_p^n)) = \Phi(\mathbb{Q}_p^n)$, i.e., $\langle f, \phi \rangle = 0$, for all $\phi \in \Phi(\mathbb{Q}_p^n)$. Consequently, $f \in \Phi^\perp$ (see Proposition 1). Thus the solutions of the homogeneous problem (21) are indistinguishable as elements of the space $\Phi'(\mathbb{Q}_p^n)$. □

Let $P_N(z) = \sum_{k=0}^N a_k z^k$ be a polynomial, where $a_k \in \mathbb{C}$ are constants. Let us consider the equation

$$P_N(D_x^\alpha)f = g, \quad g \in \Phi'(\mathbb{Q}_p^n), \tag{22}$$

where $(D_x^\alpha)^k \stackrel{\text{def}}{=} D_x^{\alpha k}$, $\alpha \in \mathbb{C}$ and $f$ is the desired distribution.

**Theorem 2.** *If $P_N(z) \neq 0$ for all $z > 0$ then equation (22) has the unique solution*

$$f(x) = F^{-1}\left[\frac{F[g](\xi)}{P_N(|\xi|_p^\alpha)}\right](x) \in \Phi'(\mathbb{Q}_p^n). \tag{23}$$

*In particular, the unique solution of the equation*

$$D_x^\alpha f = g, \quad g \in \Phi'(\mathbb{Q}_p^n),$$

*is given by the formula* $f = D_x^{-\alpha} g \in \Phi'(\mathbb{Q}_p^n)$.

*Proof.* According to formulas (4)–(6), (10)–(12), $F[\kappa_\alpha(x)] = |\xi|_p^{-\alpha}$, $\alpha \in \mathbb{C}$ in $\Phi'(\mathbb{Q}_p^n)$. Consequently, applying the Fourier transform to the left-hand and right-hand sides of relation (22), we obtain (23). Here we must take into account the fact that $\frac{1}{P_N(|\xi|_p^\alpha)}$ is a multiplier in $\Psi(\mathbb{Q}_p^n)$. Thus (23) is the solution of the problem (22).

In view of the proof of Theorem 1, the homogeneous problem (22) has only a trivial solution. □

## ACKNOWLEDGMENTS

The authors would like to thank S. V. Kozyrev and I. V. Volovich for fruitful discussions. The authors also use this occasion to congratulate Branko Dragovich with his 60th birthday.

## REFERENCES

1. S. Albeverio, A.Yu. Khrennikov, V.M. Shelkovich, Associated homogeneous $p$-adic distributions, *J. Math. An. Appl.* **313**, 64–83 (2006).
2. S. Albeverio, A.Yu. Khrennikov, V.M. Shelkovich, Associated homogeneous $p$-adic generalized functions, *Dokl. Ross. Akad. Nauk* **393**, no. 3, 300–303 (2003). English transl. in *Russian Doklady Mathematics* **68**, no. 3, 354–357 (2003).
3. Ya. Aref'eva, B. Dragovich, and I.V. Volovich On the adelic string amplitudes, *Phys. Lett. B* **209**, no. 4, 445–450 (1998).
4. B. Dragovich, $p$-adic and adelic quantum mechanics, *Tr. Mat. Inst. Steklova* **245**, 72–85 (2004); Izbr. Vopr. $p$-adich. Mat. Fiz. i Anal.; translation in Proc. Steklov Inst. Math., **245**, no. 2, 64–77 (2004).
5. B. Dragovich, I.V. Volovich, $p$-adic strings and noncommutativity. *Noncommutative structures in mathematics and physics* (Kiev, 2000), NATO Sci. Ser. II Math. Phys. Chem., 22, Kluwer Acad. Publ., Dordrecht, pp. 391–399 (2001).
6. I.M. Gel'fand and G.E. Shilov, *Generalized functions. vol 1: Properties and operations.* New York, Acad. Press, 1964.
7. I.M. Gel'fand, M.I. Graev and I.I. Piatetskii-Shapiro, *Generalized functions, Vol **6**: Representation theory and automorphic functions*, Nauka, Moscow, 1966.
8. A. Khrennikov, Fundamental solutions over the field of $p$-adic numbers, *St. Petersburg Math. J.* **4**, no. 3, 613–628 (1993).
9. A. Khrennikov, *p-Adic valued distributions in mathematical physics*, Kluwer Academic Publ., Dordrecht, 1994.
10. A. Khrennikov, *Non-archimedean analysis: quantum paradoxes, dynamical systems and biological models*, Kluwer Academic Publ., Dordrecht, 1997.
11. A. Khrennikov, *Information dynamics in cognitive, psychological, social and anomalous phenomena*, Kluwer Academic Publ., Dordrecht, 2004.
12. A.N. Kochubei, *Pseudo-differential equations and stochastics over non-archimedean fields*, Marcel Dekker. Inc. New York, Basel, 2001.
13. P.I. Lizorkin, Generalized Liouville differentiation and the functional spaces $L_p^r(E_n)$, Imbedding theorems, *(Russian) Mat. Sb. (N.S.)* **60** (102), 613–628 (1963).
14. P.I. Lizorkin, Operators connected with fractional differentiation, and classes of differentiable functions, (Russian) *Studies in the theory of differentiable functions of several variables and its applications*, IV. Trudy Mat. Inst. Steklov. Vol. **117**, 613–628 (1972).
15. S.G. Samko, *Hypersingular integrals and their applications.* Taylor & Francis, London, 2002.
16. S.G. Samko, A.A. Kilbas, and O.I. Marichev, *Fractional integrals and derivatives and some of their applications*, Minsk, Nauka i Tekhnika, 1987 (in Russian); English translation: *Fractional integrals and derivatives. Theory and applications*, Gordon and Breach, London, 1993.
17. M.H. Taibleson, Harmonic analysis on $n$-dimensional vector spaces over local fields. I. Basic results on fractional integration, *Math. Annalen* **176**, 191–207 (1968).
18. M.H. Taibleson, *Fourier analysis on local fields*, Princeton University Press, Princeton, 1975.
19. V.S. Vladimirov, I.V. Volovich and E.I. Zelenov, *p-Adic analysis and mathematical physics*, World Scientific, Singapore, 1994.
20. V.S. Vladimirov, I.V. Volovich, $p$-Adic quantum mechanics, *Commun. Math. Phys.* 659–676 **123**, (1989).
21. I.V. Volovich, $p$-Adic string, *Class. Quant. Grav.* **4**, L83–L87 (1987).

# Sequence-Spaces and Applications

Nicole De Grande – De Kimpe

*Vrije Universiteit Brussels, Brussels, BELGIUM*

**Abstract.** We give a survey of the general theory of $p$-adic sequence spaces (special attention is paid to Köthe spaces) and show how this theory can be applied to the solution of various problems in $p$-adic functional analysis.

## 1. THE SPACE $c_0$

$\mathbb{K}$ will denote a field with non-archimedean valuation $|\cdot|$, which is non-trivial. We assume that $\mathbb{K}$ is complete for the metric induced by this valuation.

Let us introduce the following notations

$$c_0 = \{a = (\alpha_n)_n \mid \alpha_n \in \mathbb{K}, \forall n, \lim_n \alpha_n = 0\}, \qquad \|a\| = \max_n |\alpha_n|.$$

The following facts are well-known (see [9]):
- $(c_0, \|\cdot\|)$ *is a Banach space.*
- $(c_0)' = \ell^\infty$ *where* $(c_0)'$ *is the topological dual of* $c_0$ *and*
  $\ell^\infty = \{b = (\beta_n)_n \mid \beta_n \in \mathbb{K}, \forall n, \sup_n |\beta_n| < \infty\}$.
- $\sigma(c_0, \ell^\infty)$, *the weak topology on* $c_0$, *is coarser than the norm-topology but they have the same convergent sequences.*
- $(c_0, \sigma(c_0, \ell^\infty))$ *is sequentially complete.*
- $(c_0, \|\cdot\|)$ *is of countable type (i.e. it contains a countable subset whose linear span is dense in X). Indeed, the vectors* $e_n = (0, \ldots, 0, 1, 0, \ldots)$ *with 1 on place n form a Schander basis for* $(c_0, \|\cdot\|)$.
- $(c_0, \|\cdot\|)$ *is the "only" Banach space of countable type, i.e. every Banach space of countable type is linearly homeomorphic to* $c_0$ *([12], p. 66).*

We only consider infinite dimensional spaces.

## 2. PERFECT SEQUENCE SPACES

Everything in this section comes from [1] (if not specified otherwise) and remains true if $\mathbb{K}$ is not spherically complete.

A sequence space $\Lambda$ is a vector subspace of $\mathbb{K}^\mathbb{N}$. Its *Köthe dual* $\Lambda^\times$ is defined by

$$\Lambda^\times = \{b = (\beta_n)_n \mid \beta_n \in \mathbb{K}, \forall n, \text{ and } \lim_n \alpha_n \beta_n = 0, \forall a = (\alpha_n) \in \Lambda\}$$

We obviously have $\Lambda \subset \Lambda^{\times\times}$.

$\Lambda$ is called *perfect* if $\Lambda = \Lambda^{\times\times}$. Let us mention that $\Lambda^\times$ is always perfect. We only consider perfect sequence spaces.

$(\Lambda, \Lambda^\times)$ form a dual pair of vector spaces for the bilinear form

$$\Lambda \times \Lambda^\times \longrightarrow \mathbb{K}: \quad (a,b) \longrightarrow \sum_n \alpha_n \beta_n$$

The locally convex topologies of this dual pair we are interested in, at the moment are

- the weak topology $\sigma(\Lambda, \Lambda^\times)$ generated by the family of semi-norms

$$p_b(a) = \left|\sum_n \alpha_n \beta_n\right|, \quad a \in \Lambda, \ b \in \Lambda^\times$$

- the (finer) natural topology $\eta(\Lambda, \Lambda^\times)$ generated by the family of semi-norms

$$q_b(a) = \max_n |\alpha_n \beta_n|, \quad a \in \Lambda, \ b \in \Lambda^\times$$

The topologies $\sigma(\Lambda, \Lambda^\times)$ and $\eta(\Lambda, \Lambda^\times)$ yield the same convergent sequences. $(\Lambda, \eta(\Lambda, \Lambda^\times))$ is complete and $(\Lambda, \sigma(\Lambda, \Lambda^\times))$ is sequentially complete. Moreover $(\Lambda, \eta(\Lambda, \Lambda^\times))' = \Lambda^\times$.

The sequences $e_n$, $n = 1, 2, \ldots$ form a *Schander basis* for $(\Lambda, \eta(\Lambda, \Lambda^\times))$ and $(\Lambda, \eta(\Lambda, \Lambda^\times))' = \Lambda^\times$ (the same holds for $\sigma(\Lambda, \Lambda^\times)$).

Let $E$ be a locally convex space and $p$ a continuous semi-norm in $E$. We denote by $E_p$ the space $E/\text{Ker}\, p$ normed by $(\pi_p : E \longrightarrow E_p)$

$$\|\pi_p(x)\|_p = p(x), \quad x \in E.$$

Of course if $p$ is a norm then $E_p = E$.

The space $E$ is said to be *countable type* if $E_p$ is of countable type for each $p$.

Every locally convex space with a Schander basis is of countable type.

Every locally convex space of countable type is *strongly polar* (i.e. for every $U$, zero-neighbourhood (closed convex) in $E$ we have $U = U^{00}$). If $\mathbb{K}$ is spherically complete then every locally convex space over $\mathbb{K}$ is strongly polar (see [10]).

A subset $B$ of $E$ is called *compactoid* if for every $U \in \mathscr{U}$ ($\mathscr{U}$ = a basis of closed, convex, zero-neighbourhood of $0$ in $E$) then exists a finite subset $X$ of $E$ such that $B \subset C_0(X) + U$, where $C_0(X)$ is the convex hull of $X$.

It is clear that compactoid subsets of $E$ are bounded in $E$.

A subset $B$ of $(\Lambda, \eta(\Lambda, \Lambda^\times))$ is compactoid iff there exists $a \in \Lambda$ such that

$$B \subset \{(\gamma_n, \alpha_n) \mid |\gamma_n| \leq 1\}.$$

# 3. EXAMPLES

A) $\mathbb{K}^\mathbb{N}$.

$\Lambda = \mathbb{K}^\mathbb{N}$ is a perfect sequence space.
$\Lambda^\times = \bigoplus_n \mathbb{K}$ = the space of the finitely non-zero sequences.

$\sigma(\Lambda, \Lambda^\times) = \eta(\Lambda, \Lambda^\times)$ = the product topology on $\Lambda$ (for which it is a Fréchet space). It is clear that on $\Lambda$ (for any of these topologies) there is no continuous norm.

Conversely we have ( [3] Prop. 2.6).

*Let $E$ be a Fréchet space on which there is no continuous norm then*
  *(i) $E$ has a subspace which is linearly homeomorphic to $\mathbb{K}^\mathbb{N}$.*
  *(ii) If $\mathbb{K}$ is spherically complete, this subspace is complemented in $E$.*

B) **The space $c_0$.**

$c_0$ is a perfect sequence space, $(c_0)^\times = \ell^\infty$ and $\eta(c_0, c_0^\times)$ coincides with norm-topology on $c_0$.

About the role of $c_0$ in this theory we mention two results.
  (i) *Let $b = (b_n)$ be a sequence of strictly positive real numbers. We denote by $\Lambda_b$ the sequence space (see [1] and [8])*

$$\Lambda_b = \{a = (\alpha_n) \mid \lim_n |\alpha_n| \beta_n = 0\}.$$

*Then $\Lambda_b$ is a perfect sequence space*

$$\Lambda_b^\times = \{a = (\alpha_n) \mid \sup_n \frac{|\alpha_n|}{\beta_n} < \infty\}.$$

*and the space $\Lambda_b$ is linearly homeomorphic to $c_0$ when $\Lambda_b$ is equipped with $\eta(\Lambda_b, \Lambda_b^\times)$.*

  (ii) *An operator ( = a continuous linear mapping) $T : E \longrightarrow F$ ($E, F$ locally convex spaces) is called* compact *when there exists a zero-neighbourhood $U$ in $E$ such that $T(U)$ is compactoid in $F$.*

*A locally convex space $E$ is called* nuclear *if for every continuous semi-norm $p$ on $E$ there exists a continuous semi-norm $q$ on $E$ with $q \geq p$ such that continuous map $\varphi_{pq} : E_q \longrightarrow E_p$ is compact.*

*Every nuclear space is of countable type. "Conversely" we have (see [2])*

*Suppose $E$ is of countable type. If every operator from $E$ to $c_0$ is compact then $E$ is nuclear (see also [10]).*

208

Other characterizations of nuclear spaces are given in [5].
More examples of perfect sequence spaces follow.

## 4. KÖTHE SPACES

Let $B = (b_n^k)_{n,k}$ be an infinite matrix consisting of strictly positive real numbers satisfying the condition

$$b_n^k \leq b_n^{k+1} \quad \text{for all } n \text{ and all } k.$$

We denote by $K(B)$ the sequence space

$$K(B) = \{a = (\alpha_n) \mid \lim_n |\alpha_n| b_n^k = 0, \ k = 1, 2, \ldots\}.$$

Then $\Lambda = K(B)$ is a perfect sequence space. The topology determined by the increasing sequence of norm

$$p_b(a) = \max_n |\alpha_n| b_n^k$$

is exactly $\eta(\Lambda, \Lambda^\times)$. Much more information about spaces of this type can be found in [7].

We concentrate on $\Lambda = K(B)$ with the above topology. Note that $K(B)$ is a Fréchet space and that the vectors $e_n$ ($n = 1, 2, \ldots$) form a Schander basis for $\Lambda$. Also note that $c_0$ is a Köthe space for $b_n^k = 1$, $\forall n, \forall k$ and that

$$K(B)^\times = \{(\beta_n) \mid \exists k \text{ with } \sup_n \frac{|\beta_n|}{b_n^k} < \infty\}.$$

Conversely we have:

*Every countably normed Fréchet space with a Schander basis is linearly homeomorphic to a Köthe space (see [3]).*

The nuclearity of a Köthe space can easily be characterized. Indeed (see [3]):

$\Lambda(B)$ is nuclear iff $\forall k, \exists k_1 > k$ with $\dfrac{b_n^k}{b_n^{k_1}} \longrightarrow 0$ *(when $n$ goes to $\infty$).*

It was also proved in [3] that

*the Köthe dual $\Lambda^\times$ of a Köthe space $\Lambda$ is always nuclear for its natural topology $\eta(\Lambda^\times, \Lambda)$.*

We finish this section with the following remark (also proved in [3]). There exist nuclear Köthe spaces which are not linearly homeomorphic (compare with $c_0$).

# 5. APPLICATIONS

**1. The space $A$ of entire functions (see [3]).**

$f: \mathbb{K} \longrightarrow \mathbb{K}$ is an entire function if can be written as

$$f(x) = \sum_{n=1}^{\infty} \alpha_n x^n, \alpha_n \in \mathbb{K}, \forall n \quad \text{and} \quad \lim_n \alpha_n x^n = 0, \forall x \in \mathbb{K}.$$

A (canonical) locally convex topology on $A$ is given by the sequence of norms

$$p_k(f) = \max_n |\alpha_n| k^n, \qquad k = 1, 2, 3, \ldots$$

Identifying $f$ with its sequence of coefficients we see that $A = K(B)$ (algebraically and topologically) where

$$b_n^k = k^n.$$

It follows immediately that $A$ is nuclear.

**2. The space $A_1$ of functions that are analytic in the unit ball of $\mathbb{K}$ (see [3]).**

If $f \in A_1$ then it can be written as

$$f(x) = \sum_{n=1}^{\infty} \alpha_n x^n, \alpha_n \in \mathbb{K}, \forall n \quad \text{and} \quad \lim_n \alpha_n x^n = 0, \forall x \in \mathbb{K}, |x| < 1.$$

$A_1$ has a (canonical) locally convex topology given by the sequence of norms:

$$p_k(f) = \max_n |\alpha_n| \left(\frac{k}{k+1}\right)^n, \qquad k = 1, 2, 3, \ldots$$

Identifying $f \equiv (\alpha_n)$ we see again that $A_1 = K(B)$ with

$$b_n^k = \left(\frac{k}{k+1}\right)^n.$$

It follows that $A_1$ is nuclear.

**3. The space $C^\infty(\mathbb{Z}_p)$ (see [4]).**

The space $s$ of rapidly decreasing sequences is the Köthe space $\Lambda(B)$ associated with the matrix $B = (b_n^k)$ where $b_n^k = n^k$, i.e.

$$s = \{a = (\alpha_n) \mid \lim_n |\alpha_n| n^k = 0, \ k = 1, 2, \ldots\}$$

with norm

$$\|a\|_k = \max_n |\alpha_n| n^k, \qquad k = 1, 2, \ldots$$

It is easy to see that this space is nuclear.

Let now $f \in C(\mathbb{Z}_p \longrightarrow \mathbb{Q}_p)$ (the continuous functions from $\mathbb{Z}_p$ to $\mathbb{Q}_p$) and with $f$ in its expansion with respect to the Mahler basis, i.e. for $x \in \mathbb{Z}_p$

$$f(x) = \sum_{n=0}^{\infty} \alpha_n \binom{x}{n} \quad \text{where} \quad \binom{x}{n} = \frac{x(x-1)\cdots(x-n+1)}{n!}. \tag{5.1}$$

Then apply the following results (proved in [10]),

*Let $f$ be written as in (5.1) then*

(i) $f \in C^k(\mathbb{Z}_p) \iff \lim_n |\alpha_n| n^k = 0$.
(ii) *The norms* $\max_{0 \leq j \leq k} \|\Phi_j f\|_\infty$ *and* $|\alpha_0| \vee \sup_{n \geq 1} |\alpha_n| n^k$ *are equivalent.*

Since

$$C^\infty(\mathbb{Z}_p) = \bigcap_k C^k(\mathbb{Z}_p) \quad \text{it follows} \quad C^\infty(\mathbb{Z}_p) = s.$$

## 4. The Laplace transform (see [5] for details).

We assume $\mathbb{K} = \mathbb{Q}_p$, but most of the results are still valid in a more general case.

We called $A$ the *space of entire functions* (see Subsection 1 of this Section) and $A_0 = A^\times$ the *space of function that are analytic at zero* (In [5] we explained the reason for this name. Compare also with Subsection 2 of this Section). On $A^\times$ we consider the topology $\eta(\Lambda, \Lambda^\times)$.

The non-archimedean functions $e^x, \sin x, \cos x, \log(1+x), \ldots$ (see [10] for their definitions) are elements of $A_0$.

As usual the spaces $A$ and $A_0$ will be spaces of test functions. Their dual spaces $A'$ and $A'_0$ are the spaces of distributions. Note that topology on $A_0 = A^\times$, namely the inductive limit topology coincides with $\eta(\Lambda, \Lambda^\times)$.

We define, for $y \in \mathbb{Q}_p$ the function

$$\exp y_\bullet : x \longrightarrow e^{xy}.$$

Then $\exp y_\bullet \in A_0$ and it can be identified with the sequence $(y^n/n!)$.

The definition of the Laplace transform then reads (as in the classical case): for $g \in A'_0$ and $y \in \mathbb{Q}_p$

$$\mathscr{L}(g)(y) = \langle \exp y_\bullet, g \rangle,$$

where $\langle \exp y_\bullet, g \rangle$ is the value of $g \in A'_0$ at $\exp y_\bullet \in A_0$.

This Laplace transform has all the nice properties we have in classical case, e.g. $\mathscr{L}$ and its transposed $\mathscr{L}^*$ behave well with respect to derivatives and convolutions. In fact all these properties can be proved easily identifying the test functions and distributions with their corresponding sequences. So we obtain, e.g. the Laplace transform $\mathscr{L}$ is a linear homeomorphism

$$\mathscr{L} : (A'_0, \beta(A'_0, A)) = (A, \eta(A, A^\times)) \longrightarrow (A, \eta(A, A^\times)).$$

Since we have all the necessary properties the Laplace transform is an important tool in the solution of differential equations (see [5] for these applications).

## 5. A counterexample.

In a nuclear space every bounded subset is compactoid and one may ask if the property "every bounded subset is compactoid" implies the nuclearity of the space. Sometimes it does.

Let for example, $X$ be a zero-dimensional Hausdorff space and $C(X) = \{f : X \longrightarrow \mathbb{K} \mid f \text{ is continuous}\}$. On $C(X)$ we consider compact-open topology $\tau_c$. It is generated by the family of semi-norms

$$p_A(f) = \sup_{x \in A} |f(x)|, \qquad A \text{ is compact subset of } X.$$

Then the following statements are equivalent
(a) $(C(X), \tau_c)$ is nuclear.
(b) Every $\tau_c$ bounded subset of $C(X)$ is $\tau_c$ compactoid.
(c) Every compact subset of $X$ is finite.

A proof of this result and more equivalences can be found in [6].

But this is not always the case; even for Fréchet spaces with a Schander basis.

A counterexample is given in [7]. It is in fact a Köthe space $K(B)$.

Firstly, we find a condition on $B$ (note that all the information on $K(B)$ is contained in the structure of $B$) such that in $\Lambda(B)$ every bounded subset is compactoid ($K(B)$ is then called a *Montel space*).

We have (see [7]):

(*) *If for every $k$ and every subsequence $(i_n)$ of the indices, there exists $k_1 > k$ such that $(b_{i_n}^{k_1}/b_{i_n}^k)_n$ is bounded. Then $K(B)$ is a Montel space.*

On the other hand the nuclearity of $K(B)$ can be seen (Subsection 3).

It then is sufficient to find a matrix $B$ for which (*) is satisfied and such that $K(B)$ is not nuclear. Such a (kind of exotic) matrix is given in [7].

## REFERENCES

1. De Grande-De Kimpe N., Perfect locally $\mathbb{K}$-convex sequence spaces, *Proc. Kon. Ned. Akad. Wct* **A75**, 471–482 (1971).

2. ——, Structure theorems for locally $\mathbb{K}$-convex spaces, *Proc. Kon. Ned. Akad. Wct* **A80**, 11–22 (1977).
3. ——, Non-archimedean Fréchet spaces generalizing spaces of analytic functions, *Proc. Kon. Ned. Akad. Wct* **A85**, 423–439 (1982).
4. ——, The non-archimedean space $C^\infty(X)$, *Comp. Math.* **48**, 297–309 (1983).
5. De Grande-De Kimpe N. and Khrennikov A., The non-archimedean Laplace transform, *Belg. Math. Soc. Simon Stevin* **3**, 225–237 (1996).
6. De Grande-De Kimpe N. and Navarro S., Non-archimedean nuclearity and spaces of continuous functions, *Indag. Math. N. S.* **2**, 201–206 (1991).
7. De Grande-De Kimpe N., Perez-Garcia C. and Schikhof W., Non-archimedean $t-$ frames and *FM* spaces, *Can. Math. Bull.* **35**, 475–483 (1992).
8. De Grande-De Kimpe N., Kahol T., Perez-Garcia C. and Schikhof W., *p*-adic locally convex induction limits; in: *p-adic functional analysis* (Schikhof, Perez-Garcia, Kahol eds.) Marcel Dekker. *Proceedings of the 1996 Conference of p-adic functional analysis (Nymegen)*, 159–222.
9. Monna A.F., Espaces linéaires à une infinité dénormiable de coordonnées, *Proc. Kon. Ned. Akad. Wct* **A53**, 1548–1559 (1950).
10. Schikhof W., *Ultrametric calculus*, Cambridge University Press, 1984.
11. ——, Locally convex spaces over non-spherically complete fields I-II, *Bull. Soc. Math. Belg. (ser B)* **XXXVIII**, 187–224 (1986).
12. Van Rooij A., *Non-archimedean functional analysis*, Marcel Dekker, 1978.

# Ultrametric Gelfand Transforms

Alain Escassut* and Nicolas Maïnetti[†]

*Laboratoire de Mathématiques Pures, Université Blaise Pascal (Clermont-Ferrand),
63177 Aubière Cedex, FRANCE

[†]LLAIC, IUT Université d'Auvergne (Clermont-Ferrand), BP 86,
63172 Aubière Cedex, FRANCE
email: mainetti@iut.u-clermont1.fr

**Abstract.** Let $K$ be an algebraically closed field, complete for a non-trivial ultrametric absolute value, and let $A$ be a commutative normed $K$-algebra with identity. We call *multiplicative spectrum* of $A$ the set $Mult(A, \|\cdot\|)$ of continuous multiplicative semi-norms on $A$. We denote by $Mult(K[x])$ the set of multiplicative semi-norms on the polynomial algebra $K[x]$. Both sets of semi-norms are endowed with the topology of pointwise convergence. We also denote by $\mathscr{X}(A, K)$ the set of $K$-algebra homomorphism from $A$ onto $K$. Unlike in complex analysis, there might exist some maximal ideals in $A$ which are not the kernel of elements of $\mathscr{X}(A,K)$. In $A$, we can define two kinds of Gelfand transform. $G_A$ and $GM_A$. The first one, denoted by $G_A$ is similar to that in complex analysis, consisting of associating to each element $f$ of $A$ the mapping $\widehat{f}$ from $\mathscr{X}(A, K)$ to $K$ defined as $\widehat{f}(\chi) = \chi(f)$, $(\chi \in \mathscr{X}(A, K))$. The second, denoted by $GM_A$ consists of associating to each element $f$ of $A$ the mapping $f^*$ from $Mult(A, \|\cdot\|)$ to $Mult(K[x])$ defined as $f^*(\phi)(P) = \phi(P \circ f)$. This transform allows us to interpret any element of $A$ as a continuous function defined on a compact space $(Mult(A, \|\cdot\|))$ and with value in a locally compact space $(Mult(K[x]))$. We study these transforms and particularly the injectivity of the second. We deduce some spectral properties of this injectivity. Given $\phi \in Mult(A, \|\cdot\|)$, we will denote by $Z_\phi$ the mapping from $A$ into $Mult(K[x])$ defined as $Z_\phi(f) = f^*(\phi)$. We show that each function $Z_\phi$ is continuous. Moreover we put a metric topology $\delta$ on $(Mult(K[x]))$ such that the family of functions $Z_\phi$, $\phi \in Mult(A, \|\cdot\|)$ is equicontinuous.

**Keywords:** Ultrametric Functional Analysis, Normed Algebras, Gelfand Transform.
**PACS:** 02.30.Sa.

## 1. INTRODUCTION AND RESULTS

**Notation.** Throughout the paper, $K$ will denote an algebraically closed field complete for a nontrivial ultrametric absolute value and $(A, \|\cdot\|)$ will be a commutative normed $K$-algebra with unity. We will denote by $\|\cdot\|_{si}$ the spectral semi-norm of $A$ defined as $\|x\|_{si} = \lim_{n \to \infty} \|x^n\|^{\frac{1}{n}}$. We will denote by $Max(A)$ the set of maximal ideals of $A$ and by $Max_a(A)$ the set of maximal ideals of codimension 1 of $A$.

We will denote by $Mult(A)$ the set of non identically zero multiplicative semi-norms of $A$, by $Mult_m(A)$ the set of the $\varphi \in Mult(A)$ such that $Ker(\varphi) \in Max(A)$, by $Mult_a(A)$ the set of the $\varphi \in Mult(A)$ such that $Ker(\varphi) \in Max_a(A)$.

We will denote by $Mult(A, \|\cdot\|)$ the set of $\varphi \in Mult(A)$ which are continuous with respect to the norm $\|.\|$ and we put $Mult_m(A,\|.\|) = Mult(A,\|.\|) \cap Mult_m(A)$ and $Mult_a(A,\|.\|) = Mult(A,\|.\|) \cap Mult_a(A)$.

We can define several usual semi-multiplicative semi-norms:

$\|x\|_{sa} = \sup\{\varphi(x) \,|\, \varphi \in Mult_a(A,\|.\|)\}$,

$\|x\|_{si} = \lim_{n \to \infty} \|x^n\|^{\frac{1}{n}}$.

Moreover, we put $\tau(x) = \sup\{|\lambda| \,|\, \lambda \in sp(x) \cup \{0\}\}$.

**Remark 1.** It is well known that $\|x\|_{si} = \sup\{\varphi(x) \,|\, \varphi \in Mult(A,\|.\|)\}$ ([10], Theorem 6.25).

**Remark 2.** In certain cases, $sp(x)$ might be empty (for instance in a field extension of $K$). This is why $\tau(x)$ involves $sp(x) \cup \{0\}$.

Proposition A is immediate:

**Proposition A.** *Assume $Max_a(A) \neq \emptyset$. Then $\|x\|_{sa} \leq \tau(x) \leq \|x\|_{sm} \leq \|x\|_{si}$ $\forall x \in A$. Moreover, $\|x\|_{sa} = \sup\{|\chi(x)| \,|\, \chi \in \mathscr{X}(A,K)\}$ $\forall x \in A$.*

**Remark 3.** $\tau$ is not a semi-norm. Indeed, as we will see later, there might exist $\chi \in \mathscr{X}(A) \setminus \mathscr{X}(A,K)$ and $x,y \in A$ such that $\chi(x) = 1+t$, $\chi(y) = -t$, with $t \notin K$, and $\max(\tau(x), \tau(y)) < 1$. However we have $\chi(x+y) = 1$, hence $\tau(x+y) \geq 1$.

**Notation.** In $A$ we will denote by o), p), q) these properties:

o) $\|x\|_{sa} = \tau(x)$, $\forall x \in A$,

p) $\|x\|_{sa} = \|x\|_{si}$, $\forall x \in A$,

q) $\tau(x) = \|x\|_{si}$, $\forall x \in A$.

**Notation.** In $A$ we can define two kinds of Gelfand transform $G_A$ and $GM_A$. The first one, denoted by $G_A$ is similar to that in complex analysis in complex analysis, consisting of associating to each element $f$ of $A$ the mapping $\hat{f}$ from $\mathscr{X}(A,K)$ to $K$ defined as $\hat{f}(\chi) = \chi(f)$, $(\chi \in \mathscr{X}(A,K))$. The second, denoted by $GM_A$ consists of associating to each element $f$ of $A$ the mapping $f^*$ from $Mult(A,\|.\|)$ to $Mult(K[x])$ defined as $f^*(\phi)(P) = \phi(P \circ f)$.

**Remark 4.** By definition, $G_A$ is injective if and only if the intersection of all maximal ideals of codimension 1 is null.

As in complex analysis, Theorem B is immediate:

**Theorem B.** *$\mathscr{X}(A,K)$ being provided with the topology of simple convergence, for every $f \in A$, $\hat{f}$ belongs to $\mathscr{C}(\mathscr{X}((A,K),K)$. Moreover, $G_A$ satisfies $\|\hat{f}\| = \|f\|_{sa}$ for every $f \in A$*

**Theorem C.** *Assume that $A$ satisfies Property p). Then $\mathscr{C}(\mathscr{X}(A,K),K)$ being provided with the norm of uniform convergence, then $G_A$ is an isometry if and only if $\|x^2\| = \|x\|^2$ $\forall x \in A$.*

**Theorem D.** *Assume that A satisfies Property p). The following two properties are equivalent on A:*

*(i1) A is semi-simple and $G_A(A)$ is closed in $\mathscr{C}(\mathscr{X}(A,K),K)$,*
*(i2) A is uniform.*

Now, we shall examine the mapping $GM_A$.

**Theorem E.** *Given $f \in A$, the mapping $f^*$ from $Mult(A, \|\cdot\|)$ to $Mult(K[x])$ is continuous with respect to the topology of simple convergence.*

**Theorem F.** *Assume that the intersection of maximal ideals of codimension 1 is null. Then $GM_A$ is injective.*

Now, we have to recall results of [2] and [6].

We call *circular filter of center a and diameter r on K* the filter $\mathscr{F}$ which admits as a generating system the family of sets $\Gamma(\alpha, r', r'')$ with $\alpha \in d(a,r), r' < r < r''$, i.e. $\mathscr{F}$ is the filter which admits for base the family of sets of the form $\bigcap_{i=1}^{q} \Gamma(\alpha_i, r'_i, r''_i))$ with $\alpha_i \in d(a,r), r'_i < r < r''_i$ ($1 \leq i \leq q$, $q \in \mathbb{N}$).

In order to characterizing all absolute values of $K(x)$, we call *circular filter with no center, of diameter r of canonical base* $(D_n)_{n \in \mathbb{N}}$ a filter admitting for base a sequence $(D_n)_{n \in \mathbb{N}}$ where each $D_n$ is a disk $d(a_n, r_n)$, such that $\bigcap_{n=1}^{\infty} d(a_n, r_n) = \emptyset$ and $\lim_{n \to \infty} r_n = r$.

Finally the filter of neighbourhoods of a point $a \in K$ is called *circular filter of the neighbourhoods of a*. It will also be named *circular filter of center a and diameter 0* or *Cauchy circular filter of limit a*.

A circular filter is said to be *large* if it has diameter different from 0, and to be *punctual* if it is a Cauchy circular filter.

Given a circular filter $\mathscr{F}$, its diameter will be denoted by $diam(\mathscr{F})$ and the set of its centers is denoted by $\mathscr{Q}(\mathscr{F})$. So, if $\mathscr{F}$ is a circular filter of center $a$ and diameter $r$, then $\mathscr{Q}(\mathscr{F}) = d(a,r)$.

The set of circular filters on $K$ secant with a subset $D$ of $K$ will be denoted by $\Phi(D)$ and the set of large circular filters will be denoted by $\Phi'(D)$. Given two circular filters $\mathscr{F}$ and $\mathscr{G}$, $\mathscr{F}$ is said *to surround* $\mathscr{G}$ if either $\mathscr{G}$ is secant with $\mathscr{Q}(\mathscr{F})$, or if $\mathscr{F} = \mathscr{G}$.

We will denote by $\preceq$ the relation on the set of circular filters defined as $\mathscr{F} \preceq \mathscr{G}$ if $\mathscr{G}$ surrounds $\mathscr{F}$ and by $\prec$ the relation defined as $\mathscr{F} \prec \mathscr{G}$ if $\mathscr{F} \preceq \mathscr{G}$ and $\mathscr{F} \neq \mathscr{G}$.

**Theorem G ([2]).** *The relation $\preceq$ is an order relation on $\Phi(K)$ and $\prec$ is the strict order associated to this order relation. Moreover, if $\mathscr{F}, \mathscr{G}, \mathscr{H}$ satisfy $\mathscr{F} \preceq \mathscr{G}$ and $\mathscr{F} \preceq \mathscr{H}$ then $\mathscr{G}$ and $\mathscr{H}$ are comparable.*

**Proposition H.** *Let $\mathscr{F}, \mathscr{G}$ be two circular filters which are not comparable for $\preceq$. Given disks $F \in \mathscr{F}$ and $G \in \mathscr{G}$ such that $F \in \mathscr{F}, G \in \mathscr{G}$, $F \cap G = \emptyset$ then $\delta(F,G)$ does not depend on the choice of F and G.*

**Notation.** In the hypothesis of Proposition G, we put $\lambda(\mathscr{F},\mathscr{G}) = \delta(F,G)$ with $F$, $G$ disks such that $F \in \mathscr{F}, G \in \mathscr{G}, F \cap G = \emptyset$.

**Theorem I ([2]).** *Let $\mathscr{F},\mathscr{G} \in \Phi(K)$. There exists $\sup(\mathscr{F},\mathscr{G}) \in \Phi(K)$ and it is the unique circular filter of diameter $\lambda(\mathscr{F},\mathscr{G})$ which surrounds both $\mathscr{F}$, $\mathscr{G}$.*

**Notation.** Let $\mathscr{F}$, $\mathscr{G}$ be two circular filters and let $\mathscr{S} = \sup(\mathscr{F},\mathscr{G})$. We put $\delta(\mathscr{F},\mathscr{G}) = \max(diam(\mathscr{S}) - diam(\mathscr{F}), diam(\mathscr{S}) - diam(\mathscr{G}))$.

**Theorem J ([2]).** *$\delta$ is a distance on $\Phi(K)$.*

**Theorem K (G. Garandel [8], [9]).** *For every circular filter $\mathscr{F}$ on $K$, for every polynomial $P(x) \in K[x], |P(x)|$ has a limit $\varphi_{\mathscr{F}}(P)$ along the filter $\mathscr{F}$.*

**Theorem L (Garandel-Guennebaud [8], [9]).** *The mapping $\Psi$ from $\Phi(K)$ into $Mult(K[x])$ defined as $\Psi(\mathscr{F}) = \varphi_{\mathscr{F}}$ is a bijection. Moreover, the restriction $\Psi'$ of $\Psi$ to $\Phi'(K)$ is a bijection from $\Phi'(K)$ onto the set of multiplicative norms on $K[x]$, and therefore onto the set of multiplicative norms on $K(x)$.*

**Notation.** Let $\mathscr{F}$, $\mathscr{G}$ be two circular filters. Since we have a one to one correspondence between $Mult(K[x])$ and $\Phi(K)$, we can also put $\delta(\varphi_{\mathscr{F}}, \varphi_{\mathscr{G}}) = \delta(\mathscr{F},\mathscr{G})$, giving $Mult(K[x])$ a metric topology.

**Theorem M ([2]).** *The $\delta$-topology is stronger than the topology of simple convergence on $Mult(K[x])$. The restrictions of both topologies to a totally ordered subset of $Mult(K[x])$ are identical.*

We are now able to claim Theorem N:

**Theorem N.** *Let $A$ be uniform. Let $\phi \in Mult(A, \|\,.\,\|)$, let $f, g \in A$ be such that $f^*(\phi) \neq g^*(\phi)$. Let $\varphi_{\mathscr{F}} = f^*(\phi)$, $\varphi_{\mathscr{G}} = g^*(\phi)$. Then $diam(\sup(\mathscr{F},\mathscr{G})) = \phi(f-g)$.*

**Corollary N1.** *Let $A$ be uniform. Let $\phi \in Mult(A, \|\,.\,\|)$, let $f, g \in A$. Let $\varphi_{\mathscr{F}} = f^*(\phi)$, $varphi_{\mathscr{G}} = g^*(\phi)$. Then $\delta(f^*(\phi), g^*(\phi))) \leq \phi(f-g) \leq \|f-g\|$.*

**Notation.** Given $\phi \in Mult(A, \|\,.\,\|)$, we will denote by $Z_\phi$ the mapping from $A$ into $Mult(K[x])$ defined as $Z_\phi(f) = f^*(\phi)$.

**Corollary N2.** *Let $A$ be uniform. The family of functions $Z_\phi$, $\phi \in Mult(A, \|\,.\,\|)$ is uniformly equicontinuous with respect to the $\delta$-topology on $Mult(K[x])$.*

**Corollary N3.** *Let $A$ be uniform. Each function $Z_\phi$, $\phi \in Mult(A, \|\,.\,\|)$ is continuous with respect to the topology of simple convergence on $Mult(K[x])$.*

**Theorem O.** *Let $A$ be uniform and complete. If $A$ does not satisfy Property q), then $GM_A$ is not injective.*

**Corollary O1.** *Let A be uniform, having at least one maximal ideal of codimension* 1 *and such that the intersection of all maximal ideals of codimension* 1 *is equal to* $\{0\}$. *Then A satisfies Property q).*

Indeed, by Theorem F $GM_A$ is injective, hence by Theorem I, $A$ satisfies Property q).

## 2. THE PROOFS

**Proof of Theorem C.** Suppose $\|x^2\| = \|x\|^2$ is true for all $x \in A$, let $x \in A$ and let $\rho = \frac{\|x\|_{si}}{\|x\|}$. Then we check that $\frac{\|x^{2^n}\|_{si}}{\|x^{2^n}\|} = \rho^{(2^n)}$ hence $\|x\|_{si} = \rho \lim_{n\to\infty} \|x^n\|^{\frac{1}{n}}$, and therefore $\rho = 1$. So, $G_A$ is an isometry. The converse is trivial. □

**Proof of Theorem D.** First, suppose that $A$ is uniform. Let $x$ belong to the Jacobson radical of $A$. In particular, $x$ belongs to the intersection of maximal ideals of codimension 1 of $A$. So, we have $\|x\|_{sa} = 0$, so, by Property p), $\|x\|_{si} = 0$, and therefore $x = 0$, hence $A$ is semi-simple. Let $h$ belong to the closure of $G_A(A)$ in $\mathscr{C}(\mathscr{X}(A,K),K)$. Since $\mathscr{C}(\mathscr{X}(A,K),K)$ is provided with the norm of uniform convergence, we just have to consider a sequence $\widehat{f_n}$ of $G_A$ converging to $h$. Thus, $\|h - \widehat{f_n}\| = \sup\{|\chi(f_n) - h(\chi)| \,|\, \chi \in \mathscr{X}(A,K)\}$. The sequence $(f_n)_{n \in \mathbb{N}}$ is a Cauchy sequence in $A$, with respect to $\|x\|_{si} = 0$, and therefore with respect to the norm of $A$ because $A$ is uniform. Let $f = \lim_{n\to\infty} f_n$. Then we can check that $\widehat{f} = \lim_{n\to\infty} \widehat{f_n}$ in $\mathscr{C}(\mathscr{X}(A,K),K)$ and therefore $h = \widehat{f}$. Conversely, suppose that $A$ is semi-simple and that $G_A(A)$ is closed in $\mathscr{C}(\mathscr{X}(A,K),K)$. Since $A$ is semi-simple and satisfies Property p), $\| \cdot \|_{sa}$ is a norm equal to $\|x\|_{si}$. Consider a Cauchy sequence $(f_n)$ with respect to the norm $\| \cdot \|_{si}$. Then the sequence $(\widehat{f_n})$ is a Cauchy sequence in $\mathscr{C}(\mathscr{X}(A,K),K)$, and has a limit $h$ which actually lies in $G_A(A)$ because $G_A(A)$ is closed in $\mathscr{C}(\mathscr{X}(A,K),K)$. Hence, there exists $f \in A$ such that $\widehat{f} = h$. Consequently, $f$ is the limit of the sequence $(f_n)$ with respect to the norm $\| \cdot \|_{si}$. Thus, $A$ is a $K$-algebra complete for both $\| \cdot \|_{si}$ and $\| \cdot \|$. And since $\|x\|_{si} \leq \|x\| \; \forall x \in A$, by Banach's Theorem the two norms are equivalent. □

**Proof of Theorem E.** Let $\mathscr{G} = f^*(\phi)$. By [2] the family of sets $Mult(H(E), \| \cdot \|_E)$, where $E$ is an infraconnected affinoid subset of $K$ lying in $\mathscr{G}$, makes a base of neighborhoods of $\phi_{\mathscr{G}}$. So, we take an arbitrary $\mathscr{G}$-affinoid $B$ and will show that there exists a neighborhood $W$ of $\phi$ inside $Mult(A, \| \cdot \|)$ whose image by $f^*$ is included in $Mult(H(B), \| \cdot \|_B)$. If $\mathscr{G}$ has center $b$ and diameter $r > 0$, we can take an arbitrary $\mathscr{G}$-affinoid $B$ of the form $d(b, r+\varepsilon) \setminus \left(\bigcup_{j=1}^{q} d(b_j, (r-\varepsilon^-))\right)$, and if $\mathscr{G}$ has no center and but has diameter $r$, or is a Cauchy filter, we can take an arbitrary $\mathscr{G}$-affinoid $B$ of the form $d(b, r+\varepsilon)$. Putting $b = b_0$, we consider now the neighborhood of $\phi$ $W = \{\psi \in Mult(A, \| \cdot \|) \,|\, |\psi(f - b_j) - \phi(f - b_j)| \leq \varepsilon, (0 \leq j \leq q)\}$. Then we can check that $f^*(W) \subset Mult(H(B), \| \cdot \|_B)$, which proves the claim. □

Lemma 1 is easy:

**Lemma 1.** *Let $\chi \in \mathscr{X}(A,K)$ and let $f \in A$. Then $f^*(|\chi|) = \varphi_{\chi(f)}$.*

Proof. Let $\mathscr{F}$ be the circular filter such that $\varphi_{\mathscr{F}} = f^*(|\chi|)$. Then $\varphi_{\mathscr{F}}$ clearly belongs to $Mult_m(K[x])$ and $\mathscr{F}$ is the Cauchy filter of neighborhoods of $\chi(f)$ because $\varphi_{\mathscr{F}}(P) = |P(\chi(f))|\ \forall P \in K[x]$. □

**Proof of Theorem F.** Suppose $f^* = g^*$ and $f \neq g$. Since the intersection of maximal ideals of codimension 1 is null, there exists ($\chi \in \mathscr{X}(A,K)$ such that $\chi(g-f) \neq 0$. We have $f^*(|\chi|) = \varphi_{\chi(f)}$ and $g^*(|\chi|) = \varphi_{\chi(g)}$. But since $f^* = g^*$, we have $f^*(|\chi|) = g^*(|\chi|)$. Let $\mathscr{F}$ be the circular filter such that $\varphi_{\mathscr{F}} = f^*(|\chi|) = g^*(|\chi|)$. Then $\varphi_{\mathscr{F}}$ clearly belongs to $Mult_m(K[x])$ and $\mathscr{F}$ is the Cauchy filter of neighborhoods of $\chi(f)$ because $\varphi_{\mathscr{F}}(P) = |P(\chi(f))|$. Similarly, it is the Cauchy filter of neighborhoods of $\chi(g)$. Consequently, $\chi(f) = \chi(g)$, hence $\chi(f-g) = 0$, a contradiction to the hypothesis. □

From Lemma 2.11 [4] we can extract the following lemma:

**Lemma 2.** *Let $\mathscr{F}$ be a circular filter and let $a \in K$. There exists a unique $r > 0$ such that $\mathscr{F}$ is approaching $C(a,r)$. Moreover, for every $s > diam(\mathscr{F})$ there exists a unique disk E of the form $d(b,s)$ which belongs to $\mathscr{F}$.*

**Proof of Theorem N.** Suppose that $f^*(\phi) \neq g^*(\phi)$. Let $r = diam(\mathscr{F})$, $s = diam(\mathscr{G})$. Suppose first that $\mathscr{F} \prec \mathscr{G}$, hence $r < s$. Let us take $l \in ]r,s[$. By Lemma 2 there exists a unique disk $d(a,l)$ which belongs to $\mathscr{F}$ Then $a$ is a center of $\mathscr{G}$. Thus, we have $\phi(f-a) \leq l$, $\phi(g-a) = s$, hence, since semi-multiplicative semi-norms are ultrametric, $\phi(g-f) = s = diam(\sup(\mathscr{F},\mathscr{G}))$. Now, suppose that $\mathscr{F}$ and $\mathscr{G}$ are incomparable. Let $\mathscr{S} = \sup(\mathscr{F},\mathscr{G})$. By a classical result [5] we can find disks $F = d(a,r') \in \mathscr{F}$, $G = d(b,s') \in \mathscr{G}$ such that $\delta(F,G) = \lambda(\mathscr{F},\mathscr{G}) = |a-b| = diam(\sup(\mathscr{F},\mathscr{G}))$. Of course we have $|b-a| > r'$ and $|b-a| > s'$. Then, $\phi(g-a) = |b-a|$ and $\phi(f-a) < |b-a|$, hence $\phi(g-f) = |b-a|$, and therefore $\phi(g-f) = diam(\sup(\mathscr{F},\mathscr{G}))$ which ends the proof. □

**Lemma 3.** *Let $x \in A$ be such that $\tau(x) < \|x\|$. There exists $\phi \in Mult(A,\|\cdot\|)$, a number $\sigma > 0$ an affinoid set F which belongs to the filter $\mathscr{F} = \Psi^{-1}(\phi_x)$, such that $diam(\Psi^{-1}(\psi_x)) \geq \sigma$, for all $\psi \in Mult(A,\|\cdot\|)$ such that $\Psi^{-1}(\psi_x) \in \Phi(F)$.*

Proof. We assume Lemma is not true. Let $r = \tau(x)$ and let $s = \|x\|$, let $m \in ]r,s[$, let $\mathscr{O}$ be the strict x-spectral partition, and let $\phi_0 \in Mult(A,\|\cdot\|)$ be such that $\Psi^{-1}((\phi_0)_x)$ is secant with $\Gamma(0,m,s)$. Let $\mathscr{F}_0 = \Psi^{-1}((\phi_0)_x)$ and let $s_0 = diam(\mathscr{F}_0)$. Suppose we have already constructed a finite sequence $(\phi_n)_{n \in \mathbb{N}}$ for $n = 0,...,q$, satisfying $diam(\Psi^{-1}((\phi_n)_x)) < \dfrac{s}{n+1}$ and $\delta(\Psi^{-1}((\phi_n)_x),\Psi^{-1}((\phi_{n+1})_x)) < \dfrac{2s}{n+1}$. Let $\mathscr{F}_n = \Psi^{-1}((\phi_n)_x)$. Since the Lemma is supposed to be false, there exists no affinoid $F \in \mathscr{F}_n$ such that $\mathscr{F}_n$ is the only circular filter of the form $\Psi^{-1}(\psi_x)$, $\psi \in Mult(A,\|\cdot\|)$, which is secant with F. Suppose that we can't find $\phi_{n+1} \in Mult(A,\|\cdot\|)$ such that the filter $\mathscr{F}_{n+1} = \Psi^{-1}((\phi_{n+1})_x))$ satisfies $diam(\mathscr{F}_{n+1}) < \dfrac{s}{n+2}$, and $\delta(\mathscr{F}_{n+1},\mathscr{F}_n) < \dfrac{2s}{n+1}$. This means that for every $\psi \in$

$Mult(A, \|\,.\,\|)$ such that $\delta(\Psi^{-1}(\psi_x), \mathscr{F}_n) < \dfrac{2s}{n+1}$, we have $diam(\Psi^{-1}(\psi_x)) \geq \dfrac{1}{n+1}$. Let $F$ be the unique disk of diameter $\dfrac{2s}{n+1}$ such that $\mathscr{F}_n \in \Phi(F)$. Then for all $\mathscr{G} \in \Phi(F)$ we have $\delta(\mathscr{F}, \mathscr{G}) \leq \dfrac{2s}{n+1}$, hence $diam(\Psi^{-1}(\psi_x)) \geq \dfrac{s}{n+1}$ for all $\psi \in Mult(A, \|\,.\,\|)$ such that $\Psi^{-1}(\psi_x) \in \Phi(F)$, which proves the conclusion of the Lemma. Thus, since we have supposed that the conclusion is false, we can find $\phi_{n+1} \in Mult(A, \|\,.\,\|)$ such that the filter $\mathscr{F}_{n+1} = \Psi^{-1}((\phi_{n+1})_x))$ satisfies $diam(\mathscr{F}_{n+1}) < \dfrac{s}{n+2}$, and $\delta(\mathscr{F}_{n+1}, \mathscr{F}_n) < \dfrac{2s}{n+1}$. And this is true for all $n \in \mathbb{N}$, hence the sequence $((\phi_n)_x)_{n \in \mathbb{N}}$ is a Cauchy sequence with respect to the $\delta$-topology. Consequently, its limit in $Mult(K[x])$ is of the form $\varphi_\alpha$, with $\alpha \in \Gamma(0, r, s)$. On the other hand, since $Mult(A, \|\,.\,\|)$ is compact for the topology of simple convergence, the sequence $(\phi_n)_{n \in \mathbb{N}}$ admits a point of adherence $\theta$ with respect to the topology of simple convergence. Consequently, $\theta_x$ is a point of adherence of the sequence $((\psi_n)_x)_{n \in \mathbb{N}}$. Next, since $Mult(K[x])$ is sequentially compact, we can extract from the sequence $(\psi_n)_x$ a subsequence converging to $\theta_x$ with respect to the topology of simple convergence. Thus, without loss of generality, we can assume that the sequence $(\psi_n)_x$ converges to $\theta_x$ with respect to the topology of simple convergence in $Mult(K[x])$. But since the sequence $(\psi_n)_x$ converges to $\varphi_\alpha$ for the $\delta$-topology, so much the more does it converge to $\varphi_\alpha$ for the topology of simple convergence, hence $\varphi_\alpha = \theta_x$. Thus, $\theta_\alpha$ is punctual, a contradiction to the fact that $\alpha \notin sp(x)$. Consequently, our hypothesis "the conclusion is false" is wrong, and this finishes proving Lemma 3. $\square$

**Proposition T.** Let $\mathscr{O} = (d(b_i, r_i^-))_{i \in J}$ be a classic partition of $d(a, r) \setminus D$, and suppose that $\mathscr{O}$ admits an increasing (resp. a decreasing) idempotent T-sequence $(T_n)$ of center $b$ and diameter $l$. Let $s \in ]0, l[$ (resp. $s \in ]l, r[$) and let $\varepsilon \in ]0, 1[$. There exists $f \in H'(D, \mathscr{O}, (b_i)_{i \in J})$ satisfying

(1) $|\psi(f-1)|_\infty < \varepsilon \;\; \forall \psi \in Mult(H(D, \mathscr{O}), \|\,.\,\|_{D, \mathscr{O}})$ such that $\psi(x-b) \leq s$ and $\psi(f) = 0 \;\; \forall \psi \in Mult(H(D, \mathscr{O}), \|\,.\,\|_{D, \mathscr{O}})$ such that $\psi(x-b) \geq l$,

(2) $\psi(f) \neq 0 \;\; \forall \psi \in Mult(H(D, \mathscr{O}), \|\,.\,\|_{D, \mathscr{O}}) \setminus Mult_a(H(D, \mathscr{O}), \|\,.\,\|_{D, \mathscr{O}})$ such that $s < \psi(f) < l$. resp.

(1) $\psi(f-1) \leq \varepsilon \;\; \forall \psi \in Mult(H(D, \mathscr{O}), \|\,.\,\|_{D, \mathscr{O}})$ such that $\psi(x-b) \geq s$ and $\psi(f) = 0 \;\; \forall \psi \in Mult(H(D, \mathscr{O}), \|\,.\,\|_{D, \mathscr{O}})$ such that $\psi(x-b) \leq l$

(2) $\psi(f) \neq 0 \;\; \forall \psi \in Mult(H(D, \mathscr{O}), \|\,.\,\|_{D, \mathscr{O}}) \setminus Mult_a(H(D, \mathscr{O}), \|\,.\,\|_{D, \mathscr{O}})$ such that $l < \psi(f) < s$.)

**Proof of Theorem O.** Suppose that $A$ does not satisfy Property q) and let $x \in A$ be such that $\tau(x) < \|x\|$. Consider now the strict $x$-spectral partition $\mathscr{O}$ of $d(0, s)^{(x)}$. Let $r = \tau(x)$ and let $s = \|x\|$. By Lemma 3 there exists $\phi \in Mult(A, \|\,.\,\|)$, a number $\sigma > 0$ and an affinoid set $F$ which belongs to the filter $\mathscr{F} = \Psi^{-1}(\phi_x)$, such that $diam(\Psi^{-1}(\psi_x)) \geq \sigma$, for all $\psi \in Mult(A, \|\,.\,\|)$ such that $\Psi^{-1}(\psi_x) \in \Phi(F)$. Whether or not $\mathscr{F}$ has a center, we are going to construct an element $G \in A$ such that, given $\psi \in Mult(A, \|\,.\,\|)$, either $\phi(G) = 0$, or $diam(\Psi^{-1}(\psi_x)) \geq \sigma$. Suppose first that $\mathscr{F}$ has no center. By Lemma 2 there exists disks $d(a,t) \in \mathscr{F}$, $d(a,l) \in \mathscr{F}$ included in $F$, with $l < t$. By Proposition T. there exists $g \in H(sp(x), \mathscr{O})$ such that $\varphi_\mathscr{G}(g) = 1 \ \forall \mathscr{G}$ secant with $d(a,l)$ and $\varphi_\mathscr{G}(g) = 0 \ \forall \mathscr{G}$ secant with $d(0,s) \setminus d(a,t^-)$. Suppose now that $\mathscr{F}$ has center $a$ and diameter $m$. Without loss of generality we can assume that $F$ is of the form $d(a, m+\varepsilon) \setminus \bigcup_{j=1}^{q} d(a_j, (m-\varepsilon)^-)$, with $m + \varepsilon < s$ when $m < s$. For each $j = 1,...,q$, by Proposition T there exists $g_j \in H(sp(x), \mathscr{O})$ such that $\varphi_\mathscr{G}(g_j) = 1 \ \forall \mathscr{G}$ secant with $d(0,s) \setminus d(a,m)^-$ and $\varphi_\mathscr{G}(g_j) = 0 \ \forall \mathscr{G}$ secant with $d(a_j, m - \varepsilon)$. Moreover, if $m = s$ we put $g_0 = 1$, and if $m < s$, by Proposition T there exists $g_0 \in H(sp(x), \mathscr{O})$ such that $\varphi_\mathscr{G}(g) = 1 \ \forall \mathscr{G}$ secant with $d(a,m)$ and $\varphi_\mathscr{G}(g) = 0 \ \forall \mathscr{G}$ secant with $d(0,s) \setminus d(a, (m+\varepsilon)^-)$. Then we put $g = \prod_{j=0}^{q} g_j$. In both cases, we now put $G = \Theta_x(g)$ and we can check that given $\psi \in Mult(A, \| \cdot \|)$, either $\phi(G) = 0$, or $diam(\Psi^{-1}(\psi_x)) \geq \sigma$. Let $\xi \in K$ satisfy $\|\xi G\| < \sigma$ and let $y = x + \xi G$. Then $\|y - x\| < \sigma$. Consider $\psi \in Mult(A, \| \cdot \|)$, let $\varphi_\mathscr{R} = x^*(\psi)$, $\varphi_\mathscr{T} = y^*(\psi)$. If $\psi(G) = 0$, then $\psi(y-x) = 0$, hence by Theorem N we have $x^*(\psi) = y^*(\psi)$. And now, if $\psi(G) \neq 0$, then we can see that $\psi(y-x) < \sigma \leq diam(\mathscr{R}) \leq diam(\sup(\mathscr{R}, \mathscr{T}))$, therefore $x^*(\psi) = y^*(\psi)$ again. Consequently, $x^*(\psi) = y^*(\psi) \ \forall \psi \in Mult(A, \| \cdot \|)$. However, $x \neq y$ because $\phi(G) \neq 0$. □

# REFERENCES

1. Boussaf K., Image of circular filters, *International Journal of Mathematics, Game Theory and Algebra*, Vol **10**, no. 5, 365–372.
2. Boussaf K., Mainetti N. and Hemdahoui M., Tree structure on the set of multiplicative semi-norms of Krasner algebras $H(D)$, *Revista Matematica Complutense*, Vol **XIII**, no. 1, 85–109 (2000).
3. Boussaf K., *An interpretation of analytic functions. p-adic functional analysis*, Lecture Notes in Pure and Applied Mathematics **222**, Marcel Dekker, (2001).
4. Boussaf K., Escassut A. and Mainetti N., Analytic mappings in the tree $Mult(K[x])$, Preprint.
5. Escassut A., *Analytic Elements in p-adic Analysis*, World Scientific Publishing Inc., Singapore 1995.
6. Garandel G., Les semi-normes multiplicatives sur les algèbres d'éléments analytiques au sens de Krasner, *Indag. Math.* **37**, no. 4, 327–341, (1975).
7. Guennebaud B., *Sur une notion de spectre pourles algèbres normées ultramétriques*, thèse Université de Poitiers, 1973.
8. Mainetti N., *Spectral properties of p-adic Banach algebras*, Lecture Notes in Pure and Applied Mathematics **207**, 189–210, (1999).
9. Rivera-Letelier J., Espace hyperbolique p-adique et dynamique des fonctions rationnelles, to appear in Compositio.
10. Van Rooij A.C.M., *Non-Archimedean Functional Analysis*, Marcel Dekker, inc., 1978.

// # $p$-Adic Multiple Zeta Values: a Précis

Hidekazu Furusho

*Graduate School of Mathematics, Nagoya University,*
*Nagoya, 464-8602, JAPAN*
*email: furusho@math.nagoya-u.ac.jp*

**Abstract.** This is an expose of the theory of $p$-adic multiple zeta value developed by the author. This theory is almost parallel to the story of the complex case. $p$-adic multiple polylogarithm, which is a $p$-adic function, is constructed by Coleman's $p$-adic iterated integration theory. $p$-adic multiple zeta value is defined to be its special value at 1. Many nice properties of $p$-adic multiple zeta value is shown. $p$-adic KZ equation is also introduced and the $p$-adic Drinfel'd associator is constructed from two fundamental solutions of the $p$-adic KZ equation. A relation of $p$-adic multiple zeta value with the $p$-adic Drinfel'd associator as well as its relation with the fundamental group of the projective line minus three points is established.

**Keywords:** $p$-adic KZ equation, $p$-adic Drinfel'd associator, $p$-adic multiple zeta value, $p$-adic multiple polylogarithm, $p$-adic integration.
**PACS:** 02.10.De.

## 1. INTRODUCTION

This paper is a survey of the theory of $p$-adic multiple zeta values ($p$: a prime) developed in [8] and [9]. The notions of (one-variable) $p$-adic multiple polylogarithms, $p$-adic KZ equation and $p$-adic Drinfel'd associator are introduced in [8]. They are $p$-adic analogues of basic materials in the complex case. They are used to define $p$-adic multiple zeta value and show its nice properties. In [9] a tannakian origin of the $p$-adic Drinfel'd associator is revealed. This established a close relationship of the $p$-adic multiple zeta values with the unipotent fundamental group of the projective line minus three points.

Below we explain that it is not a priori to give a definition of a $p$-adic analogue of multiple zeta values. Let $k_1, \cdots, k_m \in \mathbf{N}$. The (usual) *multiple zeta value* is the real number defined by the following series

$$\zeta(k_1, \cdots, k_m) = \sum_{\substack{0 < n_1 < \cdots < n_m \\ n_i \in \mathbf{N}}} \frac{1}{n_1^{k_1} \cdots n_m^{k_m}}. \tag{1}$$

Especially in the case when $m = 1$, the multiple zeta value coincides with the Riemann zeta value $\zeta(k)$. We can check easily that this series converges in the topology of $\mathbf{R}$ if and only if $k_m > 1$, however, this series never converges in the topology of $\mathbf{Q}_p$! Thus it is not so easy and not so straightforward to give a definition of $p$-adic version of multiple zeta value. To give a nice definition, we need another interpretation of multiple zeta values.

Suppose $z \in \mathbf{C}$. The (one variable) *multiple polylogarithm* is a function defined by the following series

$$Li_{k_1,\cdots,k_m}(z) = \sum_{\substack{0<n_1<\cdots<n_m \\ n_i \in \mathbf{N}}} \frac{z^{n_m}}{n_1^{k_1}\cdots n_m^{k_m}} \ . \tag{2}$$

Especially in the case when $m = 1$, the multiple polylogarithm coincides with the classical polylogarithm $Li_k(z)$. Easily we see that this series converges for $|z| < 1$. In §3.2, we will define the *p-adic multiple polylogarithm* to be the function defined by the above series just replacing $z \in \mathbf{C}$ by $z \in \mathbf{C}_p$. We remark that especially in the case when $m = 1$, the $p$-adic multiple polylogarithm is equal to the $p$-adic polylogarithm $\ell_k(z)$ which was studied by Coleman [4]. What is interesting is that this $p$-adic multiple polylogarithm converges for $|z|_p < 1$ similarly to the above complex case. Here $|\cdot|_p$ means the standard multiplicative valuation of $\mathbf{C}_p$.

An important relationship between the multiple polylogarithm and the multiple zeta value is the following formula:

$$\zeta(k_1,\cdots,k_m) = \lim_{\substack{z \to 1 \\ |z|<1}} Li_{k_1,\cdots,k_m}(z) \ . \tag{3}$$

In this paper, we will define the $p$-adic multiple zeta value à la formula (3) instead of à la (1). But we note that here happens a serious problem because the open unit disk centered at 0 on $\mathbf{C}_p$ and the one centered at 1 on $\mathbf{C}_p$ are disjoint! Thus we cannot consider $\lim_{z \to 1}$ of $p$-adic multiple polylogarithms which are functions defined on $|z|_p < 1$, i.e. on the open unit disk centered at 0. To give a meaning of this limit, we will make an analytic continuation of $p$-adic multiple polylogarithms by Coleman's $p$-adic iterated integration theory [4] and then define $p$-adic multiple zeta values to be a limit value at 1 of analytically continued $p$-adic multiple polylogarithms.

The organization of this paper is as follows. §2 is devoted to a review of known results on complex case. §3 is explanation of our results, including construction of $p$-adic multiple zeta values and their many nice properties. Appendix is a short discussion on weights of polylogarithm both in the étale side and in the rigid side.

## 2. THE COMPLEX CASE

We shall review definitions of multiple polylogarithms and multiple zeta values in §2.1, shall recall notions of the KZ equation and the Drinfel'd associator briefly in §2.2 and shall see a tannakian interpretation of the Drinfel'd associator in §2.3.

### 2.1. Multiple polylogarithms and multiple zeta values

We review briefly definitions and properties of multiple polylogarithms and multiple zeta values. For more details, consult [7] and [11] for example.

Let $k_1, \cdots, k_m \in \mathbf{N}$ and $z \in \mathbf{C}$.

**2.1 Definition.** *The (one variable) multiple polylogarithm (MPL for short) $Li_{k_1,\cdots,k_m}(z)$ is defined by (2).*

Easily we can check the following.

**2.2 Lemma.** *The MPL $Li_{k_1,\cdots,k_m}(z)$ converges for $|z| < 1$.*

**2.3 Lemma.** *Suppose that $|z| < 1$. Then*

$$\frac{d}{dz}Li_{k_1,\cdots,k_m}(z) = \begin{cases} \frac{1}{z}Li_{k_1,\cdots,k_m-1}(z) & k_m \neq 1, \\ \frac{1}{1-z}Li_{k_1,\cdots,k_m-1}(z) & k_m = 1, \end{cases}$$

$$\frac{d}{dz}Li_1(z) = \frac{1}{1-z}.$$

By Lemma 2.3, for $|z| < 1$, we get $Li_1(z) = -\log(1-z)$ and the following,

$$Li_{k_1,\cdots,k_m}(z) = \begin{cases} \int_0^z \frac{1}{t} Li_{k_1,\cdots,k_m-1}(t)dt & k_m \neq 1, \\ \int_0^z \frac{1}{1-t} Li_{k_1,\cdots,k_m-1}(t)dt & k_m = 1, \end{cases}$$

from which we get an expression of the MPL by iterated integral of $\frac{dt}{t}$ and $\frac{dt}{1-t}$. Since $\frac{dt}{t}$ and $\frac{dt}{1-t}$ admit poles at $t = 0, 1$ and $\infty$, we cannot give an analytic continuation of the MPL to the whole complex plane due to monodromies around $0$, $1$ and $\infty$. However we can say that

**2.4 Lemma.** *The MPL $Li_{k_1,\cdots,k_m}(z)$ can be analytically continued to the universal unramified covering $\widetilde{\mathbf{P}^1(\mathbf{C})\backslash\{0,1,\infty\}}$ of $\mathbf{P}^1(\mathbf{C})\backslash\{0,1,\infty\}$.*

Since the simply-connected Riemann surface $\widetilde{\mathbf{P}^1(\mathbf{C})\backslash\{0,1,\infty\}}$ is an infinite covering of $\mathbf{P}^1(\mathbf{C})\backslash\{0,1,\infty\}$, each MPL admits (*countably*) infinite branches. The following are well-known (for example, see [11]).

**2.5 Lemma.** $\lim\limits_{\substack{z \to 1 \\ |z|<1}} Li_{k_1,\cdots,k_m}(z)$ *converges if $k_m > 1$.*

**2.6 Lemma.** $\lim\limits_{\substack{z \to 1 \\ |z|<1}} Li_{k_1,\cdots,k_m}(z)$ *diverges if $k_m = 1$.*

**2.7 Definition.** *For $k_1, \cdots, k_m \in \mathbf{N}$, $k_m > 1$, the multiple zeta value (MZV for short) is defined to be*

$$\zeta(k_1,\cdots,k_m) = \lim_{\substack{z \to 1 \\ |z|<1}} Li_{k_1,\cdots,k_m}(z) \left( = \sum_{0 < n_1 < \cdots < n_m} \frac{1}{n_1^{k_1} \cdots n_m^{k_m}} \right).$$

Since MPL's are **C**-valued functions, MZV's lie in **C**. However we can say more.

**2.8 Lemma.** *For $k_1, \cdots, k_m \in \mathbf{N}$, $k_m > 1$, $\zeta(k_1, \cdots, k_m) \in \mathbf{R}$.*

**2.9 Notation.** *For each natural number $w$, let $Z_w$ be the $\mathbf{Q}$-vector subspace of $\mathbf{R}$ generated by all MZV's $\zeta(k_1, \cdots, k_m)$ with*
$$k_1 + \cdots + k_m = w, \text{ i.e. } Z_w := \langle \zeta(k_1, \cdots, k_m) \mid k_1 + \cdots + k_m = w \rangle_{\mathbf{Q}} \subseteq \mathbf{R},$$
*and put $Z_0 = \mathbf{Q}$. Denote $Z_{\bullet}$ to be the formal direct sum of $Z_w$ for all $w \geqslant 0$: $Z_{\bullet} := \bigoplus_{w \geqslant 0} Z_w$.*

The following is one of the fundamental properties of MZV's.

**2.10 Proposition.** *The graded $\mathbf{Q}$-vector space $Z_{\bullet}$ forms a graded $\mathbf{Q}$-algebra, i.e. $Z_a \cdot Z_b \subseteq Z_{a+b}$ for $a, b \geqslant 0$.*

*Proof.* At least we have two proofs [11]. The first one is given by the *harmonic product formulae*, from which it follows, for example,
$$\zeta(m) \cdot \zeta(n) = \zeta(m,n) + \zeta(n,m) + \zeta(m+n).$$

The other one is given by the *shuffle product formulae*, from which it follows, for example,
$$\zeta(m) \cdot \zeta(n) = \sum_{i=0}^{m-1} \binom{n-1+i}{i} \zeta(m-i, n+i) + \sum_{j=0}^{n-1} \binom{m-1+j}{j} \zeta(n-j, m+j).$$

$\square$

*Double shuffle relations* are combinations of the above two formulae. Besides them, MZV satisfies the other type relation called *regularization relations* (cf. [12]).

## 2.2. The KZ equation and the Drinfel'd associator

In this subsection, we will briefly review the definition of the KZ equation and the Drinfel'd associator. For more detailed information on the KZ equation and the Drinfel'd associator, see [6], [7] and [13].

Let $A_{\mathbf{C}}^{\wedge} = \mathbf{C}\langle\langle A, B \rangle\rangle$ be the non-commutative formal power series ring generated by two elements $A$ and $B$ with complex number coefficients.

**2.11 Definition.** *The (formal) Knizhnik-Zamolodchikov equation* [1] *(KZ equation for short) is the differential equation*
$$\frac{\partial G}{\partial u}(u) = \left(\frac{A}{u} + \frac{B}{u-1}\right) \cdot G(u) \quad , \tag{KZ}$$

---

[1] This is a special case of the KZ equation for $\mathbf{P}^1(\mathbf{C}) \setminus \{0, 1, \infty\}$ in [13].

where $G(u)$ is an analytic function in complex variable $u$ with values in $A_{\mathbf{C}}^{\wedge}$ where 'analytic' means each of whose coefficient is analytic.

The equation (KZ) has singularities only at $0, 1$ and $\infty$. Let $\mathbf{C}'$ be the complement of the union of the real half-lines $(-\infty, 0]$ and $[1, +\infty)$ in the complex plane. This is a simply-connected domain. The equation (KZ) has a unique analytic solution on $\mathbf{C}'$ having a specified value at any given points on $\mathbf{C}'$. Moreover, for the singular points $0$ and $1$, there exist unique solutions $G_0(u)$ and $G_1(u)$ of (KZ) such that

$$G_0(u) \approx u^A \ (u \to 0), \qquad G_1(u) \approx (1-u)^B \ (u \to 1),$$

where $\approx$ means that $G_0(u) \cdot u^{-A}$ (resp. $G_1(u) \cdot (1-u)^{-B}$) has an analytic continuation in a neighborhood of $0$ (resp. $1$) with value $1$ at $0$ (resp. $1$). Here

$$u^A := exp(A \log u) := 1 + \frac{A \log u}{1!} + \frac{(A \log u)^2}{2!} + \frac{(A \log u)^3}{3!} + \cdots \quad \text{and} \quad \log u := \int_1^u \frac{dt}{t}$$

in $\mathbf{C}'$. In the same way, $(1-u)^B$ is well-defined on $\mathbf{C}'$. Since $G_0(u)$ and $G_1(u)$ are both non-zero unique solutions of (KZ) with the specified asymptotic behaviors, they must coincide with each other up to multiplication from the right by an invertible element of $A_{\mathbf{C}}^{\wedge}$.

**2.12 Definition.** *The Drinfel'd associator is the element $\Phi_{KZ}(A, B)$ of $A_{\mathbf{C}}^{\wedge}$ which is defined by*

$$G_0(u) = G_1(u) \cdot \Phi_{KZ}(A, B) \ .$$

We note that $\Phi_{KZ}(A, B)$ is independent of $u$.

**2.13 Proposition.** *MZV's appear at each coefficient of the Drinfel'd associator as follows:*

$$\Phi_{KZ}(A, B) = 1 + \cdots + (-1)^m \zeta(k_1, \ldots, k_m) A^{k_m - 1} B \cdots A^{k_1 - 1} B + \cdots .$$

For its explicit formulae, see [14] and [7]Proposition 3.2.3.

## 2.3. Tannakian interpretation

We explain that MZV is related to the fundamental group of the projective line minus three points.

Let $X = \mathbf{P}^1 \backslash \{0, 1, \infty\}$ and $\pi_1^{DR}(X, \overrightarrow{01})$ denote the de Rham fundamental group of $X$ with the tangential base point $\overrightarrow{01}$ at $0$. This is a pro-algebraic group defined over $\mathbf{Q}$ (cf. [5]). Their set of $\mathbf{Q}$-valued points is embedded into the non-commutative formal power series ring

$$i: \pi_1^{DR}(X, \overrightarrow{01}) \hookrightarrow \mathbf{Q}\langle\langle A, B \rangle\rangle \tag{4}$$

where $A$ and $B$ correspond to the loops around 0 and 1 respectively. Let $\pi_1^{DR}(X,\vec{01},\vec{10})$ (resp. $\pi_1^{Be}(X,\vec{01},\vec{10})$ ) be the de Rham (resp. Betti) fundamental torsor of $X$ from $\vec{01}$ to $\vec{10}$. They are pro-algebraic variety defined over $\mathbf{Q}$. There is a Hodge-type comparison isomorphism (cf. loc.cit) between them

$$\pi_1^{DR}(X,\vec{01},\vec{10})(\mathbf{C}) \cong \pi_1^{Be}(X,\vec{01},\vec{10})(\mathbf{C}).$$

In LHS a canonical de Rham path $d$ is constructed in loc.cit. In contrast there is a canonical path $b$ in RHS which is provided by the standard path going from 0 to 1 on the real axis (cf. [9] §2.2). Going via this Betti path $b$ and coming back via this de Rham path $d$, we get a de Rham loop $d^{-1}b$ in $\pi_1^{DR}(X,\vec{01})(\mathbf{C})$.

**2.14 Proposition** ([9], §2.2). *The de Rham loop* $d^{-1}b \in \pi_1^{DR}(X,\vec{01})(\mathbf{C})$ *corresponds to the Drinfel'd associator by the embedding* $i$, *i.e.*

$$i(d^{-1}b) = \Phi_{KZ}(A,B).$$

Since MZV's appear as coefficients of the Drinfel'd associator by Proposition 2.13, we may say that the loop $d^{-1}b$ which is a combination of a de Rham path and a Betti path is a tannakian origin of MZV's.

## 3. THE $p$-ADIC CASE

§3.1 is a review of Coleman's $p$-adic integration theory [4]. §3.2 and §3.3 are summary of [8]. The $p$-adic multiple polylogarithms and the $p$-adic multiple zeta values will be introduced §3.2. Notions of the $p$-adic KZ equation and the $p$-adic Drinfel'd associator will be established in §3.3. §3.4 is an expose of [9]§2.2. A tannakian interpretation of the $p$-adic Drinfel'd associator will be given. The reader will find interesting analogies between §2 and §3.

### 3.1. Review of Coleman's $p$-adic iterated integration theory

We will review the $p$-adic iterated integration theory by R. Coleman [4]. This theory will be employed in the analytic continuation of $p$-adic multiple polylogarithms in §3.2. Suppose that $X/\mathscr{O}_{\mathbf{C}_p}$ is a smooth projective and surjective scheme over the ring $\mathscr{O}_{\mathbf{C}_p}$ of integers of $\mathbf{C}_p$, of relative dimension 1 with its generic fiber $X_{\mathbf{C}_p}$ and its special fiber $X_{\overline{\mathbf{F}_p}}$. Let $Y = X - D$ where $D$ is a closed subscheme of $X$ which is relatively etale over $\mathscr{O}_{\mathbf{C}_p}$. Let $j : Y_{\overline{\mathbf{F}_p}} \hookrightarrow X_{\overline{\mathbf{F}_p}}$ to be the associated open embedding. Put $A^\dagger := \Gamma(]X_{\overline{\mathbf{F}_p}}[, j^\dagger \mathscr{O}_{]X_{\overline{\mathbf{F}_p}}[})$ and $\Omega^\dagger := \Gamma(]X_{\overline{\mathbf{F}_p}}[, j^\dagger \Omega^1_{]X_{\overline{\mathbf{F}_p}}[})$ (for these meanings, see [1]). Fix $a \in \mathbf{C}_p$. It determines a branch of the $p$-adic logarithm $\log^a : \mathbf{C}_p^\times \to \mathbf{C}_p$ which is characterized by $\log^a(p) = a$. We call this $a \in \mathbf{C}_p$ the *branch parameter* of the $p$-adic logarithm.

In [4], Coleman constructed an $A^\dagger$-algebra $A^a_{\text{Col}}$ called *the ring of Coleman functions* attached to a branch parameter $a \in \mathbf{C}_p$. This is a subalgebra of the algebra of locally analytic functions over $Y(\mathbf{C}_p)$. He also constructed a $\mathbf{C}_p$-linear map

$$\int_{(a)} : A^a_{\text{Col}} \otimes_{A^\dagger} \Omega^\dagger \longrightarrow A^a_{\text{Col}} \Big/ \mathbf{C}_p \cdot 1 \quad \text{satisfying} \quad d \circ \int_{(a)} = id_{A^a_{\text{Col}} \otimes \Omega^\dagger}$$

called *the p-adic Coleman integration* attached to a branch parameter $a \in \mathbf{C}_p$. We often drop the subscript $_{(a)}$. He showed basic properties of Coleman functions including the uniqueness principle and the functorial property (see [4]).

**3.1 Notation.** Let $a \in \mathbf{C}_p$ and $\omega \in A^a_{\text{Col}} \otimes_{A^\dagger} \Omega^\dagger$. Then by Coleman's integration theory, there exists (uniquely modulo constant) a Coleman function $F_\omega$ such that $\int \omega \equiv F_\omega$ (modulo constant). For $x, y \in ]Y(\overline{\mathbf{F}_p})[\,^2$, we define $\int_x^y \omega$ to be $F_\omega(y) - F_\omega(x)$. It is clear that $\int_x^y \omega$ does not depend on any choice of $F_\omega$ (although it may depend on $a \in \mathbf{C}_p$). If $F_\omega(x)$ and $F_\omega(y)$ make sense for some $x, y \in X(\mathbf{C}_p)$, we also denote $F_\omega(y) - F_\omega(x)$ by $\int_x^y \omega$. When we let $y$ vary, we regard $\int_x^y \omega$ as the Coleman function which is characterized by $dF_\omega = \omega$ and $F_\omega(x) = 0$.

### 3.2. *p*-adic multiple polylogarithms and *p*-adic multiple zeta values

We will define *p*-adic multiple polylogarithms to be Coleman functions which admits an expansion around 0 similar to the complex case. Then we will define *p*-adic multiple zeta value to be its 'limit' value at 1.

We fix a branch parameter $a \in \mathbf{C}_p$ and employ Coleman's *p*-adic integration theory attached to this branch parameter $a \in \mathbf{C}_p$ for $X = \mathbf{P}^1_{\mathscr{O}_{\mathbf{C}_p}}$ and $Y = \text{Spec}\,\mathscr{O}_{\mathbf{C}_p}[t, \frac{1}{t}, \frac{1}{1-t}]$. Let $k_1, \cdots, k_m \in \mathbf{N}$ and $z \in \mathbf{C}_p$. Consider $Li_{k_1,\cdots,k_m}(z)$ defined by (2).

**3.2 Lemma.** *The series* $Li_{k_1,\cdots,k_m}(z)$ *converges on the open unit disk* $|z|_p < 1$.

**3.3 Lemma.** *Let* $z \in \mathbf{C}_p$ *such that* $|z|_p < 1$. *Then* $Li_{k_1,\cdots,k_m}(z)$ *satisfies the same differential equation to Lemma 2.3.*

The proofs are easy. Lemma 3.2 and Lemma 3.3 are *p*-adic analogue of Lemma 2.2 and Lemma 2.3 respectively.

**3.4 Definition.** *We define recursively the (one variable) p-adic multiple polylogarithm (p-adic MPL for short)* $Li^a_{k_1,\cdots,k_m}(z) \in A^a_{\text{Col}}$ *attached to* $a \in \mathbf{C}_p$ *which is the Coleman*

---

[2] For a subset $S \subset X(\overline{\mathbf{F}_p})$ we denote its tubular neighborhood (see [1]) in $X(\mathbf{C}_p)$ by $]S[$. As a set-theoretically it is the set of points in $X(\mathbf{C}_p)$ reducing to $S$.

*function characterized below:*

$$Li^a_{k_1,\cdots,k_m}(z) := \begin{cases} \int_0^z \frac{1}{t} Li^a_{k_1,\cdots,k_m-1}(t)dt & k_m \neq 1, \\ \int_0^z \frac{1}{1-t} Li^a_{k_1,\cdots,k_m-1}(t)dt & k_m = 1, \end{cases}$$

$$Li^a_1(z) := -log^a(1-z) := \int_0^z \frac{dt}{1-t} \quad .$$

It is easy to see that $Li^a_{k_1,\cdots,k_m}(z) = \sum_{0<n_1<\cdots<n_m} \frac{z^{n_m}}{n_1^{k_1}\cdots n_m^{k_m}}$ if $|z|_p < 1$. The following is a *p*-adic version of Lemma 2.4.

**3.5 Proposition** ([8], Proposition 2.11). *The p-adic MPL* $Li^a_{k_1,\cdots,k_m}(t)$ *is locally analytic on* $\mathbf{P}^1(\mathbf{C}_p)\setminus\{1,\infty\}$. *More precisely,* $Li^a_{k_1,\cdots,k_m}(t)\big|_{]0[} \in A(]0[)$, $Li^a_{k_1,\cdots,k_m}(t)\big|_{]1[} \in A(]1[)[log^a(t-1)]$ *and* $Li^a_{k_1,\cdots,k_m}(t)\big|_{]\infty[} \in A(]\infty[)[log^a(\frac{1}{t})]$.

*Proof.* By construction, the claim is proved inductively. □

This proposition means that *p*-adic MPL can be analytically continued to $\mathbf{P}^1(\mathbf{C}_p)\setminus\{1,\infty\}$, although the complex MPL cannot be analytically continued to $\mathbf{P}^1(\mathbf{C})\setminus\{0,1,\infty\}$ but only to its universal unramified covering $\widetilde{\mathbf{P}^1(\mathbf{C})\setminus\{0,1,\infty\}}$ instead by Lemma 2.4. We note that the *p*-adic MPL admits *uncountably* infinite branches which correspond to branches $a \in \mathbf{C}_p$ of *p*-adic logarithms $log^a(z)$ although complex MPL admits *countably* infinite branches. We call $Li^a_{k_1,\cdots,k_m}(z)$ the branch of *p*-adic MPL corresponding to $a \in \mathbf{C}_p$.

**3.6 Notation.** *Let* $\alpha \in \mathbf{C}_p$ *and let* $f(z)$ *be a function defined on* $\mathbf{C}_p$. *We denote* $\lim'_{z \to \alpha} f(z)$ *to be* $\lim_{n \to \infty} f(z_n)$ *if this limit converges to the same value for any sequence* $\{z_n\}_{n=1}^\infty$ *which satisfies* $z_n \to \alpha$ *in* $\mathbf{C}_p$ *and* $e(\mathbf{Q}_p(z_1,z_2,\cdots)/\mathbf{Q}_p) < \infty$ *(which means that the field generated by* $z_1, z_2, \cdots$ *over* $\mathbf{Q}_p$ *is a finitely ramified (possibly infinite) algebraic extension field over* $\mathbf{Q}_p$). *If the latter limit converges (resp. does not converge) to the same value, we call* $\lim'_{z \to \alpha} f(z)$ *converges (resp. diverges).*

**3.7 Theorem** ([8], Theorem 2.22). *If* $k_m > 1$, $\lim'_{\substack{z \to 1 \\ z \in \mathbf{C}_p - \{1\}}} Li^a_{k_1,\cdots,k_m}(z)$ *always converges on* $\mathbf{C}_p$. *Moreover this limit value is independent of any choice of branch parameter* $a \in \mathbf{C}_p$.

It is striking that $\lim'_{\varepsilon \to 0} Li^a_{k_1,\cdots,k_m}(1-\varepsilon)$ does not depend on any choice of branch parameter $a \in \mathbf{C}_p$ although $Li^a_{k_1,\cdots,k_m}(1-\varepsilon)$ takes whole values on $\mathbf{C}_p$ if we fix $\varepsilon$ ($0 < |\varepsilon|_p < 1$) and let $a$ vary on $\mathbf{C}_p$.

**3.8 Definition.** *For any index* $(k_1,\cdots,k_m)$ *with* $k_m > 1$, *we define the corresponding p-adic multiple zeta vale (p-adic MZV for short)* $\zeta_p(k_1,\cdots,k_m)$ *to be its limit in* $\mathbf{C}_p$, *i.e.*

$$\zeta_p(k_1,\cdots,k_m) := \lim'_{\substack{z \to 1 \\ z \in \mathbf{C}_p - \{1\}}} Li^a_{k_1,\cdots,k_m}(z) \in \mathbf{C}_p.$$

We note that this definition of $p$-adic MZV is independent of any choice of branches by Theorem 3.7. As for a $p$-adic analogue of Lemma 2.6, at present, we have nothing to say except the following.

**3.9 Note.** *The limit* $\lim\limits_{\substack{z \to 1 \\ z \in \mathbf{C}_p - \{1\}}}' Li^a_{k_1,\cdots,k_m}(z)$ *sometimes converges and sometimes diverges on* $\mathbf{C}_p$ *for* $k_m = 1$.

**3.10 Examples.** *(a) By the Coleman's calculation in [4], $\zeta_p(n) = \frac{p^n}{p^n-1} L_p(n, \omega^{1-n})$ for $n > 1$. So we have the equality $\zeta_p(2k) = 0$.*
*(b) $\lim\limits_{z \to 1}' Li^a_{2,1}(z)$ converges to $-2\zeta_p(1,2)$, i.e. $\zeta_p(2,1) = -2\zeta_p(1,2)$.*
*(c) $\lim\limits_{z \to 1}' Li^a_{3,1}(z)$ diverges if and only if $\zeta_p(3) \neq 0$, equivalently if and only if $p$ satisfies the 3rd Leopoldt conjecture $(L_3)$ see [8]Remark 2.20. Suppose that $(L_3)$ fails at a prime $p$. Then we get $\zeta_p(3,1) = -2\zeta_p(1,3) - \zeta_p(2,2)$ for this prime $p$.*
*(d) $\zeta_p(3) = \zeta_p(1,2)$ and $\zeta_p(1,3) = \zeta_p(2,2) = \zeta_p(1,1,2) = 0$.*

Those $p$-adic MZV's *were* defined to be elements of $\mathbf{C}_p$, but actually we can say more.

**3.11 Theorem** ([8], Theorem 2.25). *All $p$-adic MZV's are $p$-adic numbers, i.e. $\zeta_p(k_1,\cdots,k_m) \in \mathbf{Q}_p$.*

*Proof.* By Theorem 3.7 we may assume $a \in \mathbf{Q}_p$. Suppose that $\lim\limits_{\substack{z \to 1 \\ z \in \mathbf{C}_p-\{1\}}}' Li^a_{k_1,\cdots,k_m}(z)$ converges. Recall that $p$-adic MPL $Li^a_{k_1,\cdots,k_m}(z)$ ($a \in \mathbf{C}_p$) is an iterated integral of $\frac{dt}{t}$ and $\frac{dt}{1-t}$ which is a rational 1-form defined over $\mathbf{Q}_p$ and notice that $Li^a_{k_1,\cdots,k_m}(z) \in \mathbf{Q}_p$ for all $z \in p\mathbf{Z}_p$. Then from the Galois equivariancy of Coleman integration it follows that $Li^a_{k_1,\cdots,k_m}(z)$ is $\mathrm{Gal}(\overline{\mathbf{Q}_p}/\mathbf{Q}_p)$-invariant for $z \in \mathbf{P}^1(\mathbf{Q}_p)\backslash\{1,\infty\}$. Therefore in this case, we get $Li^a_{k_1,\cdots,k_m}(z) \in \mathbf{Q}_p$ for $z \in \mathbf{P}^1(\mathbf{Q}_p)\backslash\{1,\infty\}$. Thus we get $\lim\limits_{\substack{z \to 1 \\ z \in \mathbf{Q}_p-\{1\}}} Li^a_{k_1,\cdots,k_m}(z) \in \mathbf{Q}_p$, which yields the theorem. $\square$

It may be better to say that this theorem is a $p$-adic version of Lemma 2.8.

**3.12 Definition.** *For each natural number $w$, let $Z_w^{(p)}$ be the finite dimensional $\mathbf{Q}$-linear subspace of $\mathbf{Q}_p$ generated by all $p$-adic MZV's of indices with weight $w$, and put $Z_0^{(p)} = \mathbf{Q}$. Define $Z_\bullet^{(p)}$ to be the formal direct sum of $Z_w^{(p)}$ for all $w \geq 0$: $Z_\bullet^{(p)} := \bigoplus\limits_{w \geq 0} Z_w^{(p)}$.*

By Theorem 3.11, we see that $Z_w^{(p)} = \langle \zeta_p(k_1,\cdots,k_m) \mid k_1 + \cdots + k_m = w, k_m > 1, m \in \mathbf{N}\rangle_\mathbf{Q} \subset \mathbf{Q}_p$

**3.13 Theorem** ([8], Theorem 2.28). *The graded $\mathbf{Q}$-vector space has a structure of $\mathbf{Q}$-algebra, i.e. $Z_a^{(p)} \cdot Z_b^{(p)} \subseteq Z_{a+b}^{(p)}$ for $a, b \geq 0$.*

This is a *p*-adic analogue of Proposition 2.10. The proof in loc. cit is based on showing the *shuffle product formulae* for *p*-adic MZV. On the other hand the validity of the *harmonic product formulae* for *p*-adic MZV is non-trivial because we do not have a series expansion of *p*-adic MZV such as (1). However in [3] it is proved and by using this *double shuffle relations* for *p*-adic MZV are deduced. As for the extra relations, the *regularization relations* for *p*-adic MZV, the validity is a main theorem of [10].

## 3.3. The *p*-adic KZ equation and the *p*-adic Drinfel'd associator

Notions of the *p*-adic KZ equation, its fundamental solutions and the *p*-adic Drinfel'd associator will be introduced. They are essential to prove all theorems in the previous subsection.

Let $A^\wedge_{\mathbf{C}_p} = \mathbf{C}_p\langle\langle A, B\rangle\rangle$ be the non-commutative formal power series ring with $\mathbf{C}_p$-coefficients generated by two elements $A$ and $B$.

**3.14 Definition.** *The (formal) p-adic Knizhnik-Zamolodchikov equation (p-adic KZ equation for short) is the differential equation*

$$\frac{\partial G}{\partial u}(u) = \left(\frac{A}{u} + \frac{B}{u-1}\right) \cdot G(u) \quad , \tag{KZ$^p$}$$

*where $G(u)$ is an analytic function in variable $u \in \mathbf{P}^1(\mathbf{C}_p)\backslash\{0,1,\infty\}$ with values in $A^\wedge_{\mathbf{C}_p}$ where 'analytic' means each of whose coefficient is locally p-adic analytic.*

Unfortunately, because $\mathbf{P}^1(\mathbf{C}_p)\backslash\{0,1,\infty\}$ is topologically totally disconnected, the equation (KZ$^p$) does not have a unique solution on $\mathbf{P}^1(\mathbf{C}_p)\backslash\{0,1,\infty\}$ even locally as in the complex analytic function case. But fortunately we get the following nice property on Coleman functions.

**3.15 Theorem** ([8], Theorem 3.3). *Fix $a \in \mathbf{C}_p$. Then there exists unique (invertible) solutions $G_0^a(u)$ and $G_1^a(u) \in A^a_{\mathrm{Col}} \hat\otimes A^\wedge_{\mathbf{C}_p}$ of (KZ$^p$) which is defined and locally analytic on $\mathbf{P}^1(\mathbf{C}_p)\backslash\{0,1,\infty\}$ and satisfies*

$$G_0^a(u) \approx u^A (u \to 0) \text{ and } G_1^a(u) \approx (1-u)^B (u \to 1).$$

For the definition of $\approx$, consult [8]§3.1. As in the same way to the complex case we see that $G_0^a(u)$ and $G_1^a(u)$ are both invertible and they must coincide with each other up to a multiplication from the right by an invertible element $\Phi_{KZ}^{(a),p}(A,B) \in A^\wedge_{\mathbf{C}_p} = \mathbf{C}_p\langle\langle A,B\rangle\rangle$ (which is independent of $u \in \mathbf{P}^1(\mathbf{C}_p)\backslash\{0,1,\infty\}$). Namely $G_0^a(u) = G_1^a(u) \cdot \Phi_{KZ}^{(a),p}(A,B)$.

**3.16 Theorem** ([8], Theorem 3.10). *Actually $\Phi_{KZ}^{(a),p}(A,B)$ is independent of any choice of branch parameter $a \in \mathbf{C}_p$.*

*Proof.* Put $z_0 \in \,]\mathbf{P}^1_{\overline{\mathbf{F}_p}}\backslash\{0,1,\infty\}[$. Since $]\mathbf{P}^1_{\overline{\mathbf{F}_p}}\backslash\{0,1,\infty\}[$ is a branch independent region, the special value of $G_0^a(u)$ at $u = z_0$ actually does not depend on any choice of branch parameter $a \in \mathbf{C}_p$. Similarly we see that the value of $G_1^a(u)$ at $u = z_0$ does not depend on any choice of branch parameter. Therefore $\Phi_{KZ}^{(a),p}(A,B) = G_1^a(z_0)^{-1} \cdot G_0^a(z_0)$ is actually independent of any choice of branch parameter $a \in \mathbf{C}_p$. □

**3.17 Definition.** *The $p$-adic Drinfel'd associator $\Phi_{KZ}^p(A,B)$ is the element of $\mathbf{C}_p\langle\langle A,B\rangle\rangle^\times$, which is defined by*

$$G_0^a(u) = G_1^a(u) \cdot \Phi_{KZ}^p(A,B).$$

This definition of the $p$-adic Drinfel'd associator $\Phi_{KZ}^p(A,B)$ is independent of $u \in \mathbf{P}^1(\mathbf{C}_p)\backslash\{0,1,\infty\}$ and any choice of branch parameter $a \in \mathbf{C}_p$ by Theorem 3.16.

**3.18 Theorem** ([8], Theorem 3.30). *$p$-adic MZV's appear at each coefficient of the $p$-adic Drinfel'd associator as follows:*

$$\Phi_{KZ}^p(A,B) = 1 + \cdots + (-1)^m \zeta_p(k_1,\ldots,k_m) A^{k_m-1} B \cdots A^{k_1-1} B + \cdots.$$

The following is a low degree part of the $p$-adic Drinfel'd associator.

**3.19 Examples.**

$$\begin{aligned}\Phi_{KZ}^p(A,B) = &\ 1 - \zeta_p(2)AB + \zeta_p(2)BA - \zeta_p(3)A^2B + 2\zeta_p(3)ABA \\ &+ \zeta_p(1,2)AB^2 - \zeta_p(3)BA^2 - 2\zeta_p(1,2)BAB + \zeta_p(1,2)B^2A \\ &- \zeta_p(4)A^3B + 3\zeta_p(4)A^2BA + \zeta_p(1,3)A^2B^2 - 3\zeta_p(4)ABA^2 \\ &+ \zeta_p(2,2)ABAB - (2\zeta_p(1,3) + \zeta_p(2,2))AB^2A - \zeta_p(1,1,2)AB^3 \\ &+ \zeta_p(4)BA^3 - (2\zeta_p(1,3) + \zeta_p(2,2))BA^2B + (4\zeta_p(1,3) + \zeta_p(2,2))BABA \\ &+ 3\zeta_p(1,1,2)BAB^2 - \zeta_p(1,3)B^2A^2 - 3\zeta_p(1,1,2)B^2AB \\ &+ \zeta_p(1,1,2)B^3A + \cdots.\end{aligned}$$

## 3.4. Tannakian interpretation

The $p$-adic MZV will be related to the fundamental group of the projective line minus three points. It is one of the main results in [9].

Let $\pi_1^{\text{rig},p}(X_{\mathbf{F}_p}, \overrightarrow{01}, \overrightarrow{10})$ be the rigid fundamental torsor of $X$ from $\overrightarrow{01}$ to $\overrightarrow{10}$. This is a pro-algebraic variety defined over $\mathbf{Q}_p$. There is a Berthelot-Ogus-type comparison isomorphism between the de Rham and the rigid

$$\pi_1^{\text{DR}}(X, \overrightarrow{01}, \overrightarrow{10})(\mathbf{Q}_p) \cong \pi_1^{\text{rig},p}(X_{\mathbf{F}_p}, \overrightarrow{01}, \overrightarrow{10})(\mathbf{Q}_p)$$

(see [9]§2.1). In LHS there is a canonical de Rham path $d$ (cf. §2.3). In contrast Besser [2] and Vologodsky [16] constructed a canonical path $c$ in RHS, which is a unique Frobenius invariant path. Going via this rigid path $c$ and coming back via this de Rham path $d$, we get a de Rham loop $d^{-1}c$ in $\pi_1^{\mathrm{rig},p}(X,\overrightarrow{01})(\mathbf{Q}_p)$.

**3.20 Theorem** ([9], §2.1). *The de Rham loop* $d^{-1}c \in \pi_1^{DR}(X,\overrightarrow{01})(\mathbf{Q}_p)$ *corresponds to the p-adic Drinfel'd associator by the embedding i* (4), *i.e.*

$$i(d^{-1}c) = \Phi_{KZ}^p(A,B).$$

Since $p$-adic MZV's appear as coefficients of the $p$-adic Drinfel'd associator by Theorem 3.18, we may say that the loop $d^{-1}c$ which is a combination of a de Rham path and a rigid path is a tannakian origin of $p$-adic MZV's.

In [9] a tannakian origin of $p$-adic MPL's is also revealed. There is another notion of $p$-adic MZV's introduced by P.Deligne. An explicit relationship between our $p$-adic MZV's and his $p$-adic MZV's is also established in loc.cit.

# A. ON WEIGHTS OF *l*-ADIC POLYLOGARITHMS AND *p*-ADIC POLYLOGARITHMS

This is a report a $l$-adic behavior of the Wojtkowiak's $l$-adic polylogarithm and a $p$-adic behavior of the (Coleman's) $p$-adic polylogarithm [3] with respect to weights.

## A.1. *l*-adic polylogarithms

We employ notations in [15]. Let $l$ be a prime and $F$ be a subfield of $\mathbf{C}$. Put $\sigma \in \mathrm{Gal}(\overline{F}/F)$ and $z \in \mathbf{P}^1(F)\setminus\{0,1,\infty\}$. The distribution $\{\kappa_{l^n}^{z,(a)}(\sigma) \in \mathbf{Z}_l^\times\}_{n=0,1,2,\ldots}^{a\in\mathbf{Z}/l^n}$ (loc. cit. Definition 2) which determines a measure $\mu_F^{z,\sigma}$ of $\mathbf{Z}_l$ satisfies the following:

$$\kappa_{l^n}^{z,(a)}(\sigma) = \kappa_{l^{n+1}}^{z^l,(la)}(\sigma). \tag{5}$$

Let us note that actually the definition of $\kappa_{l^n}^{z,(a)}(\sigma)$ in LHS depends on a choice of a topological path in $\pi_1^{\mathrm{top}}(\mathbf{P}^1(\mathbf{C})\setminus\{0,1,\infty\};\overrightarrow{01},z)$ and we take its corresponding image in $\pi_1^{\mathrm{top}}(\mathbf{P}^1(\mathbf{C})\setminus\{0,1,\infty\};\overrightarrow{01},z^l)$ by the endomorphism of $\mathbf{P}^1(\mathbf{C})\setminus\{0,1,\infty\}$ sending $x \mapsto x^l$ when we consider the definition of $\kappa_{l^{n+1}}^{z^l,(la)}(\sigma)$ in RHS.

---

[3] This is the special case of $p$-adic MPL in Definition 3.4 with $m=1$.

For $m \geq 1$, the $l$-adic generalized Soulé element $\tilde{\chi}_m^z : Gal(\overline{F}/F) \to \mathbf{Z}_l$ is defined by the following congruence property (loc. cit. Definition 3):

$$\tilde{\chi}_m^z(\sigma) \equiv \sum_{a=0}^{l^n-1} a^{m-1} \kappa_{l^n}^{z,(a)}(\sigma) \mod l^n \mathbf{Z}_l \quad (n \geq 1),$$

which gives its integral expression

$$\tilde{\chi}_m^z(\sigma) = \int_{\mathbf{Z}_l} x^{m-1} d\mu_F^{z,\sigma}.$$

Particularly when $\sigma \in Gal\left(\overline{F}/F(z^{\frac{1}{l^\infty}})\right)$, we have $\ell_m^z(\sigma) = (-1)^{m-1} \frac{\tilde{\chi}_m^z(\sigma)}{(m-1)!}$ (loc. cit. Corollary) where $\ell_m^z(\sigma)$ stands for the $l$-adic polylogarithm studied by Wojtkowiak. Put $\chi_m^{(l),\sigma}(z) := \tilde{\chi}_m^{zl}(\sigma) - l^{m-1} \tilde{\chi}_m^z(\sigma) \in \mathbf{Z}_l$.

**A.1 Proposition.** *Fix $\sigma \in Gal(\overline{F}/F)$ and $z \in \mathbf{P}^1(F) \setminus \{0, 1, \infty\}$. Let $m, m', M \in \mathbf{N}$. If $m \equiv m' \mod (l-1)l^M$, then*

$$\chi_m^{(l),\sigma}(z) \equiv \chi_{m'}^{(l),\sigma}(z) \mod l^{M+1} \mathbf{Z}_l.$$

*Proof.* By (5), we compute

$$\tilde{\chi}_m^z(\sigma) = \lim_{n \to \infty} \sum_{0 \leq a \leq l^n-1} a^{m-1} \kappa_{l^n}^{z,(a)}(\sigma) = \lim_{n \to \infty} \frac{1}{l^{m-1}} \sum_{0 \leq a \leq l^n-1} (la)^{m-1} \kappa_{l^{n+1}}^{zl,(la)}(\sigma)$$

$$= \lim_{n \to \infty} \frac{1}{l^{m-1}} \sum_{\substack{0 \leq a \leq l^{n+1}-1 \\ l|a}} a^{m-1} \kappa_{l^{n+1}}^{zl,(a)}(\sigma) = \frac{1}{l^{m-1}} \int_{l\mathbf{Z}_l} x^{m-1} d\mu_F^{zl,\sigma}.$$

Then we have

$$\chi_m^{(l),\sigma}(z) = \tilde{\chi}_m^{zl}(\sigma) - l^{m-1} \tilde{\chi}_m^z(\sigma) = \int_{\mathbf{Z}_l^\times} x^{m-1} d\mu_F^{zl,\sigma}.$$

The equality $x^{m-1} \equiv x^{m'-1} \mod l^{M+1} \mathbf{Z}_l$ for $x \in \mathbf{Z}_l^\times$ by $m \equiv m' \mod (l-1)l^M$ gives an expression $x^{m-1} - x^{m'-1} = l^{M+1} h(x)$ for $h(x) \in \text{Cont}(\mathbf{Z}_l^\times, \mathbf{Z}_l)$. By $\int_{\mathbf{Z}_l^\times} h(x) d\mu_F^{zl,\sigma} \in \mathbf{Z}_l$,

$$\int_{\mathbf{Z}_l^\times} x^{m-1} d\mu_F^{zl,\sigma} \equiv \int_{\mathbf{Z}_l^\times} x^{m'-1} d\mu_F^{zl,\sigma} \mod l^{M+1} \mathbf{Z}_l,$$

from which we get the claim. $\square$

Especially by restricting into $\sigma \in Gal\left(\overline{F}/F(z^{\frac{1}{l^\infty}})\right)$, we see that a modification

$$\ell_m^{(l),\sigma}(z) := \left\{(m-1)! l^m \ell_m^{zl}(\sigma)\right\} - \frac{1}{l^{m-1}} \left\{(m-1)! l^m \ell_m^z(\sigma)\right\} \tag{6}$$

of the $l$-adic polylogarithm $\ell_m^z(\sigma)$ admits a nice $l$-adic behavior with respect to the weight $m$.

## A.2. *p*-adic polylogarithms

Let $p$ be a prime. Put $m \in \mathbf{N}$. The $p$-adic polylogarithm $Li_m(z) = \sum_{n=1}^{\infty} \frac{z^n}{n^m}$, converges on $\{z \in \mathbf{C}_p : |z|_p < 1\}$. As is explained in §3.2, this function is analytically continued to $\mathbf{P}^1(\mathbf{C}_p)\setminus\{1,\infty\}$ as (not a rigid analytic function but) a Coleman function of $\mathbf{P}^1(\mathbf{C}_p)\setminus\{0,1,\infty\}$. Consider its modification

$$\ell_m^{(p)}(z) := Li_m(z) - \frac{1}{p^m} Li_m(z^p) \tag{7}$$

which has an expansion $\ell_m^{(p)}(z) = \sum_{(n,p)=1} \frac{z^n}{n^m}$ on $|z|_p < 1$. It is shown in [4] Proposition 6.2 that actually this function extends to a rigid analytic function of $B := \mathbf{P}^1(\mathbf{C}_p) - \left\{|z-1|_p < p^{\frac{-1}{p-1}}\right\}$.

**A.2 Proposition.** *Let $m, m', M \in \mathbf{N}$. If $m \equiv m' \mod (p-1)p^M$, then*

$$\ell_m^{(p)}(z) \equiv \ell_{m'}^{(p)}(z) \mod p^{M+1} \mathbf{Z}_p[[z]].$$

*Proof.* It is clear by the congruence $n^m \equiv n^{m'} \mod p^{M+1} \mathbf{Z}$ for $(n,p) = 1$. □

Since the ring of rigid analytic functions on $B$ is embedded into $\mathbf{C}_p[[z]]$ by taking expansions at $z = 0$, we see that the modification $\ell_m^{(p)}(z) = Li_m(z) - \frac{1}{p^m} Li_m(z^p)$ of the $p$-adic polylogarithm $Li_m(z)$ admits a nice $p$-adic behavior with respect to the weight $m$ as a rigid analytic function.

One notices that the adjustment factor is $\frac{1}{p^m}$ in (7) while it is $\frac{1}{l^{m-1}}$ in (6). The author thinks that there might be a theoretical reasoning why such gap between étale side and rigid side happens.

## ACKNOWLEDGMENTS

The author thanks Branko Dragovich for a hospitality during his stay at Institute of Physics, Belgrade, Serbia and Montenegro. He is also grateful to Hiroshi Fujiwara who helped him to type this paper.

## REFERENCES

1. Berthelot P., *Cohomologie rigide et cohomologie rigide à support propre, Première partie*, Prépublication IRMAR 96-03, 89 pages (1996).

2. Besser A., Coleman integration using the Tannakian formalism, *Math. Ann.* **322**, no. 1, 19–48 (2002).
3. Besser A. and Furusho H., The double shuffle relation for *p*-adic multiple zeta values, math.NT/0310177.
4. Coleman R., Dilogarithms, regulators and *p*-adic *L*-functions, *Invent. Math.* **69**, no. 2, 171–208 (1982).
5. Deligne P., *Le groupe fondamental de la droite projective moins trois points, Galois groups over Q* (Berkeley, CA, 1987), 79–297, Math. S. Res. Inst. Publ., 16, Springer, New York-Berlin, 1989.
6. Drinfel'd V.G., On quasitriangular quasi-Hopf algebras and a group closely connected with $\text{Gal}(\overline{Q}/Q)$, *Leningrad Math. J.* **2**, no. 4, 829–860 (1991).
7. Furusho H., The multiple zeta value algebra and the stable derivation algebra, to appear in *Publ. Res. Inst. Math. Sci.* **39**, no. 4, math.NT/0011261.
8. _____, *p*-adic multiple zeta values I — *p*-adic multiple polylogarithms and the *p*-adic KZ equation, *Invent. Math.* **155**, no. 2, 253–286(2004), math.NT/0304085.
9. _____, *p*-adic multiple zeta values II — tannakian interpretations, math.NT/0506117.
10. _____, and Jafari A., Regularization and generalized double shuffle relations for *p*-adic multiple zeta values, math.AG/0510681.
11. Goncharov A.B., Multiple polylogarithms, cyclotomy and modular complexes, *Math. Res. Lett.* **5**, no. 4, 497–516 (1998).
12. Ihara K., Kaneko M. and Zagier D., Derivation and double shuffle relations for multiple zeta values, to appear in *Compositio Math.*
13. Kassel C., *Quantum groups*, Graduate Texts in Mathematics **155**, Springer-Verlag, New York, 1995.
14. Le T.T.Q. and Murakami J., Kontsevich's integral for the Kauffman polynomial, *Nagoya Math. J.* **142**, 39–65 (1996).
15. Nakamura H. and Wojtkowiak Z., On explicit formulae for *l*-adic polylogarithms, Arithmetic fundamental groups and noncommutative algebra, 285–294, *Proc. Sympos. Pure Math.*, **70**, Amer. Math. Soc., Providence, RI, 2002.
16. Vologodsky V., Hodge structure on the fundamental group and its application to *p*-adic integration, *Mosc. Math. J.* **3**, no. 1, 205–247 (2003).

# Aspects of *p*-Adic Non-Linear Functional Analysis

Helge Glöckner

*TAU Darmstadt, FB Mathematik AG 5, Schlossgartenstr. 7,
64289 Darmstadt, GERMANY
email: gloeckner@mathematik.tu-darmstadt.de*

**Abstract.** The article provides an introduction to infinite-dimensional differential calculus over topological fields and surveys some of its applications, notably in the areas of infinite-dimensional Lie groups and dynamical systems.

**Keywords:** Ultrametric calculus, non-archimedian analysis, non-linear functional analysis, fixed point theorem, inverse function theorem, implicit function theorem, Lie group, invariant manifold, stable manifold, pseudo-stable manifold, non-linear mapping, dependence on parameters.
**MSC 2000:** 26E30, 46S10, 22E65, 26E15, 26E20, 37D10, 46T20, 47J07, 58C15, 58C20.

## 1. INTRODUCTION

We describe aspects of non-linear functional analysis over topological fields, considered as the study of non-linear mappings between topological vector spaces, their fixed points and differentiability properties. Our first aim is to give an introduction to the differential calculus of smooth and $C^k$-maps between topological vector spaces over topological fields developed in [6]. This approach generalizes Schikhof's single variable calculus over complete ultrametric fields (as in [51]) to multi- and infinite-dimensional situations. As concerns mappings between real locally convex spaces, it is equivalent to the usual locally convex calculus (Keller's $C_c^k$-theory, as in [38] or [46]). The second aim is to survey some recent applications of differential calculus over topological fields, and the specific techniques underlying them. In particular, we discuss the following topics:

- The existence of fixed points and their $C^k$-dependence on parameters (as in [28]);
- An implicit function theorem for $C^k$-maps from arbitrary topological vector spaces over valued fields to Banach spaces (established in [25] and [28]);
- The construction of the main types of infinite-dimensional Lie groups (linear Lie groups, mapping groups, diffeomorphism groups, direct limit groups) over topological fields (carried out in [27], also [21] and [24]);
- The construction of invariant manifolds around hyperbolic fixed points for dynamical systems modelled on Banach spaces over valued fields (as in [29] and [30]), by an adaptation of Irwin's method (developed in [35] and [36], also [13] and [58]);
- Some special tools of calculus used in the preceding constructions (results ensuring differentiability properties of non-linear mappings between function spaces [27]

and spaces of sequences [29]; exponential laws for function spaces [27]).

While analytic mappings between Banach spaces over complete valued fields and the corresponding analytic Banach-Lie groups are classical objects of study (see [9], [11]; also [52] for the finite-dimensional case), we are interested just as well in mappings between general topological vector spaces, which have hardly been investigated so far. This is essential for infinite-dimensional Lie theory, since many interesting groups cannot be modelled on Banach spaces. It is also essential to work with smooth (and $C^k$-) maps, rather than analytic ones. In fact, already in the real case, typical examples of infinite-dimensional Lie groups (like diffeomorphism groups) are smooth Lie groups, but cannot be given an analytic Lie group structure. Even worse, over each local field of positive characteristic, finite-dimensional smooth Lie groups exist which do not admit an analytic Lie group structure compatible with their topological group structure [23]. For this reason, we shall use smooth maps as the basis of Lie theory.[1]

We mention that many important techniques usually applied in finite-dimensional non-archimedian analysis do not possess infinite-dimensional counterparts: Neither the techniques of algebraic geometry, which help to analyze the most prominent examples of finite-dimensional $p$-adic Lie groups (linear algebraic groups); nor the techniques of rigid analytic geometry (see [8], [17]), the non-archimedean analogue of complex geometry and function theory. However, the ideas of *real* differential calculus turn out to be extremely robust: surprisingly large parts of ordinary differential calculus carry over to mappings between open subsets of topological vector spaces over topological fields, once they are reformulated in an appropriate way. Schikhof's textbook [51] bears witness of this phenomenon in the case of functions of a single variable (see also [1], [3], [14], [15], [37], [40], [41] and [56], part of which include functions of several variables). The present article is based on the infinite-dimensional extension of Schikhof's ultrametric calculus developed in [6] (cf. [44] and [45] for a different, earlier approach). We hope to illustrate that the techniques of real differential calculus remain powerful also when dealing with mappings between topological vector spaces over topological fields.

Our illustrating examples are mostly taken from Lie theory and dynamical systems. In the theory of non-archimedean dynamical systems, a wealth of results is available for one-dimensional polynomial or analytic systems, inspired by classical complex dynamics (see, e.g., [2], [4], [39], [43]). Motivated by Lie-theoretic applications, we here describe complementary results dealing with higher-dimensional (and infinite-dimensional) dynamical systems, which have hardly been investigated before. Also the study of infinite-dimensional Lie groups over topological fields is of recent origin. It started with Ludkovsky's discussions of irreducible representations and invariant measures for certain infinite-dimensional $p$-adic groups (see [44], [45] and the references therein). In the meantime, all major constructions of real infinite-dimensional Lie groups have been adapted to more general topological fields ([27], [24]).

---

[1] Another problem is that a convincing notion of analytic map between (non-normable) locally convex spaces over ultrametric fields does not seem to exist yet, in contrast to the real and complex cases.

Outside the realms of non-linear analysis and Lie theory discussed in this article, the differential calculus over topological fields (and suitable commutative topological rings) has been used to give an essentially algebraic approach to differential geometry, which is entirely based on differentiation and does not involve integrals, nor flows [5]. Jordan theoretic applications have been explored in [7].

## 2. DIFFERENTIAL CALCULUS OVER TOPOLOGICAL FIELDS

The basic idea of differential calculus over topological fields is to call a map $C^1$ if one can pass from directional difference quotients to directional derivatives continuously. This idea, and its implications, will be described in more detail now. Throughout this section, $\mathbb{K}$ denotes a topological field (i.e., a field, equipped with a non-discrete Hausdorff topology which turns the field operations into continuous mappings). Topological $\mathbb{K}$-vector spaces are defined as in the real case, and are assumed Hausdorff. As the default, $E, F$ and $E_1, E_2, \ldots$ denote topological $\mathbb{K}$-vector spaces, and $U \subseteq E$ an open subset. To define continuous differentiability, let $E$ and $F$ be topological $\mathbb{K}$-vector spaces, and $f \colon U \to F$ be a map on an open subset $U \subseteq E$. Then the directional difference quotient

$$f^{]1[}(x,y,t) := \frac{f(x+ty)-f(x)}{t}$$

makes sense for all $(x,y,t)$ in the subset $U^{]1[} := \{(x,y,t) \in U \times E \times \mathbb{K}^{\times} : x+ty \in U\}$ of $E \times E \times \mathbb{K}$. To incorporate directional derivatives, we must enlarge this set by allowing also the value $t=0$. Hence, we consider now

$$U^{[1]} := \{(x,y,t) \in U \times E \times \mathbb{K} : x+ty \in U\}.$$

Thus $U^{[1]} = U^{]1[} \cup (U \times E \times \{0\})$, as a disjoint union.

**Definition 2.1.** $f \colon U \to F$ is called *continuously differentiable* (or $C^1$) if $f$ is continuous and there exists a continuous map $f^{[1]} \colon U^{[1]} \to F$ which extends $f^{]1[} \colon U^{]1[} \to F$.

Thus, we assume the existence of a continuous map $f^{[1]} \colon U^{[1]} \to F$ such that

$$f^{[1]}(x,y,t) = \frac{f(x+ty)-f(x)}{t} \quad \text{for all } (x,y,t) \in U^{[1]} \text{ such that } t \neq 0.$$

Given a $C^1$-map $f \colon U \to F$ as before, its directional derivative

$$df(x,y) := D_y f(x) := \lim_{0 \neq t \to 0} \tfrac{1}{t}(f(x+ty)-f(x)) = \lim_{0 \neq t \to 0} f^{[1]}(x,y,t) = f^{[1]}(x,y,0)$$

at $x \in U$ in the direction $y \in E$ exists, by continuity of $f^{[1]}$. The map $df \colon U \times E \to F$ is continuous, being a partial map of $f^{[1]}$, and it can be shown that the "differential"

$f'(x) := df(x, \bullet) : E \to F$ of $f$ at $x \in U$ is a continuous $\mathbb{K}$-linear map [6, Proposition 2.2]. Since $f^{[1]}(x,y,0)$ is a limit of difference quotients, $f^{[1]}$ is uniquely determined by $f$.

**Example 2.2.** To prove that a given map is $C^1$, usually one first writes down directional difference quotients and then tries to guess a continuous extension. To illustrate this strategy, let us show that every continuous linear map $\lambda : E \to F$ between topological $\mathbb{K}$-vector spaces is continuously differentiable. For $x, y \in E$ and $t \in \mathbb{K}^\times$, we have

$$\frac{\lambda(x+ty) - \lambda(x)}{t} = \lambda(y), \tag{1}$$

exploiting the linearity of $\lambda$. We now note that the right hand side of (1) makes sense just as well for $t = 0$, and defines a continuous function

$$\lambda^{[1]} : E \times E \times \mathbb{K} \to F, \qquad \lambda^{[1]}(x,y,t) := \lambda(y). \tag{2}$$

Thus $\lambda$ is $C^1$, with $\lambda^{[1]}$ as just described (which is again a continuous linear map). In particular, $\lambda'(x) = \lambda$ for each $x \in E$.

The same idea can be used to prove the Chain Rule for $C^1$-maps.

**Proposition 2.3.** *If $f$ and $g$ are composable $C^1$-maps, then also $f \circ g$ is $C^1$, and*

$$(f \circ g)^{[1]}(x,y,t) = f^{[1]}\bigl(g(x), g^{[1]}(x,y,t), t\bigr).$$

*In particular, $d(f \circ g)(x,y) = df\bigl(g(x), dg(x,y)\bigr)$.*

**Proof.** For $t \neq 0$, we have

$$\frac{f(g(x+ty)) - f(g(x))}{t} = \frac{f\bigl(g(x) + t \frac{g(x+ty) - g(x)}{t}\bigr) - f(g(x))}{t} = f^{[1]}\bigl(g(x), g^{[1]}(x,y,t), t\bigr).$$

Since the final expression makes sense just as well for $t = 0$ and defines a continuous function, we see that $f \circ g$ is $C^1$, with $(f \circ g)^{[1]}$ as asserted. □

Since $U^{[1]}$ is an open subset of the topological $\mathbb{K}$-vector space $E \times E \times \mathbb{K}$, it is possible to define $C^k$-maps by recursion.

**Definition 2.4.** Given an integer $k \geq 2$, a map $f : E \supseteq U \to F$ is called $C^k$ if it is $C^1$ and $f^{[1]} : U^{[1]} \to F$ is $C^{k-1}$. As usual, $f$ is called $C^\infty$ (or *smooth*) if it is $C^k$ for each $k \in \mathbb{N}$.

The present recursive definition of $C^k$-maps is particularly well-adapted to inductive arguments. In [44] and [45], a different approach to higher differentiability was used.

**Example 2.5.** We have seen above that $\lambda^{[1]}$ is again continuous linear if $\lambda : E \to F$ is a continuous linear map. Hence a straightforward induction shows that $\lambda$ is $C^k$ for each $k \in \mathbb{N}$ and thus smooth. A similar argument shows that every continuous $n$-linear map $\beta : E_1 \times \cdots \times E_n \to F$ is smooth (use [6, §3.3] and induction).

**Remark 2.6.** $C^k$-maps have many of the properties familiar from the real case.
(a) Compositions of composable $C^k$-maps are $C^k$ (see [6, Proposition 4.5]).
(b) Higher differentials can be defined and have the usual properties: If $f: E \subseteq U \to F$ is $C^k$, then the iterated directional derivative

$$d^j f(x, y_1, \ldots, y_j) := (D_{y_j} \cdots D_{y_1} f)(x)$$

exists for all $j \in \mathbb{N}$ such that $j \leq k$ and all $x \in U$, $y_1, \ldots, y_j \in E$. The mapping $d^j f: U \times E^j \to F$ so defined is continuous, and $d^j f(x, \bullet): E^j \to F$ is a continuous, symmetric, $j$-linear map, for each $x \in U$ (see [6, Chapter 4]).
(c) Finite order Taylor expansions are available in the following form: If $f: E \supseteq U \to F$ is $C^k$, with $k$ finite, then there are functions $a_j: U \times E \to F$ for $j \in \{0, 1, \ldots, k\}$ and a continuous map $R_k: U^{[1]} \to F$ with $R_k(x, y, 0) = 0$ for all $(x, y) \in U \times E$, such that

$$f(x+ty) = \sum_{j=0}^{k} t^j a_j(x, y) + t^k R_k(x, y, t) \quad \text{for all } (x, y, t) \in U^{[1]}.$$

The functions $a_1, \ldots, a_k$ and $R_k$ are uniquely determined. Furthermore, $a_j$ is $C^{k-j}$, homogeneous of degree $j$ in the second argument, and $j! a_j(x, y) = d^j f(x, y, \ldots, y)$. If $\mathbb{K}$ has characteristic 0, we can divide by $j!$ and recover Taylor's formula in its familiar form with $a_j(x, y) = \frac{1}{j!} d^j f(x, y, \ldots, y)$ (see [6, Chapter 5] for details).
(d) Despite its appearance, being $C^k$ is a local property: If $(U_i)_{i \in I}$ is an open cover of $U$ and $f|_{U_i}$ is $C^k$ for each $i \in I$, then $f: U \to F$ is $C^k$ (see [6, Lemma 4.9]).

The preceding facts make it possible to define $C^k$-manifolds in the usual way.

**Definition 2.7.** A $C^k$-*manifold* modelled on a topological $\mathbb{K}$-vector space $E$ is a Hausdorff topological space $M$, equipped with a set $\mathscr{A}$ of homeomorphisms $\phi: U \to V$ from open subsets of $M$ onto open subsets of $E$, such that the domains $U$ cover $M$ and the transition maps $\phi \circ \psi^{-1}$ are $C^k$ on their domain, for all $\phi, \psi \in \mathscr{A}$.

Now $C^k$-maps between $C^k$-manifolds can be defined as usual (such that the $C^k$-property can be tested in charts). Also the direct product of two manifolds can be defined as usual.

**Definition 2.8.** A *Lie group* over a topological field $\mathbb{K}$ is a group $G$, equipped with a smooth manifold structure modelled on a topological $\mathbb{K}$-vector space which turns the group inversion $G \to G$ and the group multiplication $G \times G \to G$ into smooth mappings.

$C^k$-Lie groups for finite $k$ are defined analogously.

**Remark 2.9.** The $C^k$-maps considered here are related to traditional concepts as follows.
(a) First of all, a map $f: \mathbb{R}^n \to \mathbb{R}^m$ is $C^k$ in our sense if and only if it is $C^k$ in the usual sense. In fact, if $f$ is an ordinary $C^1$-map, then

$$f^{[1]}(x, y, t) := \int_0^1 df(x+ty, y) \, dt$$

defines a map $f^{[1]}\colon \mathbb{R}^n \times \mathbb{R}^n \times \mathbb{R} \to \mathbb{R}^m$ which is continuous as a parameter-dependent integral, and produces the desired directional difference quotient for $t \neq 0$, by the Mean Value Theorem. Conversely, existence of $f^{[1]}$ easily entails that $f$ is $C^1$ in the usual sense. For higher $k$, one argues by a simple induction.

(b) More generally, a map $f\colon E \supseteq U \to F$ to a real locally convex space $F$ is $C^k$ in our sense if and only if it is a "Keller $C_c^k$-map," i.e. $f$ is continuous, the iterated directional derivatives $d^j f(x, y_1, \ldots, y_j)$ (as in Remark 2.6 (b)) exist for all $j \in \mathbb{N}$ such that $j \leq k$, and define continuous maps $d^j f\colon U \times E^j \to F$ (see [6, Proposition 7.4]).

(c) Every $k$ times continuously Fréchet differentiable map ($FC^k$-map) between real Banach spaces is $C^k$, and conversely every $C^{k+1}$-map is $FC^k$ (cf. (b) and [38]).

(d) A map $E \supseteq U \to F$ to a complex locally convex space is $C^\infty$ if and only if it is complex analytic, i.e., $f$ is continuous and given locally around each point by a pointwise convergent series of continuous homogeneous polynomials [6, Proposition 7.7].

(e) Finally, a map $\mathbb{K} \supseteq U \to \mathbb{K}$ on an open subset of a complete ultrametric field $\mathbb{K}$ is $C^k$ in our sense if and only if it is a $C^k$-map in the sense of [51, Definition 29.1], as usually considered in non-archimedian analysis (see [6, Proposition 6.9]).

Of course, we are mainly interested in differential calculus over a valued field $(\mathbb{K}, |.|)$; thus $\mathbb{K}$ is a field and $|.|\colon \mathbb{K} \to [0, \infty[$ an "absolute value" with properties analogous to the modulus of real or complex numbers. We shall always assume that $|.|$ is non-trivial in the sense that the metric $d(x, y) := |x - y|$ defines a non-discrete (field) topology on $\mathbb{K}$. A valued field $(\mathbb{K}, |.|)$ is called an *ultrametric field* if its absolute value $|.|$ satisfies the ultrametric inequality, viz. $|x + y| \leq \max\{|x|, |y|\}$ for all $x, y \in \mathbb{K}$. Besides $\mathbb{R}$ and $\mathbb{C}$, the most prominent examples of valued fields are local fields (i.e., totally disconnected, locally compact topological fields). Any such is known to admit an ultrametric absolute value defining its topology. Up to isomorphism, every local field either is a finite extension of the field $\mathbb{Q}_p$ of $p$-adic numbers, or a field $\mathbb{F}((X))$ of formal Laurent series over a finite field $\mathbb{F}$ (see [57]). If $(\mathbb{K}, |.|)$ is a valued field, then we can speak of norms and seminorms on $\mathbb{K}$-vector spaces, as in the real and complex cases. A normed space $(E, \|.\|)$ over $\mathbb{K}$ which is complete in the metric determined by $\|.\|$ is called a *Banach space*. If $(\mathbb{K}, |.|)$ is an ultrametric field here and $\|.\|$ satisfies the ultrametric inequality, then $(E, \|.\|)$ is called an *ultrametric* Banach space. See also [10], [47] and [50].

**Remark 2.10.** We mention that a certain range of strengthened differentiability properties is available for mappings between topological vector spaces over a valued field $(\mathbb{K}, |.|)$. In this case, also $k$ times strictly differentiable mappings ($SC^k$-maps) and $k$ times Lipschitz differentiable mappings ($LC^k$-maps) can be defined [28, §2 and §3]. Then

$$C^{k+1} \Rightarrow LC^k \Rightarrow SC^k \Rightarrow C^k$$

(see [28, Remark 3.17]). The strengthened differentiability properties are useful for some refined results. Strict differentiability at a point for mappings on normed spaces is a classical concept (see [9, 1.2.2]). Mappings between real Banach spaces are strictly differentiable if and only if they are once continuously Fréchet differentiable [9, 2.3.3].

If $\mathbb{K}$ is locally compact and $E$ a finite-dimensional $\mathbb{K}$-vector space, then a map $f: U \to F$ on an open subset $U \subseteq E$ is $C^k$ if and only if it is $SC^k$ (cf. [28, Lemma 3.11]).

## 3. FIXED POINTS, INVERSE AND IMPLICIT FUNCTIONS

In this section, we present an implicit function theorem for $C^k$-maps from arbitrary topological vector spaces over valued fields to Banach spaces. We also describe some of the tools used in its proof, notably a theorem ensuring the $C^k$-dependence of fixed points on parameters, which may be of independent interest. The results are taken from [28] (for slightly less general versions, see [25]). Similar implicit function theorems in the real and complex cases, in different settings of analysis, were also obtained in [33], [34] (for Keller's $C^k_\Pi$-theory) and [53] (in the Convenient Setting of Analysis, [18], [42]).

While classical implicit function theorems are restricted to mappings between Banach spaces, the following result (a special case of [28, Theorem 5.2]) only requires that the range space be a Banach space (for the case $k = 1$, see Remark 3.7).

**Theorem 3.1 (Generalized Implicit Function Theorem).** *Let $\mathbb{K}$ be a valued field, $E$ be a topological $\mathbb{K}$-vector space, $F$ be a Banach space over $\mathbb{K}$, and $f: U \times V \to F$ be a $C^k$-map, where $U \subseteq E$ and $V \subseteq F$ are open subsets and $k \in \mathbb{N} \cup \{\infty\}$, $k \geq 2$. Assume that $f(x_0, y_0) = 0$ for some $(x_0, y_0) \in U \times V$ and $f'_{x_0}(y_0) \in \mathrm{GL}(F) = \mathscr{L}(F)^\times$, where $f_x := f(x, \bullet): V \to F$ for $x \in U$. Then there exist open neighbourhoods $U_0 \subseteq U$ of $x_0$ and $V_0 \subseteq V$ of $y_0$ such that $\{(x, y) \in U_0 \times V_0 : f(x, y) = 0\}$ is the graph of a $C^k$-map $\lambda: U_0 \to V_0$.*

**Strategy of the proof.** The idea is to deduce the implicit function theorem from an "Inverse Function Theorem with Parameters." Since $f'_{x_0}(y_0) \in \mathrm{GL}(F)$, where $\mathrm{GL}(F)$ is open in $\mathscr{L}(F)$ and the map $U \to \mathscr{L}(F)$, $x \mapsto f'_x(y_0)$ is continuous, we see that $f'_x(y_0) \in \mathrm{GL}(F)$ for $x$ close to $x_0$. Then $f_x^{-1}$ exists locally around $f_x(y_0)$, by a suitable Inverse Function Theorem (e.g., [9, 1.5.1] or Proposition 3.2 below). Now the essential point is that the map

$$\psi: (x, z) \mapsto (f_x)^{-1}(z) \tag{3}$$

actually makes sense on a whole neighbourhood $U_0 \times W$ of $(x_0, 0)$ in $U \times F$, and is $C^k$ there (further explanations will be given below). Once we have this, the rest is easy: The map $\lambda: U_0 \to F$, $\lambda(x) := (f_x)^{-1}(0) = \psi(x, 0)$ is $C^k$, and $f(x, \lambda(x)) = 0$. □

To see that the map $\psi$ in (3) is defined on a whole neighbourhood, one exploits that suitable versions of the inverse function theorem provide *quantitative* information on the size of the domain of the local inverse. For example, one can use the following Lipschitz version of the inverse function theorem, [28, Theorem 5.3] (adapted from the real case in [58]). If $E$ is a Banach space over $\mathbb{K}$ and $f: E \supseteq X \to E$ a map, we set $\mathrm{Lip}(f) := \sup\{\|f(y) - f(x)\| \cdot \|y - x\|^{-1} : x \neq y \in X\}$ and call $f$ *Lipschitz* if $\mathrm{Lip}(f) < \infty$.

**Proposition 3.2 (Lipschitz Inverse Function Theorem).** *Let $(E, \|.\|)$ be a Banach space over a valued field $(\mathbb{K}, |.|)$. Let $x \in E$, $r > 0$ and $f: B_r(x) \to E$ be a map on the ball $B_r(x) := \{y \in E: \|y - x\| < r\}$ of the form*

$$f = A + \tilde{f},$$

*where $A \in \mathrm{GL}(E)$ and $\tilde{f}: B_r(x) \to E$ is a Lipschitz map with $\mathrm{Lip}(\tilde{f}) < \|A^{-1}\|^{-1}$. Then $f$ has open image and is a homeomorphism onto its image. Furthermore, the inverse map $f^{-1}: f(B_r(x)) \to B_r(x)$ is Lipschitz with $\mathrm{Lip}(f^{-1}) \leq (\|A^{-1}\|^{-1} - \mathrm{Lip}(\tilde{f}))^{-1}$, and*

$$B_{ar}(f(x)) \subseteq f(B_r(x)) \subseteq B_{br}(f(x))$$

*with $a := \|A^{-1}\|^{-1} - \mathrm{Lip}(\tilde{f})$ and $b := \|A\| + \mathrm{Lip}(\tilde{f})$.*

The proof of Proposition 3.2 is based on a simple Newton-type iteration. Accordingly, $\psi(x,z)$ in (3) is the fixed point of a contraction $g_{x,z}$ of a (subset of a) Banach space. The following general result ensures that the fixed point $\psi(x,z)$ of $g_{x,z}$ is a $C^k$-function of $(x,z)$. It applies to so-called "uniform" families of contractions.

**Definition 3.3.** Let $F$ be a Banach space over a valued field $\mathbb{K}$, and $U \subseteq F$ be a subset. A family $(f_p)_{p \in P}$ of mappings $f_p: U \to F$ is called a *uniform family of contractions* if there exists $\theta \in [0,1[$ such that $\|f_p(x) - f_p(y)\| \leq \theta \|x - y\|$ for all $x, y \in U$ and $p \in P$.

We now formulate the technical backbone of the generalized implicit function theorem (a special case of [28, Theorem 4.7]). It was stimulated by the discussion of fixed points in real Banach spaces and their dependence on parameters in *Banach* spaces in [36].

**Theorem 3.4 (on the Parameter-Dependence of Fixed Points).** *Let $\mathbb{K}$ be a valued field, $E$ be a topological $\mathbb{K}$-vector space, $F$ be a Banach space over $\mathbb{K}$ and $f: P \times U \to F$ be a $C^k$-map, where $P \subseteq E$ and $U \subseteq F$ are open and $k \in \mathbb{N}_0 \cup \{\infty\}$. Assume that the maps $f_p := f(p, \cdot): U \to F$ define a is a uniform family of contractions $(f_p)_{p \in P}$. Then*

$$Q := \{p \in P: f_p \text{ has a fixed point } x_p\}$$

*is open in $P$, and $\phi: Q \to F$, $p \mapsto x_p$ is a $C^k$-map.*

Exploiting the $C^k$-dependence of fixed points on parameters, one also obtains the following analogue of the classical Inverse Function Theorem (see [28, Theorem 5.1]):

**Theorem 3.5 (Inverse Function Theorem).** *Let $E$ be a Banach space over a valued field $\mathbb{K}$ and $f: U \to E$ be a $C^k$-map on an open subset $U \subseteq E$, where $k \in \mathbb{N} \cup \{\infty\}$ and $k \geq 2$. If $f'(x) \in \mathrm{GL}(E)$ for some $x \in U$, then there exists an open neighbourhood $V \subseteq U$ of $x$ such that $f(V)$ is open in $E$ and $f|_V: V \to f(V)$ is a $C^k$-diffeomorphism.*

In the ultrametric case, closer inspection of the mapping $\psi$ in (3) leads to the following useful result (which is a special case of [28, Theorem 5.17]).

**Theorem 3.6 (Ultrametric Inverse Function Theorem with Parameters).** *Let $(\mathbb{K}, |.|)$ be an ultrametric field, $E$ be a topological $\mathbb{K}$-vector space, and $F$ be an ultrametric Banach space over $\mathbb{K}$. Let $P_0 \subseteq E$ and $U \subseteq F$ be open, and $f: P_0 \times U \to F$ be a $C^k$-map, where $k \in \mathbb{N} \cup \{\infty\}$ and $k \geq 2$. Let $(p_0, x_0) \in P_0 \times U$ be given such that $A := f'_{p_0}(x_0) \in \mathrm{GL}(F)$, where $f_p := f(p, \cdot): U \to F$ for $p \in P_0$. Then there exists an open neighbourhood $P \subseteq P_0$ of $p_0$ and $r > 0$ such that $B := B_r(x_0) \subseteq U$ and the following holds:*

(a) $f_p(B) = f(p_0, x_0) + A.B_r(0) =: V$, for each $p \in P$, and $\phi_p: B \to V$, $\phi_p(y) := f(p, y)$ is a $C^k$-diffeomorphism;

(b) $f_p(B_s(y)) = f_p(y) + A.B_s(0)$ for all $p \in P$, $y \in B$ and $s \in \,]0, r]$;

(c) The map $\psi: P \times V \to B$, $\psi(p, v) := \phi_p^{-1}(v)$ is $C^k$;

(d) $\xi: P \times B \to P \times V$, $\xi(p, y) := (p, f(p, y))$ is a $C^k$-diffeomorphism, with inverse given by $\xi^{-1}(p, v) = (p, \psi(p, v))$.

A similar result is available for non-ultrametric Banach spaces [28, Theorem 5.13].

**Remark 3.7.** Theorems 3.1 and 3.6 remain valid for $k = 1$ if $F$ is finite-dimensional, and Theorem 3.5 remains valid for finite-dimensional $E$ (see [28]). In the infinite-dimensional case, $C^1$-maps need not be approximated good enough by their linearization around a given point to guarantee the hypotheses of Proposition 3.2. This problem disappears for $SC^k$-maps or $LC^k$-maps. All of the results presented so far in this section have analogues for $SC^k$-maps and $LC^k$-maps, valid for all $k \in \mathbb{N} \cup \{\infty\}$ (see [28]).

**Applications.** We mention various applications of the preceding theorems in non-archimedean analysis. For applications in the real and complex cases, see [25] and [32].

- As described in more detail in Section 4, the implicit function theorem can be used to construct stable manifolds around hyperbolic fixed points.
- In [27], the ultrametric inverse function theorem with parameters in a Fréchet space is used to prove that the inversion map $\mathrm{Diff}(M) \to \mathrm{Diff}(M)$, $\gamma \mapsto \gamma^{-1}$ of the diffeomorphism group of a paracompact, finite-dimensional smooth manifold over a local field is smooth (see also Section 5 below).
- Varying an idea from [12], $C^{k+n}$-solutions to (systems of) $p$-adic differential equations of the form $y^{(k)} = f(x, y, y', \ldots, y^{(k-1)})$ for $f$ an $LC^n$-map and $k \in \mathbb{N}$, $n \in \mathbb{N}_0$ can be constructed using our inverse function theorems (with and without parameters), which depend on initial conditions and parameters in a controlled way (work in progress; cf. [51, §65] for $C^1$-solutions to scalar-valued first order equations). This is a first step towards the study of equations which are not accessible by the existing highly-developed techniques for linear or analytic equations (cf. [16], [54]).
- The ultrametric inverse function theorem with parameters is also used in the construction of a $C^k$-compatible analytic Lie group structure on each finite-dimensional $p$-adic $C^k$-Lie group in [26] (for $k \in \mathbb{N} \cup \{\infty\}$).

# 4. INVARIANT MANIFOLDS AROUND FIXED POINTS

We briefly discuss invariant manifolds over valued fields and an application.

**Definition 4.1.** Let $E$ be a Banach space over a valued field $(\mathbb{K}, |.|)$, and $a \in \,]0,1]$. A (bicontinuous) linear automorphism $\alpha \in \mathrm{GL}(E)$ is called *a-hyperbolic* if $E = E_1 \oplus E_2$ for certain $\alpha$-invariant closed vector subspaces $E_1, E_2$;

$$\|x_1 + x_2\| = \max\{\|x_1\|, \|x_2\|\} \quad \text{for all } x_1 \in E_1 \text{ and } x_2 \in E_2$$

holds for a norm $\|.\|$ on $E$ equivalent to the original one; and $\alpha = \alpha_1 \oplus \alpha_2$ with $\|\alpha_1\| < a$ and $\|\alpha_2^{-1}\|^{-1} > a$. The 1-hyperbolic automorphisms are simply called *hyperbolic*.

**Remark 4.2.** If $E$ is an ultrametric Banach space, then the norm $\|.\|$ described in the definition can be chosen ultrametric as well.

**Remark 4.3.** If $\mathbb{K}$ is a local field and $\dim_\mathbb{K}(E) < \infty$, then $\alpha$ is $a$-hyperbolic if and only if $a \neq |\lambda|$ for each eigenvalue $\lambda \in \overline{\mathbb{K}}$ of $\alpha \otimes_\mathbb{K} \mathrm{id}_{\overline{\mathbb{K}}} \in \mathrm{GL}(E \otimes_\mathbb{K} \overline{\mathbb{K}})$, where $\overline{\mathbb{K}}$ is an algebraic closure of $\mathbb{K}$ ([29]; cf. also [19, proof of Lemma 3.3]).

We begin with the global Stable Manifold Theorem (see [29]).

**Theorem 4.4.** Let $M$ be a $C^\infty$-manifold modelled on a Banach space over a valued field, $f: M \to M$ be a $C^\infty$-diffeomorphism, and $p \in M$ be a hyperbolic fixed point of $f$, i.e. $f(p) = p$ and the differential $T_p f: T_p M \to T_p M$ is a hyperbolic automorphism of the tangent space $T_p M$. Then

$$W_s := \{x \in M : f^n(x) \to p \text{ as } n \to \infty\}$$

is an immersed $C^\infty$-submanifold of $M$.

**Remark 4.5.** Analogous conclusions are available for manifolds and diffeomorphisms which are $C^k$ (with $k \geq 2$), resp., $SC^k$ and $LC^k$ (with $k \in \mathbb{N}$). The theorem also holds for $\mathbb{K}$-analytic diffeomorphism of $\mathbb{K}$-analytic manifolds, in the sense of [9] (see [29]).

Using standard arguments, the global stable manifold theorem follows from corresponding local results. Therefore, we now briefly discuss the construction of local $a$-stable manifolds, for $a \in \,]0,1]$. For simplicity, we only consider smooth maps, and concentrate on the case where $E$ is an ultrametric Banach space over an ultrametric field $(\mathbb{K}, |.|)$.

Let $\alpha \in \mathrm{GL}(E)$ be $a$-hyperbolic, say $E = E_1 \oplus E_2$ with $E_1$ and $E_2$ as in Definition 4.1. Let $r > 0$ and $f: U \to E$ be a smooth map on the ball $U := B_r^E(0) = U_1 \times U_2$, where $U_j := B_r^{E_j}(0)$, such that $f(0) = 0$ and $f'(0) = \alpha$. The $a$-stable set of $f$ is defined as

$$W_{s,a} := \{z \in U : f^n(z) \text{ is defined for all } n \in \mathbb{N}_0, a^{-n}\|f^n(z)\| < r \text{ and } f^n(z) = o(a^n)\}.$$

Then $f(W_{s,a}) \subseteq W_{s,a}$. If $z \in W_{s,a}$, then the orbit $\omega := (f^n(z))_{n \in \mathbb{N}_0}$ is an element of the Banach sequence space

$$\mathscr{S}_a(E) := \{z = (z_n)_{n \in \mathbb{N}_0} \in E^{\mathbb{N}_0} : z_n = o(a^n)\}$$

with norm $\|z\|_a := \max\{a^{-n}\|z_n\| : n \in \mathbb{N}_0\}$. After shrinking $r$, we may assume that $\mathrm{Lip}(f - \alpha) < \min\{1, \|\alpha_2^{-1}\|^{-1}\}$. Then the following holds:

**Theorem 4.6.** *$W_{s,a}$ is the graph of a $C^\infty$-map $\phi: U_1 \to U_2$ with $\phi(0) = 0$ and $\phi'(0) = 0$. Thus $W_{s,a}$ is a $C^\infty$-submanifold of $E$, which is tangent to the a-stable subspace $E_1$ at $0$.*

The proof given in [29] follows Irwin's method (see [35] and [58]). The idea is to construct not $\phi(x)$ itself, but the orbit $\omega$ of $(x, \phi(x))$, which is an element of the ball $\mathscr{U} := \{z \in \mathscr{S}_a(E) : \|z\|_a < r\}$. The orbit $\omega$ turns out to solve an equation $\Phi(x, \omega) = 0$, for a suitable smooth map

$$\Phi: U_1 \times \mathscr{U} \to \mathscr{S}_a(E),$$

to which the implicit function theorem can be applied. This idea, used by Irwin in the real case, works just as well in the present situation. Essentially, one simply uses Theorem 3.1 instead of the implicit function theorem for real Banach spaces; the desired differentiability properties of $\Phi$ are ensured by the next proposition (from [29]).

**Proposition 4.7.** *Let $E$ and $F$ be Banach spaces over a valued field, $f: B_r^E(0) \to F$ be a mapping such that $f(0) = 0$, $a \in {]0,1]}$ and $\mathscr{U} := \{w \in \mathscr{S}_a(E) : \|w\|_a < r\}$. If $f$ is $C^k$, $SC^k$, $LC^k$ with $k \in \mathbb{N} \cup \{\infty\}$, resp., $\mathbb{K}$-analytic, then also the map*

$$\mathscr{S}_a(f): \mathscr{U} \to \mathscr{S}_a(F), \qquad (x_n)_{n \in \mathbb{N}_0} \mapsto (f(x_n))_{n \in \mathbb{N}_0}$$

*is $C^k$, $SC^k$, $LC^k$, resp., $\mathbb{K}$-analytic.*

We mention that also Irwin's construction of pseudo-stable manifolds (see [36] and [13]) can be adapted to general valued fields [30]. Combining these constructions, all main types of invariant manifolds (stable and unstable manifolds, center-stable and center-unstable manifolds, as well as center manifolds) become available also over valued fields. Using these invariant manifolds, the following result (from [31]) can be obtained, which provided the original stimulus for the author's studies compiled in this section.

**Theorem 4.8.** *Given $k \in \mathbb{N} \cup \{\infty, \omega\}$, let $G$ be a finite-dimensional $C^k$-Lie group over a local field $\mathbb{K}$ and $\alpha: G \to G$ be a $C^k$-automorphism whose contraction group*

$$U_\alpha := \{x \in G : \alpha^n(x) \to 1 \text{ as } n \to \infty\}$$

*is closed in $G$. Then $U_\alpha$, $U_{\alpha^{-1}}$ and $M_\alpha := \{x \in G : \alpha^{\mathbb{Z}}(x) \text{ is relatively compact}\}$ are closed Lie subgroups, and the product map $U_\alpha \times M_\alpha \times U_{\alpha^{-1}} \to G$, $(x, y, z) \mapsto xyz$ is a $C^k$-diffeomorphism onto an open, $\alpha$-stable identity neighbourhood of $G$.*

For $\mathbb{K} = \mathbb{Q}_p$, this is a classical result by Wang [55, Theorem 3.5]. In this case, $U_\alpha$ is always closed, and $\alpha$ simply looks like a linear automorphism in an exponential chart, which facilitates to reduce the proof to the case of linear automorphisms of vector spaces. By contrast, automorphisms of Lie groups over local fields of positive characteristic need not be linear in any chart (cf. [23]), whence non-linear analysis cannot be avoided.

## 5. EXAMPLES OF INFINITE-DIMENSIONAL LIE GROUPS

In this section, we discuss the main examples of Lie groups over topological fields, in parallel with some of the specific techniques of non-linear functional analysis needed to construct their Lie group structures. In particular, we describe various results concerning non-linear mappings between function spaces.

### Linear Lie groups

Among the easiest examples of real or complex Lie groups are linear Lie groups, i.e., unit groups of unital Banach algebras (or other well-behaved topological algebras) and their Lie subgroups. If $\mathbb{K}$ is a general topological field, then a good class of topological algebras to look at are the *continuous inverse algebras*, i.e., unital associative topological $\mathbb{K}$-algebras $A$ such that the group of units $A^\times$ is open and the inversion map $\eta: A^\times \to A$, $a \mapsto a^{-1}$ is continuous. An elementary argument shows (see [27, Proposition 2.2]):

**Proposition 5.1.** *If $A$ is a continuous inverse algebra, then the inversion map $\eta: A^\times \to A$ is smooth and thus $A^\times$ is a Lie group.*

For example, $\mathbb{K}$ is a continuous inverse algebra, and more generally every finite-dimensional unital associative $\mathbb{K}$-algebra when equipped with the canonical vector topology ($\cong \mathbb{K}^d$), see [27, Proposition 2.6]. If $A$ is a continuous inverse algebra over $\mathbb{K}$, then so is the matrix algebra $M_n(A)$, for each $n \in \mathbb{N}$ (see [27, Proposition 2.3]). Furthermore, $A \otimes_\mathbb{K} \mathbb{L}$ is a continuous inverse algebra over $\mathbb{L}$, for each finite extension $\mathbb{L}$ of $\mathbb{K}$, by [27, Corollary 2.8]. If $A$ a continuous inverse algebra over a locally compact topological field $\mathbb{K}$, and $K$ a compact $C^r$-manifold over $\mathbb{K}$ (where $r \in \mathbb{N}_0 \cup \{\infty\}$), then also the algebra $C^r(K, A)$ of $A$-valued $C^r$-maps is a continuous inverse algebra, with respect to pointwise operations and the natural topology on this function space described below (see [27, Proposition 5.7]). For further examples over $\mathbb{R}$ or $\mathbb{C}$, see [20].

Summing up, we always have a certain supply of continuous inverse algebras over each topological field, whose unit groups provide a certain supply of Lie groups.

### Mapping groups and related constructions

The second widely studied class of infinite-dimensional real Lie groups are the mapping groups, for example, loop groups $C(\mathbb{S}^1, G)$ and $C^\infty(\mathbb{S}^1, G)$, where $\mathbb{S}^1$ is the unit circle and

$G$ a finite-dimensional real Lie group ([46], [49]). The classical constructions of mapping groups can be generalized to a large extent to the case of Lie groups over topological fields. For example, $C(K,G)$ can be made a Lie group for each compact topological space $K$ and Lie group $G$ over a topological field $\mathbb{K}$ (see [27, Proposition 5.1]). We now discuss groups of differentiable maps in more detail. [27, Proposition 5.1] subsumes:

**Proposition 5.2.** *Let $K$ be a compact (and hence finite-dimensional) $C^k$-manifold over a locally compact topological field $\mathbb{K}$, where $k \in \mathbb{N}_0 \cup \{\infty\}$, and $G$ a Lie group, modelled on a topological $\mathbb{K}$-vector space $E$. Then there is a uniquely determined smooth manifold structure on $C^k(K,G)$ making it a Lie group modelled on $C^k(K,E)$, and such that*

$$\Phi \colon C^k(K,U) \to C^k(K,V), \quad \gamma \mapsto \phi \circ \gamma$$

*defines a chart of $C^k(K,G)$ around $1$, for some chart $\phi \colon G \supseteq U \to V \subseteq E$ of $G$.*

Here $C^k(K,U) := \{\gamma \in C^k(K,G) \colon \gamma(K) \subseteq U\}$ and $C^k(K,V) := \{\gamma \in C^k(K,E) \colon \gamma(K) \subseteq V\}$, which is an open subset of the topological $\mathbb{K}$-vector space $C^k(K,E)$ of $E$-valued $C^k$-maps on $K$. The topology on the latter is defined as follows.

**Definition 5.3.** Given topological $\mathbb{K}$-vector spaces $X$ and $E$ over a topological field $\mathbb{K}$, and a $C^k$-map $\gamma \colon U \to E$ on an open subset $U \subseteq X$, where $k \in \mathbb{N}_0 \cup \{\infty\}$, we recursively define $U^{[j]} := (U^{[1]})^{[j-1]}$ and $\gamma^{[j]} := (\gamma^{[1]})^{[j-1]} \colon U^{[j]} \to E$ for each $j \in \mathbb{N}$ such that $j \leq k$ (with $U^{[0]} := U$ and $f^{[0]} := f$). We equip $C^k(U,E)$ with the initial topology with respect to the family of mappings

$$C^k(U,E) \to C(U^{[j]},E), \quad \gamma \mapsto \gamma^{[j]}$$

such that $j \in \mathbb{N}_0$ and $j \leq k$, where $C(U^{[j]},E)$ carries the compact-open topology. If $M$ is a $C^r$-manifold modelled on $X$, with $C^k$-atlas $\mathscr{A}$, we equip $C^k(M,E)$ with the initial topology with respect to the family of maps $C^k(M,E) \to C^k(V,E)$, $\gamma \mapsto \gamma \circ \phi^{-1}$, for $\phi \colon M \supseteq U \to V \subseteq X$ ranging through $\mathscr{A}$ (see [27, §4] for details).

This is a very natural definition, which provides spaces with the expected completeness, metrizability and local convexity properties in relevant situations, and which produces the conventional topologies in the real locally convex case [27, Proposition 4.19].

The proof of Proposition 5.2 is based on the following result (and variants for mappings between spaces of $C^k$-maps, [27, Proposition 4.20 and Corollary 4.21]). It ensures smoothness of the relevant non-linear mappings between spaces of smooth functions:

**Proposition 5.4.** *Let $K$ be a compact smooth manifold over a locally compact topological field $\mathbb{K}$, $E$ and $F$ be topological $\mathbb{K}$-vector spaces, and $U \subseteq E$ be open. Then*

$$C^\infty(K,f) \colon C^\infty(K,U) \to C^\infty(K,F), \quad \gamma \mapsto f \circ \gamma$$

*is a smooth map, for each smooth map $f \colon U \to F$. More generally,*

$$f_* \colon C^\infty(K,U) \to C^\infty(K,F), \quad \gamma \mapsto (x \mapsto f(x,\gamma(x)))$$

*is smooth, for each smooth map $f \colon K \times U \to F$.*

Now the $C^\infty$-version of Proposition 5.2 easily follows. For example, assuming $U = U^{-1}$, the inversion map of $C^\infty(K,G)$ is smooth on $C^\infty(K,U)$ (considered as a smooth manifold with global chart $\Phi$), by the following argument: The inversion map of $G$ restricts to a smooth map $i: U \to U$, corresponding to the smooth map $j := \phi \circ i \circ \phi^{-1}: V \to V$ in the local chart $\phi$. The restriction of the inversion map of $C^\infty(K,G)$ to $C^\infty(K,U)$ is $C^\infty(K,i)$. This map is smooth, since $\Phi \circ C^\infty(K,i) \circ \Phi^{-1} = C^\infty(K,j)$ is smooth by Proposition 5.4.

**Remark 5.5.** We mention that an analogue of Proposition 5.4 is available for mappings between spaces of compactly supported vector-valued $C^k$-functions on a paracompact finite-dimensional manifold $M$ over a locally compact topological field $\mathbb{K}$ (see [27, Proposition 8.22 and Corollary 8.23]). For each Lie group $G$ modelled on a topological $\mathbb{K}$-vector space $E$, this facilitates to turn the "test function group"

$$C_c^k(M,G) = \{\gamma \in C^k(M,G): \gamma(x) = 1 \text{ for all } x \text{ outside some compact set}\}$$

of compactly supported $G$-valued $C^k$-maps into a Lie group modelled on $C_c^k(M,E)$, equipped with a suitable vector topology (see [27, Proposition 9.1]).

As a tool for the discussion of diffeomorphism groups, it is useful to know that the *weak direct product*

$$\textstyle\prod^*_{i \in I} G_i := \{(x_i)_{i \in I} \in \prod_{i \in I} G_i : x_i = 1 \text{ for all but finitely many } i\}$$

of a family $(G_i)_{i \in I}$ of Lie groups over a valued field $\mathbb{K}$ can always be turned into a Lie group, modelled on the direct sum $E := \bigoplus_{i \in I} E_i$ of the respective modelling spaces, equipped with the box topology (see [27, Proposition 7.1]). Thus, sets of the form $\bigoplus_{i \in I} U_i := E \cap \prod_{i \in I} U_i$ are taken as a basis of open 0-neighbourhoods for $E$, where each $U_i$ is an open 0-neighbourhood in $E_i$. The following result concerning typical non-linear mappings between direct sums ([27, Proposition 6.9]) is used to construct the Lie group structure on weak direct products.

**Proposition 5.6.** *Let $(E_i)_{i \in I}$ and $(F_i)_{i \in I}$ be families of topological $\mathbb{K}$-vector spaces indexed by a set $I$. Let $k \in \mathbb{N}_0 \cup \{\infty\}$, and $f_i: U_i \to F_i$ be a $C^k$-map, for $i \in I$, defined on an open 0-neighbourhood $U_i \subseteq E_i$. Then $\bigoplus_{i \in I} f_i: \bigoplus_{i \in I} U_i \to \bigoplus_{i \in I} F_i$, $(x_i)_{i \in I} \mapsto (f_i(x_i))_{i \in I}$ is a $C^k$-map on the open subset $\bigoplus_{i \in I} U_i$ of $\bigoplus_{i \in I} E_i$.*

For $\mathbb{K} \in \{\mathbb{R}, \mathbb{C}\}$, analogous results can be obtained using locally convex direct sums [22].

## Diffeomorphism groups

Let $M$ be a finite-dimensional, paracompact smooth manifold over a local field $\mathbb{K}$. We explain some ideas used in [27, §13] to define a Lie group structure on $\text{Diff}(M)$, the group of all $C^\infty$-diffeomorphisms of $M$:

**Proposition 5.7.** *$\text{Diff}(M)$ is a Lie group modelled on the space $C_c^\infty(M,TM)$ of compactly supported smooth vector fields on $M$.*

To construct the Lie group structure, one exploits that $M$ is a disjoint union $M = \bigcup_{i \in I} B_i$ of balls, i.e., open subsets $B_i \subseteq M$ diffeomorphic to $\mathbb{O}^d$, where $\mathbb{O}$ is the maximal compact subring of $\mathbb{K}$ and $d$ the dimension of the modelling space of $M$. The main step (explained more closely below) is to make each $\operatorname{Diff}(B_i)$ a Lie group. Then the weak direct product $\prod_{i \in I}^* \operatorname{Diff}(B_i)$ can be made a Lie group, as described above. Here $\prod_{i \in I}^* \operatorname{Diff}(B_i)$ can be identified with a subgroup of $\operatorname{Diff}(M)$ in an apparent way. In a third step, one verifies that $\operatorname{Diff}(M)$ can be given a Lie group structure making $\prod_{i \in I}^* \operatorname{Diff}(B_i)$ an open subgroup.

The following "exponential law" (covered by [27, Lemma 12.1 and Proposition 12.2]) is essential (cf. also [27, Proposition 12.6] for a variant for metrizable manifolds).

**Proposition 5.8.** *Let $\mathbb{K}$ be a topological field, $M$ and $N$ be smooth manifolds modelled on topological $\mathbb{K}$-vector spaces, and $E$ be a topological $\mathbb{K}$-vector space.*
  (a) *For every smooth mapping $f: M \times N \to E$, also the mapping $f^\vee: M \to C^\infty(N, E)$, $f^\vee(x) := f(x, \bullet)$ is smooth.*
  (b) *If $\mathbb{K}$ is locally compact and $N$ is finite-dimensional, then a map $g: M \to C^\infty(N, E)$ is smooth if and only if $g^\wedge: M \times N \to E$, $g^\wedge(x, y) := g(x)(y)$ is smooth.*

*Furthermore, the map $C^\infty(M \times N, E) \to C^\infty(M, C^\infty(N, E))$, $f \mapsto f^\vee$ is an isomorphism of topological vector spaces in the situation of* (b), *with inverse $g \mapsto g^\wedge$.*

We now explain the essential first step of the above construction of the Lie group structure on diffeomorphism groups in more detail, namely the construction of the Lie group structure on $\operatorname{Diff}(M)$ for $M = \mathbb{O}^d$ a ball. Then $P := \operatorname{Diff}(M)$ is an open subset of the topological vector space $C^\infty(M, \mathbb{K}^d)$ (as above).

*Smoothness of inversion.* The inclusion map $i: P \to C^\infty(M, \mathbb{K}^d)$, $\gamma \mapsto \gamma$ being smooth, the second half of the exponential law (Proposition 5.8 (b)) ensures that also

$$f: P \times M \to \mathbb{K}^d, \quad f(\gamma, x) := i^\wedge(\gamma, x) := i(\gamma)(x) = \gamma(x)$$

is smooth, using that $M$ is finite-dimensional. Then $f_\gamma := f(\gamma, \bullet) = \gamma$ for each $\gamma \in P$, whence $f'_\gamma(x) = \gamma'(x) \in \operatorname{GL}(\mathbb{K}^d)$ for each $x \in M$. Applying the inverse function with parameters (Theorem 3.6) with the diffeomorphism $\gamma$ as the parameter, we see that

$$g: P \times M \to M, \quad g(\gamma, x) := (f_\gamma)^{-1}(x) = \gamma^{-1}(x)$$

is smooth. Now the first half of the exponential law (Proposition 5.8 (a)) shows that $\operatorname{Diff}(M) \to C^\infty(M, \mathbb{K}^d)$, $\gamma \mapsto g^\vee(\gamma) := g(\gamma, \bullet) = \gamma^{-1}$ is smooth. But this is the inversion map of the group $\operatorname{Diff}(M)$.

*Smoothness of composition.* Since $\operatorname{Diff}(M)$ is open in $C^\infty(M, \mathbb{K}^d)$, we only need to show that $\Gamma: C^\infty(M, \mathbb{K}^d \times \mathbb{K}^d) \cong C^\infty(M, \mathbb{K}^d) \times C^\infty(M, \mathbb{K}^d) \to C^\infty(M, \mathbb{K}^d)$, $\Gamma(\gamma, \eta) := \gamma \circ \eta$ is smooth. Because the evaluation map $\operatorname{ev}: C^\infty(M, \mathbb{K}^d) \times M \to \mathbb{K}^d$, $\operatorname{ev}(\gamma, x) := \gamma(x)$ is smooth by [27, Proposition 11.1], we deduce from the formula

$$\Gamma^\wedge((\gamma, \eta), x) = \gamma(\eta(x)) = \operatorname{ev}(\gamma, \operatorname{ev}(\eta, x)) \quad \text{for } \gamma, \eta \in C^\infty(M, \mathbb{K}^d), x \in M$$

that $\Gamma^\wedge$ is smooth and hence also $\Gamma$, by Proposition 5.8 (b).

**Remark 5.9.** For differentiability properties of the composition map between spaces of $C^k$-functions, see [27, §11]. Diffeomorphism groups can also be found in [45].

## Direct limit groups

Consider an ascending sequence $G_1 \subseteq G_2 \subseteq \cdots$ of finite-dimensional Lie groups over a locally compact topological field $\mathbb{K}$, such that each inclusion map $G_n \to G_{n+1}$ is a smooth immersion. Then $G := \bigcup_{n \in \mathbb{N}} G_n$ can be given a Lie group structure modelled on the direct limit topological $\mathbb{K}$-vector space $\varinjlim E_n$ of the respective finite-dimensional modelling spaces, which makes $G$ the direct limit of the given directed sequence in the category of $\mathbb{K}$-Lie groups and smooth homomorphisms [24] (cf. also [48] and [21]).

## REFERENCES

1. Araujo J. and Schikhof W., The Weierstrass-Stone approximation theorem for $p$-adic $C^n$-functions *Ann. Math. Blaise Pascal* **1**, 61–74 (1994).
2. Arrowsmith D.K. and Vivaldi F., Geometry of $p$-adic Siegel discs, *Physica D* **71**, 222–236 (1994).
3. Barsky D., Fonctions $k$-Lipschitziennes sur un anneau local et polynômes à valeurs entières, *Bull. Soc. math. France* **101**, 397–411 (1973).
4. Benedetto R.L., $p$-adic dynamics and Sullivan's no wandering domains theorem, *Compositio Math.* **122** (2000), 281–298.
5. Bertram W., Differential geometry, Lie groups and symmetric spaces over general base fields and rings, to appear in *Memoirs of the AMS*, math.DG/0502168.
6. Bertram W., Glöckner H. and Neeb K.H., Differential calculus over general base fields and rings, *Expo. Math.* **22**, 213–282 (2004).
7. Bertram W. and Neeb K.H., Projective completions of Jordan pairs. Part II: Manifold structures and symmetric spaces, *Geom. Dedicata* **112** (2005), 73–113.
8. Bosch S., Güntzer U. and Remmert R., *Non-Archimedian Analysis*, Springer, 1984.
9. Bourbaki N., *Variétés différentielles et analytiques. Fascicule de résultats*, Hermann, Paris, 1967.
10. Bourbaki N., *Topological Vector Spaces* Chapters 1–5, Springer-Verlag, 1987.
11. Bourbaki N., *Lie Groups and Lie Algebras* Chapters 1–3, Springer-Verlag, 1989.
12. Chow S.N. and Hale J.K., *Methods of Bifurcation Theory*, Springer-Verlag, 1982.
13. de la Llave R. and Wayne C.E., On Irwin's proof of the pseudostable manifold theorem, *Math. Z.* **219**, 301–321 (1995).
14. De Smedt S., $p$-adic continuously differentiable functions of several variables, *Collect. Math.* **45**, 137–152 (1994).
15. De Smedt S., Local invertibility of non-archimedian vector-valued functions, *Ann. Math. Blaise Pascal* **5**, 13–23 (1998).
16. Dwork B., *Lectures on p-adic Differential Equations*, Springer-Verlag, 1982.
17. Fresnel J. and van der Put M., *Rigid Analytic Geometry and its Applications*, Birkhäuser, 2004.
18. Frölicher A. and Kriegl A., *Linear Spaces and Differentiation Theory*, John Wiley, 1988.
19. Glöckner H., Scale functions on $p$-adic Lie groups, *Manuscr. Math.* **97**, 205–215 (1998).
20. Glöckner H., Algebras whose groups of units are Lie groups, *Studia Math.* **153**, 147–177 (2002).
21. Glöckner H., Direct limit Lie groups and manifolds, *J. Math. Kyoto Univ.* **43**, 1–26 (2003).
22. Glöckner H., Lie groups of measurable mappings, *Canadian J. Math.* **55**, 969–999 (2003).
23. Glöckner H., Smooth Lie groups over local fields of positive characteristic need not be analytic, *J. Algebra* **285**, 356–371 (2005).

24. Glöckner H., Fundamentals of direct limit Lie theory, *Compositio Math.* **141**, 1551–1577 (2005).
25. Glöckner H., Implicit functions from topological vector spaces to Banach spaces, to appear in *Israel J. Math.*, math.GM/0303320.
26. Glöckner H., Every smooth *p*-adic Lie group admits a compatible analytic structure, to appear in *Forum Math.*, math.GR/0312113.
27. Glöckner H., Lie groups over non-discrete topological fields, preprint, math.GM/0303320.
28. Glöckner H., Finite order differentiability properties, fixed points and implicit functions over valued fields, preprint, math.FA/0511218.
29. Glöckner H., Stable manifolds for dynamical systems over valued fields, in preparation.
30. Glöckner H., Pseudo-stable manifolds for dynamical systems over valued fields, in preparation.
31. Glöckner H., Scale functions on Lie groups over local fields of positive characteristic, in preparation.
32. Glöckner H. and Neeb K.H., *Infinite-Dimensional Lie Groups*, book in preparation.
33. Hiltunen S., Implicit functions from locally convex spaces to Banach spaces, *Studia Math.* **134**, 235–250 (1999).
34. Hiltunen S., A Frobenius theorem for locally convex global analysis, *Monatsh. Math.* **129**, 109–117 (2000).
35. Irwin M.C., On the stable manifold theorem, *Bull. London Math. Soc.* **2**, 196–198 (1970).
36. Irwin M.C., A new proof of the pseudostable manifold theorem, *J. London Math. Soc.* **21**, 557–566 (1980).
37. Katok S., *p*-adic analysis in comparison with real, pp. 11–87 in: *MASS selecta*, Amer. Math. Soc. Providence, 2003.
38. Keller H.H., *Differential Calculus in Locally Convex Spaces*, Springer Verlag, 1974.
39. Khrennikov A. and Nilsson M., Behaviour of Hensel perturbations of *p*-adic monomial dynamical systems, *Anal. Math.* **29**, 107–133 (2003).
40. Kim H.S. and Kim T., On *p*-adic differentiability and bounded functions, *Kyungpook Math. J.* **35**, 171–178 (1995).
41. Kim T., S.D. Kim and Park D.W., On uniform differentiability and *q*-Mahler expansions, *Adv. Stud. Contemp. Math. (Kyungshang)* **4**, 35–41 (2001).
42. Kriegl A. and Michor P.W., *The Convenient Setting of Global Analysis*, AMS, Providence, 1997.
43. Lubin J., Non-Archimedian dynamical systems, *Compositio Math.* **94**, 321–346 (1994).
44. Ludkovsky S.V., Quasi-invariant measures on non-Archimedian groups and semigroups of loops and paths, their representations I, II, *Ann. Math. Blaise Pascal* **7**, 19–80 (2000).
45. Ludkovsky S.V., A structure and representations of diffeomorphism groups of non-Archimedian manifolds, *Southeast Asian Bull. Math.* **26**, 975–1004 (2003).
46. Milnor J., Remarks on infinite-dimensional Lie groups, pp. 1008–1057 in: DeWitt B., and R. Stora (Eds.), *Relativity, Groups and Topology II*, North Holland, 1983.
47. Monna A.F., *Analyse Non-Archimédienne*, Springer-Verlag, 1979.
48. Natarajan L., Rodríguez-Carrington E. and Wolf J.A., Differentiable structure for direct limit groups, *Letters in Math. Phys.* **23**, 99–109 (1991).
49. Pressley A., and Segal G.B., *Loop Groups*, Clarendon Press, Oxford, 1986.
50. Rooij A.C.M., *Non-Archimedian Functional Analysis*, Marcel Dekker, 1978.
51. Schikhof W.H., *Ultrametric Calculus*, Cambridge University Press, 1984.
52. Serre J.P., *Lie Algebras and Lie Groups*, Springer-Verlag, 1992.
53. Teichmann J., A Frobenius theorem on convenient manifolds, *Monatsh. Math.* **134**, 159–167 (2001).
54. van der Put M. and Singer M., *Galois Theory of Linear Differential Equations*, Springer-Verlag, 2003.
55. Wang J.S.P., The Mautner phenomenon for *p*-adic Lie groups, *Math. Z.* **185**, 403–412 (1984).
56. Weisman S.S., On *p*-adic differentiability, *J. Number Theory* **9**, 79–86 (1997).
57. Weil A., *Basic Number Theory*, Springer-Verlag, 1973.
58. Wells J.C., Invariant manifolds on non-linear operators, *Pacific J. Math.* **62**, 285–293 (1976).

# Umbral Calculus and Holonomic Modules in Positive Characteristic

Anatoly N. Kochubei

*Institute of Mathematics, National Academy of Sciences of Ukraine,
Tereshchenkivska 3, Kiev, 01601 UKRAINE
email: kochubei@i.com.ua*

**Abstract.** In the framework of analysis over local fields of positive characteristic, we develop algebraic tools for introducing and investigating various polynomial systems. In this survey paper we describe a function field version of umbral calculus developed on the basis of a relation of binomial type satisfied by the Carlitz polynomials. We consider modules over the Weyl-Carlitz ring, a function field counterpart of the Weyl algebra. It is shown that some basic objects of function field arithmetic, like the Carlitz module, Thakur's hypergeometric polynomials, and analogs of binomial coefficients arising in the positive characteristic version of umbral calculus, generate holonomic modules.

**Keywords:** Function fields, differential equations with Carlitz derivatives, umbral calculus, holonomic modules.
**PACS:** 02.10.De; 02.10.Hh.

## 1. INTRODUCTION AND PRELIMINARIES

Classical umbral calculus [24] and the theory of holonomic modules over rings of differential operators [7] belong to methods of algebraic analysis widely used in various applications including mathematical physics; see, for example, [3, 5, 9, 13, 21]. Within the contemporary tendency to find non-Archimedean, in particular, the positive characteristic, counterparts of all major structures of analysis and mathematical physics, it was natural to try to introduce and study positive characteristic analogs of these two theories. This paper is a brief survey of recent results in this direction; for a detailed exposition see [18, 19].

It is well known that any non-discrete locally compact topological field of a positive characteristic $p$ is isomorphic to the field $K$ of formal Laurent series with coefficients from the Galois field $\mathbb{F}_q$, $q = p^\nu$, $\nu \in \mathbb{Z}_+$. Denote by $|\cdot|$ the non-Archimedean absolute value on $K$; if $z \in K$,

$$z = \sum_{i=m}^{\infty} \zeta_i x^i, \quad m \in \mathbb{Z}, \zeta_i \in \mathbb{F}_q, \zeta_m \neq 0,$$

then $|z| = q^{-m}$. This valuation can be extended onto the field $\overline{K}_c$, the completion of an algebraic closure of $K$. Let $O = \{z \in K : |z| \leq 1\}$ be the ring of integers in $K$. The ring

$\mathbb{F}_q[x]$ of polynomials (in the indeterminate $x$) with coefficients from $\mathbb{F}_q$ is dense in $O$ with respect to the topology defined by the metric $d(z_1, z_2) = |z_1 - z_2|$.

It is obvious that standard notions of analysis do not make sense in the characteristic $p$ case. For instance, $n! = 0$ if $n \geq p$, so that one cannot define a usual exponential function on $K$, and $\frac{d}{dt}(t^n) = 0$ if $p$ divides $n$. On the other hand, some well-defined functions have unusual properties. In particular, there are many functions with the $\mathbb{F}_q$-linearity property

$$f(t_1 + t_2) = f(t_1) + f(t_2), \quad f(\alpha t) = \alpha f(t),$$

for any $t_1, t_2, t \in K$, $\alpha \in \mathbb{F}_q$. Such are, for example, all power series $\sum c_k t^{q^k}$, $c_k \in \overline{K}_c$, convergent on some region in $K$ or $\overline{K}_c$.

The analysis on $K$ taking into account the above special features was initiated in a seminal paper by Carlitz [4] who introduced, for this case, the appropriate notions of a factorial, an exponential and a logarithm, a system of polynomials $\{e_i\}$ (now called the Carlitz polynomials), and other related objects. In subsequent works by Carlitz, Goss, Thakur, and many others (see references in [12, 28]) analogs of the gamma, zeta, Bessel and hypergeometric functions were introduced and studied. A difference operator $\Delta$,

$$(\Delta u)(t) = u(xt) - x u(t)$$

(an inner derivation of composition rings of $\mathbb{F}_q$-linear polynomials or more general $\mathbb{F}_q$-linear functions) acting on functions over $K$ or its subsets, which was introduced in [4], became (as an analog of the operator $t\frac{d}{dt}$) the main ingredient of the calculus and the analytic theory of differential equations on $K$ [15, 16, 17]. It appears also in a characteristic $p$ analog of the canonical commutation relations of quantum mechanics found in [14]. In fact, the counterpart of the classical derivative is the nonlinear ($\mathbb{F}_q$-linear) operator $d = \sqrt[q]{\ } \circ \Delta = \tau^{-1} \Delta$, where $\tau u = u^q$ is the $\mathbb{F}_q$-linear Frobenius operator.

The definition of the Carlitz polynomials is as follows. Let $e_0(t) = t$,

$$e_i(t) = \prod_{\substack{m \in \mathbb{F}_q[x] \\ \deg m < i}} (t - m), \quad i \geq 1 \tag{1}$$

(we follow the notation in [11] used in the modern literature; the initial formulas from [4] have different signs in some places). It is known [4, 11] that

$$e_i(t) = \sum_{j=0}^{i} (-1)^{i-j} \begin{bmatrix} i \\ j \end{bmatrix} t^{q^j} \tag{2}$$

where

$$\begin{bmatrix} i \\ j \end{bmatrix} = \frac{D_i}{D_j L_{i-j}^{q^j}},$$

$D_i$ is the Carlitz factorial

$$D_i = [i][i-1]^q \ldots [1]^{q^{i-1}}, \quad [i] = x^{q^i} - x \ (i \geq 1), \ D_0 = 1, \tag{3}$$

the sequence $\{L_i\}$ is defined by
$$L_i = [i][i-1]\ldots[1] \ (i \geq 1); \ L_0 = 1. \tag{4}$$

It follows from (3), (4) that
$$|D_i| = q^{-\frac{q^i-1}{q-1}}, \quad |L_i| = q^{-i}.$$

The normalized polynomials $f_i(t) = \dfrac{e_i(t)}{D_i}$ form an orthonormal basis in the Banach space $C_0(O, \overline{K}_c)$ of all $\mathbb{F}_q$-linear continuous functions $O \to \overline{K}_c$, with the supremum norm $\|\cdot\|$. Thus every function $\varphi \in C_0(O, \overline{K}_c)$ admits a unique representation as a uniformly convergent series
$$\varphi = \sum_{i=0}^{\infty} a_i f_i, \quad a_i \in \overline{K}_c, \ a_i \to 0,$$

satisfying the orthonormality condition
$$\varphi = \sup_{i \geq 0} |a_i|.$$

For several different proofs of this fact see [6, 8, 14, 31]. Note that we consider functions with values in $\overline{K}_c$ defined on a compact subset of $K$.

The sequences $\{D_i\}$ and $\{L_i\}$ are involved in the definitions of the Carlitz exponential and logarithm:
$$e_C(t) = \sum_{n=0}^{\infty} \frac{t^{q^n}}{D_n}, \quad \log_C(t) = \sum_{n=0}^{\infty} (-1)^n \frac{t^{q^n}}{L_n}, \quad |t| < 1. \tag{5}$$

It is seen from (2) and (5) that the above functions are $\mathbb{F}_q$-linear on their domains of definition.

The most important object connected with the Carlitz polynomials is the Carlitz module
$$C_s(z) = \sum_{i=0}^{\deg s} f_i(s) z^{q^i} = \sum_{i=0}^{\deg s} \frac{e_i(s)}{D_i} z^{q^i}, \quad s \in \mathbb{F}_q[x]. \tag{6}$$

Note that by (1) $e_i(s) = 0$ if $s \in \mathbb{F}_q[x]$, $\deg s < i$.

The function $C_s$ appears in the functional equation for the Carlitz exponential,
$$C_s(e_C(t)) = e_C(st).$$

Its main property is the relation
$$C_{ts}(z) = C_t(C_s(z)), \quad s, t \in \mathbb{F}_q[x], \tag{7}$$

which obtained a far-reaching generalization in the theory of Drinfeld modules, the principal objects of the function field arithmetic (see [12, 28]).

## 2. UMBRAL CALCULUS

Classical umbral calculus [25, 24] is a set of algebraic tools for obtaining, in a unified way, a rich variety of results regarding structure and properties of various polynomial sequences. There exist many generalizations extending umbral methods to other classes of functions. However there is a restriction common to the whole literature on umbral calculus – the underlying field must be of zero characteristic. An attempt to mimic the characteristic zero procedures in the positive characteristic case [10] revealed a number of pathological properties of the resulting structures. More importantly, these structures were not connected with the existing analysis in positive characteristic based on a completely different algebraic foundation.

A version of umbral calculus implementing such a connection was developed by the author [18], and we summarize it in this section. Its basic notion is motivated by the following identity for the non-normalized Carlitz polynomials $e_i = D_i f_i$:

$$e_i(st) = \sum_{n=0}^{i} \binom{i}{n}_K e_n(t) \{e_{i-n}(s)\}^{q^n} \tag{8}$$

where the "$K$-binomial coefficients" $\binom{i}{n}_K$ are defined as

$$\binom{i}{n}_K = \frac{D_i}{D_n D_{i-n}^{q^n}}.$$

Computing the absolute values of the Carlitz factorials directly from their definition (3), it is easy to show that

$$\left| \binom{i}{n}_K \right| = 1, \quad 0 \le n \le i.$$

In fact, $\binom{i}{n}_K \in \mathbb{F}_q(x)$, and we can consider also other places of $\mathbb{F}_q(x)$, that is other non-equivalent absolute values. It can be proved [19] that $\binom{i}{n}_K$ belongs to the ring of integers for any finite place of $\mathbb{F}_q(x)$.

We see the relation (8) as a function field counterpart of the classical binomial identity [25, 24] satisfied by many classical polynomials. Now, considering a sequence $u_i$ of $\mathbb{F}_q$-linear polynomials with coefficients from $\overline{K}_c$, we call it *a sequence of K-binomial type* if $\deg u_i = q^i$ and for all $i = 0, 1, 2, \ldots$

$$u_i(st) = \sum_{n=0}^{i} \binom{i}{n}_K u_n(t) \{u_{i-n}(s)\}^{q^n}, \quad s, t \in K. \tag{9}$$

As in the conventional umbral calculus, the dual notion is that of a delta operator. However, in contrast to the classical situation, here the delta operators are only $\mathbb{F}_q$-linear, not linear.

Denote by $\rho_\lambda$ the operator of multiplicative shift, $(\rho_\lambda u)(t) = u(\lambda t)$. We call a linear operator $T$, on the $\overline{K}_c$-vector space $\overline{K}_c\{t\}$ of all $\mathbb{F}_q$-linear polynomials, *invariant* if it commutes with $\rho_\lambda$ for each $\lambda \in K$.

A $\mathbb{F}_q$-linear operator $\delta = \tau^{-1}\delta_0$, where $\delta_0$ is a linear invariant operator on $\overline{K}_c\{t\}$, is called *a delta operator* if $\delta_0(t) = 0$ and $\delta_0(f) \neq 0$ for $\deg f > 1$. A sequence $\{P_n\}_0^\infty$ of $\mathbb{F}_q$-linear polynomials is called *a basic sequence* corresponding to a delta operator $\delta = \tau^{-1}\delta_0$, if $\deg P_n = q^n$, $P_0(1) = 1$, $P_n(1) = 0$ for $n \geq 1$,

$$\delta P_0 = 0, \quad \delta P_n = [n]^{1/q} P_{n-1}, \; n \geq 1, \tag{10}$$

or, equivalently,

$$\delta_0 P_0 = 0, \quad \delta_0 P_n = [n] P_{n-1}^q, \; n \geq 1. \tag{11}$$

It is clear that $d = \tau^{-1}\Delta$ is a delta operator. It follows from well-known identities for the Carlitz polynomials $e_i$ [11] that the sequence $\{e_i\}$ is basic with respect to the operator $d$.

**Theorem 1.** *For any delta operator $\delta = \tau^{-1}\delta_0$, there exists a unique basic sequence $\{P_n\}$, which is a sequence of K-binomial type. Conversely, given a sequence $\{P_n\}$ of K-binomial type, define the action of $\delta_0$ on $P_n$ by the relations (11), extend it onto $\overline{K}_c\{t\}$ by linearity and set $\delta = \tau^{-1}\delta_0$. Then $\delta$ is a delta operator, and $\{P_n\}$ is the corresponding basic sequence.*

The analogs of the higher Carlitz difference operators

$$\left(\Delta^{(n)} u\right)(t) = \Delta^{(n-1)} u(xt) - x^{q^{n-1}} \Delta^{(n-1)} u(t), \quad n \geq 2; \quad \Delta^{(1)} = \Delta,$$

in the present general context are the operators $\delta_0^{(l)} = \tau^l \delta^l$. The identity

$$\delta_0^{(l)} P_j = \frac{D_j}{D_{j-l}^{q^l}} P_{j-l}^{q^l} \tag{12}$$

holds for any $l \leq j$. If $f$ is a $\mathbb{F}_q$-linear polynomial, $\deg f \leq q^n$, then a generalized Taylor formula

$$f(st) = \sum_{l=0}^n \frac{\left(\delta_0^{(l)} f\right)(s)}{D_l} P_l(t) \tag{13}$$

holds for any $s,t \in K$. For the Carlitz polynomials $e_i$, the formulas (12) and (13) are well known [11]. It is important that, in contrast to the classical umbral calculus, the linear operators involved in (13) are not powers of a single linear operator.

Any linear invariant operator $T$ on $\overline{K}_c\{t\}$ admits a representation

$$T = \sum_{l=0}^\infty \sigma_l \delta_0^{(l)}, \quad \sigma_l = \frac{(TP_l)(1)}{D_l}. \tag{14}$$

The infinite series in (14) becomes actually a finite sum if both sides of (14) are applied to any $\mathbb{F}_q$-linear polynomial. Conversely, any such series defines a linear invariant operator on $\overline{K}_c\{t\}$.

Let us consider the case where $\delta = d$, so that $\delta_0^{(l)} = \Delta^{(l)}$. The next result leads to new delta operators and basic sequences.

**Theorem 2.** *The operator* $\theta = \tau^{-1}\theta_0$, *where*

$$\theta_0 = \sum_{l=1}^{\infty} \sigma_l \Delta^{(l)},$$

*is a delta operator if and only if*

$$S_n \stackrel{\text{def}}{=} \sum_{l=1}^{n} \frac{\sigma_l}{D_{n-l}^{q^l}} \neq 0 \text{ for all } n = 1, 2, \ldots. \tag{15}$$

**Example 1.** Let $\sigma_l = 1$ for all $l \geq 1$, that is

$$\theta_0 = \sum_{l=1}^{\infty} \Delta^{(l)}. \tag{16}$$

Estimates of $|D_n|$ which follow directly from (3) show that $|S_n| = q^{\frac{q^n-q}{q-1}}$, so that (15) is satisfied. Comparing (16) with a classical formula from [25] we may see the polynomials $P_n$ for this case as analogs of the Laguerre polynomials.

**Example 2.** Let $\sigma_l = \frac{(-1)^{l+1}}{L_l}$. For this case it can be shown [18] that $S_n = D_n^{-1}$, $n = 1, 2, \ldots$; $\theta_0(t^{q^j}) = t^{q^j}$ for all $j \geq 1$ (of course, $\theta_0(t) = 0$), and $P_0(t) = t$, $P_n(t) = D_n \left( t^{q^n} - t^{q^{n-1}} \right)$ for $n \geq 1$.

As in the $p$-adic case [29, 30, 23], the umbral calculus can be used for constructing new orthonormal bases in $C_0(O, \overline{K}_c)$.

Let $\{P_n\}$ be the basic sequence corresponding to a delta operator $\delta = \tau^{-1}\delta_0$,

$$\delta_0 = \sum_{l=1}^{\infty} \sigma_l \Delta^{(l)}. \tag{17}$$

The sequence $Q_n = \frac{P_n}{D_n}$, $n = 0, 1, 2, \ldots$, called the normalized basic sequence, satisfies the identity

$$Q_i(st) = \sum_{n=0}^{i} Q_n(t) \{Q_{i-n}(s)\}^{q^n},$$

another form of the K-binomial property. Though it resembles its classical counterpart, the presence of the Frobenius powers is a feature specific for the case of a positive characteristic.

**Theorem 3.** *If $|\sigma_1| = 1$, $|\sigma_l| \leq 1$ for $l \geq 2$, then the sequence $\{Q_n\}_0^\infty$ is an orthonormal basis of the space $C_0(O, \overline{K}_c)$ – for any $f \in C_0(O, \overline{K}_c)$ there is a uniformly convergent expansion*

$$f(t) = \sum_{n=0}^{\infty} \psi_n Q_n(t), \quad t \in O,$$

*where $\psi_n = \left(\delta_0^{(n)} f\right)(1)$, $|\psi_n| \to 0$ as $n \to \infty$,*

$$\|f\| = \sup_{n \geq 0} |\psi_n|.$$

By Theorem 3, the Laguerre-type polynomial sequence from Example 1 is an orthonormal basis of $C_0(O, \overline{K}_c)$. The sequence from Example 2 does not satisfy the conditions of Theorem 3.

Note that the conditions of Theorem 3 imply that $S_n \neq 0$ for all $n$, so that the series (17) considered in Theorem 3 always correspond to delta operators.

In [18] recursive formulas and generating functions for normalized basic sequences are also given.

## 3. HOLONOMIC MODULES OVER THE WEYL-CARLITZ RING

The theory of holonomic modules over the Weyl algebra and more general algebras of differential or $q$-difference operators is becoming increasingly important, both as a crucial part of the general theory of D-modules and in view of various applications. Usually, the holonomic property of the module corresponding to a system of differential equations is a sign of its "regular" behavior. Most of the classical special functions are associated (see [5]) with holonomic modules, which helps to investigate their properties.

In the positive characteristic case a natural counterpart of the Weyl algebra is, for the case of a single variable, the ring $\mathfrak{A}_1$ generated by $\tau, d$, and scalars from $\overline{K}_c$, with the relations

$$d\tau - \tau d = [1]^{1/q}, \quad \tau\lambda = \lambda^q \tau, \quad d\lambda = \lambda^{1/q} d \; (\lambda \in \overline{K}_c). \tag{18}$$

The ring consists of finite sums

$$a = \sum_{i,j} \lambda_{ij} \tau^i d^j, \quad \lambda_{ij} \in \overline{K}_c, \tag{19}$$

and the representation of an element in the form (19) is unique.

Basic algebraic properties of $\mathfrak{A}_1$ [16, 2] are similar to those of the Weyl algebra in characteristic 0 and quite different from the case of the algebra of usual differential operators over a field of positive characteristic [22].

The ring $\mathfrak{A}_1$ is left and right Noetherian, without zero divisors. $\mathfrak{A}_1$ possesses no non-trivial two-sided ideals stable with respect to the mapping

$$\sum_{i,j} \lambda_{ij} \tau^i d^j \mapsto \sum_{i,j} \lambda_{ij}^q \tau^i d^j.$$

The center of $\mathfrak{A}_1$ is described explicitly in [2]; it contains countably many elements (this corrects an erroneous statement from [16]). In fact, $\mathfrak{A}_1$ belongs to the class of generalized Weyl algebras [1]. A well-developed theory available for them enabled Bavula [2] to classify ideals in $\mathfrak{A}_1$, as well as all simple modules over $\mathfrak{A}_1$.

A generalization of $\mathfrak{A}_1$ to the case of several variables is not straightforward because the Carlitz derivatives $d_s$ and $d_t$ do not commute on a monomial $f(s,t) = s^{q^m} t^{q^n}$, if $m \neq n$. Moreover, if $m > n$, then $d_s^m f$ is not a polynomial, nor even a holomorphic function in $t$ (since the action of $d$ is not linear and involves taking the $q$-th root).

A reasonable generalization is inspired by Zeilberger's idea (see [5]) to study holonomic properties of sequences of functions making a transform with respect to the discrete variables, which reduces the continuous-discrete case to the purely continuous one (simultaneously in all the variables). In our situation, if $\{P_k(s)\}$ is a sequence of $\mathbb{F}_q$-linear polynomials with $\deg P_k \leq q^k$, we set

$$f(s,t) = \sum_{k=0}^{\infty} P_k(s) t^{q^k}, \tag{20}$$

and $d_s$ is well-defined. In the variable $t$, we consider not $d_t$ but the linear operator $\Delta_t$. The latter does not commute with $d_s$ either, but satisfies the commutation relations

$$d_s \Delta_t - \Delta_t d_s = [1]^{1/q} d_s, \quad \Delta_t \tau - \tau \Delta_t = [1] \tau,$$

so that the resulting ring $\mathfrak{A}_2$ resembles a universal enveloping algebra of a solvable Lie algebra.

More generally, denote by $\mathfrak{F}_{n+1}$ the set of all germs of functions of the form

$$f(s, t_1, \ldots, t_n) = \sum_{k_1=0}^{\infty} \cdots \sum_{k_n=0}^{\infty} \sum_{m=0}^{\min(k_1,\ldots,k_n)} a_{m,k_1,\ldots,k_n} s^{q^m} t_1^{q^{k_1}} \ldots t_n^{q^{k_n}} \tag{21}$$

where $a_{m,k_1,\ldots,k_n} \in \overline{K}_c$ are such that all the series are convergent on some neighbourhoods of the origin. We do not exclude the case $n = 0$ where $\mathfrak{F}_1$ will mean the set of all $\mathbb{F}_q$-linear power series $\sum_m a_m s^{q^m}$ convergent on a neighbourhood of the origin. $\widehat{\mathfrak{F}}_{n+1}$ will denote the set of all polynomials from $\mathfrak{F}_{n+1}$, that is the series (21) in which only a finite number of coefficients is different from zero.

The ring $\mathfrak{A}_{n+1}$ is generated by the operators $\tau, d_s, \Delta_{t_1}, \ldots \Delta_{t_n}$ on $\mathfrak{F}_{n+1}$, and the operators of multiplication by scalars from $\overline{K}_c$. To simplify the notation, we write $\Delta_j$ instead of $\Delta_{t_j}$

and identify a scalar $\lambda \in \overline{K}_c$ with the operator of multiplication by $\lambda$. The operators $\Delta_j$ are $\overline{K}_c$-linear, so that

$$\Delta_j \lambda = \lambda \Delta_j, \quad \lambda \in \overline{K}_c, \tag{22}$$

while the operators $\tau, d_s$ satisfy the commutation relations (18). In the action of each operator $d_s, \Delta_j$ (acting in a single variable), other variables are treated as scalars. The operator $\tau$ acts simultaneously on all the variables and coefficients. We have the relations involving $\Delta_j$:

$$\Delta_j \tau - \tau \Delta_j = [1]\tau, \quad d_s \Delta_j - \Delta_j d_s = [1]^{1/q} d_s, \quad j = 1,\ldots,n. \tag{23}$$

Using the commutation relations (18), (22), and (23), we can write any $a \in \mathfrak{A}_{n+1}$, in a unique way, as a finite sum

$$a = \sum c_{l,\mu,i_1,\ldots,i_n} \tau^l d_s^\mu \Delta_1^{i_1} \ldots \Delta_n^{i_n}. \tag{24}$$

Let us introduce a filtration in $\mathfrak{A}_{n+1}$ (an analog of the Bernstein filtration) denoting by $\Gamma_\nu$, $\nu \in \mathbb{Z}_+$, the $\overline{K}_c$-vector space of operators (24) with $\max\{l + \mu + i_1 + \cdots + i_n\} \leq \nu$ where the maximum is taken over all the terms of (24). Then $\mathfrak{A}_{n+1}$ is a left and right Noetherian filtered ring.

In a standard way (see [7]) we define filtered left modules over $\mathfrak{A}_{n+1}$. All the basic notions regarding a filtered module $M$ (like those of the graded module $\mathrm{gr}(M)$, dimension $d(M)$, multiplicity $m(M)$, good filtration etc) are introduced just as their counterparts in the theory of modules over the Weyl algebra.

If we consider $\mathfrak{A}_{n+1}$ as a left module over itself, then

$$d(\mathfrak{A}_{n+1}) = n+2, \quad m(\mathfrak{A}_{n+1}) = 1. \tag{25}$$

For any finitely generated left $\mathfrak{A}_{n+1}$-module $M$, we have $d(M) \leq n+2$. By (25), this bound cannot be improved in general. However, if $I$ is a non-zero left ideal in $\mathfrak{A}_{n+1}$, then

$$d(\mathfrak{A}_{n+1}/I) \leq n+1. \tag{26}$$

For the module $\widehat{\mathfrak{F}}_{n+1}$ of $\mathbb{F}_q$-linear polynomials (21), we have

$$d\left(\widehat{\mathfrak{F}}_{n+1}\right) = n+1, \quad m\left(\widehat{\mathfrak{F}}_{n+1}\right) = n!$$

The proofs of all these results, as well as the ones given in this section below, can be found in [19].

It is natural to call an $\mathfrak{A}_{n+1}$-module $M$ *holonomic* if $d(M) = n+1$. Thus, $\widehat{\mathfrak{F}}_{n+1}$ is an example of a holonomic module.

The next theorem demonstrates, already for the case of $\mathfrak{A}_1$-modules, a sharp difference from the case of modules over the Weyl algebras. In particular, we see that an analog of the Bernstein inequality (see [7]) does not hold here without some additional assumptions.

**Theorem 4.** (i) *For any $k = 1, 2, \ldots$, there exists such a nontrivial $\mathfrak{A}_1$-module $M$ that $\dim M = k$ (dim means the dimension over $\overline{K}_c$), that is $d(M) = 0$.*

(ii) *Let $M$ be a finitely generated $\mathfrak{A}_1$-module with a good filtration. Suppose that there exists a "vacuum vector" $v \in M$, such that $d_s v = 0$ and $\tau^m(v) \neq 0$ for all $m = 0, 1, 2, \ldots$. Then $d(M) \geq 1$.*

Let us consider the case of holonomic submodules of the $\mathfrak{A}_{n+1}$-module $\mathfrak{F}_{n+1}$, consisting of $\mathbb{F}_q$-linear functions (21) polynomial in $s$ and holomorphic near the origin in $t_1, \ldots, t_n$. Let $0 \neq f \in \mathfrak{F}_{n+1}$,
$$I_f = \{\varphi \in \mathfrak{A}_{n+1} : \varphi(f) = 0\}.$$

$I_f$ is a left ideal in $\mathfrak{A}_{n+1}$. The left $\mathfrak{A}_{n+1}$-module $M_f = \mathfrak{A}_{n+1}/I_f$ is isomorphic to the submodule $\mathfrak{A}_{n+1} f \subset \mathfrak{F}_{n+1}$ – an element $\varphi(f) \in \mathfrak{A}_{n+1} f$ corresponds to the class of $\varphi \in \mathfrak{A}_{n+1}$ in $M_f$. A natural good filtration in $M_f$ is induced from that in $\mathfrak{A}_{n+1}$.

As we know (see (26)), if $I_f \neq \{0\}$, then $d(M_f) \leq n+1$. We call a function $f$ *holonomic* if the module $M_f$ is holonomic, that is $d(M_f) = n+1$. The condition $I_f \neq \{0\}$ means that $f$ is a solution of a non-trivial "differential equation" $\varphi(f) = 0$, $\varphi \in \mathfrak{A}_{n+1}$. The case $n = 0$ is quite simple.

**Theorem 5.** *If a non-zero function $f \in \mathfrak{F}_1$ satisfies an equation $\varphi(f) = 0$, $0 \neq \varphi \in \mathfrak{A}_1$, then $f$ is holonomic.*

In particular, any $\mathbb{F}_q$-linear polynomial of $s$ is holonomic, since it is annihilated by $d_s^m$, with a sufficiently large $m$.

If $n > 0$, the situation is more complicated. We call the module $M_f$ (and the corresponding function $f$) *degenerate* if $d(M_f) < n+1$ (by the Bernstein inequality, there is no degeneracy phenomena for modules over the complex Weyl algebra). The simplest example of a degenerate function (for $n = 1$) is $f(s, t_1) = g(st_1) \in \mathfrak{F}_2$ where the function $g$ belongs to $\mathfrak{F}_1$ and satisfies an equation $\varphi(g) = 0$, $\varphi \in \mathfrak{A}_1$. It can be shown that $d(M_f) = 1$.

In order to exclude the degenerate case, we introduce the notion of a non-sparse function. A function $f \in \mathfrak{F}_{n+1}$ of the form (21) is called *non-sparse* if there exists such a sequence $m_l \to \infty$ that, for any $l$, there exist sequences $k_1^{(i)}, k_2^{(i)}, \ldots, k_n^{(i)} \geq m_l$ (depending on $l$), such that $k_\nu^{(i)} \to \infty$ as $i \to \infty$ ($\nu = 1, \ldots, n$), and $a_{m, k_1^{(i)}, \ldots, k_n^{(i)}} \neq 0$.

**Theorem 6.** *If a function $f$ is non-sparse, then $d(M_f) \geq n+1$. If, in addition, $f$ satisfies an equation $\varphi(f) = 0$, $0 \neq \varphi \in \mathfrak{A}_{n+1}$, then $f$ is holonomic.*

We use Theorem 6 to prove that the functions (21) obtained via the sequence-to-function transform (20) or its multi-index generalizations, from some well-known sequences of polynomials over $K$ are holonomic. In all the cases below the non-sparseness is evident, and we have only to prove that the corresponding function satisfies a non-trivial Carlitz differential equation.

a) *The Carlitz polynomials.* The transform (20) of the sequence $\{f_k\}$ is the Carlitz module function $C_s(t)$; see (6). It is easy to check that $d_s C_s(t) = C_s(t)$. Therefore the Carlitz module function is holonomic, jointly in both its variables.

b) *Thakur's hypergeometric polynomials.* We consider the polynomial case of Thakur's first analog of the hypergeometric function [26, 27, 28], that is

$$_1F_\lambda(-a_1,\ldots,-a_l;-b_1,\ldots,-b_\lambda;z) = \sum_m \frac{(-a_1)_m \ldots (-a_l)_m}{(-b_1)_m \ldots (-b_\lambda)_m D_m} z^{q^m} \quad (27)$$

where $a_1,\ldots,a_l,b_1,\ldots,b_\lambda \in \mathbb{Z}_+$, and the Pochhammer type symbols are defined as

$$(a)_n = \begin{cases} D_{n+a-1}^{q^{-(a-1)}}, & \text{if } a \geq 1; \\ L_{-a-n}^{-q^n}, & \text{if } a \leq 0, n \leq -a; \\ 0, & \text{if } a \leq 0, n > -a. \end{cases}$$

It is clear that the terms in (27), which make sense and do not vanish, are those with $m \leq \min(a_1,\ldots,a_l,b_1,\ldots,b_\lambda)$. Let the function $f \in \mathfrak{F}_{l+\lambda+1}$ be given by

$$f(s,t_1,\ldots,t_l,u_1,\ldots,u_\lambda)$$
$$= \sum_{k_1=0}^{\infty} \ldots \sum_{k_l=0}^{\infty} \sum_{v_1=0}^{\infty} \ldots \sum_{v_\lambda=0}^{\infty} {}_1F_\lambda(-k_1,\ldots,-k_l;-v_1,\ldots,-v_\lambda;s)$$
$$\times t_1^{q^{k_1}} \ldots t_l^{q^{k_l}} u_1^{q^{v_1}} \ldots u_\lambda^{q^{v_\lambda}}.$$

It is known ([28], Sect. 6.5) that

$$d_{sl} F_\lambda(-k_1,\ldots,-k_l;-v_1,\ldots,-v_\lambda;s)$$
$$= {}_1F_\lambda(-k_1+1,\ldots,-k_l+1;-v_1+1,\ldots,-v_\lambda+1;s) \quad (28)$$

if all the parameters $k_1,\ldots,k_l,v_1,\ldots,v_\lambda$ are different from zero. If at least one of them is equal to zero, then the left-hand side of (28) equals zero. This property implies the identity $d_s f = f$, the same as that for the Carlitz module function. Thus, $f$ is holonomic.

c). *K-binomial coefficients.* It can be shown [19] that the $K$-binomial coefficients $\binom{k}{m}_K$ satisfy the Pascal-type identity

$$\binom{k}{m}_K = \binom{k-1}{m-1}_K^q + \binom{k-1}{m}_K^q D_m^{q-1} \quad (29)$$

where $0 \leq m \leq k$ and it is assumed that $\binom{k}{-1}_K = \binom{k-1}{k}_K = 0$.

Consider a function $f \in \mathfrak{F}_2$ associated with the $K$-binomial coefficients, that is

$$f(s,t) = \sum_{k=0}^{\infty} \sum_{m=0}^{k} \binom{k}{m}_K s^{q^m} t^{q^k}. \quad (30)$$

The identity (29) implies the equation

$$d_s f(s,t) = \Delta_t f(s,t) + [1]^{1/q} f(s,t)$$

for the function (30). Therefore $f$ is holonomic.

Another non-trivial example of a holonomic module over $\mathfrak{A}_{n+1}$ is given by the theory [20] of evolution type equations over $\overline{K}_c$, that is equations of the form

$$\{P(\Delta_1,\ldots,\Delta_n) + Q(\Delta_1,\ldots,\Delta_n)d_s\} u = 0$$

where $P,Q$ are non-zero polynomials with coefficients from $\overline{K}_c$. Denote by $R$ the operator in the left-hand side:

$$R = P(\Delta_1,\ldots,\Delta_n) + Q(\Delta_1,\ldots,\Delta_n)d_s$$

where $P,Q$ are non-zero polynomials. Let $I = \mathfrak{A}_{n+1}R$.

**Theorem 7.** *The module $M = \mathfrak{A}_{n+1}/I$ is holonomic.*

## ACKNOWLEDGMENTS

This work was supported in part by the Ukrainian Foundation for Fundamental Research, Grant 10.01/004.

## REFERENCES

1. V. Bavula, *St. Petersburg Math. J.* **4**, no. 1, 71–92 (1993).
2. V. Bavula, The Carlitz algebras, math.RA/0505397.
3. Yu. Berest and A. Kasman, *Lett. Math. Phys.* **43**, 279–294 (1998).
4. L. Carlitz, *Duke Math. J.* **1**, 137–168 (1935).
5. P. Cartier, *Astérisque* **206**, 41–91 (1992).
6. K. Conrad, *J. Number Theory* **84**, 230–257 (2000).
7. S.C. Coutinho, *A Primer of Algebraic D-modules*, Cambridge University Press, 1995.
8. B. Diarra, *Contemporary Math.* **319**, 75–97 (2003).
9. A. Dimakis, F. Müller-Hoissen, and T. Striker, *J. Phys. A* **29**, 6861–6876 (1996).
10. L. Ferrari, *Comp. Math. Appl.* **41**, 1099–1108 (2001).
11. D. Goss, *K-Theory* **1**, 533–555 (1989).
12. D. Goss, *Basic Structures of Function Field Arithmetic*, Springer, Berlin, 1996.
13. M.A. Guest, *Topology* **44**, 263–282 (2005).
14. A.N. Kochubei, *Lett. Math. Phys.* **45**, 11–20 (1998).
15. A.N. Kochubei, *J. Number Theory* **76**, 281–300 (1999).
16. A.N. Kochubei, *J. Number Theory* **83**, 137–154 (2000).
17. A.N. Kochubei, *Finite Fields Appl.* **9**, 250–266 (2003).
18. A.N. Kochubei, *Adv. Appl. Math.* **34**, 175–191 (2005).
19. A.N. Kochubei, Holonomic Modules in Positive Characteristic, math.RA/0503398.
20. A.N. Kochubei, Evolution Equations and Functions of Hypergeometric Type over Fields of Positive Characteristic, math.NT/0510481.
21. D. Levi, P. Tempesta, and P. Winternitz, *J. Math. Phys.* **45**, 4077–4105 (2004).

22. M. van der Put, *Compositio Math.* **97**, 227–251 (1995).
23. A.M. Robert, *A Course in p-Adic Analysis*, Springer, New York, 2000.
24. S. Roman, *The Umbral Calculus*, Academic Press, London, 1984.
25. G.C. Rota, D. Kahaner, and A. Odlyzko, *J. Math. Anal. Appl.* **42**, 684–760 (1973).
26. D.S. Thakur, *Finite Fields and Their Appl.* **1**, 219–231 (1995).
27. D.S. Thakur, *J. Ramanujan Math. Soc.* **15**, 43–52 (2000).
28. D.S. Thakur, *Function Field Arithmetic*, World Scientific, Singapore, 2004.
29. L. Van Hamme, Continuous operators, which commute with translations, on the space of continuous functions on $\mathbb{Z}_p$, in *p-Adic Functional Analysis* (J.M. Bayod et al., eds.), Lect. Notes Pure Appl. Math. **137**, Marcel Dekker, New York, pp. 75–88 (1992).
30. A. Verdoodt, Umbral calculus in non-Archimedean analysis, in *p-Adic Functional Analysis* (A.K. Katsaras et al., eds.), Lect. Notes Pure Appl. Math. **222**, Marcel Dekker, New York, pp. 309–322 (2001).
31. C.G. Wagner, *J. Reine Angew. Math.* **251**, 153–160 (1971).

# Point on Curves Whose Coordinates are $p$-Adic $U$-Numbers

Hamza Menken* and Khanlar R. Mamedov*

*Mersin University, Mathematics Department 33343 Mersin, TURKEY
emails: hmenken@mersin.edu.tr, hanlar@mersin.edu.tr

**Abstract.** In the present paper it is shown that if a curve $\Gamma$ in $\mathbf{Q}_p^n$ has parametrization by non constant rational functions with rational coefficients, then there exist infinitely many points on $\Gamma$ whose coordinates are $p$-adic $U$-numbers.

**Keywords:** $p$-adic algebraic numbers, $p$-adic $U$-numbers.
**MSC 2000:** 11J61, 11K60.

## 1. INTRODUCTION

A classification of the set of all complex numbers into four disjoint aggregates, termed $A$-, $S$-, $T$-, $U$-numbers, was introduced by Mahler [10] in 1932, and it has proved to be of considerable value in the general development of the subject. The first classification of this kind was outlined by Maillet [12] in 1906, and others were described by Perna and Morduchai-Boltovskoj; but to Mahler's classification attaches by far the most interest (see [2]).

Everywhere below, we use the following notation. A prime $p \geq 2$ is assumed to be fixed. Denote by $\mathbf{Q}_p$ the field of $p$-adic numbers, by $|x|_p$ the $p$-adic norm on $x \in \mathbf{Q}_p$ and by $H(\alpha)$ the height of a $p$-adic algebraic number $\alpha$, i.e. the maximum of the $p$-adic values of the coefficients of the minimal polynomial of $\alpha$.

As in the real case, Mahler [11] had characterized the set of $p$-adic numbers as follows:
Let $\xi$ be a $p$-adic number, and for each pair of positive integers $n, H$, let $P(x)$ be a polynomial with integer coefficients; and define $\omega_n(H, \xi)$ by the equation

$$\omega_n(H,\xi) = \min\left\{|P(\xi)|_p: \ \deg P \leq n, \ H(P) \leq H, \ P(\xi) \neq 0\right\}. \tag{1}$$

It is clear that $0 \leq \omega_n(H,\xi) \leq 1$, since if $P(x) = 1$ then $|P(\xi)|_p = 1$. Further define

$$\omega_n(\xi) = \lim_{H \longrightarrow \infty} \sup \frac{-\log \omega_n(H,\xi)}{\log H}, \tag{2}$$

$$\omega(\xi) = \lim_{n \longrightarrow \infty} \sup \frac{\omega_n(\xi)}{n}. \tag{3}$$

It is clear that $\omega_n(\xi)$ does not decreasing as a function of $n$. One has, $0 \leq \omega_n(\xi) \leq \infty$ and $0 \leq \omega(\xi) \leq \infty$. Let $\mu(\xi)$ be the least positive integer $n$ for which $\omega_n(\xi) = \infty$, writing $\mu(\xi) = \infty$ if, in fact, $\omega_n(\xi) < \infty$ for all $n$. The number $\xi$ is called $p$-adic

$$A-\text{number if } \omega(\xi) = 0 \text{ and } \mu(\xi) = \infty, \tag{4}$$

$$S-\text{number if } 0 < \omega(\xi) < \infty \text{ and } \mu(\xi) = \infty, \tag{5}$$

$$T-\text{number if } \omega(\xi) = \infty \text{ and } \mu(\xi) = \infty, \tag{6}$$

$$U-\text{number if } \omega(\xi) = \infty \text{ and } \mu(\xi) < \infty. \tag{7}$$

We note that the $p$-adic $A$-numbers are just the $p$-adic algebraic numbers. Thus, a $p$-adic transcendental number $\xi$ is an $S$-number if $\omega_n(H,\xi)$ is uniformly bounded for all $n, H$, a $p$-adic $U$-number if, for some $n$, $\omega_n(H,\xi)$ is unbounded, a $T$-number otherwise.

On the other hand, Schlickewei [15] introduced a classification of p-adic numbers closely to Mahler's classification: Let $\xi$ be a $p$-adic number, and for each pair of positive integers $n, H$, let $\alpha$ be a $p$-adic algebraic number; and define $\omega_n^*(H,\xi)$ by the equation

$$\omega_n^*(H,\xi) = \min\left\{|\xi - \alpha|_p : \deg \alpha \leq n, \ H(\alpha) \leq H, \ \xi \neq \alpha\right\}. \tag{8}$$

Further define

$$\omega_n^*(\xi) = \lim_{H \to \infty} \frac{-\log \omega_n^*(H,\xi)}{\log H}, \tag{9}$$

$$\omega^*(\xi) = \lim_{n \to \infty} \sup \frac{\omega_n^*(\xi)}{n}. \tag{10}$$

It is clear that the inequalities $0 \leq \omega_n^*(\xi) \leq \infty$ and $0 \leq \omega^*(\xi) \leq \infty$ hold. Let $\mu^*(\xi)$ be the least positive integer $n$ for which $\omega_n^*(\xi) = \infty$, writing $\mu^*(\xi) = \infty$ if, in fact, $\omega_n^*(\xi) < \infty$ for all $n$. So, $\mu^*(\xi)$ is uniquely determined and neither $\mu^*(\xi)$ nor $\omega^*(\xi)$ can be finite. There are the following four possibilities for $\xi$. The number $\xi$ is called $p$-adic

$$A^* - \text{number if } \omega^*(\xi) = 0 \text{ and } \mu^*(\xi) = \infty, \tag{11}$$

$$S^* - \text{number if } 0 < \omega^*(\xi) < \infty \text{ and } \mu^*(\xi) = \infty, \tag{12}$$

$$T^* - \text{number if } \omega^*(\xi) = \infty \text{ and } \mu^*(\xi) = \infty, \tag{13}$$

$$U^* - \text{number if } \omega^*(\xi) = \infty \text{ and } \mu^*(\xi) < \infty. \tag{14}$$

$\xi$ is called a $U^*$-number of degree $m$ $(m \geq 1)$ if $\mu^*(\xi) = m$. The set of $p$-adic $U^*$-numbers of degree $m$ is denoted by $U_m^*$. Thus $U^* = \bigsqcup_{m=1}^{\infty} U_m^*$ holds. The $p$-adic set $U_1^*$ is called $p$-adic Liouville numbers. For instance, it can be easily shown that the number

$$1 + p + p^{2!} + p^{3!} + \cdots + p^{n!} + \cdots \tag{15}$$

is a $p$-adic Liouville number. Long [9] proved that $U_m = U_m^*$ for any natural number $n$. We note that there is an exact correspondence between the two classifications, $A^*$-, $S^*$-, $T^*$- and $U^*$-classes being in fact that identical with the $A$-, $S$-, $T$- and $U$-classes, respectively.

It is well known that a $p$-adic number is a rational if and only if its $p$-adic expansion is periodical. In general, to determinate the $p$-adic expansion of a $p$-adic number is not easy. For instance, the series

$$\sum_{n=1}^{\infty} n! = 1 + 2! + 3! + \cdots + n! + \cdots \tag{16}$$

converges in every $Q_p$ since $\lim_{n\to\infty} n! = 0$, but we do not know if its canonical expansion is periodic or not for some prime $p$? This problem has been open since 1971 [8]. Furthermore, we do not know the place of this sum in Mahler's classification of the $p$-adic numbers?

In the present paper we consider an approximation problem in the theory of Diophantine approximations in $p$-adic numbers. The theory of Diophantine approximations is developed for field of real numbers, but there are a few investigations in the field of $p$-adic numbers, for this we refer to [3], [4] and [5].

From the above classifications, the following definition is well known.

**Definition 1.** *Let $\gamma \in Q_p$ and $m \in Z^+$. The number $\gamma$ is called a $p$-adic $U_m$-number if for every $\omega > 0$, there exists a $p$-adic algebraic number $\alpha$ of degree $m$ with*

$$0 < |\gamma - \alpha|_p < H(\alpha)^{-\omega} \tag{17}$$

*and if there exist real constants $C, K > 0$ depending only on $\gamma$ and $m$ such that the relation*

$$0 < |\gamma - \beta|_p < CH(\beta)^{-K} \tag{18}$$

*holds for every $p$-adic algebraic number $\beta$ of degree $< m$.*

In [1] Alniacik has shown that if a curve $C$ in $R^n$ has a parametrization by non constant rational functions having rational coefficients then there exist infinitely many points $P(x_1, ..., x_n)$ on $C$ where $x_i \in U_m$ $(i = 1, ..., n)$. In [13] the similar problem in $Q_p$ was considered for polynomials. In the present work we improve this problem in $Q_p$ for rational functions. Since the field of $p$-adic numbers $Q_p$ is not ordered, a curve $\Gamma$ in $Q_p^n$ is defined on an open ball $K(a, r)$ where $a \in Q_p$ and $r > 0$, and it is considered a function from the open ball $K(a, r)$ to $Q_p^n$ as $\Gamma : K(a, r) \longrightarrow Q_p^n$, $\Gamma(x) = (f_1(x), ..., f_n(x))$. We assume that the functions $f_1(x), ..., f_n(x)$, the components of $\Gamma$, are rational functions with rational coefficients. We will show that if the functions $f_i$ parameterizing the curve $\Gamma$ satisfy the mean-value theorem, then there exist infinitely many points on $\Gamma$ whose coordinates are $p$-adic $U$-numbers. It is well known that for rational functions or even polynomial functions, the mean-value theorem does not hold without some restrictions [14].

We need the following lemma.

**Lemma 1** ([7]). *Let $\alpha_1, ..., \alpha_k$ $(k \geq 1)$ be algebraic numbers in $Q_p$ with $[Q(\alpha_1, ..., \alpha_k) : Q] = g$ and let $F(y, x_1, ..., x_k)$ be a polynomial with integer coefficients, whose degree in*

$y$ is at least one. If $\eta$ is an algebraic number such that $F(\eta, \alpha_1, ..., \alpha_k) = 0$, then the degree of $\eta \leq dg$ and

$$H(\eta) \leq 3^{2dg+(l_1+...+l_k)g} H^g H(\alpha_1)^{l_1 g} ... H(\alpha_k)^{l_k g} \qquad (19)$$

where $H(\eta)$ is the height of $\eta$, $H(\alpha_i)$ $(i = 1, ..., k)$ is the height of $\alpha_i$ and, $H$ is the maximum of absolute values of the coefficients of $F$, $l_i$ $(i = 1, ..., k)$ is the degree of $F$ in $x_i$ and $d$ is the degree of $F$ in $y$.

To prove our results we use the following theorem.

**Theorem 1** ([6]). *Let $\{\alpha_i\}$ be a sequence of p-adic algebraic numbers with*
*(1)* $\deg \alpha_i = m_i \leq l$, $\lim_{i \to \infty} H(\alpha_i) = \infty$ $(l \in \mathbb{Z}^+)$,
*(2)* $|\alpha_{i+1} - \alpha_i|_p = H(\alpha_i)^{-\omega_i}$ *where* $\lim_{i \to \infty} \omega_i = \infty$,
*(3)* $], 0 < |\alpha_{i+1} - \alpha_i|_p < H(\alpha_{i+1})^{-\delta}$ *for some fixed* $\delta > 0$.
*Then,* $\lim_{i \to \infty} \alpha_i \in U_m^*$ *where* $m = \lim_{i \to \infty} \inf m_i$.

## 2. MAIN RESULTS

In the present work we obtain the following main results.

**Theorem 2.** *Let* $m \in \mathbb{Z}^+$, $f_i(x) = \frac{p_i(x)}{q_i(x)}$ *and* $p_i(x), q_i(x) \in \mathbb{Z}[x]$, *where* $(p_i(x), q_i(x)) = 1$, $\min(\deg p_i(x), \deg q_i(x)) \geq 1$, $(i = 1, ..., k)$. *If the rational functions* $f_i(x)$ $(i = 1, ..., k)$ *satisfy the mean-value theorem in any open ball, then there exist infinitely many p-adic U-numbers* $\gamma$ *such that* $f_i(\gamma) \in U_m$ *for* $i = 1, ..., k$.

*Proof.* Let $\alpha$ be a $p$-adic algebraic number of degree $m$ and $\alpha^{(1)} = \alpha, \alpha^{(2)}, ..., \alpha^{(m)}$ denote the conjugates of $\alpha$. Consider the equation

$$P_i(\alpha^{(r)} + y) = P_i(\alpha^{(s)} + y) \quad (1 \leq r, s \leq m, r \neq s). \qquad (20)$$

For fixed $r, s, i$, the equation (20) is equivalent to some polynomial equation

$$c_t y^t + ... + c_1 y + c_0 = 0 \qquad (21)$$

where the coefficients $c_j$ are $p$-adic algebraic numbers. Since $\alpha^{(r)} \neq \alpha^{(s)}$ for $r \neq s$, and $c_t \neq 0$, so (20) has only finitely many solutions in $y \in \mathbb{Q}_p$. Consider $y = p^n$, then there exists a natural number $n_0$ such that $\deg P_i(\alpha + p^n) = m$ for all $n \geq n_0$. Let $\{\omega_i\}$ be a sequence of positive real numbers with $\lim_{i \to \infty} \omega_i = \infty$. We define algebraic numbers $\alpha_i$ and integers $n_i$ as

$$\deg P_i(\alpha + p^{n_1}) = m \quad (i = 1, ..., k), \quad \alpha_1 = \alpha + p^{n_1} \qquad (22)$$

$$\deg P_t(\alpha_i + p^{n_{i+1}}) = m \quad (t = 1,\ldots,k) \tag{23}$$

$$H(\alpha_i)^{\omega_i} < p^{n_{i+1}} \tag{24}$$

$$n_i^2 < n_{i+1} \quad (i \geq 1) \tag{25}$$

$$\alpha_{i+1} = \alpha_i + p^{n_{i+1}}. \tag{26}$$

From (22) and (26) it follows that $\alpha_{i+1} = \alpha + \sum_{j=1}^{i+1} p^{n_j}$. Also, the relation $F(\alpha_{i+1}, \alpha, \sum_{j=1}^{i+1} p^{n_j}) = 0$ holds for the polynomial $F(y, x_1, x_2) = y - x_1 - x_2$. Applying Lemma 1 we obtain

$$H(\alpha_{i+1}) \leq 3^{4m} H(\alpha)^{2m} H(\sum_{j=1}^{i+1} p^{n_j})^{2m}. \tag{27}$$

Using (25) we write

$$H(\sum_{j=1}^{i+1} p^{n_j}) = p^{n_1} + \ldots + p^{n_{i+1}} \leq (i+1) p^{n_{i+1}} \leq p^{2n_{i+1}}. \tag{28}$$

Since $\lim_{i \to \infty} p^{n_i} = \infty$, there exists a natural number $i_1$ such that

$$p^{n_{i+1}} \geq 3^{4m} H(\alpha)^{2m} \tag{29}$$

for all $i \geq i_1$. Thus, for all $i \geq i_1$ we can write

$$H(\alpha_{i+1}) \leq (p^{n_{i+1}})^{2m+1}. \tag{30}$$

A combination of (26) and (30) gives us

$$|\alpha_{i+1} - \alpha_i|_p = |p^{n_{i+1}}|_p = p^{-n_{i+1}} \leq H(\alpha_{i+1})^{-\frac{1}{2m+1}} \quad (i \geq i_1). \tag{31}$$

Letting a $\delta = \frac{1}{2m+1}$, we write

$$|\alpha_{i+1} - \alpha_i|_p \leq H(\alpha_{i+1})^{-\delta} \quad (i \geq i_1). \tag{32}$$

On the other hand, it follows from (24) and (26) that

$$|\alpha_{i+1} - \alpha_i|_p \leq H(\alpha_i)^{-\omega_i} \quad (i \geq i_1). \tag{33}$$

Thus, the sequence $\{\alpha_i\}$ satisfies the conditions (1), (2) and (3) of Theorem 1, and so we have $\lim_{i \to \infty} \alpha_i = \gamma \in U_m^*$.

Now we show that $f_t(\gamma) \in U_m$ $(t = 1,\ldots,k)$. Put $\beta_j = f_t(\alpha_i)$. Since we assume that the functions $f_t(x)$ $(t = 1,\ldots,k)$ satisfy the mean-value theorem we can write

$$|f_t(\alpha_{i+1}) - f_t(\alpha_i)|_p = |\alpha_{i+1} - \alpha_i|_p |f_t'(\theta)|_p. \tag{34}$$

Put $c = \max_{1 \leq t \leq k} |f'_t(\theta)|_p$. Hence, we have

$$|\beta_{i+1} - \beta_i|_p \leq |\alpha_{i+1} - \alpha_i|_p c. \tag{35}$$

For the polynomial $F(y,x) = y - f_t(x)$ the relation $F(\beta_i, \alpha_i) = 0$ holds. Applying Lemma 1 we have

$$H(\beta_i) \leq H(\alpha_i)^{qm+1} \tag{36}$$

where $q$ is a positive constant depending only $\deg p_i, \deg q_i$ ($i = 1, ..., k$). Using (33) and (36) in (35) we obtain

$$|\beta_{i+1} - \beta_i|_p \leq H(\alpha_i)^{-(\omega_i+1)/qm+1} \tag{37}$$

and combining (32) and (37) we have

$$|\beta_{i+1} - \beta_i|_p \leq |\alpha_{i+1} - \alpha_i|_p c < |\alpha_{i+1} - \alpha_i|_p^{\frac{1}{2}} \leq H(\beta_i)^{-\rho} \tag{38}$$

where $\rho = (4m+4)^{-1}(qm+1)^{-1} > 0$. Thus by Theorem 1 it follows that the number

$$\lim_{i \to \infty} \beta_i = f_t(\lim_{i \to \infty} \alpha_i) = f_t(\gamma) \tag{39}$$

is $p$-adic $U$-number ($i = 1, ..., k$). □

The following result is a consequence of the Theorem 2.

**Corollary 1.** *Let $\Gamma$ be a curve in $\mathbb{Q}_p^n$ parameterized by non constant the rational functions with rational coefficients satisfying the mean-value theorem. Then there exist infinitely many points $P(x_1, x_2, ..., x_n)$ on $\Gamma$ where $x_i$ is a $p$-adic $U$-number ($i = 1, ..., n$).*

## REFERENCES

1. K. Alniacik, The Points on Curves whose Coordinates are $U$-numbers, *Rendiconti di Matematica, Serie VII* **18**, 649–653 (1998).
2. A. Baker, *Transcendental Number Theory*, Cambridge Unv. Press, Cambridge and London, 1975.
3. V.V. Beresnevich, V.I. Bernik, E.I. Kovalevskaya, On Approximation of $p$-adic Numbers by $p$-adic Algebraic Numbers, *Journal of Number Theory* **111**, 33–56, (2005).
4. V.V. Beresnevich, E.I. Kovalevskaya, On Diophantine Approximations of Dependent Quantities in $p$-adic Case, *Mathematical Notes* **73**, no. 1, 21–25 (2003).
5. E.B. Burger, On Liouville Decompositions in Local Fields, *Proc. of American Math. Soc.* **124**, no. 11, 3305–3310 (1996).
6. H. Duru, On Semi-strong $U$−numbers in $p$−adic Fields $Q_p$, *Unv. of Istanbul Faculty of Science the Journal of Math.* **57**, 103–111 (1998).
7. O.Ş. İçen, Anhang zu den Arbeiten "Über die Functionswerte der $p$-adish Elliptischen Functionen I and II", *Unv. of Istanbul Faculty of Science the Journal of Math.* A **38**, 25–35 (1973).
8. A. Khrennikov, $p$−Adic Valued Distributions in Mathematical Physics, Kluwer Academic Pub., Dordrecht, 1994.
9. X.X. Long, On Mahler's Classification of $p$-adic Numbers, *Pure Apply. Math.* **5**, 73–80 (1989).

10. K. Mahler, Zur Approximation der Exponentialfunction und des Logaritmus I, II, *J. Reine Angew. Math.* **166**, 118–150 (1932).
11. K. Mahler, Über eine Klassen-Einteilung der $p$-adischen Zahlen, *Mathematica (Leiden)* **3**, 177–185 (1935).
12. E. Maillet, Sur la Classification des Irrationnelles, *C. R.* **143**, 26–28 (1906).
13. H. Menken, An Investigation on $p$-adic $U$-numbers, *Univ. of Istanbul Faculty of Science the Journal of Math.* **59**, 111–143 (2000).
14. A.M. Robert, *A Course in p-adic Analysis*, Spinger-Verlag New York, 2000.
15. H.P. Schlickewei, $p$-adic $T$-numbers Do Exist, *Acta Aritmatica* **XXXIX**, 181–191 (1981).

# Infinitesimals in Nonstandard Analysis versus Infinitesimals in $p$-Adic Fields

Žarko Mijajlović*, Miloš Milošević† and Aleksandar Perović†

*Faculty of Mathematics, University of Belgrade, Studentski trg 16, p.o. box 550,
11 001 Belgrade, SERBIA AND MONTENEGRO
email: zarkom@eunet.yu

†Mathematical Institute of SANU, Knez Mihailova 35, p.o. box 367,
11 001 Belgrade, SERBIA AND MONTENEGRO
emails: mionamil@eunet.yu , peramail314@yahoo.com

## 1. INTRODUCTION

In general, infinite quantities reflect certain non-Archimedean property of the underlying structure. For example, the fields of hyperreal numbers and $p$-adic numbers are non-Archimedean according to ordering and norm, respectively. We shall compare these structures in regard to the infinitesimal notions, construction and related transfer techniques.

## 2. INFINITESIMALS AND RELATED NOTIONS

In terms of a first order predicate logic, one can think of the structure of hyperreal numbers as some $\aleph_1$-saturated elementary extension of standard universe of analysis. That particulary means that our language contains enough function symbols to interpret any real function and any real functional.

Without additional set theoretic assumptions, such as CH, there can be many nonisomorphic $\aleph_1$-saturated elementary extensions of standard universe of analysis, so we shall precise the intended model: starting from standard universe $\mathbb{R}$ of analysis, we pick arbitrary nonprincipal ultrafilter $D$ over $\mathbb{N}$, and then define nonstandard universe $^*\mathbb{R}$ as an ultrapower of $\mathbb{R}$ over $D$. Since

$$x \mapsto \langle x \mid n \in \mathbb{N} \rangle_D$$

is an elementary embedding of $\mathbb{R}$ into $^*\mathbb{R}$, we may assume that $\mathbb{R}$ is a substructure of $^*\mathbb{R}$.

For the given prime number $p$, by $\mathbb{Q}_p$ we shall denote the structure of $p$-adic numbers. Regardless of the chosen construction (fraction field of the projective limit or completion in $p$-adic norm), this structure is unique up to isomorphism. Recall that each $p$-adic number ($\neq 0$) can be expressed as

$$\sum_{n=k}^{\infty} a_n p^n, \; a_n \in \{0, \ldots, p-1\}, \; a_k \neq 0, \; k \in \mathbb{Z}$$

and that $|\sum_{n=k}^{\infty} a_n p^n|_p = p^{-k}$.

One can introduce infinitesimals in variety of ways, but we shall focus on the following two:

- Suppose that $\langle X, +, \cdot, 0, 1 \rangle$ is a field of characteristic 0 and that the map $\| \; \|_X : X \longrightarrow \mathbb{R}$ is a norm over $X$. We say that $x \in X$ is an *infinitesimal* if for every positive integer $n$ the inequality

$$\|nx\|_X < 1 \tag{1}$$

holds. $x \in X$ is *finite* if $x = 0$ or $x^{-1}$ is not an infinitesimal;

- Suppose that $\langle X, +, \cdot, 0, 1, < \rangle$ is ordered field. We say that $x \in X$ is an *infinitesimal* if for every positive integer $n$ relation

$$nx < 1 \tag{2}$$

holds. $x \in X$ is *finite* if there is positive integer $n$ such that $-n \leqslant x \leqslant n$.

Regardless of approach, we shall use the following definitions and notation:

- Infinitesimals will be denoted by $\varepsilon, \varepsilon_0, \varepsilon_1$, etc. Infinitesimal $\varepsilon$ is a *proper infinitesimal* if $\varepsilon \neq 0$;
- $x \approx y$ iff $x - y$ is an infinitesimal; such $x$ and $y$ will be called *infinitely close*. The *monad* $\mu(x)$ of $x$ is the set of all elements of $X$ infinitely close to $x$. In particular, $\mu(0)$ is the set of all infinitesimals in $X$;
- The set of all finite elements of $X$ will be denoted by $X_{fin}$. The *galaxy* of $x \in X$ is the set $G(x)$ defined by

$$G(x) = \{y \in X \mid x - y \in X_{fin}\}.$$

Clearly, $G(0) = X_{fin}$.

Notice that $X_{fin}$ is a subring of $X$ and that $X_{fin} = X$ if and only if there are no proper infinitesimals in $X$. Also notice that the corresponding monads of 0 are maximal ideals of $^*\mathbb{R}_{fin}$ and $\mathbb{Q}_{p\,fin} = \mathbb{Z}_p$, respectively.

If $\langle a_n \mid n \in \mathbb{N} \rangle$ is a sequence of real numbers such that for some $S \in D$ we have that $\lim_{n \in S} a_n = 0$ and $a_n \neq 0$ for each $n \in S$, then $\langle a_n \mid n \in \mathbb{N} \rangle_D$ is a proper infinitesimal in $^*\mathbb{R}$. Recall that $D$ is nonprincipal ultrafilter over $\mathbb{N}$.

It is easy to see that each $p$-adic number $\sum_{n=k}^{\infty} a_n p^n$ such that $k > 0$ and $a_k \neq 0$ is a proper infinitesimal in $\mathbb{Q}_p$.

Infinitesimal notions plays the key role in the nonstandard analysis and its application in the real analysis. The following examples serves as an illustration of this fact:

1. Function *standard part* $st : {}^*\mathbb{R}_{fin} \longrightarrow \mathbb{R}$ is defined by

$$st(x) = \sup_{\mathbb{R}}\{y \mid y \leqslant x\}. \tag{3}$$

   Clearly, standard part is an epimorphism and $\mu(0)$ is its kernel. According to the isomorphism theorem we have that

$$ {}^*\mathbb{R}_{fin}/ker(st) \cong \mathbb{R}.$$

2. If $\varepsilon$ is an infinitesimal, then $st(a+\varepsilon) = st(a)$;
3. $f : \mathbb{R} \longrightarrow \mathbb{R}$ is continuous iff for all $a \in {}^*\mathbb{R}_{fin}$

$$st({}^*f(a)) = f(st(a));$$

4. $f : \mathbb{R} \longrightarrow \mathbb{R}$ is uniformly continuous iff for all $a,b \in {}^*\mathbb{R}$

$$a \approx b \to {}^*f(a) \approx {}^*f(b);$$

5. Let $f : \mathbb{R} \longrightarrow \mathbb{R}$ be a differentiable function and let $\varepsilon$ be a proper infinitesimal. Then $f'(x) = st\left(\frac{{}^*f(x+\varepsilon)-f(x)}{\varepsilon}\right)$. For example,

$$(x^2)' = st\left(\frac{x^2 + 2x\varepsilon + \varepsilon^2 - x^2}{\varepsilon}\right) = st(2x+\varepsilon) = 2x;$$

6. Let $S$ be an infinite subset of $\mathbb{R}$. Then ${}^*S \setminus S$ is also infinite;
7. Let $H$ be an infinite natural number, i.e. $H \in {}^*\mathbb{N} \setminus \mathbb{N}$. Then

$$E = \{S \subseteq \mathbb{N} \mid H \in {}^*S\}$$

   is a nonprincipal ultrafilter over $\mathbb{N}$;

8. $\int_0^1 f(x)dx = \frac{1}{H} \sum_{i=0}^{H} {}^*f(\frac{i}{H})$, where $H \in {}^*\mathbb{N} \setminus \mathbb{N}$ and $f$ is continuous;
9. Problem of covering the Euclidean plane with tiles (or dominoes): if there is a covering of each square, prove that we can cover entire plane. One solution is the following: by the extension principle, we can find the covering of a square which edges has length $H$, where $H \in {}^*\mathbb{N} \setminus \mathbb{N}$. This particular cover induces the covering of entire Euclidean plane.

# 3. TRANSFER TECHNIQUES

Due to Łoś's theorem (see [2]), the ultrapower construction gives us the strongest possible transfer of first order properties among the structures of real and hyperreal numbers. Even more is true, the ultrapower construction preserves $\Sigma_1^1$ sentences.

Unfortunately, the projective limit construction of $p$-adic numbers gives us much weaker transfer. Still, this construction has important consequences; some of them are listed below.

1. Using the preservation theorems (see [2] or [6]) one can easily prove the following two facts:
   - Let $\mathscr{M} = \varprojlim \mathscr{M}_j$ ($j \in J$) and let $\varphi$ be a first order sentence in the language of $\mathscr{M}$. Then
   $$\mathscr{M} \models \varphi \text{ implies for all } j \in J \; \mathscr{M}_j \models \varphi$$
   iff $\varphi$ is a positive sentence. We say that sentence $\varphi$ is positive if there are no occurrences of negation and implication in it.
   - Properties that are preserved under products and submodels are also preserved under projective limits.

2. Each infinite profinite group has cardinality $2^{\aleph_0}$.

3. If $k \leqslant K$ is a finite or infinite Galois extension and $\mathscr{A}$ is the set of all finite Galois extensions $N$ of $k$ contained in $K$, then
$$\mathscr{G}al(K/k) \cong \varprojlim \mathscr{G}al(N/k).$$

4. Let $(p_n)$ be an increasing sequence of prime numbers and let $A_n$ be the field $\mathbb{Q}(\sqrt{p_1}, \ldots, \sqrt{p_n})$. Then
$$\mathscr{G}al(\mathbb{Q}(\sqrt{p_i} \mid i \in \omega)/\mathbb{Q}) \cong \varprojlim \mathscr{G}al(A_n/\mathbb{Q}) \cong \varprojlim C_2^n \cong C_2^{\aleph_0}.$$

5. Let $k = \mathbb{Q}(\zeta_p)$ and $K = \bigcup_{n=1}^{\infty} \mathbb{Q}(\zeta_{p^n})$, where $\zeta_{p^n}$ is a primitive $p^n$-th root of unity. Then:
   - $\mathscr{G}al(K/k) \cong \mathbb{Z}/2\mathbb{Z} \times \mathbb{Z}_2$, $\mathbb{Q}_2^* \cong \mathbb{Z} \times \mathscr{G}al(K/k)$;
   - $\mathscr{G}al(K/k) \cong \mathbb{Z}_p$, $\mathbb{Q}_p^* \cong \mathbb{Z} \times \mathbb{Z}_p \times \mathbb{Z}/(p-1)\mathbb{Z}$, for $p \neq 2$.

The Hasse-Minkowski theorem gives us a kind of downward transfer: if quadratic form $f(\bar{x})$ over $\mathbb{Q}$ has nontrivial zero in each $\mathbb{Q}_p$ (including $\mathbb{Q}_\infty = \mathbb{R}$), then it also has a nontrivial zero in $\mathbb{Q}$ (the converse is trivial, since $\mathbb{Q}$ is a subfield of each $\mathbb{Q}_p$).

Observe that assertion "$f(\bar{x})$ has a nontrivial zero in $\mathbb{R}$" may be reduced to the validity of some quantifier free formula in $\mathbb{Q}$. Namely, due to quantifier elimination for RCF,

$$\mathbb{R} \models \exists x_1 \ldots \exists x_n (\sum_{i=1}^{n} a_i x_i^2 + \sum_{i<j} b_{i,j} x_i x_j = 0 \wedge \bigvee_{i=1}^{n} x_i \neq 0) \tag{4}$$

iff

$$\mathbb{R} \models \varphi(a_1,\ldots,a_n,b_{1,1},\ldots,b_{n-1,n}), \tag{5}$$

where $\varphi$ is a Boolean combination of equalities and inequalities of rational numbers, so (4) is reducible to simple decidable condition on $\mathbb{Q}$. This is particulary useful when quadratic form has a rank $\geqslant 5$, because in this case Meyer's theorem implies existence of a nontrivial zero in $\mathbb{Q}$ iff there is a nontrivial zero in $\mathbb{R}$.

The next theorem is a characterization of elementary equivalence between Henselian fields, so it may be considered as a strong form of transfer.

**Theorem 1 (Ax, Kochen, Ershov).** *Let $(K, Z, \mathrm{ord})$ and $(K', Z', \mathrm{ord}')$ be two unramified Henselian fields of characteristic 0, and $\mathbf{k}$ and $\mathbf{k}'$ their residue fields. The following are equivalent:*

- $K \equiv K'$ *(as valued fields);*
- $\mathbf{k} \equiv \mathbf{k}'$ *(as fields) and* $Z \equiv Z'$ *(as ordered groups).*

This transfer principle is crucial in the proof that the Artin's conjecture is correct for almost all $p$-adic fields $\mathbb{Q}_p$. We remind the reader of the statement of the Artin's conjecture: if $n > d^2$, then each homogenous polynomial of degree $d$ in $n$ variables over $\mathbb{Q}_p$ has a nontrivial zero.

To summarize our discussion, we give the following table:

|  | $^*\mathbb{R}$ | $\mathbb{Q}_p$ |
|---|---|---|
| non-Archimedean | $\leqslant$ | $\|\ \|_p$ |
| infinitesimals | $\varepsilon \approx 0$ | $\|\varepsilon\|_p < 1$ |
| monads | $\mu(a)$ | unit ball |
| construction: | $^*\mathbb{R} = \prod_D \mathbb{R}$ | $\mathbb{Z}_p = \varprojlim \mathbb{Z}/p^n\mathbb{Z}$ |
| transfer | Łoś's theorem | Hasse-Minkowski theorem |

## REFERENCES

1. S. Albeverio et al, *Nonstandard Methods in Stohastic Analysis and Mathematical Physics*, Academic Press Inc. 1986.
2. C.C. Chang, H.J. Keisler, *Model Theory*, North–Holland 1990.
3. G. Cherlin, *Model Theoretic Algebra - Selected Topics*, Springer 1976.
4. J. Denef, *Arithmetic and Geometric Applications of Quantifier elimination for Valued Fields*, MSRI Publications, Vol **39**, 2000.
5. H. Koch, *Galoissche theorie der p-erweiterungen*, VEB Deutscher Verlag der Wissenschaften, Berlin 1970.

6. D. Marker, *Model Theory: An Introduction*, Springer-Verlag, 2002.
7. J.P. Serre, *A Course in Arithmetic*, Springer 1996.
8. K.D. Stroyan, W.A.J. Luxemburg, *Introduction to the theory of infinitesimals*, N. Y. a. o.: Academic Press, 1976.

# Barrelledness of $p$-Adic $C^1$-Function Spaces

Wim H. Schikhof

*Department of Mathematics, University of Nijmegen, Toernooiveld,
6525 ED Nijmegen, The Netherlands
email: schikhof@math.kun.nl*

**Abstract.** [1] Let $K$ be a complete non-archimedean rank 1 valued field, let $X \subset K$ be non-empty, without isolated points. On the space $C^1(X)$ of continuously differentiable functions $X \longrightarrow K$ (§ 3) we define the topology $\tau_c^1$ of compact convergence of functions and their first order difference quotients (4.3). Our main result (6.2) is barrelledness of $(C^1(X), \tau_c^1)$. As a corollary we obtain (7.1) that $(C^1(X), \tau_c^1)$ is reflexive if and only if $K$ is *not* spherically complete.

## 1. PRELIMINARIES

Throughout, $K$ is a complete non-archimedean valued field with valuation $|\,.\,| : K \longrightarrow [0, \infty)$. For notations and facts about locally convex spaces we refer to [3]. The symbol [ ] indicates 'linear span'. By $K^{(\mathbb{N})}$ we mean the $K$-vector space of all sequences $(\lambda_1, \lambda_2, \ldots)$ where $\lambda_n \in K$ for all $n$ and $\lambda_n = 0$ for large $n$. Let $\tau$ be the strongest locally convex topology on $K^{(\mathbb{N})}$. Then each $\tau$-bounded set lies in a finite-dimensional space and the dual $(K^{(\mathbb{N})})'$ of $(K^{(\mathbb{N})}, \tau)$ is naturally isomorphic to $K^{\mathbb{N}}$.

From now on in §1, let $M$ be an ultrametric space with ultrametric $d$. The closure of $Y \subset M$ is denoted $\overline{Y}$. A subset of $M$ is called 'clopen' if it is both closed and open. The *ball* with radius $r > 0$ about $a \in M$, $B(a, r) := \{x \in M : d(x, a) \leq r\}$, is clopen. For $Y \subset M$, $Y \neq \emptyset$, and $x \in M$ we set $d(x, Y) := \inf\{d(x, y) : y \in Y\}$ and, for $r > 0$, $Y^r := \{x \in M : d(x, Y) \leq r\}$, $\emptyset^r := \emptyset$. $Y^r$ is a clopen neighbourhood of $Y$.

Let $f : M \longrightarrow K$. Then supp $f := \overline{\{x \in M : f(x) \neq 0\}}$. Let $\emptyset \neq Y \subset M$. By $f|Y$ we mean the restriction of $f$ to $Y$. We set $\|f\|_Y := \sup\{|f(x)| : x \in Y\}$ (allowing the value $\infty$), $\|f\|_\emptyset := 0$. By $\xi_Y$ we denote the $K$-valued characteristic function of $Y$.

Let $C(M)$ be the $K$-vector space of all continuous functions $M \longrightarrow K$. If $f \in C(M), Y \subset M$ compact then $\|f\|_Y < \infty$. Let $BC(M) := \{f \in C(M) : \|f\|_M < \infty\}$. $(BC(M), \|\,.\,\|_M)$ is a Banach space. If $M$ is compact then $BC(M) = C(M)$.

---

[1] **Note.** The corresponding 'classical' case is hardly a problem. Spaces of real $C^1$-functions on an interval are easily seen to be Fréchet, hence automatically barrelled.

# 2. SOME FACTS ON ULTRAMETRIC SPACES

In this section $M$ is an ultrametric space with ultrametric $d$.

**Lemma 2.1.** *Let $Y \subset U \subset M$, where $Y$ is compact and $U$ is open. Then, for some $r > 0$, $Y^r \subset U$.*

*Proof.* We may suppose $Y \neq \emptyset$. By compactness there are finitely many balls $B(x_1, r_1), \ldots, B(x_n, r_n)$ contained in $U$, covering $Y$. Let $r := \frac{1}{2} \min\{r_1, \ldots, r_n\}$, and let $z \in Y^r$. There is a $y \in Y$ with $d(z,y) < 2r$. Then $y \in B(x_i, r_i)$ for some $i \in \{1, \ldots, n\}$ and $d(z, x_i) \leq \max(d(z,y), d(y, x_i)) \leq \max(2r, r_i) = r_i$, so that $z \in B(x_i, r_i) \subset U$. It follows that $Y^r \subset U$. $\square$

**Lemma 2.2.** *Let $f \in C(M)$, $Y$ a compact subset of $M$, $\varepsilon > 0$. Then there is an $r > 0$ such that $\|f\|_{Y^r} \leq \max(\|f\|_Y, \varepsilon)$.*

*Proof.* We may assume $Y \neq \emptyset$. For each $y \in Y$ there is a ball $B_y$ about $y$ such that $|f(z) - f(y)| < \varepsilon$ for all $z \in B_y$. Then $U := \bigcup\{B_y : y \in Y\}$ is open, contains $Y$ and $\|f\|_U \leq \max(\|f\|_Y, \varepsilon)$. By the previous lemma $Y^r \subset U$ for some $r > 0$ completing the proof. $\square$

**Lemma 2.3.** *Let $Y \subset M$ be closed, but not compact. Then there exists a clopen partition $U_1, U_2, \ldots$ of $M$ such that $U_n \cap Y \neq \emptyset$ for all $n$.*

*Proof.* We distinguish two cases.

(i) $(Y, d)$ is complete. Then $Y$ is not precompact and there is an $\varepsilon > 0$ and $y_1, y_2, \ldots \in Y$ such that $B(y_1, \varepsilon), B(y_2, \varepsilon), \ldots$ are pairwise disjoint and we see that we may choose $U_1 := B(y_1, \varepsilon) \cup (M \setminus \bigcup_n B(y_n, \varepsilon))$ and, for $n \geq 2$, $U_n := B(y_n, \varepsilon)$.

(ii) $(Y, d)$ is not complete. Denote the canonical ultrametric on the completion $M^\wedge$ of $M$ again by $d$. Then $Y$ is not closed in $M^\wedge$, so there is a sequence $y_1, y_2, \ldots$ in $Y$ and a $z \in M^\wedge \setminus M$ such that $d(z, y_1) > d(z, y_2) > \ldots$, $\lim_{n \to \infty} d(z, y_n) = 0$. Putting $U_1 := \{x \in M : d(x, z) \geq d(x, y_1)\}$ and, for $n \geq 2$, $U_n := \{x \in M : d(x, y_n) \leq d(x, z) < d(x, y_{n-1})\}$ and observing that $z \notin M$ we conclude that $U_1, U_2, \ldots$ has the required properties. $\square$

**Lemma 2.4.** *Let $Y$ be a non-empty closed subset of $M$, let $\varepsilon > 0$. Then there is a map $\sigma : M \longrightarrow Y$ such that*

(i) *$\sigma$ is the identity on $Y$,*
(ii) *$d(\sigma(x), \sigma(y)) \leq (1 + \varepsilon) d(x, y)$ for all $x, y \in M$,*
(iii) *$\sigma$ is locally constant on $M \setminus Y$.*

*If $Y$ is compact we can choose $\sigma$ such that, in addition,* (ii) *holds for $\varepsilon := 0$.*

*Proof.* For the first part the proof of [2], Lemma 76.1 applies with only obvious modifications. For the second part use [2], Exercise 76.A. $\square$

From now on in this paper $X$ is a non-empty subset of $K$ without isolated points.

## 3. BANACH SPACES OF $C^1$-FUNCTIONS

For $f : X \longrightarrow K$ we set

$$\Phi_1 f(x,y) := \frac{f(x) - f(y)}{x - y} \quad (x, y \in X, \ x \neq y).$$

Recall ( [2], 27.1, 27.2 ) that $f$ is called a $C^1$-*function* if $\Phi_1 f$ can (uniquely) be extended to a continuous function on $X \times X$ or, equivalently, if $f$ is differentiable and

$$(x,y) \longmapsto \overline{\Phi}_1 f(x,y) := \begin{cases} \Phi_1 f(x,y) & \text{if } x \neq y \\ f'(x) & \text{if } x = y \end{cases}$$

is continuous on $X \times X$.

The space of all $C^1$-functions $X \longrightarrow K$ is denoted $C^1(X)$. By $BC^1(X)$ we indicate the subspace of all $f \in C^1(X)$ for which $f$ and $\overline{\Phi}_1 f$ are bounded. With the norm

$$f \longmapsto \|f\|_X^1 := \max(\|f\|_X, \|\overline{\Phi}_1 f\|_{X \times X})$$

$BC^1(X)$ is a Banach space. If $X$ is compact then $C^1(X) = BC^1(X)$.

We quote the following result on antiderivation ($P$ stands for 'primitivation').

**Proposition 3.1 ([2], 79.2, 79.B).** *There is a linear map* $P : BC(X) \longrightarrow BC^1(X)$ *such that* $(Pf)' = f$ *and* $\|Pf\|_X^1 \leq \|f\|_X$ *for all* $f \in BC(X)$.

In 6.1 we will encounter compact subsets of $K$ that may have isolated points. It is useful to extend the notion of a $C^1$-function on such domains as follows.

**Definition 3.2.** For compact $Z \subset K$, let $D^1(Z)$ be the collection of all ordered pairs $(f;g)$ where $f, g \in C(Z)$ such that

$$(x,y) \longmapsto \overline{\Phi}_1 (f;g)(x,y) := \begin{cases} \Phi_1 f(x,y) & \text{if } x \neq y \\ g(x) & \text{if } x = y \end{cases}$$

is continuous on $Z \times Z$.

**Remarks 3.3.**
1. One could, in the same spirit, define $D^1(Z)$ for *any* subset of $K$, but we will not need it here.

2. If $x \in Z$ is not isolated (in $Z$) then $g(x)$ is determined by $f$, but this conclusion does not hold in general. For example, if $Z$ consists of only a single point then $D^1(Z)$ is a two-dimensional space.

With coordinatewise operations $D^1(Z)$ is a $K$-vector space. The formula
$$\|(f;g)\|_Z^1 := \max(\|f\|_Z, \|\overline{\Phi}_1(f;g)\|_{Z \times Z})$$
defines a norm $\|\cdot\|_Z^1$ on $D^1(Z)$ for which it is a Banach space, as is easily seen. If $Z \neq \emptyset$ has no isolated points then the map $f \mapsto (f; f')$ is a linear surjective isometry $C^1(Z) \longrightarrow D^1(Z)$.

**Proposition 3.4.** *Let $Z \subset K$ be compact. Then $D^1(Z)$ is of countable type.*

*Proof.* By compactness and metrizability of $Z$ and $Z \times Z$ the spaces $C(Z)$ and $C(Z \times Z)$ are of countable type ([1], 3.T), hence so is $C(Z) \times C(Z \times Z)$. The map
$$(f;g) \longmapsto (f, \overline{\Phi}_1(f;g))$$
is a linear isometry $D^1(Z) \longrightarrow C(Z) \times C(Z \times Z)$, and the conclusion follows. □

We now prove a Tietze-Urysohn type lemma for $C^1$-functions.

**Theorem 3.5.** *Let $Z \subset X$ be compact. Then there is a linear isometry $E: D^1(Z) \longrightarrow BC^1(X)$ such that $E(f;g)|Z = f$, $E(f;g)'|Z = g$ for all $(f;g) \in D^1(Z)$.*

*Proof.* We may assume $Z \neq \emptyset$.

(i) Let $D_0^1(Z) := \{(f;g) \in D^1(Z) : g = 0\}$. We first define $E_0 := E|D_0^1(Z)$. By Lemma 2.4 there exists a map $\sigma: X \longrightarrow Z$, extending the identity on $Z$, locally constant on $X \setminus Z$ and satisfying $|\sigma(x) - \sigma(y)| \leq |x - y|$ for all $x, y \in X$. We now define for $(f;0) \in D_0^1(Z)$,
$$E_0(f;0) := f \circ \sigma.$$

To prove that $E_0$ does the job we will use the estimation
$$\left| \frac{f(\sigma(x)) - f(\sigma(y))}{x - y} \right| \leq \left| \frac{f(\sigma(x)) - f(\sigma(y))}{\sigma(x) - \sigma(y)} \right| \tag{1}$$
for all $x, y \in X$ for which $\sigma(x) \neq \sigma(y)$.

Clearly $f \circ \sigma$ extends $f$. To prove that $f \circ \sigma \in C^1(X)$ and $(f \circ \sigma)' = 0$ it suffices to show that if $x_1, x_2, \ldots$ and $y_1, y_2, \ldots$ are sequences in $X$, both converging to $a \in X$, and $x_n \neq y_n$ for all $n$, then $(f(\sigma(x_n)) - f(\sigma(y_n)))/(x_n - y_n) \to 0$. If $a \in X \setminus Z$ then by local constantness of $\sigma$ on $X \setminus Z$ we have $\sigma(x_n) = \sigma(y_n)$ for large $n$ and we are done. If $a \in Z$ we may suppose that $\sigma(x_n) \neq \sigma(y_n)$ for all $n$. Then by (1) we have
$$\left| \frac{f(\sigma(x_n)) - f(\sigma(y_n))}{x_n - y_n} \right| \leq \left| \frac{f(\sigma(x_n)) - f(\sigma(y_n))}{\sigma(x_n) - \sigma(y_n)} \right|. \tag{2}$$

Now $\sigma(x_n) \to \sigma(a) = a$, $\sigma(y_n) \to \sigma(a) = a$ and since $(f;0) \in D^1(Z)$ the limit of the right hand side of (2) equals 0 and we are done.

Clearly $E_0$ is linear. Let $(f;0) \in D_0^1(Z)$. We have $\|f \circ \sigma\|_X = \|f\|_Z$ and, by using (1), $\|\overline{\Phi}_1(f \circ \sigma)\|_{X \times X} \le \|\overline{\Phi}_1(f;0)\|_{Z \times Z}$, so $f \circ \sigma = E_0(f;0)$ belongs to $BC^1(X)$ and $\|E_0(f;0)\|_X^1 \le \|(f;0)\|_Z^1$. The opposite inequality is trivial as $f \circ \sigma$ is an extension of $f$. Thus, we conclude that $E_0$ maps $D_0^1(Z)$ linearly and isometrically into $BC^1(X)$, $E_0(f;0)|Z = f$, $E_0(f;0)'|Z = 0$ for all $(f;0) \in D_0^1(Z)$.

(ii) To attack the general case we use two maps $T, P$ as follows. By the non-archimedean version of the Tietze-Urysohn extension lemma ([1], 5.24) there is a linear isometry $T : (C(Z), \|\cdot\|_Z) \longrightarrow (BC(X), \|\cdot\|_X)$ such that $Tf|Z = f$ for all $f \in C(Z)$. Further, let $P : BC(X) \longrightarrow BC^1(X)$ be the antiderivation map of 3.1. Let $(f;g) \in D^1(Z)$. Then $PTg \in BC^1(X)$, $(PTg)' = Tg$, hence $(PTg|Z;g) \in D^1(Z)$ so that $(f - PTg|Z;0) \in D_0^1(Z)$ and we can define

$$E(f;g) := PTg + E_0(f - PTg|Z;0).$$

Straightforward verifications show that $E$ extends $E_0$ and satisfies the conditions. □

## 4. THE TOPOLOGY OF COMPACT CONVERGENCE

We need the following as a preliminary.

**Definition 4.1.** Let $M$ be an ultrametric space. The *topology of compact convergence* $\tau_c$ on $C(M)$ is the locally convex topology induced by the seminorms $f \mapsto \|f\|_Y$ where $Y \subset M$ is compact.

It is easily seen that $(C(M), \tau_c)$ is Hausdorff and complete.

**Proposition 4.2.** $(C(M), \tau_c)$ *is of countable type.*

*Proof.* Let $Y \subset M$ be compact. Since $(C(Y), \|\cdot\|_Y)$ is of countable type there is a countable set $S \subset C(Y)$ whose linear hull is $\|\cdot\|_Y$-dense in $C(Y)$. Now let $T : C(Y) \longrightarrow BC(M)$ be a linear isometry such that $Tf|Y = f$ for all $f \in C(Y)$ ([1], 5.24). Then, obviously, the linear hull of $TS$ is dense in $C(M)$ with respect to the seminorm $\|\cdot\|_Y$. □

Our main concern will be the following $C^1$-version.

For a subset $Y$ of $X$ and $f \in C^1(X)$ we define $\|f\|_Y^1 := 0$ if $Y = \emptyset$; otherwise

$$\|f\|_Y^1 := \max(\|f\|_Y, \|\overline{\Phi}_1 f\|_{Y \times Y})$$

(allowing the value $\infty$). If $Y$ is compact then $\|f\|_Y^1 < \infty$, which leads to

**Definition 4.3.** The *topology of compact convergence* $\tau_c^1$ on $C^1(X)$ is the locally convex topology induced by the seminorms $f \mapsto \|f\|_Y^1$ where $Y \subset X$ is compact.

Again, $(C^1(X), \tau_c^1)$ is Hausdorff and complete, and also:

**Proposition 4.4.** $(C^1(X), \tau_c^1)$ *is of countable type.*

*Proof.* The map $f \mapsto (f, \overline{\Phi}_1 f)$ is a linear homeomorphism: $(C^1(X), \tau_c^1) \longrightarrow (C(X), \tau_c) \times (C(X \times X), \tau_c)$. By 4.2 $(C(X), \tau_c)$ and $(C(X \times X), \tau_c)$ are of countable type, hence so are their product and its subspaces, in particular $(C^1(X), \tau_c^1)$. □

We will need the next two facts on $\tau_c^1$-convergence.

**Proposition 4.5.** *Let $\{U_i : i \in I\}$ be a clopen covering of $X$. For each finite subset $J$ of $I$ let $V_J := \bigcup\{U_i : i \in J\}$. Then, for each $f \in C^1(X)$, the net $J \mapsto f \, \xi_{V_J}$ (where the collection of finite subsets of $I$ is directed by inclusion) converges to $f$ with respect to $\tau_c^1$.*

*Proof.* Let $Y \subset X$ be compact; we prove $\|f - f\, \xi_{V_J}\|_Y^1 \to 0$. By compactness, $Y \subset V_{J_0}$ for some finite subset $J_0$ of $I$. Now let $J$ be a finite subset of $I$, $J \supset J_0$. Then $f - f\, \xi_{V_J} = 0$ on $V_{J_0}$ so $\|f - f\, \xi_{V_J}\|_Y^1 \le \|f - f\, \xi_{V_J}\|_{V_{J_0}}^1 = 0$ and we are done. □

**Corollary 4.6.** *Let $\{U_n : n \in \mathbb{N}\}$ be a clopen partition of $X$ and let, for each $n$, $f_n \in C^1(X)$, supp $f_n \subset U_n$. Then, for each $(\lambda_1, \lambda_2, \ldots) \in K^{\mathbb{N}}$ the series $\sum_{n=1}^{\infty} \lambda_n f_n$ converges with respect to $\tau_c^1$.*

*Proof.* Obviously the series converges pointwise to an $f : X \longrightarrow K$ and since $f\, \xi_{U_n} = \lambda_n f_n$ for each $n$ we have $f \in C^1(X)$. By 4.5 the sequence $m \mapsto \sum_{n=1}^m f\, \xi_{U_n} = \sum_{n=1}^m \lambda_n f_n$ converges to $f$ in the $\tau_c^1$-sense. □

The next result concerns approximation of $\|\cdot\|_Y^1$ by $\|\cdot\|_U^1$ where $U$ is a neighbourhood of $Y$.

**Proposition 4.7.** *Let $f \in C^1(X)$, $Y \subset X$, $Y$ compact, $\varepsilon > 0$. Then there is a clopen $U \subset X$, containing $Y$, such that $\|f\|_U^1 \le \max(\|f\|_Y^1, \varepsilon)$.*

*Proof.* By 2.2 there is an $r_1 > 0$ such that $\|f\|_{Y^{r_1}} \le \max(\|f\|_Y, \varepsilon)$. The same lemma, applied to $Y \times Y \subset X \times X$ (with the canonical product metric) and $\overline{\Phi}_1 f$ yields an $r_2 > 0$ such that $\|\overline{\Phi}_1 f\|_{(Y \times Y)^{r_2}} \le \max(\|\overline{\Phi}_1 f\|_{Y \times Y}, \varepsilon)$. Without loss, assume $r_1 = r_2 := r$. Now $(Y \times Y)^r = Y^r \times Y^r$ so by choosing $U := Y^r$ we obtain

$$\|f\|_U^1 = \max(\|f\|_{Y^r}, \|\overline{\Phi}_1 f\|_{Y^r \times Y^r}) \le \max(\|f\|_Y, \|\overline{\Phi}_1 f\|_{Y \times Y}, \varepsilon) = \max(\|f\|_Y^1, \varepsilon). \quad \square$$

# 5. THE DUAL OF $(C^1(X), \tau_c^1)$: $C^1$-DISTRIBUTIONS

In this section we study the elements of $(C^1(X), \tau_c^1)'$, sometimes also called $C^1$-distributions. Simple examples are $f \mapsto f(a)$ and $f \mapsto f'(a)$ where $a \in X$. In general, a linear map $\varphi : C^1(X) \longrightarrow K$ is a $C^1$-distribution if and only if there exists a compact $Z \subset X$ and a $C > 0$ such that $|\varphi(f)| \leq C \|f\|_Z^1$ for all $f \in C^1(X)$. We will see that among such compact sets $Z$ there exists a smallest one, the *support of* $\varphi$, supp $\varphi$, which we will define now.

**Definition 5.1.** Let $\varphi \in (C^1(X), \tau_c^1)'$. A clopen subset $U$ of $X$ is called $\varphi$-*null* if for all $f \in C^1(X)$ with supp $f \subset U$ we have $\varphi(f) = 0$. Let $N_\varphi$ be the union of all $\varphi$-null sets, and supp $\varphi := X \setminus N_\varphi$.

**Proposition 5.2.** *Let* $\varphi \in (C^1(X), \tau_c^1)'$, *let* $Z \subset X$ *be compact and* $C > 0$ *be such that* $|\varphi(f)| \leq C \|f\|_Z^1$ *for all* $f \in C^1(X)$. *Then we have the following.*

*(i) Clopen subsets of $\varphi$-null sets are $\varphi$-null.*

*(ii) Any clopen union of $\varphi$-null sets is $\varphi$-null.*

*(iii)* supp $\varphi \subset Z$.

*(iv)* supp $\varphi$ *is compact.*

*Proof.* (i) is immediate and (iv) follows from (iii) and the fact that supp $\varphi$ is closed. To prove (ii), first consider two $\varphi$-null sets $U_1$ and $U_2$ and let $f \in C^1(X)$, supp $f \subset U_1 \cup U_2$. Then $f \, \xi_{U_1} \in C^1(X)$, supp $f \, \xi_{U_1} \subset U_1$ so $\varphi(f \, \xi_{U_1}) = 0$, but also supp $(f - f \, \xi_{U_1}) \subset U_2$ so that $\varphi(f - f \, \xi_{U_1}) = 0$. We see that $\varphi(f) = 0$ so that $U_1 \cup U_2$ is $\varphi$-null. Inductively one proves that finite unions of $\varphi$-null sets are $\varphi$-null. Now let $I$ be infinite, $U := \bigcup\{U_i : i \in I\}$ be clopen where each $U_i$ is $\varphi$-null. Let $f \in C^1(X)$, supp $f \subset U$. Applying 4.5 to the covering formed by $X \setminus U$ and the $U_i$ ($i \in I$) we find that $f = f \, \xi_U = \lim_J f \, \xi_{V_J}$, where $V_J := \bigcup\{U_i : i \in J\}$ for each finite set $J \subset I$. By the first part of the proof we have $\varphi(f \, \xi_{V_J}) = 0$ for all finite subsets $J$ of $I$, so by continuity, $\varphi(f) = 0$. We see that $U$ is $\varphi$-null.

Next we prove (iii). Let $U \subset X$ be clopen, $U \cap Z = \emptyset$, $f \in C^1(X)$, supp $f \subset U$. Then $f = 0$ on the clopen set $X \setminus U$ so that $\|f\|_{X \setminus U}^1 = 0$. But then, since $Z \subset X \setminus U$ we have $|\varphi(f)| \leq C \|f\|_Z^1 \leq C \|f\|_{X \setminus U}^1 = 0$. Thus, $U$ is $\varphi$-null and we see that $X \setminus Z$ is a union of $\varphi$-null sets implying $X \setminus Z \subset N_\varphi$ i.e. supp $\varphi \subset Z$. $\square$

**Corollary 5.3.** *Every clopen subset of $X$ that is contained in $N_\varphi$ is $\varphi$-null.*

*Proof.* Follows directly from (ii) above. $\square$

Now we arrive at the crucial result. Together with 5.2 (iii) it yields the minimality property of supp $\varphi$ announced in the beginning of this section.

**Theorem 5.4.** *Let $\varphi \in (C^1(X), \tau_c^1)'$. Then there is a $C > 0$ such that $|\varphi(f)| \leq C\|f\|_{\text{supp } \varphi}^1$ for all $f \in C^1(X)$.*

*Proof.* There are a compact set $Z \subset X$ and a $C > 0$ such that $|\varphi(f)| \leq C\|f\|_Z^1$ for all $f \in C^1(X)$.

Now let $f \in C^1(X)$; we prove that $|\varphi(f)| \leq C\|f\|_{\text{supp } \varphi}^1$. To this end, let $\varepsilon > 0$. By 4.7 there is a clopen $U \supset \text{supp } \varphi$ such that

$$\|f\|_U^1 \leq \max(\|f\|_{\text{supp } \varphi}^1, \varepsilon). \tag{3}$$

By 2.4 there is a map $\sigma : X \longrightarrow U$ that is the identity on $U$, locally constant on $X \setminus U$ (hence, $\sigma \in C^1(X)$) and such that $|\sigma(x) - \sigma(y)| \leq (1+\varepsilon)|x-y|$ for all $x, y \in X$.

We have $f \circ \sigma \in C^1(X)$ and $f = f \circ \sigma$ on $U$ so that $\text{supp}(f - f \circ \sigma) \subset X \setminus U$ which a $\varphi$-null set by 5.3. Thus, $\varphi(f - f \circ \sigma) = 0$ and we have $|\varphi(f)| = |\varphi(f \circ \sigma)| \leq C\|f \circ \sigma\|_Z^1 \leq C\|f \circ \sigma\|_X^1$ so that

$$|\varphi(f)| \leq C \max(\|f \circ \sigma\|_X, \|\overline{\Phi}_1(f \circ \sigma)\|_{X \times X}). \tag{4}$$

Clearly

$$\|f \circ \sigma\|_X = \|f\|_U. \tag{5}$$

Now let $x, y \in X$.

(i) If $x \neq y$ but $\sigma(x) = \sigma(y)$ then $\Phi_1(f \circ \sigma)(x, y) = 0$, whereas, if $\sigma(x) \neq \sigma(y)$

$$|\Phi_1(f \circ \sigma)(x, y)| = \left|\frac{f(\sigma(x)) - f(\sigma(y))}{\sigma(x) - \sigma(y)}\right| \left|\frac{\sigma(x) - \sigma(y)}{x - y}\right| \leq (1+\varepsilon)\|\overline{\Phi}_1 f\|_{U \times U}.$$

(ii) If $x = y$ then $|\overline{\Phi}_1(f \circ \sigma)(x, y)| = |(f \circ \sigma)'(x)|$ which equals $|f'(x)|$ if $x \in U$ and 0 if $x \in X \setminus U$.

Putting (i),(ii) together we obtain

$$\|\overline{\Phi}_1(f \circ \sigma)\|_{X \times X} \leq (1+\varepsilon)\|\overline{\Phi}_1 f\|_{U \times U} \tag{6}$$

and combining (4), (5), (6) we get

$$|\varphi(f)| \leq C \max(\|f\|_U, (1+\varepsilon)\|\overline{\Phi}_1 f\|_{U \times U}) \leq C(1+\varepsilon)\|f\|_U^1,$$

so that, applying (3),

$$|\varphi(f)| \leq C(1+\varepsilon) \max(\|f\|_{\text{supp } \varphi}^1, \varepsilon),$$

which holds for every $\varepsilon > 0$ and we are done. $\square$

**Corollary 5.5.** *Let $f \in C^1(X)$, $\varphi \in (C^1(X), \tau_c^1)'$. If $f$ and $f'$ are 0 on $\text{supp } \varphi$ then $\varphi(f) = 0$.*

**Remark 5.6.** If supp $\varphi$ has no isolated points then $f = 0$ on supp $\varphi$ implies $f' = 0$ on supp $\varphi$, but this conclusion is not true in general. In fact, let $a \in X$, take $\varphi : f \mapsto f'(a)$. Then supp $\varphi = \{a\}$ and the function $g : x \mapsto x - a$ is 0 on supp $\varphi$ but $\varphi(g) = g'(a) = 1$.

## 6. BARRELLEDNESS OF $(C^1(X), \tau_c^1)$

For barrelledness we have to prove that for any pointwise bounded collection $\mathscr{P}$ of continuous seminorms, the seminorm $\sup\{p : p \in \mathscr{P}\}$ is continuous. But, since $(C^1(X), \tau_c^1)$ is of countable type (4.4) it suffices to prove a slightly weaker statement ('polar barrelledness') ([3], 6.1, 6.2, 6.3):

(*) Let $\mathscr{F}$ be a pointwise bounded family in $(C^1(X), \tau_c^1)'$. Then $f \mapsto \sup\{|\varphi(f)| : \varphi \in \mathscr{F}\}$ is a continuous seminorm on $(C^1(X), \tau_c^1)$.

To prove (*) we start with a lemma.

**Lemma 6.1.** *Let $\mathscr{F}$ be a pointwise bounded family in $(C^1(X), \tau_c^1)'$. Then*

$$Z := \overline{\bigcup \{\text{supp } \varphi : \varphi \in \mathscr{F}\}}$$

*is compact.*

*Proof.* Suppose $Z$ is not compact; we derive a contradiction. By 2.3 there is a clopen partition $U_1, U_2, \ldots$ of $X$ such that $U_n \cap Z \neq \emptyset$ for all $n$. Then there are $\varphi_1, \varphi_2, \ldots \in \mathscr{F}$ such that $U_n \cap \text{supp } \varphi_n \neq \emptyset$ for each $n$. Thus, no $U_n$ is $\varphi_n$-null so there is, for each $n$, an $f_n \in C^1(X)$ with support in $U_n$ such that $\varphi_n(f_n) \neq 0$.

For $n \in \mathbb{N}$ set
$$x_n := (\varphi_n(f_1), \varphi_n(f_2), \ldots).$$

Compactness of supp $\varphi_n$ and the fact that supp $f_m \subset U_m$ for each $m$ entails that $\varphi_n(f_m) = 0$ for large $m$ i.e. $x_n \in K^{(\mathbb{N})}$.

Let $\tau$ be the strongest locally convex topology on $K^{(\mathbb{N})}$; we proceed to show that $\{x_1, x_2, \ldots\}$ is $\tau$-bounded. Since $K^{(\mathbb{N})}$ is of countable type it suffices to show weak boundedness i.e. that for each $(\lambda_1, \lambda_2, \ldots) \in K^{\mathbb{N}}$ the set $H := \{\sum_m \lambda_m \varphi_n(f_m) : n \in \mathbb{N}\}$ is bounded. Now by 4.6 $g := \sum_{m=1}^{\infty} \lambda_m f_m \in C^1(X)$, so $H = \{\varphi_n(g) : n \in \mathbb{N}\}$ which is bounded by assumption. Thus, $\{x_1, x_2, \ldots\}$ is $\tau$-bounded, implying finite-dimensionality of $[x_1, x_2, \ldots]$, an impossibility as $\varphi_n(f_n) \neq 0$ for each $n$. □

We now have all material for our main theorem.

**Theorem 6.2.** *Let $X$ be a non-empty subset of $K$ without isolated points. Then $(C^1(X), \tau_c^1)$ is barrelled.*

*Proof.* It suffices to prove (*) at the beginning of this section. From 6.1 we infer that

$$Z := \overline{\bigcup \{\operatorname{supp} \varphi : \varphi \in \mathscr{F}\}}$$

is compact. Let $R : C^1(X) \longrightarrow D^1(Z)$ be the 'restriction map' $f \mapsto (f|Z; f'|Z)$. If $f \in \operatorname{Ker} R$ then $f = f' = 0$ on $Z$ so, for all $\varphi \in \mathscr{F}$, we have $\varphi(f) = 0$ by 5.5. Then, for each $\varphi \in \mathscr{F}$ we have the factorization

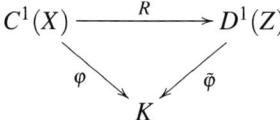

where $\tilde{\varphi}$ is linear and $\{\tilde{\varphi} : \varphi \in \mathscr{F}\}$ is pointwise bounded as $R$ is surjective by 3.5. Continuity of $\tilde{\varphi}$ follows from

$$|\tilde{\varphi}(Rf)| = |\varphi(f)| \leq C \, \|f\|^1_{\operatorname{supp} \varphi} \leq C \, \|f\|^1_Z = C \, \|Rf\|^1_Z$$

for all $f \in C^1(X)$ and some $C > 0$. Now $D^1(Z)$ is Banach, hence barrelled so $\{\tilde{\varphi} : \varphi \in \mathscr{F}\}$ is uniformly bounded i.e. there is an $\mathscr{M} > 0$ such that $|\tilde{\varphi}(Rf)| \leq \mathscr{M} \, \|Rf\|^1_Z$ (i.e. $|\varphi(f)| \leq \mathscr{M} \, \|f\|^1_Z$) for all $f \in C^1(X)$ and all $\varphi \in \mathscr{F}$. We see that the seminorm $\sup\{|\varphi| : \varphi \in \mathscr{F}\}$ is majorized by the continuous seminorm $\mathscr{M} \, \| \cdot \|^1_Z$, hence is itself continuous. □

**Remarks 6.3.**

1. It is not hard to see that, in the above manner, one can also prove that $(C(X), \tau_c)$ is barrelled. In fact, the proof has much less complications.

2. If $X$ is locally compact and separable there is a shorter proof of 6.2 by showing that in this case $X$ is the union of an increasing sequence $U_1 \subset U_2 \subset \ldots$ of open compact sets, so that the seminorms $\{\| \cdot \|^1_{U_n} : n \in \mathbb{N}\}$ induce the topology $\tau_c^1$. Then $(C^1(X), \tau_c^1)$ is metrizable, hence Fréchet and therefore automatically barrelled.

3. Let $n \in \{1, 2, \ldots\}$. On $C^n(X)$ (see [2], 29.1) one can define a topology $\tau_c^n$ in the spirit of 4.3, this time by taking into account all difference quotients up to order $n$. We do not know whether $(C^n(X), \tau_c^n)$ is barrelled or not for $n \geq 2$. The answer may depend on the characteristic of $K$.

## 7. REFLEXIVITY

As an application of the previous theory we now prove the following.

**Theorem 7.1.** *Let $X$ be a non-empty subset of $K$ without isolated points. Then $(C^1(X), \tau_c^1)$ is reflexive if and only if $K$ is not spherically complete.*

*Proof.* We have that $(C^1(X), \tau_c^1)$ is Hausdorff, complete and of countable type (4.4). Then by [3], 9.8, if $K$ is not spherically complete, $(C^1(X), \tau_c^1)$ is reflexive.

Now let $K$ be spherically complete. Let $a \in X$. Since $a$ is not isolated there are $a_1, a_2, \ldots \in X$ such that $|a_1 - a| > |a_2 - a| > \ldots$, $\lim_{n \to \infty} a_n = a$. Then $Z := \{a, a_1, a_2, \ldots\}$ is compact and $D^1(Z)$ is an infinite-dimensional Banach space, hence not reflexive by [1], 4.16. Let $R : C^1(X) \longrightarrow D^1(Z)$ be the 'restriction map', $h \mapsto (h|Z, h'|Z)$ and let $E : D^1(Z) \longrightarrow BC^1(X)$ be as in 3.5. Then $ER$ is a continuous projection $(C^1(X), \tau_c^1) \longrightarrow (C^1(X), \tau_c^1)$ and its image is linearly homeomorphic to $D^1(Z)$ yielding a linear homeomorphism of $C^1(X)$ onto $D^1(Z) \oplus \operatorname{Ker} R$. If $(C^1(X), \tau_c^1)$ were reflexive then so would be $D^1(Z)$, a contradiction. □

## REFERENCES

1. A.C.M. van Rooij, *Non-Archimedean Functional Analysis*, Marcel Dekker, New York, 1978.
2. W.H. Schikhof, *Ultrametric Calculus*, University Press, Cambridge, 1984.
3. W.H. Schikhof, Locally convex spaces over nonspherically complete valued fields I-II, *Bull. Soc. Math. Belg. Sér. B* **38**, 187–224 (1986).

# Local $p$-Adic Differential Equations

Marius van der Put* and Lenny Taelman*

*Institute of Mathematics and Computer Science, PO-BOX 800,
9700AV Groningen, THE NETHERLANDS
email: lenny@math.rug.nl

**Abstract.** This paper studies divergence in solutions of $p$-adic linear local differential equations. Such divergence is related to the notion of $p$-adic Liouville numbers. Also, the influence of the divergence on the differential Galois groups of such differential equations is explored. A complete result is given for second order equations and a conjecture for higher order equations is proposed.

**Keywords:** $p$-Adic analysis, differential equations, Liouville numbers, divergence.

## 1. INTRODUCTION

Let $k$ be a complete non-archimedean valued field containing the field of $p$-adic numbers. We suppose that $k$ is algebraically closed. Let $k\{x\}$ denote the ring of convergent power series over $k$ and let $K = k(\{x\})$ be its field of fractions. The aim of this paper is to investigate the divergence of solutions of linear differential equations and to compute differential Galois groups. This project is related in spirit to the work of J.P. Ramis and J. Martinet which is concerned with the same questions but then with base field the complex numbers (see [2], [5] and [4]). The main observation is that in the complex case, the divergence of solutions is due to the singularity of the equation and that in particular regular singular equations do not produce divergence. In contrast to this, in the $p$-adic case it is not the irregularity that produces divergence. In fact, the divergence in the $p$-adic situation comes from the regular singular case when $p$-adic Liouville numbers are present. This paper can be seen as a continuation of [1] and [3]. The analysis of the divergence for $p$-adic differential equations and the contribution of this to the differential Galois group is much more involved than in the complex case. There are two reasons for this, namely the divergence cannot be classified in levels and moreover there is at present no asymptotic theory on sectors (or the like) available in the $p$-adic case.

## 2. NOTATION

- $\mathbb{Z}_p$, $\mathbb{Q}_p$: the ring of $p$-adic integers and its quotient field
- $k$: a complete and algebraically closed field containing $\mathbb{Q}_p$
- $\widehat{K} = k((x))$, $K = k(\{x\})$: the field of formal Laurent series and the subfield of convergent Laurent series

- $k[[x]] \subset \widehat{K}$, $k\{x\} \subset K$: the subrings of functions without pole at $x = 0$
- $\delta$: the derivation $x\frac{d}{dx}$ on the above rings and fields

## 3. $p$-ADIC LIOUVILLE NUMBERS

Whereas the classical Liouville numbers are real numbers that can be rapidly approximated by rational numbers, the $p$-adic Liouville numbers are those numbers that can be rapidly approximated by *positive integers* for the $p$-adic valuation. The precise definition is the following. An element $\lambda \in k$ is called a *$p$-adic Liouville number* if $\liminf_{n \to +\infty} |\lambda - n|^{1/n} = 0$. The terminology is introduced in [C] and was rediscovered in [P80]. In the latter the above definition is given because it is better adapted to differential equations. Indeed, consider the inhomogeneous differential equation

$$(\delta - \lambda)y = \frac{1}{1-x}, \text{ with } \lambda \notin \mathbb{Z}. \tag{1}$$

This equation has a unique formal solution, namely $\sum_{n \geq 0} \frac{1}{n-\lambda} x^n$. This solution is divergent precisely when $\lambda$ is a $p$-adic Liouville number.

It seems that the set $\mathscr{L}$ of $p$-adic Liouville numbers does not have much structure. One can make the following observations

1. $\mathscr{L} \subset \mathbb{Z}_p$
2. $\mathscr{L}$ has measure 0 for the (real) Haar measure on $\mathbb{Z}_p$
3. If $a \in \mathscr{L}$ and $n, m \in \mathbb{Z}$ with $m > 0$ then $n + ma \in \mathscr{L}$
4. $\mathscr{L} \neq -\mathscr{L}$ and $\mathscr{L} \cap -\mathscr{L} \neq \emptyset$
5. Every $a \in \mathscr{L}$ is transcendent over $\mathbf{Q}$

Let a subset $\{\lambda_1, \ldots, \lambda_s\}$ of $k$ be given such that $\lambda_i - \lambda_j \in \mathbb{Z}$ implies that $i = j$ (this condition will correspond to a differential equation being non-resonant). We associate to this set an oriented graph $E$. The vertices of $E$ are $\{v_1, \ldots, v_s\}$ and $(v_i, v_j)$ (with $i \neq j$) is an oriented edge if and only if $\lambda_j - \lambda_i \in \mathscr{L}$.

**Theorem 1.** *Let $E$ be a finite oriented graph such that:*

(a) *The two ends of every oriented edge are distinct.*

(b) *For every pair of distinct vertices $a, b$ there is at most one oriented edge from $a$ to $b$.*

*There exists a finite subset $\{\lambda_1, \ldots, \lambda_s\}$ of $k$ such that $\lambda_i - \lambda_j \in \mathbb{Z}$ implies that $i = j$ and such that $E$ is its associated oriented graph.*

The proof is rather intricate and combinatorial.

# 4. DIFFERENTIAL EQUATIONS OVER $p$-ADIC FIELDS

A differential operator
$$L = \delta^n + a_{n-1}\delta^{n-1} + \ldots + a_0$$
with coefficients in $K$ (or in $\widehat{K}$) is called *regular singular* if and only if all $a_i$ are in $k\{x\}$ (or in $k[[x]]$), that is, have no pole at $x = 0$. $L$ is called *irregular* when it is not regular singular.

Irreducible irregular operators with coefficients in $K$ have the property that the formal solutions of the equation $L(y) = f$ with $f \in K$ are automatically convergent. One can, for example, easily verify this for the equation

$$(\delta - x^{-1})y = \sum_{n \geq 0} b_n x^n \qquad (2)$$

This surprising result is completely opposite to the complex case. In fact, in questions on divergence, the regular singular equations over $K$ will be the most interesting, as already can be seen in the simple example (1). Also this is very different from the complex case, where divergence in regular singular equations with convergent coefficients does not occur.

More precisely, using some $p$-adic functional analysis, one can show:

**Proposition 2.** *Let $f \in \widehat{K}$ be a divergent solution of a linear differential equation with coefficients in $K$. Then the minimal homogeneous equation satisfied by $f$ has the form*

$$(\delta^m + a_{m-1}\delta^{m-1} + \cdots + a_0)y = 0 \text{ with all } a_i \in K$$

*with all $a_i \in k\{x\}$ and such that the polynomial*

$$P(T) := T^m + a_{m-1}(0)T^{m-1} + \cdots + a_0(0) \in k[T]$$

*has a root in $\mathbf{Z}$ and has a root which is a $p$-adic Liouville number.*

Hence, from now on we will mainly consider regular singular differential equations. Using the definition of a $p$-adic Liouville number one can verify the following:

**Proposition 3.** *Consider the operator $L = \delta^m + a_{m-1}\delta^{m-1} + \cdots + a_0$ with coefficients in $k\{x\}$ and the polynomial*

$$P(T) = T^m + a_{m-1}(0)T^{m-1} + \cdots + a_0(0) \in k[T].$$

*If no zero of $P$ is a $p$-adic Liouville number, then all formal solutions of the inhomogeneous equations $L(y) = f$ with $f \in K$ are convergent.*

**Example 4.** We note that the solution $y = \sum_{n \geq 0} \frac{1}{n-\lambda} x^n$ of the inhomogeneous equation $(\delta - \lambda)y = \frac{1}{1-x}$ is also a solution of the monic homogeneous equation $(\delta^2 + \frac{-\lambda + x(1-\lambda)}{1-x}\delta + \frac{\lambda x}{1-x})y = 0$. The corresponding polynomial $P(T) = T^2 - \lambda T$ has zeros $\lambda, 0$. Therefore the propositions predict that $y$ is convergent if and only if $\lambda \in \mathscr{L}$, in accordance with the definition of $\mathscr{L}$.

# 5. REGULAR SINGULAR DIFFERENTIAL MODULES

To get a better understanding of divergence in regular singular equations, it is convenient to pass from the language of differential equations to the language of differential modules. These form a "coordinate-free" representation of differential equations and moreover, they permit the use of several constructions from linear algebra. For more details, see [4].

A differential module over $K$ (or any differential field) is a finite dimensional vector space $M$ over $K$ together with an operation $\partial : M \to M$ satisfying the Leibnitz rule $\partial f = f\partial + \delta f$. A submodule $N \subset M$ is per definition a vector subspace of $M$ that is preserved by $\partial$. The direct sum of two differential modules $M_1$ and $M_2$ is the direct sum of the underlying vector spaces on which $\partial$ acts as $\partial(m_1, m_2) = (\partial m_1, \partial m_2)$. It is also possible to define quotients, tensor products, symmetric products, internal Homs for differential modules (see [4]).

One can construct a scalar differential equation from the module if one can find a cyclic vector $e \in M$, that is, a vector $e$ such that $e, \partial e, \partial^2 e, \ldots$ form a basis. Such a cyclic vector always exists. If the dimension of $M$ is $n$, we then find that $\partial^n e$ can be written as $-a_0 e - a_1 \partial e - \cdots - a_{n-1} \partial^{n-1} e$, yielding the coefficients of a linear differential equation $L(y) = 0$ with

$$L = \delta^n + a_{n-1}\delta^{n-1} + \ldots + a_0.$$

From the above equation, one obtains a first order matrix equation by taking the vector $\mathbf{y} = (y, \delta y, \ldots, \delta^{n-1} y)$ to be the unknown. One obtains an equation of the form

$$\delta \mathbf{y} = A\mathbf{y},$$

with $A$ an $n \times n$ matrix over $K$.

Such an equation in turn yields a differential module $M$ by taking $A$ to be the matrix of the action of $\partial$ on a basis of $M$.

Now, let $M$ be a regular singular differential module, i.e., a differential module associated with a regular singular equation. Then $M$ has a $\partial$-invariant lattice $M^o = k\{x\}b_1 + \cdots + k\{x\}b_m$. With respect to this basis one can represent $M$ by a matrix differential operator $\delta + A$ with $A \in \mathrm{Matr}(m, k\{x\})$. One can expand $A = A_0 + A_1 x + \cdots$ with all $A_i \in \mathrm{Matr}(m, k)$. The term $A_0$ is also the matrix of the operator $\partial$ acting upon $M^o/xM^o$. For a suitable lattice, the eigenvalues $\lambda_1, \ldots, \lambda_s$ of $A_0$ have the property that $\lambda_i - \lambda_j \notin \mathbb{Z}$ for $i \neq j$. The $\lambda_1, \ldots, \lambda_s$ are uniquely determined by $M$, up to translation over integers. For convenience, we will call them the *eigenvalues of M*.

In the complex case $\delta + A$ is equivalent to $\delta + A_0$ by a convergent transformation. Moreover, the conjugacy class of the monodromy matrix $e^{2\pi i A_0}$ classifies the regular singular differential module. In our situation this is not the case, due to "small denominators" and $p$-adic Liouville numbers.

One defines an oriented graph $E$ attached to $M$ as follows. The vertices of $E$ are called $v_1, \ldots, v_s$. There is an oriented edge $(v_i, v_j)$ with $i \neq j$ if and only if $\lambda_j - \lambda_i$ is a $p$-adic

Liouville number. Let $E_1, \ldots, E_r$ denote the connected components of the oriented graph $E$.

**Proposition 5.** *The regular singular differential module $M$ has a unique direct sum decomposition $M = \oplus_{i=1}^{r} M_i$ such that $M_i$ is a regular singular differential module with associated oriented graph $E_i$ (and with the eigenvalues corresponding to this subset of $E$).*
*In particular, suppose that the set of eigenvalues $\{\lambda_1, \ldots, \lambda_s\}$ of $M$ has the property that $\lambda_i - \lambda_j \notin \mathscr{L}$ for all $i \neq j$. Then $M$ has a basis over $K$ such that the matrix of $\partial$ w.r.t. this basis is constant (i.e., has coefficients in $k$).*

*Proof.* We sketch the proof. The formal decomposition (over $\widehat{K}$) always exists, moreover this decomposition is convergent if there are no $p$-adic Liouville numbers involved. Hence the decomposition according to the connected components $E_i$ of $E$ converges. □

The above proposition reduces the study of regular singular differential modules to the case where the associated oriented graph is connected. Our next aim is to produce submodules according to the structure of the connected oriented graph. The following proposition describes a collection of submodules of $M$ in terms of the graph $E(M)$. For a *generic* $M$ having a prescribed graph $E(M) = E$, the result is optimal, that is, the proposition describes *all* submodules.

**Proposition 6.** *Let the connected oriented graph $E$ be associated to the regular singular differential module $M$. Suppose that the set of the vertices $V$ of $E$ is a disjoint union of two sets $V_1, V_2$ such that $(v_2, v_1)$ is not an oriented edge for any $v_1 \in V_1$, $v_2 \in V_2$. Then there is a unique submodule $M_1 \subset M$ corresponding to the eigenvalues belonging to $V_1$.*

## 6. EXAMPLES OF DIFFERENTIAL GALOIS GROUPS

Consider a linear differential equation $L(y) = 0$ of order $n$ and with coefficients in $K$. Then, of course, not all solutions of the equation need to exist in $K$, but over a suitably large differential field extension of $K$, the solution space becomes $n$-dimensional. In fact, there is a smallest such differential field extension, called the Picard-Vessiot field of the equation. The differential Galois group of $L$ is defined to be the group of $K$-linear differential field automorphisms of the Picard-Vessiot field. It acts on the solution space $V$ of the equation, and one knows that it is an algebraic subgroup of $GL(V)$. We refer the reader to [4] for more information on differential Galois gruops.

In this section we compute the subgroups of $SL_2$ that are realizable as differential Galois group of a regular singular differential module over $K$.

Over $\widehat{K}$ - or equivalently, over $\mathbb{C}((x))$, since to the formal power series the topology of the field of constants does not matter - these have to be one of the following: finite cyclic, $\mathbf{G}_a$ (additive), $\mathbf{G}_m$ (multiplicative) or $\mathbf{G}_a \times \{\pm 1\}$.

Over $K$ more subgroups are possible. Let $M$ be a two-dimensional differential module over $K$. Take a $\delta$-invariant lattice $M^o = k\{x\}e_1 + k\{x\}e_2$ and let $\lambda_1, \lambda_2$ denote the eigenvalues of $\delta$ operating on $M^o/xM^o$. Assume that $\lambda_2 - \lambda_1 \notin \mathbb{Z}$. The condition that the differential Galois group $G$ is contained in $\mathrm{SL}_2(k)$ is easily seen to be equivalent to $\lambda_1 + \lambda_2 \in \mathbb{Z}$. There are several cases:

1. $\lambda_2 - \lambda_1 \notin \mathscr{L} \cup -\mathscr{L}$. So $E(M) = * \ *$
   In this case, by Proposition 5, $M$ is a direct sum of one-dimensional modules and the differential Galois groups of $M$ and of $\widehat{M}$ coincide. The list of possible subgroups is:
   $$\mathbf{G}_m, \text{ finite cyclic}, \ \mathbf{G}_a \text{ and } \{\pm 1\} \times \mathbf{G}_a .$$

2. $\alpha := \lambda_2 - \lambda_1 \in \mathscr{L} \setminus -\mathscr{L}$ and $M$ "generic". $(E(M) = * \to *)$
   Let $\lambda_1 + \lambda_2 = n \in \mathbb{Z}$. Then $\lambda_1 = -\alpha/2 + n/2$ and $\lambda_2 = \alpha/2 + n/2$. Since $\alpha$ is not rational one finds that the formal differential Galois group is $\mathbf{G}_m$. This is a subgroup of $G$. Since $M$ is generic, it has only one non trivial submodule (by 6). One concludes that $G$ is (conjugated to) the Borel subgroup $B \subset \mathrm{SL}_2(k)$.

3. $\lambda_2 - \lambda_1 \in \mathscr{L} \cap -\mathscr{L}$ and $M$ "generic". $(E(M) = * \leftrightarrow *)$
   The formal differential Galois group is again $\mathbf{G}_m$. Since $M$ is generic it has only trivial submodules (by 6). The differential Galois group $G$ can only be $\mathrm{SL}_2$ or the infinite dihedral group $D_\infty$. In the latter case $N := K(x^{1/2}) \otimes M$, as differential module over $K(x^{1/2})$, has differential Galois group $\mathbf{G}_m$. Therefore $N$ is a direct sum of two 1-dimensional subspaces. This is, however, excluded by Remark 7. We conclude that the differential Galois group of $M$ is $\mathrm{SL}_2$.

**Remark 7.** If the regular singular module $M$ over $K = k(\{x\})$ has no submodules then the same holds for the module $M \otimes_K k(\{x^{1/n}\})$ over $k(\{x^{1/n}\})$, for any $n \geq 2$. The reason for this is that the eigenvalues get multiplied by $n$ after this base change, and a positive integral multiple of a Liouville number is again a Liouville number (see §1).

Hence, we conclude that the list of realizable subgroups of $\mathrm{SL}_2$ for regular singular equations is: $\mathrm{SL}_2$, $B$, $\mathbf{G}_m, \mathbf{G}_a$, $\{\pm 1\} \times \mathbf{G}_a$, finite cyclic .
The subgroups of $\mathrm{SL}_2$ that are realizable as differential Galois groups of irregular singular modules are $\mathbf{G}_m$ and the infinite dihedral group $D_\infty$. It is interesting to compare this with the list of the realizable subgroups of $\mathrm{SL}_2$ for equations over $\mathbb{C}(\{x\})$, which is

$$\mathrm{SL}_2, \ B, \ D_\infty, \ \mathbf{G}_m, \mathbf{G}_a, \ \{\pm 1\} \times \mathbf{G}_a, \text{ finite cyclic}.$$

In the complex case, the differential Galois group is generated, as algebraic group, by the formal differential Galois group and the Stokes matrices. The groups $\mathrm{SL}_2$, $B$ and $D_\infty$ occur only for irregular singularities. $\mathrm{SL}_2$ occurs when there are at least two distinct non-trivial Stokes matrices, and $B$ occurs when there is only one non-trivial Stokes matrix. For the $D_\infty$ group, there are no Stokes matrices. The group $\mathbf{G}_m$ occurs both in the irregular case with trivial Stokes matrices, or in the regular singular case. The last three groups only occur in the regular singular case.

One observes from the above that divergence due to *p*-adic Liouville numbers is the parallel of divergence caused by the complex Stokes matrices. The reader will note, however, that the former occurs for *regular* singular *p*-adic differential equations, while the latter occurs for *irregular* singular complex differential equations.

All this leads naturally to the following conjecture.

**Conjecture 8.** *Given a graph E as before, and a generic regular singular differential module M over K with $E(M) = E$. Let S be the solution space of M and $S = \oplus S_i$ the formal decomposition corresponding to the eigenvalues $\lambda_i$. The differential Galois group of M is the smallest algebraic group G such that*

- *G contains the formal differential Galois group, i.e. the differential Galois group of $M \otimes_K \widehat{K}$ over $\widehat{K}$,*
- *for every oriented edge $(v_i, v_j)$ in E, the Lie algebra $\mathrm{Lie}(G)$ contains all linear maps*

$$S \overset{proj}{\to} S_j \overset{\alpha}{\to} S_i \subset S$$

*with $\alpha \in \mathrm{Hom}(S_j, S_i)$.*

## REFERENCES

1. D. Clark, A note on the *p*-adic convergence of solutions of linear differential equations, *Proc. Amer. Math. Soc.* **17**, 262–269 (1966).
2. J. Martinet and J.P. Ramis, Elementary acceleration and multisummability, *Annales de l'Institut Henri Poincaré, Physique Théorique* **54**, no. 4, 331–401, (1991).
3. M. van der Put, Meromorphic differential equations over valued fields, *Proceedings A* **83**, no. 3, (1980); *Indagationes Mathematicae* **42**, (1980).
4. M. van der Put and M.F. Singer, *Galois theory of linear differential equations*, Ergebnisse der Mathematik, Springer Verlag, 2003.
5. J.P. Ramis, About the inverse problem in differential Galois theory: The differential Abhyankar conjecture - In: Braaksma et. al. editors, *The Stokes Phenomenon and Hilbert's 16th Problem*, World Scientific, pp. 261–278 (1996).

# RELATED TOPICS

# Nonlocal String Tachyon as a Model for Cosmological Dark Energy

Irina Ya. Aref'eva

*Steklov Mathematical Institute, Russian Academy of Sciences,*
*Gubkin st. 8, Moscow, 119991, RUSSIA*
*email:* arefeva@mi.ras.ru

**Abstract.** There are many different phenomenological models describing the cosmological dark energy and accelerating Universe by choosing adjustable functions. In this paper we consider a specific model of scalar tachyon field which is derived from the NSR string field theory and study its cosmological applications. We find that in the effective field theory approximation the equation of state parameter $w < -1$, i.e. one has a phantom Universe. It is shown that due to nonlocal effects there is no quantum instability that the usual phantom models suffer from. Moreover due to a flip effect of the potential the Universe does not enter to a future singularity.

**Keywords:** Strings, D-branes, p-adic strings, cosmology, dark energy.

## 1. INTRODUCTION

It was suggested by Ia Supernova observations that the Universe is presently accelerating[1, 2]. The basic sets of experiments now includes also the Cosmic Microwave Background (CMB) anisotropies, X-ray data from galaxy clusters, large scale structure and age estimates of global clusters, for a review see [3, 4, 5]. It is believed that a new particle and/or gravitational physics is required to explain the acceleration of the expansion of the Universe. The observations suggest that the bulk of energy density in the Universe is gravitationally repulsive and appears like an unknown form of energy (dark energy) with negative pressure. It is believed that 2/3 of the total density of the universe is in a form of dark energy.

There exist many different models of dark energy. It is convenient to describe them by using the equation of state parameter $w = p/\rho$, where $p$ is a pressure and $\rho$ is the energy density. The analysis of the current observation data shows that $w$ lies in the range $-1.61 < w < -0.78$ [6, 3, 7].

A list of dark energy models includes (see [4]-[23] and refs. therein)

- $w = -1$: the cosmological constant;
- $w = const \neq -1$: the cosmic strings, domain walls, etc.;
- $w \neq const$: quintessence scalar field, chaplygin gas, k-essence, Dirac-Born-Infeld(DBI) action, braneworlds, etc.;
- $w < -1$: phantom models.

The most challenging for theoretical physics would be the case $w < -1$ [12]-[17] (see [19, 5, 20] for reviews). In this case the weak energy condition ($\rho > 0, \rho + p > 0$) is violated and some strange phenomena as negative entropy and temperature appear. In many models the phantom Universe in finite time ends up in the singularity called the Big Rip [14]).

In this paper we consider string field theories nonlocal tachyon where the condition $w < -1$ is realized. We show that due to a peculiar properties of nonlocal tachyon dynamics we can get a phantom universe without problems with unstability. A nonlocal dynamics for the tachyon field is obtained by truncation of the covariant string field equations. We study a nonlocal dynamics of a string tachyon in the cosmological Friedmann metric. The string theory that we have in mind is the NSR string theory compactified on a six dimensional compact manifold. Moreover we assume that all moduli are frozen and unstable non-BPS brane extends along the three large spatial dimensions. The tension of the 3-dimensional non-BPS brane acts as the 4-dimensional universe cosmological constant and the tachyon dynamics describes a deviation from a pure cosmological constant regime.

A presence of tachyons was considered as a main drawback of corresponding strings theory. Bosonic string has a tachyon and its absence in superstrings was a main motivation to introduce superstrings. However few years ego tachyon has found an application in the context of brane scenarios. The open bosonic string and GSO – NSR string tachyons were used to describe D-brane decays. According to the Sen conjecture in the perturbative vacuum there are instable branes filling space-time and all these branes disappear in the true vacuum (for review see [24],[25]). Rolling tachyon solutions describe transitions from perturbative vacuum to non-perturbative one.

Here we use the covariant open string field theory(SFT) approach [24] to tachyon dynamics (about the effective DBI approach to tachyon dynamics see [26],[27],[25] and refs. therein). A crucial feature of the rolling tachyon solution for non-BPS brane [30] is that it interpolates between two non-perturbative vacua with the same energy. We argue that at large time the dynamics is governed by an effective action that contains a ghost kinetic term, in spite of absence of ghosts in the nonlocal action. We emphasize that the ghost here is a result of an approximation to the exact nonlocal action. There is no ghost in the nonlocal action. So we get a phantom universe as an approximation to the true nonlocal dynamics. This is the reason why there are no pathologies in this scenario with the thermodynamical instabilities.

The rolling tachyon solution passes the perturbative vacuum with non-zero velocity. Therefore, to reach the true vacuum starting from the perturbative one we have to supply the tachyon with a large enough initial velocity . This is a rather non-standard situation from the local field theory point of view, where one does not have to push strongly the tachyon to make it reach the true vacuum, it is need just an infinitesimal small perturbation. This effect occurs due to the presence of derivative terms in the interaction. These terms may be studied in an effective action approximation, which corresponds to keeping only few terms of an expansion of an nonlocal operator. We note that to reach the nonperturbative vacuum one has to add to the action a brane tension which is larger that is required by the Sen hypothesis. This large brane tension can be interpreted as an effect of the closed string excitations.

We find that in the effective field theory approximation the equation of state parameter $w < -1$, i.e. one has a phantom Universe. In the DBI approach the equation of state parameter interpolates between -1 at early time and 0 at a later time. Within application of the DBI action to cosmology there are problems with large density perturbations, reheating and caustics formation [28, 29].

The paper is organized as follows. In Sect. 2.1. the nonlocal tachyon action is written in the Friedmann background. In Sect. 2.2 we shortly remind a construction of the rolling tachyon solution for non-BPS brane [30]. In Sect. 2.3 we study the tachyon dynamics in the Friedmann metric in the effective action approximation. In this approximation one gets $w < -1$. In Sect. 2.4 we show that to reach a nonperturbative vacuum one has to add to the action a brane tension which is larger that is required by the Sen hypothesis. In Sect. 3 we show the stability of the nonlocal tachyon model in the true vacuum.

It is my pleasure to dedicate this contribution to Professor Branko Dragovich, team-work with whom on p-adics strings and other subjects I very appreciate.

## 2. NON-BPS TACHYON IN FRIEDMANN SPACE-TIME

### 2.1. General set up

We consider a non-BPS tachyon leaving on 3-brane and interacting with gravity with the following action

$$S = \frac{M_p^2}{2}\int \sqrt{-g}d^4x R + S_{tach}. \qquad (2.1)$$

where

$$S_{tach} = \int \sqrt{-g}d^4x \left(-\frac{q^2}{2}g^{\mu\nu}\partial_\mu\phi\partial_\nu\phi + \frac{1}{2}\phi^2 - \frac{1}{4}\Phi^4\right), \qquad (2.2)$$

$\Phi = \exp(\frac{1}{2}\Box_g)\phi$, $\Box_g = \frac{1}{\sqrt{-g}}\partial_\mu \sqrt{-g}g^{\mu\nu}\partial_\nu$, $q^2 = const < 1$. Here we assume that all constants are absorbed into $M_p^2$. The action (2.2) generalizes the non-BPS tachyon action obtained from low level truncated SFT to the case of a non-flat metric [30].

On space homogeneous configurations in the Friedmann metric

$$ds^2 = -dt^2 + a^2(t)(dr^2 + r^2(d\theta^2 + \sin^2\theta d\phi^2)) \qquad (2.3)$$

the action (2.2) takes the form

$$S_{tach}[\phi] = \int \sqrt{-g}dt\left[\frac{1}{2}\phi^2(t) + \frac{q^2}{2}\dot\phi(t)^2 - \frac{1}{4}\Phi^4(t)\right], \qquad (2.4)$$

where $\Phi = \exp(\frac{1}{2}\mathcal{D})\phi$, $\mathcal{D} = -\partial_t^2 - 3H(t)\partial_t$ and $H(t) = \dot{a}/a$, $\dot{a} = \partial_t a$. The Einstein equations have the form

$$3H^2 = \frac{1}{M_p^2}\rho \qquad (2.5)$$

$$H^2 + 2\ddot{a}/a = -\frac{1}{M_p^2}p \qquad (2.6)$$

with the energy and pressure densities are given by [30]

$$\rho = \frac{q^2}{2}(e^{-\frac{1}{2}\mathcal{D}}\dot{\Phi})^2 - \frac{1}{2}(e^{-\frac{1}{2}\mathcal{D}}\Phi)^2 + \frac{1}{4}\Phi^4 + \mathcal{E}_1 + \mathcal{E}_2] \qquad (2.7)$$

$$p = \frac{q^2}{2}(e^{-\frac{1}{2}\mathcal{D}}\dot{\Phi})^2 + \frac{1}{2}(e^{-\frac{1}{2}\mathcal{D}}\Phi)^2 - \frac{1}{4}\Phi^4 - \mathcal{E}_1 + \mathcal{E}_2. \qquad (2.8)$$

where

$$\mathcal{E}_1 = -\frac{1}{2}\int_0^1 d\rho(e^{\frac{1}{2}\tau\mathcal{D}}\Phi^3)\mathcal{D}e^{-\frac{1}{2}\tau\mathcal{D}}\Phi \qquad (2.9)$$

$$\mathcal{E}_2 = -\frac{1}{2}\int_0^1 d\tau(\partial_t e^{\frac{1}{2}\tau\mathcal{D}}\Phi^3)\partial_t e^{-\frac{1}{2}\tau\mathcal{D}}\Phi \qquad (2.10)$$

Equation of motion for the scalar field is

$$\left(q^2\mathcal{D}+1\right)e^{-\mathcal{D}}\Phi = \Phi^3. \qquad (2.11)$$

## 2.2. Rolling solution in flat space-time

Taking $H = 0$ in (2.11) we get the following equation in the flat space

$$\left(-q^2\partial_t^2 + 1\right)e^{\partial_t^2}\Phi = \Phi^3. \qquad (2.12)$$

This equation contains infinite number of time derivatives, and actually can be written in the integral form. It has been shown numerically that for $q^2$ small enough there is a solution that interpolates between non trivial vacua $\Phi(\pm\infty) = \pm 1$ and $\Phi(0) = 0$ [32]. One can get an approximation to this solution expanding the exponent in (2.12) in powers of derivatives and keeping only the second derivatives,

$$\left((1-q^2)\partial_t^2 + 1\right)\Phi(t) = \Phi^3(t). \qquad (2.13)$$

This equation describes a particle moving in the potential $V = \frac{(\Phi^2-1)^2}{4(q^2-1)} + const$. For $q^2 < 1$ the factor $q^2 - 1$ flips the potential, Fig.1.

Equation (2.13) for $q^2 < 1$ has the kink solution $\Phi_{kink}$. Kink interpolates between two vacua during infinitely long time and it is represented in Fig.2a by a thin line.

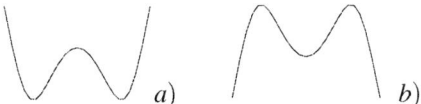

**FIGURE 1.** Flip of the potential for $q^2 < 1$

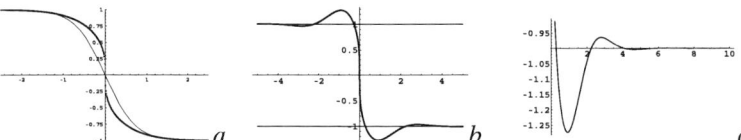

**FIGURE 2.** a) kink $\Phi_{kink}(t)$ (thin line) and $\Phi_0(t)$ (think line); b) $\Phi(t)$ for $q^2 = 0.96$; c) oscillations of $\Phi(t)$ with decreasing amplitudes around $\Phi_0$

Equation (2.12) for $q = 0$ (the p-adic string equation of motion for $p = 3$) also has a interpolating solution [33, 34, 32]. We denote it $\Phi_0(t)$ and plot it in Fig. 2a by think line. Note that the function $\Phi_0(t)$ is monotonic. From Fig.2a we see that $\Phi_{kink}$ and $\Phi_0$ have different profiles, but this difference is not too big for large times. There is an essential difference at small time. $\Phi_{kink}$ has the finite first derivative at $t = 0$, meanwhile the first derivative of $\Phi_0(t)$ becomes infinite at $t = 0$. Note, that the derivative of the initial scalar field $\phi$ related with $\Phi$ via $\phi = e^{-\frac{1}{2}\mathcal{D}}\Phi$ is finite at $t = 0$. Therefore, higher derivatives in (2.12) change the profile of $\Phi_{kink}(t)$ only at small time and do not change the asymptotic behavior at large time.

Note, that small $q^2$ also does not change too much a profile of a solution to (2.12) interpolating between two vacua. This solution is plotted in Fig.2b. The profile of this solution is not a monotonic function. It can be presented as $\Phi(t) = \Phi_0(t) + \phi(t)$, where $\phi(t)$ describes oscillations around $\Phi_0$ with decreasing amplitude. These oscillations are presented in Fig.2c.

## 2.3. Approximate solution of system of equations for Non-BPS tachyon in Friedmann space-time

Motivated by the flat case we make in (2.11) an approximation

$$\exp(\partial_t^2 + 3H(t)\partial_t)\Phi \approx (1 + \partial_t^2 + 3H(t)\partial_t)\Phi \quad (2.14)$$

and keep only terms linear on $(1 + \partial_t^2 + 3H(t)\partial_t)$. It is evident that this equation can be obtained from the action

$$S'_{scalar} = \int \sqrt{-g}d^4x \left(\frac{1-q^2}{2}g^{\mu\nu}\partial_\mu\Phi\partial_\nu\Phi - V(\Phi)\right). \quad (2.15)$$

We see that for $q^2 < 1$ we get the ghost sign in front of the kinetic terms. Assuming that $q^2 < 1$ we take for simplicity in the following formula $q^2 = 0$ ($q^2 < 1$ can be achieved just by rescaling of time). The corresponding Einstein equations have the form (2.7), (2.8) with

$$\rho = -\frac{1}{2}\dot{\Phi}^2 + V(\Phi), \quad (2.16)$$

$$p = -\frac{1}{2}\dot{\Phi}^2 - V(\Phi) \quad (2.17)$$

and the equation for $\Phi$ field read

$$\ddot{\Phi} + 3H\dot{\Phi} = V'_\Phi \quad (2.18)$$

Excluding $H$ from (2.7) and (2.18) one gets the following equation

$$\ddot{\Phi} + \frac{1}{M_p}\dot{\Phi}\sqrt{3[-\frac{1}{2}\dot{\Phi}^2 + V(\Phi)]} - V'_\Phi = 0 \quad (2.19)$$

We see that an equation similar to the usual scalar field equation in the Friedmann metric. There are only two different signs, one in front of the kinetic energy in the square root, and the second in front of the derivative of the potential. It is evident that solutions of this equation in the slow-roll regime are the same as in the usual case with "-" potential. However there are differences in the fast roll regime. The equation state parameter $w$

$$w = \frac{p}{\rho} = \frac{\frac{1}{2}\dot{\Phi}^2 + V(\Phi)}{\frac{1}{2}\dot{\Phi}^2 - V(\Phi)} \quad (2.20)$$

is always less then -1, since $w$ can be represented also as

$$w = -\frac{3H^2 + 2\dot{H}}{3H^2} = -1 - \frac{2}{3}\frac{\dot{H}}{H^2} \quad (2.21)$$

and from the equation of motions follows

$$\dot{H} = \frac{1}{2M_p^2}\dot{\Phi}^2, \quad (2.22)$$

i.e. $\dot{H}$ is positive.

## 2.4. Numerical solutions

Let us examine numerically solution of the system of equations (2.16) and (2.18) for the potential

$$V(\Phi) = \frac{1}{4}\left(\Phi^2 - 1\right)^2 \quad (2.23)$$

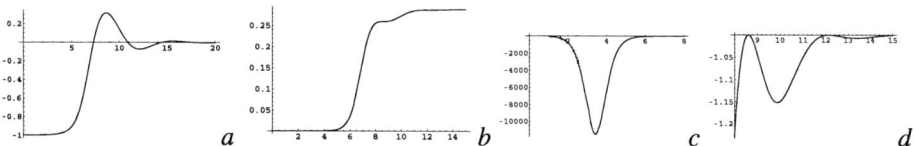

**FIGURE 3.** a) Plot of $\Phi = \Phi(t)$ with $\Phi(0) = -1$ and $\dot\Phi \simeq 0$; b) plot of $H = H(t)$; plot of $w = w(t)$ for c) $0 < t < 8$ and d) for $8 < t < 15$

There are two independent initial conditions for $\Phi(0)$ and $\dot\Phi(0)$. If the initial position $\Phi(0)$ is on the top of the hill (for the flip potential, Fig.1.b), $\Phi(0) = -1$, and the initial velocity is very small $\dot\Phi(0) \simeq 0$ (this corresponds to $H(0) \simeq 0$) then after some time $\Phi$ reaches the largest position and goes back to the bottom, and then performs few oscillations and stops at the bottom. The final value of $H$ is $1/2\sqrt{3}$. The evolutions of the scalar field and log-derivative of the scale factor are represented in Fig.3.a and Fig.3.b. The evolution of the state equation parameter $w$ is plotted in Fig.3c,d. It starts from -1, becomes a very big negative number when the field passes the bottom of the flip potential Fig.3c and goes with small fluctuations to $-1$ at large times. Fig.3.d shows that these fluctuations do not exceed $-1$.

To reach the top of the hill $\Phi = 1$ one has to increase the velocity, but since there is a restriction on the initial velocity $\dot\Phi(0)^2 \leq 2V(0)$, (the initial energy should be positive), one has to add a positive constant $V_0$ to the potential to be able to increase the initial velocity.

For large $M_p$ and a suitable $V_0$ there is a solution that starts from the top of one hill with a non-zero velocity and reach the top of other hill during an infinite time, Fig.4. In this case during the initial stages of evolution the field is near the top of the hill, $\Phi = -1$ and the acceleration is small. At later times the field begins to evolve more rapidly towards the local minimum of the flip potential and the equation state parameter $w$ becomes rather big. Finally, in very late time the field comes closed to the top of other hill, $\Phi = 1$

$$\Phi = 1 - Ae^{-\alpha t}, \qquad (2.24)$$

where $A$ is an arbitrary constant, $\alpha = (\frac{\sqrt{3V_0}}{M_p} + \sqrt{\frac{3V_0}{M_p^2} + 8})/2$ and a period of $w \simeq -1$, $w < -1$ begins. This period is infinitely long because the flip potential has the maximum at $\Phi = 1$.

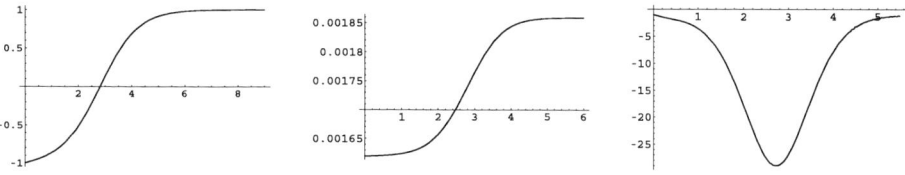

**FIGURE 4.** Plots of $\phi(t)$, $H(t)$ and $w(t)$ for $\phi(0) = -1, \dot\phi(0) = 0.1, V_0 = 0.02$

## 3. STABILITY

In this section we show that fluctuations of the scalar field around the true vacua are stable for suitable $q^2$, namely we show that the Euclidian action is positive defined.

For the usual Goldstone model in the flat space-time (2.23) fluctuations around one of minima $\Phi_0 = \pm 1$ correspond to massive excitations. In our case due to nonlocal factors in the interaction we get a different picture. Depending on parameter $q^2$ we get two excitations with different positive square mass or do not get any excitation at all.

Indeed, for fluctuations around $\Phi_0 = 1$

$$\Phi = \Phi_0 + \delta\Phi \tag{3.25}$$

we get the quadratic part of the action

$$S_0 = \int \delta\Phi \left( (q^2 \Box + 1)e^{-\Box} - 3 \right) \delta\Phi dx \tag{3.26}$$

To get the particle spectrum we are looking for solutions of the equation

$$-q^2 m^2 - 1 + 3e^{m^2} = 0 \tag{3.27}$$

with $= -k_0^2 + \vec{k}^2 \equiv k^2 = -m^2 < 0$. To find solutions of this equation we plot in Fig. 5 the function

$$f(x) = -q^2 x - 1 + 3e^x \tag{3.28}$$

and find its zeros.

We see that for $q^2 < q_{cr2}^2$ this curve has no positive zeros. For $q^2 > q_{cr2}^2$ there are two zeros, $f(m_i^2) = 0$ with $m_i^2 > 0$, $i = 1, 2$. Therefore there are no massive excitations for $q^2 < q_{cr2}^2$, and there are two massive particles for $q^2 > q_{cr2}^2$. There is one massive particle for $q_{cr}^2$.

For $q^2 < q_{cr1}$ the function $f(x)$ has also zero for a negative argument (see "cross-lines" on Fig.5) that corresponds to an appearance of tachyons.

Let us note that our propagator around new vacuum is non-standard one. If we expand $\exp(-\Box)$ in power of derivatives keeping only the first order terms on $\Box$ we get

$$S_0 \approx -\int \delta\Phi(k) \left( (q^2 - 1)k^2 + 2 \right) \delta\Phi(-k) dk \tag{3.29}$$

and we see that the case $q^2 < 1$ looks like as if the ghosts are appearing. In particular, this means that if we performing the Euclidean rotations $k^2 = -\omega^2 + \vec{k} \to k_E^2$, then our approximated Euclidean propagator $((q^2 - 1)k_E^2 + 2)$ for $q^2 < 1$ it is not positively defined. However, the full propagator that one gets from $S_0$ after the Euclidean rotations

$$S_0 = -\int \delta\Phi(k) \left( (q^2 k_E^2 - 1)e^{k_E^2} + 3 \right) \delta\Phi(-k) dk \tag{3.30}$$

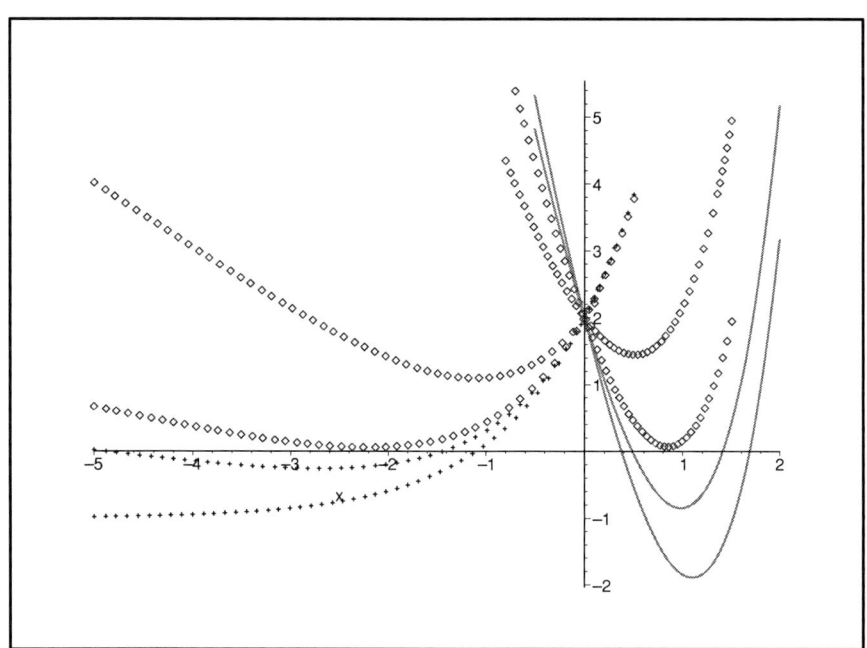

**FIGURE 5.** Plots of function $f(x) = -q^2 x - 1 + 3e^{2x}$ for different values of $q^2$: $q^2 < q_{cr1}^2$ (cross lines), $q_{cr1}^2 < q^2 < q_{cr2}^2$ (diamond lines), $q^2 > q_{cr2}^2$ (solid lines); $q_{cr1}^2 \simeq 0.33$, $q_{cr2}^2 \simeq 7$

is positive defined for $q_{cr1}^2 < q^2 < q_{cr2}^2$. Indeed,

$$\left((q^2 k_E^2 - 1)e^{k_E^2} + 3\right) = f(-k_E^2)e^{k_E^2} \tag{3.31}$$

and we see form Fig.5 that the function f(x) for negative arguments is positive for all negative arguments (positive $k_E^2$) if $q^2 > q_{cr1}^2$. Therefore, we see that the inverse propagator is strictly positive for $q_{cr2}^2 > q^2 > q_{cr1}^2$. It is interesting to note that for $q^2 = 0$ the inverse propagator $f(-k_E^2)e^{2k_E^2}$ is not positive.

## 4. CONCLUSION

We have studied the evolution of open GSO – NSR string tachyon in the Friedmann space-time. The corresponding solution in the flat space-time is known as a rolling tachyon and it describes the decay of the space filling D3 brane corresponding to the unstable perturbative vacuum to the local stable vacuum. We have performed calculations under the following approximations and assumptions:

- the level truncation and an approximation of a slow varying axillary field;

- a direct generalization of the tachyon nonlocal action to the Friedmann space-time;
- an effective local action approximation.

We have found that in the effective field theory approximation the equation of state parameter $w < -1$, i.e. one has a phantom Universe, but there is no problem with quantum instability without this approximation. We have found that to reach the nonperturbative vacuum one has to add to the action a brane tension which is larger that is required by the Sen hypothesis. This large brane tension can be interpreted as an effect of the closed string excitations.

It would be interesting to try to find an analog of this solution to a full SFT theory (without level truncation), in particular with a framework of Vacuum String Field Theory. VSFT is a version of usual SFT that is supposed to describe the theory at the minimum of the tachyon potential. Corresponding solutions for bosonic VSFT have been found recently in [35]. It would be also interesting to see if one has to shift the tension of D-brane to get a rolling solution in the Friedmann metric.

## ACKNOWLEDGMENTS

I would like to thank L. Bonora, B. Dragovich, A. Koshelev, S. Mukhanov, S. Vernov and I. Volovich for useful discussions. This work is supported in part by RFBR grant 05-01-00758 and INTAS grant 03-51-6346. Presented on: 1) QUARK-2004, Pushkinskie Gory, Russia, May 2004; 2) International Workshop Supersymmetries and Quantum Symmetries, Dubna, Russia, July 2005; 3) 2-nd conference on p-adic $M \bigcap \Phi$, Belgrade, Serbia, September, 2005; 4) FRG Workshop, Penn University, May, 2005; 5)2nd Vienna Central European Seminar, Frontiers in Astroparticle Physics, Vienna, November, 2005.

## REFERENCES

1. S.J. Perlmutter et al., Measurements of Omega and Lambda from 42 High-Redshift Supernovae, *Astroph. J.* **517** 565 (1999), `astro-ph/9812133`.
2. A. Riess et al., Observational Evidence from Supernovae for an Accelerating Universe and a Cosmological Constant, *Astron. J.* **116** 1009 (1998), `astro-ph/9805201`.
3. D.N. Spergel et al., First Year Wilkinson Microwave Anisotropy Probe (WMAP) Observations: Determination of Cosmological Parameters, *Astroph. J. Suppl.* **148** 175 (2003), `astro-ph/0302209`.
4. V. Sahni and A.A. Starobinsky, The Case for a Positive Cosmological Lambda-term, *Int. J. Mod. Phys.* **D9** 373 (2000), `astro-ph/9904398`.
5. V. Sahni, Dark Matter and Dark Energy, `astro-ph/0403324`.
6. R.A. Knop et al., New constraints on $\omega_m$, $\omega_\lambda$, and w from an independent set of eleven high - redshift supernovae observed with HST, `astro-ph/0309368`.
7. M. Tegmark al., The 3-d power spectrum of galaxies from the SDSS, *Astroph. J.* **606**, 702–740 (2004), `astro-ph/0310723`.
8. C. Armendariz-Picon, V. Mukhanov and P.J. Steinhardt, A Dynamical Solution to the Problem of a Small Cosmological Constant and Late-time Cosmic Acceleration, *Phys. Rev. Lett.* **85**, 4438 (2000), `astro-ph/0004134`.

9. Deffayet C., Dvali G. and Gabadadze G., Accelerated Universe from Gravity Leaking to Extra Dimensions, *Phys. Rev.* D **65**, 044023 (2002), astro-ph/0105068.
10. U. Alam, V. Sahni and A.A. Starobinsky, The case for dynamical dark energy revisited, astro-ph/0403687.
11. T. Padmanabhan, Accelerated expansion of the universe driven by tachyonic matter, *Phys. Rev.* **D66**, 021301 (2002), hep-th/0204150; T. Padmanabhan, T. Roy Choudhury, Can the clustered dark matter and the smooth dark energy arise from the same scalar field?, *Phys. Rev.* **D66**, 081301 (2002), hep-th/0205055.
12. R.R. Caldwell, A Phantom Menace? Cosmological consequences of a dark energy component with super-negative equation of state, *Phys. Lett.* B **545**, 23 (2002), astro-ph/9908168.
13. B. McInnes, The dS/CFT Correspondence and the Big Smash, *JHEP* **0208**, 029 (2002), hep-th/0112066.
14. R.R Caldwell, M. Kamionkowski M. and N.N. Weinberg, Phantom Energy and Cosmic Doomsday, *Phys. Rev. Lett.* **91**, 071301 (2003), astro-ph/0302506.
15. V.K. Onemli, R.P. Woodard, Super-Acceleration from Massless, Minimally Coupled $\phi^4$, *Class. Quant. Grav.* **19**, 4507 (2002), gr-qc/0204065; V.K. Onemli, R.P. Woodard, Quantum effects can render w<-1 on cosmological scales, *Phys. Rev.* D **70**, 107301 (2004), gr-qc/0406098.
16. S.M. Carroll, M. Hoffman and M. Trodden, Can the dark energy equation-of-state parameter w be less than -1?, *Phys. Rev.* D **68**, 023509 (2003), astro-ph/0301273.
17. A. Melchiorri, L. Mersini L, C.J. Odman and M. Trodden, The State of the Dark Energy Equation of State, *Phys. Rev.* D **68**, 043509 (2003), astro-ph/0211522.
18. J. Krakochvil, A. Linde, E. Linder and M. Shmakova, Testing the Cosmological Constant as a Candidate for Dark Energy, astro-ph/0312183.
19. P. Frampton, Dark energy - a pedagogic review, hep-th/0409166.
20. T. Padmanabhan, Cosmological Constant - the Weight of the Vacuum, *Phys. Rept.* **380**, 235–320 (2003), hep-th/0212290.
21. B. Feng, X. Wang, X. Zhang, Dark Energy Constraints from the Cosmic Age and Supernova, astro-ph/0404224; B. Feng, M. Li, Y.S. Piao, X. Zhang, Oscillating Quintom and the Recurrent Universe, astro-ph/0407432.
22. S. Nojiri and S. Odintsov, The final state and thermodynamics of dark energy universe, hep-th/0408170.
23. W. Fang, H.Q. Lu, Z.G. Huang and K.F. Zhang, Phantom Cosmology with Born-Infeld Type Scalar Field, hep-th/0409080.
24. K. Ohmori, A Review on Tachyon Condensation in Open String Field Theories, hep-th/0102085; I.Ya. Aref'eva, D.M. Belov, A.A. Giryavets, A.S. Koshelev, P.B. Medvedev, Noncommutative Field Theories and (Super)String Field Theories, hep-th/0111208; W. Taylor, Lectures on D-branes, tachyon condensation and string field theory, hep-th/0301094.
25. A. Sen, Tachyon Dynamics in Open String Theory, hep-th/0410103.
26. A. Sen, Rolling Tachyon, *JHEP* **0204**, 048 (2002), hep-th/0203211; A. Sen, Tachyon Matter, *JHEP* **0207**, 065 (2002), hep-th/0203265.
27. G.W. Gibbons, *Class. Quant. Grav.* **20**, S321–S346 (2003), hep-th/0301117.
28. G.N. Felder, L. Kofman, A. Starobinsky, *JHEP* **0209**, 026 (2002), hep-th/0208019.
29. A.V. Frolov, L. Kofman, A.A. Starobinsky, *Phys. Lett.* **B545**, 8–16 (2002), hep-th/0204187.
30. I.Ya. Aref'eva, L.V. Joukovskaya and A.S. Koshelev, Time Evolution in Superstring Field Theory on non-BPS brane. Rolling Tachyon and Energy-Momentum Conservation, hep-th/0301137.
31. I.Ya. Aref'eva, D.M. Belov, A.S. Koshelev, P.B. Medvedev, *Nucl. Phys.* B **638**, 3–20 2002, hep-th/0011117.
32. Ya. Volovich, Numerical study of nonlinear equations with infinite number of derivatives, *J. Phys. A* **36**, 8685–8702, (2003), math-ph/0301028.
33. L. Brekke, P.G.O. Freund, M. Olson, E. Witten, Non-archimedian string dynamics, *Nucl. Phys.* **B302**, 365 (1988).
34. N. Moeller and B. Zwiebach, Dynamics with infinitely many time derivatives and rolling tachyons, hep-th/0207107; H. Yang, Stress tensors in p-adic string theory and truncated OSFT, hep-th/0209197.
35. L. Bonora, C. Maccaferri, R.J. Scherer Santos, D.D. Tolla, Exact time-localized solutions in Vacuum String Field Theory, hep-th/0409063.

# A Note on Weil's Explicit Formula

Muharem Avdispahić* and Lejla Smajlović*

*Department of Mathematics, University of Sarajevo, Zmaja od Bosne 35,
71 000 Sarajevo, BOSNIA AND HERZEGOVINA
emails: mavdispa@pmf.unsa.ba, lejlas@pmf.unsa.ba*

**Abstract.** Taking Weil's point on adelic nature of explicit formulas, we extend the class of test functions to which his formula in Barner-Burnol version holds.

**Keywords:** Explicit formulas, Weil's functional, Hecke $L$-function.

## 1. INTRODUCTION

Let $k$ be a complete number field, $I$ its multiplicative idele group and $S_\infty$ the set of all archimedean prime spots in $k$. We shall assume that there are $r_1$ real and $r_2$ complex archimedean prime spots with local degrees $N_v = 1$ and $N_v = 2$ respectively. For $v \in S_\infty$, the valuation $\|a\|_v$ of the element $a \in k_v$ is normalized through $\|a\|_v = |a|_v^{N_v}$, where $|\cdot|_v$ is the absolute value induced on $\mathbb{R}$ or $\mathbb{C}$. By $U_v$ we denote the set of units in $k_v$ and by $\mathfrak{O}_v$ its ring of integers. Every $v \notin S_\infty$ comes from the unique prime ideal $\mathfrak{p}$ of $k$ such that $\|a\|_v = (\mathbf{N}\mathfrak{p})^{ord_\mathfrak{p} a}$, $a \in k$, where $\mathbf{N}\mathfrak{p}$ is the cardinality of $\mathfrak{O}_v/\mathfrak{p}$.

For a Hecke character $\chi$ of the idele classes, we have the equation $\chi(a) = \prod \chi_v(a_v)$, where $a = (a_v) \in I$ and $\chi_v$ are characters on local multiplicative groups $k_v^*$. The latter product is well defined, since $\chi_v(a_v) \neq 1$ for only finitely many places $v$. By $S_\chi$ we will denote the finite set of all non-archimedean prime spots of $k$ that are ramified for $\chi$, i.e. such that $\chi_v(U_v) \neq 1$ for $v \in S_\chi$. However, for each $v \in S_\chi$ there exists a natural number $m$ such that $\chi_v(1 + \mathfrak{p}^m) = 1$. The smallest such number $f_v$ is called the ramification degree of $\chi_v$, and the ideal $\mathfrak{f}_{\chi_v} = \mathfrak{p}^{f_v}$ is the conductor of $\chi_v$. The conductor of $\chi$ is the ideal $\mathfrak{f}_\chi = \prod_{v \in S_\chi} \mathfrak{f}_{\chi_v}$.

If $v \notin S_\chi$, the value of $\chi(\pi)$ does not depend on the choice of the uniformizer $\pi$. We shall denote it by $\chi(\mathfrak{p})$.

For $v \in S_\infty$, the local character $\chi_v$ is given by $\chi_v(a) = \left(\frac{a}{|a|_v}\right)^{m_v} \|a\|_v^{i\varphi_v}$. The character $\chi$ is normalized so that $\sum_{v \in S_\infty} N_v \varphi_v = 0$, i.e. $\chi$ takes value 1 on $\mathbb{R}^+$ canonically embedded in the group of idele classes.

Let $d_\chi = \mathbf{N}(\mathscr{D}\mathfrak{f}_\chi)$, where $\mathscr{D}$ denotes the local different and put $A = 2\pi^{-2r_2}\pi^{-r_1}d_\chi$. The number $\delta$ is given by $\mathbf{N}\mathscr{D} = (\mathbf{N}\mathfrak{p})^\delta$. The additive and the multiplicative Haar measure on

$k_v$ and $k_v^*$ ($v \notin S_\infty$), respectively, are chosen so that $\int_{\mathfrak{D}_v} da = (\mathbf{N}\mathscr{D})^{-\frac{1}{2}}$ and $\int_{U_v} d^*a = \log \mathbf{N}\mathfrak{p}$.

The Hecke $L$-function, $L(s,\chi)$, $s = \sigma + it$, $t \in \mathbb{R}$ is defined for $\sigma > 1$ as an absolutely convergent product

$$L(s,\chi) = \prod_{\mathfrak{p} \notin S_\chi} \left(1 - \frac{\chi(\mathfrak{p})}{\mathbf{N}\mathfrak{p}^s}\right)^{-1}.$$

Some authors (e. g., [1, pp. 5-6]) find it convenient to consider the Hecke $L$-function as valued in the distributions on the adeles $\mathbb{A}$, and use results of Tate concerning its analytical continuation. This approach yields a new formulation of the Weil explicit formula [1] for the Hecke $L$-function, valid for the class of smooth test functions with compact support. According to Burnol, the latter formulation signals a link with quantum fields, while the Hilbert-Polya idea leads rather to quantum mechanics. In Section 4 we shall consider the relation between our explicit formula and the explicit formulas obtained by Haran [2] and Burnol.

In [3], K. Barner took a path of enlarging the class of test functions to which the Weil explicit formula applies and gave a new, concise form to Weil's functional. J. Jorgenson and S. Lang [4] have proved that the Weil explicit formula is a special case of the explicit formula for the fundamental class of functions, introduced in [4].

Conditions posed on a test function in the explicit formula can be divided into two classes: conditions at zero and conditions at infinity. A condition posed at zero is concerned with the evaluation of Weil's functional, while those posed at infinity control the asymptotic behavior of the Mellin transform. In [5], we proved that Barner's conditions [3] at infinity can be weakened to include the class of functions of bounded $\phi$-variation. Similar to [6], here we shall allow test functions not necessarily continuous at zero (in contrast to Jorgenson-Lang's [4, FOU 3, on p. 62]).

## 2. THE MAIN RESULT

We will denote test functions by $f$ and $F$, assuming that they are related by $f(x) = F(-\log x)$, $x > 0$.

The main result of the paper is the following

**Theorem 2.1.** *Let $F$ be a regularized function such that*

1. $F(x) \in \phi BV(\mathbb{R}) \cap L^1(\mathbb{R})$
2. $F(x) e^{(\frac{1}{2}+\varepsilon)|x|} \in \phi BV(\mathbb{R}) \cap L^1(\mathbb{R})$, *for some $\varepsilon > 0$ and*
3. $F(x) e^{i\varphi_v x} + F(-x) e^{-i\varphi_v x} - 2F(0) = O\left(|\log|x||^{-\alpha}\right)$, *as $x \to 0$, for some $\alpha > 2$ and all $v \in S_\infty$.*

Then, the explicit formula

$$\lim_{T \to \infty} \sum_{|\operatorname{Im} \rho| < T} \operatorname{ord}(\rho) M_{\frac{1}{2}} f(\rho) = \delta_\chi \int_1^\infty \left( f(u) + f\left(\frac{1}{u}\right) \right) \left( u^{\frac{1}{2}} + u^{-\frac{1}{2}} \right) \frac{du}{u} \quad (1)$$

$$- \sum_{v \in S_\chi} \left[ \int_{\|u\|_v = 1} \left( f_v\left(\frac{1}{u}\right) - f_v(1) \right) \frac{d^*u}{\|1-u\|_v} - \delta_v f_v(1) \log N\mathfrak{p} \right]$$

$$- \sum_{v \notin S_\chi \cup S_\infty} \int_{\|u\|_v \neq 1} f_v\left(\frac{1}{u}\right) \frac{d^*u}{\|1-u\|_v} - \sum_{v \in S_\infty} N_v f_v(1) (\gamma + \log N_v \pi)$$

$$+ \sum_{v \in S_\infty} N_v \int_1^\infty \left[ \frac{2 f_v(1)}{u^2 - 1} - \frac{f_v\left(\frac{1}{u}\right) u^{2-|m_v|-N_v} + f_v(u) u^{2-|m_v|}}{u^2 - 1} \right] \frac{du}{u}$$

holds.

Here, $f_v(u) = f(\|u\|_v) \chi_v^{-1}(u) \|u\|_v^{-\frac{1}{2}}$, for all valuations $v$, $\gamma = -\Gamma'(1)$ is the Euler constant and

$$\delta_\chi = \begin{cases} 1 & \text{if } \chi \text{ is a principal character,} \\ 0 & \text{if } \chi \text{ is a non-principal character.} \end{cases}$$

$M_{\frac{1}{2}} f$ denotes the translate by $\frac{1}{2}$ of the Mellin transform of the function $f$.

The sum on the left hand side of (1) is taken over all zeros $\rho$ of the Hecke $L$-function in the critical strip, where $\operatorname{ord}(\rho)$ denotes the order of $\rho$.

The class $\phi BV(\mathbb{R})$ is the class of functions of bounded $\phi$-variation on $\mathbb{R}$, where $\phi$ is a continuous, convex function on $[0, \infty)$, increasing from zero to infinity, such that

$$\lim_{x \to 0} \frac{\phi(x)}{x} = 0, \quad \lim_{x \to \infty} \frac{\phi(x)}{x} = \infty, \quad \text{and} \quad \sum_n \phi^{-1}\left(\frac{1}{n}\right) \left(\frac{1}{n}\right)^{\frac{1}{p}} < \infty,$$

for some $p > 1$. To ensure $\phi BV$ to be a linear space, we also assume that $\phi$ satisfies the Orlicz $(\Delta_2)$ condition, i.e. there exist positive constants $x_0$ and $C$ ($C \geq 2$) such that for $0 \leq x \leq x_0$ the inequality $\phi(2x) \leq C\phi(x)$ holds (see [7]).

## 3. PROOF

Functions $L(s, \chi)$ and $L(s, \overline{\chi})$ are meromorphic functions of finite order. The logarithmic derivative of the function $L(s, \chi)$ (see e.g.[8]) has an Euler sum representation

$$-d \log L(s, \chi) = -\frac{L'}{L}(s, \chi) = \sum_{\mathfrak{p} \notin S_\chi} \sum_m (\log N\mathfrak{p}) \chi(\mathfrak{p}^m) (N\mathfrak{p})^{-ms}$$

that converges absolutely in the half plane $\mathrm{Re}(s) > 1$ and uniformly in any half plane of the form $\mathrm{Re}(s) \geq 1+\varepsilon > 1$.

In [8], the following functional equation is proved

$$W(\chi)G(s,\chi)L(s,\chi) = G(1-s,\overline{\chi})L(1-s,\overline{\chi}),$$

where $W(\chi)$ is a constant of modulus 1,

$$G(s,\chi) = A^{\frac{s}{2}} \prod_{v \in S_\infty} \Gamma\left(\frac{s_v}{2}\right), \quad G(1-s,\overline{\chi}) = A^{\frac{1-s}{2}} \prod_{v \in S_\infty} \Gamma\left(\frac{\widehat{s}_v}{2}\right),$$

$s_v = s_v(\chi) = N_v(s + i\varphi_v(\chi)) + |m_v(\chi)|$ and $\widehat{s}_v = \widehat{s}_v(\chi) = N_v(1 - s - i\varphi_v(\chi)) + |m_v(\chi)|$.
If, for a fixed $\chi$, we put $G(s) = G(s,\chi)$, $\widetilde{G}(s) = G(s,\overline{\chi})$, $Z(s) = L(s,\chi)$ and $\widetilde{Z}(s) = L(s,\overline{\chi})$ the functional equation becomes:
$Z(s)\Phi(s) = \widetilde{Z}(1-s)$, where

$$\Phi(s) = W(\chi)A^{s-\frac{1}{2}} \prod_{v \in S_\infty} \frac{\Gamma\left(\frac{s_v}{2}\right)}{\Gamma\left(\frac{\widehat{s}_v}{2}\right)}.$$

In the terminology of [4, pp. 36-37] and [9], the fudge factor $\Phi$ of the last equation is of regularized product type of the reduced order $(0,0)$.

This shows that the triple $\left(Z, \widetilde{Z}, \Phi\right)$ is in the fundamental class of functions ([4, pp. 45-46]) with $\sigma_0 = \sigma_0' = 1$, and the functions $Z$ and $\widetilde{Z}$ are of reduced order at most $(0,1)$. Now, it is possible to follow the lines of the proof of [5, Th. 6.1.] to obtain

$$\lim_{T \to \infty} \sum_{|\mathrm{Im}\,\rho| < T} \mathrm{ord}(\rho) M_{\frac{1}{2}} f(\rho) = \delta_\chi \int_{-\infty}^{\infty} F(x)\left(e^{\frac{x}{2}} + e^{-\frac{x}{2}}\right) dx$$

$$- \sum_{\mathfrak{p} \notin S_\infty} \sum_n \frac{\log N\mathfrak{p}}{N\mathfrak{p}^{\frac{n}{2}}}\left[\chi(\mathfrak{p})^n f(N\mathfrak{p}^n) + \chi(\mathfrak{p})^{-n} f\left(\frac{1}{N\mathfrak{p}^n}\right)\right] + W_\Phi(F),$$

where

$$W_\Phi(F) = \lim_{T \to \infty} \frac{1}{\sqrt{2\pi}} \int_{-T}^{T} \widehat{F}(t) \frac{\Phi'}{\Phi}\left(\frac{1}{2} + it\right) dt$$

is the Weil functional. Since

$$\frac{\Phi'}{\Phi}(s) = \log A + \sum_{v \in S_\infty} \frac{N_v}{2}\left(\frac{\Gamma'}{\Gamma}\left(\frac{s_v}{2}\right) + \frac{\Gamma'}{\Gamma}\left(\frac{\widehat{s}_v}{2}\right)\right),$$

the application of the Fourier inversion theorem for the class $HBV$, proved in [10, Th. 3.1.], yields

$$\lim_{T \to \infty} \frac{1}{\sqrt{2\pi}} \int_{-T}^{T} \widehat{F}(t) \log A \, dt = F(0) \log A.$$

To avoid posing a condition on the function $F$ at zero, we notice that

$$\lim_{T\to\infty} \frac{1}{\sqrt{2\pi}} \int_{-T}^{T} \widehat{F}(t) \left( \frac{\Gamma'}{\Gamma}\left(\frac{s_v}{2}\right) + \frac{\Gamma'}{\Gamma}\left(\frac{\hat{s}_v}{2}\right) \right) dt =$$

$$= \lim_{T\to\infty} \frac{1}{\sqrt{2\pi}} \int_{-T+\varphi_v}^{T+\varphi_v} \widehat{F}_{\chi_v}(t) \left( \frac{\Gamma'}{\Gamma}\left(a + i\frac{N_v t}{2}\right) + \frac{\Gamma'}{\Gamma}\left(a - i\frac{N_v t}{2}\right) \right) dt$$

$$= \lim_{T\to\infty} \frac{1}{\sqrt{2\pi}} \int_{-T}^{T} \left( \widehat{F}_{\chi_v}(t) + \widehat{F}_{\chi_v}(-t) \right) \frac{\Gamma'}{\Gamma}\left(a + i\frac{N_v t}{2}\right) dt,$$

where $\quad F_{\chi_v}(x) = F(x)e^{i\varphi_v x} \quad$ and $\quad a = \frac{N_v}{4} + \frac{|m_v|}{2}.$

(Obviously, the integral over $(-T-\varphi_v, -T+\varphi_v)$ is $o(1)$, as $T \to \infty$.)

Application of the general Parseval formula proved in [10, Th.6.1.] implies that under conditions 1. and 3. posed on the test function we have

$$W_\Phi(F) = F(0)\log d_\chi - \sum_{v\in S_\infty} N_v F(0) \log N_v \pi +$$

$$+ \sum_{v\in S_\infty} \int_0^\infty \left[ \frac{N_v F(0)}{x} - (F_{\chi_v}(x) + F_{\chi_v}(-x)) \cdot \frac{e^{\left(\frac{2-|m_v|}{N_v} - \frac{1}{2}\right)x}}{1 - e^{-\frac{2x}{N_v}}} \right] e^{-\frac{2x}{N_v}} dx. \quad (2)$$

Now, we observe that

$$F(0)\log d_\chi = \sum_{v\in S_\chi} \left[ \int_{\|u\|_v = 1} \left( f_v(1) - f_v\left(\frac{1}{u}\right) \right) \frac{d^*u}{\|1-u\|_v} + \delta f_v(1) \log N\mathfrak{p} \right].$$

Similarly, the equality

$$\sum_{\mathfrak{p}\notin S_\chi} \sum_n \frac{\log N\mathfrak{p}}{N\mathfrak{p}^{\frac{n}{2}}} \left[ \chi(\mathfrak{p})^n f(N\mathfrak{p}^n) + \chi(\mathfrak{p})^{-n} f\left(\frac{1}{N\mathfrak{p}^n}\right) \right] = \sum_{v\notin S_\chi \cup S_\infty} \int_{\|u\|_v \neq 1} f_v\left(\frac{1}{u}\right) \frac{d^*u}{\|1-u\|_v}$$

holds once the test function satisfies conditions 1. and 2. of the theorem.

It is left to rewrite the sum over $v \in S_\infty$ on the right hand side of (2). First, notice that

$$N_v \gamma F(0) = \int_0^\infty \left( \frac{F(0)}{1 - e^{-\frac{2x}{N_v}}} - \frac{N_v F(0)}{x} \right) e^{-\frac{2x}{N_v}} dx.$$

Therefore, after the change of variable $u = e^{\frac{x}{N_v}}$, we obtain

$$\int_0^\infty \left[\frac{N_v F(0)}{x} - (F_{\chi_v}(x) + F_{\chi_v}(-x)) \cdot \frac{e^{\left(\frac{2-|m_v|}{N_v} - \frac{1}{2}\right)x}}{1 - e^{\frac{-2x}{N_v}}}\right] e^{\frac{-2x}{N_v}} dx + N_v \gamma F(0)$$

$$= N_v \int_1^\infty \left[\frac{2f_v(1)}{u^2 - 1} - \frac{f_v\left(\frac{1}{u}\right) u^{2-|m_v|-N_v} + f_v(u) u^{2-|m_v|}}{u^2 - 1}\right] \frac{du}{u}.$$

This completes the proof of the theorem.

## 4. CONCLUDING REMARKS

The most important special case of the Weil explicit formula is the case $k = \mathbb{Q}$, where the corresponding $L$–function, with $\chi = 1$, is the Riemann zeta function $\zeta$. In this case, the triple $(\zeta, \zeta, \eta)$ belongs to the fundamental class of functions in a symmetric case, considered in [6]. Here,

$$\eta(s) = \pi^{-s+\frac{1}{2}} \frac{\Gamma\left(\frac{s}{2}\right)}{\Gamma\left(\frac{1-s}{2}\right)}.$$

The explicit formula in a symmetric case, proved in [6, Th. 4.1.], yields the following corollary.

**Corollary 4.1.** *Let a function $F$ be a regularized function such that*
1. $(F(x) + F(-x)) e^{\left(\frac{1}{2}+\varepsilon\right)|x|} \in \phi BV(\mathbb{R}) \cap L^1(\mathbb{R})$, *for some $\varepsilon > 0$ and*
2. $F(x) + F(-x) - 2F(0) = O\left(|\log|x||^{-\alpha}\right)$, *as $x \to 0$, for some $\alpha > 2$.*

*Then, the formula*

$$\lim_{T \to \infty} \sum_{|\operatorname{Im}\rho| < T} \operatorname{ord}(\rho) M_{\frac{1}{2}} f(\rho) = \int_1^\infty \left(f(u) + f\left(\frac{1}{u}\right)\right)\left(u^{\frac{1}{2}} + u^{-\frac{1}{2}}\right) \frac{du}{u}$$

$$- \sum_p \int_{\|u\|_p \neq 1} f_p\left(\frac{1}{u}\right) \frac{d^*u}{\|1-u\|_v} - f_\infty(1)(\gamma + \log \pi) \quad (3)$$

$$+ \int_1^\infty \left[\frac{2f_\infty(1)}{u^2 - 1} - \frac{f_\infty\left(\frac{1}{u}\right) u + f_\infty(u) u^2}{u^2 - 1}\right] \frac{du}{u}$$

*holds.*

Here, the sum on the left is taken over all zeros $\rho = \beta + i\gamma$ of the Riemann zeta function, the sum on the right is taken over all primes $p$, where $\|\cdot\|_p$ denotes $p-$ adic valuation and $d^*u$ is induced by the multiplicative Haar measure $\mu_p$ on $\mathbb{Q}_p$, such that the measure of units is $\log p$. The function $f_p$ is given by

$$f_p(u) = f\left(\|u\|_p\right) \|u\|_p^{-\frac{1}{2}}$$

(note that for $p = \infty$ $\|\cdot\|_\infty$ is the classical absolute value on $\mathbb{R}$).

This form of the explicit formula differs slightly from the formula obtained by S. Haran in [2], for a class of smooth, compactly supported functions on $\mathbb{R}^+$. Actually, this formula differs from the Haran's one in the term at infinity. Simple calculations show that if condition 2. on a test function is replaced by the stronger one

$$F(x) - F(0) = O\left(|\log|x||^{-\alpha}\right), \quad \text{as } x \to 0 \ (\alpha > 2), \tag{4}$$

the last integral on the right hand side of (3) can be expressed as

$$-\left[\int_1^\infty \frac{u^2 f_\infty(u) - f_\infty(1)}{u^2 - 1} \frac{du}{u} + \int_0^1 \frac{f_\infty(u) - u f_\infty(1)}{1 - u^2} du\right],$$

what is the term at infinity given in [2, p. 701].

Note that (4) implies continuity of the test function $F$ at zero, while Cor. 4.1 is more general.

J-F. Burnol [11] has shown that Haran's term at infinity can be written as

$$-\left[\log 2 + \int_{\frac{1}{2}}^\infty \frac{f_\infty\left(\frac{1}{u}\right) - f_\infty(1)}{|1 - u|} d^*u + \int_{-\infty}^{\frac{1}{2}} \frac{f_\infty\left(\frac{1}{u}\right)}{|1 - u|} d^*u\right], \tag{5}$$

where

$$d^*u = \frac{du}{2|u|}.$$

It is easy to see that our term at infinity can also be written in this form, under assumption (4).

An analogous observation is true in the case of a general number field $k$ and a Hecke character $\chi$. It is possible, under assumption (4), to rewrite the Weil functional from (1) and express it as the sum of local terms analogous to (5), see [1, pp. 13-14]. Note that our class of test functions (even the one with the condition at zero replaced by (4)) is still larger than the one considered in [1].

# REFERENCES

1. J.F. Burnol, The explicit formula and a propagator, electronic manuscript, math.NT/9809119.
1. A. Weil, Sur les "formules explicites" de la theorie des nombres premiers, *Comm. Sem. Math. Univ. Lund*, 252–265 (1952).
2. S. Haran, Riesz potentials and explicit sums in arithmetic, *Invent. math.* **101**, 697–703 (1990).
3. K. Barner, On A. Weil's explicit formula, *J. reine angew. Math.* **323**, 139–152 (1981).
4. J. Jorgenson and S. Lang, *Explicit formulas for regularized products and series*, Springer Lecture Notes in Mathematics **1593**, 1994.
5. M. Avdispahić, L. Smajlović, Explicit formula for a fundamental class of functions, *Bull. Belg. Math. Soc. Simon Stevin* **12**, no. 4, 569–587 (2005).
6. M. Avdispahić, L. Smajlović, An explicit formula and its application to the Selberg trace formula, to appear in *Monatsh. fuer Math.*
7. J. Musielak, W. Orlicz, On generalized variations (I), *Studia Math.* **18**, 11–41 (1959).
8. S. Lang, *Algebraic numbers*, Adison-Wesley, Reading, Mass. 1964.
9. J. Jorgenson, S. Lang, *Basic analysis of regularized series and products*, Springer Lecture Notes in Mathematics, **1564**, 1993.
10. M. Avdispahić, L. Smajlović, $\phi$-variation and Barner-Weil formula, *Math. Balkanica* **17**, no. 3–4, 267–289, (2003).
11. J.F. Burnol, The explicit formula in simple terms, electronic manuscript, math.NT/9810169.

# The Use of Path Integral Ideals: Deriving the Euler Summation Formula for Path Integrals

Aleksandar Bogojević*, Antun Balaž* and Aleksandar Belić*

*Institute of Physics, P. O. Box 57, 11001 Belgrade, SERBIA AND MONTENEGRO
emails: alex@phy.bg.ac.yu, antun@phy.bg.ac.yu, abelic@phy.bg.ac.yu

**Abstract.** We present and comment on a new quantity that we have recently introduced: the path integral ideal. The new quantity governs the flow of a discrete quantum theory to its continuum limit. Path integral ideals satisfy a unique integral equation – the distinction between different quantum theories being in the boundary conditions. An asymptotic expansion of this equation has led to the derivation of a generalization of Euler's summation formula for path integrals. The new analytical method has brought about a systematic improvement of the convergence of path integrals. Applied to numerical procedures, the new analytical input has resulted in the speedup of numerical simulations by many orders of magnitude. On the analytical side, the integral equation for ideals may turn out to be a useful setting for extending the obtained results to a wider setting – e.g. to p-adic valued theories and theories on non-commuting space-times.

**Keywords:** Path integral, Quantum theory, Effective action, Asymptotic expansion.
**PACS:** 05.30.-d, 03.65.Db, 03.65.-w .

## 1. INTRODUCTION

Path integrals present a rich and flexible formalism for dealing with quantum and statistical theories [1, 2] that has proven extremely useful for handling symmetries, deriving non-perturbative results, establishing connections between different theories [3, 4], and extending the quantization procedure to ever more complicated systems. They have served as catalysts for the exchange of key ideas between different areas of physics, most notably high energy and condensed matter physics [5, 6]. Today, analytical and numerical approaches to path integrals [7, 8, 9, 10] play important roles not only in physics but also in chemistry and materials science, and are acquiring a prominent role in mathematics and modern finance [11].

Further development of the path integral method is constrained by the small number of solvable models, as well as by our rather limited knowledge of their precise mathematical properties. In fact, most of our knowledge is negative, e.g. we know which trajectories do not contribute to the path integral rather than which do. One of the few positive statements concerning path integrals is that relevant trajectories exhibit stochastic self-similarity [1]. As a result they have non-trivial fractal dimension and jaggedness [12, 13]. Researchers working on numerical approaches to path integrals have successfully utilized these kinematic consequences of self-similarity to produce efficient path-generating algorithms [9, 10].

In a recent series of papers [14, 15] we have investigated the dynamical implications of stochastic self-similarity by studying the relation between discretizations of path integrals with different coarseness. This has resulted in a systematic analytical procedure that may be used to reduce path integral error to $O(\varepsilon_N^p)$ for arbitrary $p \in \mathbb{N}$, where $\varepsilon_N$ is the discrete time step. Note that $\varepsilon_N = T/N$, $T$ being the time of propagation and $N$ the discretization coarseness. This reduction of error brings about a substantial increase in the speed of numerical algorithms. Additional information can be found on our web site [16]. Self-similarity played a crucial role in this procedure in that it allowed us to derive an integral equation relating discretized theories viewed at different coarseness and to solve it in terms of an asymptotic series. The asymptotic expansion, however, implies that the obtained method is directly applicable only for $\varepsilon_N < 1$.

The fact that we can arbitrarily decrease the error points to the possibility that one can extend the formalism and obtain exact information (i.e not given as a power series in $\varepsilon_N$, and so valid even for large values of $\varepsilon_N$) about the continuum theory. Large $\varepsilon_N$ corresponds to long times of propagation, precisely what interests us in quantum field theory (or in modern finance). Equivalently, in condensed matter and materials science this corresponds to the physically most interesting region of small temperatures. Large $\varepsilon_N$ behavior is also central for determining the energy spectrum of a given model, and as such is applicable in many areas of physics [11] (e.g. atomic and molecular physics, quantum dots).

The central quantity we will work with in this paper is the path integral ideal which governs the flow of a generic discrete theory to the continuum. The ideal was first introduced in [17]. In that paper we showed that the flow to the continuum is classified according to the degree of divergence of the potential at spatial infinity. In addition we derived certain asymptotic properties of ideals. We will here show how the formalism of ideals may be used to derive a generalization of the Euler summation formula to path integrals [18].

## 2. EULER SUMMATION FORMULA FOR ORDINARY INTEGRALS

The current status of the development of the path integral formalism is quite similar to that of ordinary integrals before the setting up of integration theory by Riemann. In those days integrals were calculated directly from the defining formula, i.e. one looked at a specific discretization of the integral (Darboux sum), attempted to do the sum explicitly, and finally tried to calculate the continuum limit. For example,

$$I[f] \equiv \int_0^T f(t)dt = \lim_{N\to\infty} I_N[f] \text{ , where } I_N[f] = \sum_{n=1}^N f(t_n)\varepsilon_N , \qquad (1)$$

$\varepsilon_N = T/N$ and $t_n = n\varepsilon_N$. It goes without saying that done this way, even the simplest ordinary integrals presented a challenge. The mathematicians of the 18th century did not have computers at their disposal or the development of integration theory might have come much later, i.e. they might have succumbed to doing brute force numerical calculations of integrals of all but the simplest functions. The problem with these hypothetical numerical calculations would have been two fold: they would have been inefficient (the discretized sums converge slowly to the continuum value), and they would have worked (thus quite probably slowing down the further development of integration theory). Luckily, this early numerical road was not open. The last great step in the development of integration before Riemann was made by Euler.

Discretization is not unique. This makes it possible to change $f(t)$ to some other function (adding terms proportional to $\varepsilon_N$, $\varepsilon_N^2$, etc.) without changing the integral. Let us assume that $f^*(t)$ is such an equivalent function with the added property that the sums $I_N[f^*]$ do not depend on $N$. In fact we shall present a way of explicitly constructing $f^*(t)$ for any given $f(t)$. We first look at the simple case of $f(t) = 1$. Now

$$I_N[1] = \sum_{n=1}^{N} \varepsilon_N = T , \tag{2}$$

which is already $N$-independent. Hence, in this case, all the additional terms vanish. Note that $f^*(t)$ is completely determined by the original function $f(t)$ (and by $\varepsilon_N$), so that the additional terms necessarily depend only on the derivatives $f'$, $f''$, etc.

The second step is to take $f(t) = t$. In this case we get

$$I_N[t] = \sum_{n=1}^{N} t_n \varepsilon_N = \frac{N(N+1)}{2} \frac{T^2}{N^2} = \frac{T^2}{2} + \frac{T^2}{2N} . \tag{3}$$

From this it follows that $I_N[t - \frac{\varepsilon_N}{2}] = \frac{T^2}{2}$. Therefore, up to $f''$ and higher derivatives of $f$ that all vanish for linear $f(t)$, we have $f^*(t) = f(t) - \frac{\varepsilon_N}{2} f'(t)$.

We continue this procedure by looking at $f(t) = t^2$. In this case we find

$$I_N[t^2] = \sum_{n=1}^{N} t_n^2 \varepsilon_N = \frac{N(N+1)(2N+1)}{6} \frac{T^3}{N^3} = \frac{T^3}{3} + \frac{T^3}{2N} + \frac{T^3}{6N^2} . \tag{4}$$

It follows that $I_N[t^2 - \varepsilon_N t_n - \frac{2}{3}\varepsilon_N^2] = \frac{T^3}{3}$. In terms of $f^*$ this gives $f^*(t) = f(t) - \frac{\varepsilon_N}{2} f'(t) - \frac{2\varepsilon_N^2}{3} f''(t) + \ldots$. The additional terms now depend on higher powers of $\varepsilon_N$ as well as on higher derivatives and are determined by considering $I_N[t^3]$, and so on. In this way we have constructed a procedure for finding $f^*(t)$ for any given $f(t)$. Remembering that $I_N[f^*]$ does not depend on $N$ we find

$$\int_0^T f(t) dt = \sum_{n=1}^{N} f(t_n) \varepsilon_N - \frac{\varepsilon_N}{2} \sum_{n=1}^{N} f'(t_n) \varepsilon_N - \frac{2\varepsilon_N^2}{3} \sum_{n=1}^{N} f''(t_n) \varepsilon_N + \ldots . \tag{5}$$

This is the well-known Euler summation formula. We may also write it more compactly as

$$I[f] = I_N[f^{(p)}] + O(\varepsilon_N^p) , \tag{6}$$

where $f^{(p)}$ is the truncation of $f^*$ to the first $p$ terms. The Euler formula gives the analytical relation between integrals and their discretized sums. Looked at numerically, this formula allows us to increase the speed of convergence of discretized expressions to the continuum limit. For example, in the defining relation the discretized expressions differ from the continuum by a term of order $O(1/N)$. By using the Euler sum formula with $p$ terms we can reduce that error to $O(1/N^p)$. All that is needed to do this is that the integrand is differentiable $p-1$ times. the following sections we will generalize the above approach to path integrals.

## 3. GENERAL PROPERTIES OF PATH INTEGRALS

In the functional formalism the quantum mechanical amplitude $A(a,b;T) = \langle b|e^{-T\hat{H}}|a\rangle$ is given in terms of a path integral which is simply the $N \to \infty$ limit of the $(N-1)$-fold integral expression

$$A_N(a,b;T) = \left(\frac{1}{2\pi\varepsilon_N}\right)^{\frac{N}{2}} \int dq_1 \cdots dq_{N-1} e^{-S_N} . \qquad (7)$$

The Euclidean time interval $[0,T]$ has been subdivided into $N$ equal time steps of length $\varepsilon_N = T/N$, with $q_0 = a$ and $q_N = b$. $S_N$ is the naively discretized action of the theory. We focus on actions of the form

$$S = \int_0^T dt \left(\frac{1}{2}\dot{q}^2 + V(q)\right) , \qquad (8)$$

whose naive discretization is simply

$$S_N = \sum_{n=0}^{N-1} \left(\frac{\delta_n^2}{2\varepsilon_N} + \varepsilon_N V_n\right) , \qquad (9)$$

where $\delta_n = q_{n+1} - q_n$, $V_n = V(\bar{q}_n)$, and $\bar{q}_n = \frac{1}{2}(q_{n+1} + q_n)$. We use units in which $\hbar$ and particle mass equal 1.

As was the case with ordinary integrals the definition of the path integrals also makes it necessary to make the transition from the continuum to the discretized theory, a process that is far from unique. For theories described by eq. (8) we have the freedom to choose any point in $[q_n, q_{n+1}]$ in which to evaluate the potential without changing physics – the discretized amplitudes do differ, but they tend to the same continuum limit. The calculations we present turn out to be simplest in the mid-point prescription where the potential $V$ is evaluated at $\bar{q}_n$. A more important freedom related to our choice of discretized action has to do with the possibility of introducing additional terms that explicitly vanish in the continuum limit. Actions with such additional terms will be called effective. For example, the term $\sum_{n=0}^{N-1} \varepsilon_N \delta_n^2 g(\bar{q}_n)$, where $g$ is regular when $\varepsilon_N \to 0$, does not change the continuum physics since it goes over into $\varepsilon_N^2 \int_0^T dt \, \dot{q}^2 g(q)$, i.e. it vanishes as $\varepsilon_N^2$. Such terms do not change the physics, but they do affect the speed of convergence. A systematic study of the relation between different discretizations of the same path integral will allow us to explicitly construct a series of effective actions with progressively faster convergence to the continuum. Before we do this we will parallel the derivation in the previous section and derive some general properties of the best effective action.

The amplitude $A(a,b;T)$ of some theory with action $S$ satisfies

$$A(a,b;T) = \int dq_1 \cdots dq_{n-1} A(b, q_{n-1}; \varepsilon_N) \cdots A(q_1, a; \varepsilon_N), \qquad (10)$$

for all $N$. This general relation is a direct consequence of the linearity of states in a quantum theory. In analogy with ordinary integrals let us now suppose that there exists an effective action $S^*$ that is equivalent to $S$ (i.e that leads to the same continuum limit for all path integrals) with the additional property that its $N$-fold discretized amplitude $A_N^*(a,b;T)$ does not depend on $N$, i.e. that satisfies

$$A_N^*(a,b;T) = A(a,b;T). \qquad (11)$$

As was the case in the previous section we will in fact construct a general procedure for evaluating this effective action. For actions of the form given in eq. (8) we may write the amplitude as

$$A(q_{n+1}, q_n; \varepsilon_N) = \left(\frac{1}{2\pi\varepsilon_N}\right)^{\frac{1}{2}} \exp\left(-\frac{\delta_n^2}{2\varepsilon_N}\right) \mathscr{A}(q_{n+1}, q_n; \varepsilon_N), \qquad (12)$$

where the reduced amplitude $\mathscr{A} \to 1$ as $\varepsilon_N \to 0$. Writing $S_N^*$ as

$$S_N^* = \sum_{n=0}^{N-1} \left(\frac{\delta_n^2}{2\varepsilon_N} + \varepsilon_N W_n^*\right), \qquad (13)$$

and using eq. (7), (10) and (11) we find

$$\exp(-\varepsilon_N W_n^*) = \mathscr{A}(q_{n+1}, q_n; \varepsilon_N). \qquad (14)$$

Note that $W_n^*$ is reminiscent of some effective potential, so it should depend on $\bar{q}_n$, however, from the above relation we see that it must also depend on $\delta_n$. In addition, $W^*$ also has an explicit dependence on the discrete time step $\varepsilon_N$, hence

$$W_n^* = W^*(\delta_n, \bar{q}_n; \varepsilon_N). \qquad (15)$$

As we have seen, the above functional form is a direct consequence of the linearity of quantum theory. The equivalence of $S$ and $S^*$ implies that $W^* \to V(\bar{q})$ when $\varepsilon_N$ and $\delta$ go to zero. The final general property of $W^*$ follows from the reality of amplitudes in the Euclidean formalism. Using the hermiticity of the Hamiltonian we find $A(a,b;T) = A(a,b;T)^\dagger = \langle b|e^{-T\hat{H}}|a\rangle^\dagger = \langle a|e^{-T\hat{H}}|b\rangle = A(b,a;T)$. In terms of $W^*$ this gives us

$$W^*(\delta_n, \bar{q}_n; \varepsilon_N) = W^*(-\delta_n, \bar{q}_n; \varepsilon_N), \qquad (16)$$

or, said another way, only even powers of $\delta_n$ are present in the expansion of $W^*$:

$$W^*(\delta_n, \bar{q}_n; \varepsilon_N) = g_0(\bar{q}_n; \varepsilon_N) + \delta_n^2 g_1(\bar{q}_n; \varepsilon_N) + \delta_n^4 g_2(\bar{q}_n; \varepsilon_N) + \ldots . \tag{17}$$

All the functions $g_k$ are regular in the $\varepsilon_N \to 0$ limit. The link to the starting theory is now simply $g_0(\bar{q}_n; \varepsilon_N) \to V(\bar{q}_n)$ as $\varepsilon_N$ goes to zero. This concludes the general properties of $W^*$. The remaining properties will be analyzed in the following section by studying the relation of discretizations of different coarseness.

We next derive an equation for path integral ideals by studying the relation between the $2N$-fold and $N$-fold discretizations of the same theory. From eq. (7) we see that we can write the $2N$-fold amplitude as an $N$-fold amplitude given in terms of a new action $\widetilde{S}_N$ determined by

$$e^{-\widetilde{S}_N} = \left(\frac{2}{\pi \varepsilon_N}\right)^{\frac{N}{2}} \int dx_1 \cdots dx_N \, e^{-S_{2N}}, \tag{18}$$

where $S_{2N}$ is the $2N$-fold discretization of the starting action. We have written the $2N$-fold discretized coordinates $Q_0, Q_1, \ldots, Q_{2N}$ in terms of $q$'s and $x$'s in the following way: $Q_{2k} = q_k$ and $Q_{2k-1} = x_k$. Note that we have $q_0 = a$, $q_N = b$, while the $N-1$ remaining $q$'s play the role of the dynamical coordinates in the $N$-fold discretized theory. The $x$'s are the $N$ remaining intermediate points that we integrate over in eq. (18). It is not difficult to see that if we use the naively discretized action $S_N$ one obtains for $\widetilde{S}_N$ an expression that is not of the same form as $S_N$. Having in mind the results derived at the beginning of this section it is best to use the effective action

$$S_N^* = \sum_{n=0}^{N-1} \left(\frac{\delta_n^2}{2\varepsilon_N} + \varepsilon_N W^*(\delta_n, \bar{q}_n; \varepsilon_N)\right), \tag{19}$$

which gives the same result for both the $2N$-fold and $N$-fold discretizations. Therefore, in this case we get

$$e^{-S_N^*} = \left(\frac{2}{\pi \varepsilon_N}\right)^{\frac{N}{2}} \int dx_1 \cdots dx_N \, e^{-S_{2N}^*}. \tag{20}$$

From this it is straightforward to show that path integral ideals satisfy

$$\exp(-\varepsilon_N W^*(\delta, q; \varepsilon_N)) = \sqrt{\frac{2}{\pi \varepsilon_N}} \int_{-\infty}^{+\infty} dy \, e^{-2y^2/\varepsilon_N} \tag{21}$$

$$\times \exp\left(-\frac{\varepsilon_N}{2} W^*\left(\frac{\delta}{2} - y, q + \frac{\delta}{4} + \frac{y}{2}; \frac{\varepsilon_N}{2}\right) - \frac{\varepsilon_N}{2} W^*\left(\frac{\delta}{2} + y, q - \frac{\delta}{4} + \frac{y}{2}; \frac{\varepsilon_N}{2}\right)\right).$$

We end this section by briefly commenting on some general properties of the above integral equation. All quantum theories have been reduced to a single integral equation. We must first solve this equation and only then impose boundary conditions ($\varepsilon_N \to 0$ limit) that link us to a specific theory. The integral equation is easily solved for quadratic ideals, in which case we recover the usual free particle and harmonic oscillator results. Eq. (21) is an good starting point for developing various approximation schemes and for analyzing non-perturbative properties of ideals [17]. We also believe that thus integral equation will have natural extensions to fermions, higher dimensions, as well as to more complex settings such as p-adic theories and quantum theories on non-commuting space-times. The general solution of the above integral equation in the form of an asymptotic expansion will be given in the following section.

## 4. EULER SUMMATION FORMULA FOR PATH INTEGRALS

In order to solve eq. (21) in the form of an asymptotic expansion we write it as

$$e^{-\varepsilon_N W^*(\delta_n, \bar{q}_n; \varepsilon_N)} = \left(\frac{2}{\pi \varepsilon_N}\right)^{\frac{1}{2}} \int_{-\infty}^{+\infty} dy \, \exp\left(-\frac{2}{\varepsilon_N} y^2\right) F\left(\bar{q}_n + y; \frac{\varepsilon_N}{2}\right), \quad (22)$$

where

$$-\frac{2}{\varepsilon_N} \ln F(x; \varepsilon_N) = g_0\left(\frac{q_{n+1}+x}{2}; \varepsilon_N\right) + g_0\left(\frac{x+q_n}{2}; \varepsilon_N\right) \quad (23)$$

$$+ (q_{n+1}-x)^2 g_1\left(\frac{q_{n+1}+x}{2}; \varepsilon_N\right) + (x-q_n)^2 g_1\left(\frac{x+q_n}{2}; \varepsilon_N\right) + \ldots$$

Note the integral in eq. (22) is in a form that is ideal for an asymptotic expansion [19]. The time step $\varepsilon_N$ is playing the role of small parameter (in complete parallel to the role $\hbar$ plays in standard semi-classical, or loop, expansion). As is usual, the above asymptotic expansion is carried through by first Taylor expanding $F\left(\bar{q}_n + y; \frac{\varepsilon_N}{2}\right)$ around $\bar{q}_n$ and then by doing the remaining Gaussian integrals. Assuming that $\varepsilon_N < 1$ (i.e. $N > T$) we have

$$g_0(\bar{q}_n; \varepsilon_N) + \delta_n^2 g_1(\bar{q}_n; \varepsilon_N) + \delta_n^4 g_2(\bar{q}_n; \varepsilon_N) + \ldots = \quad (24)$$

$$= -\frac{1}{\varepsilon_N} \ln\left[\sum_{m=0}^{\infty} \frac{F^{(2m)}\left(\bar{q}_n; \frac{\varepsilon_N}{2}\right)}{(2m)!!} \left(\frac{\varepsilon_N}{4}\right)^m\right].$$

Note that $F^{(2m)}(x; \varepsilon_N)$ denotes the corresponding derivative with respect to $x$. All that remains is to calculate these expressions using eq. (23) and to expand all the $g_k$'s around the mid-point $\bar{q}_n$. This is a straight forward though tedious calculation. In this paper we will illustrate the general procedure for calculating $S^*$ by explicitly giving its expansion to order $\varepsilon_N^3$:

$$g_0(\bar{q}_n; \varepsilon_N) = g_0\left(\bar{q}_n; \frac{\varepsilon_N}{2}\right) + \varepsilon_N\left[\frac{1}{4} g_1\left(\bar{q}_n; \frac{\varepsilon_N}{2}\right) + \frac{1}{32} g_0''\left(\bar{q}_n; \frac{\varepsilon_N}{2}\right)\right]$$

$$+ \varepsilon_N^2 \left[\frac{3}{16} g_2\left(\bar{q}_n; \frac{\varepsilon_N}{2}\right) - \frac{1}{32} g_0'^2\left(\bar{q}_n; \frac{\varepsilon_N}{2}\right) + \frac{1}{2048} g_0^{(4)}\left(\bar{q}_n; \frac{\varepsilon_N}{2}\right)\right.$$

$$\left. + \frac{3}{128} g_1''\left(\bar{q}_n; \frac{\varepsilon_N}{2}\right)\right]$$

$$g_1(\bar{q}_n; \varepsilon_N) = \frac{1}{4} g_1\left(\bar{q}_n; \frac{\varepsilon_N}{2}\right) + \frac{1}{32} g_0''\left(\bar{q}_n; \frac{\varepsilon_N}{2}\right) \quad (25)$$

$$+ \varepsilon_N \left[ \frac{3}{8} g_2\left(\bar{q}_n; \frac{\varepsilon_N}{2}\right) + \frac{1}{1024} g_0^{(4)}\left(\bar{q}_n; \frac{\varepsilon_N}{2}\right) - \frac{1}{64} g_1''\left(\bar{q}_n; \frac{\varepsilon_N}{2}\right) \right]$$

$$g_2(\bar{q}_n; \varepsilon_N) = \frac{1}{16} g_2\left(\bar{q}_n; \frac{\varepsilon_N}{2}\right) + \frac{1}{6144} g_0^{(4)}\left(\bar{q}_n; \frac{\varepsilon_N}{2}\right) + \frac{1}{128} g_1''\left(\bar{q}_n; \frac{\varepsilon_N}{2}\right).$$

In the above relations we expanded $g_0$ up to $\varepsilon_N^2$, $g_1$ up to $\varepsilon_N$, etc. We also disregarded all the higher $g_k$'s. The reason for this is that the short time propagation of any theory satisfies $\delta_n^2 \propto \varepsilon_N$ while the $g_k$ term enters the action multiplied by $\delta_n^{2k}$. In general, if we wish to expand the effective action to $\varepsilon_N^p$ we need to evaluate only $g_0$ (up to $\varepsilon_N^{p-1}$) and the remaining $p-1$ functions $g_k$ (up to $\varepsilon_N^{p-1-k}$). The task of calculating the effective action to large powers of $\varepsilon_N$ is time-consuming and is best done with the help of a standard package for algebraic calculations such as Mathematica. Using Mathematica we determined the corresponding expressions for $p \leq 9$.

Although the above system of recursive relations is non-linear, it is in fact quite easy to solve if we remember that the system itself was derived via an expansion in $\varepsilon_N$. Having this in mind we first write all the functions as expansions in powers of $\varepsilon_N$ that are appropriate to the level $p$ we are working at. For $p = 3$, we have

$$g_0(\bar{q}_n; \varepsilon_N) = V(\bar{q}_n) + \varepsilon_N R_1(\bar{q}_n) + \varepsilon_N^2 R_2(\bar{q}_n)$$
$$g_1(\bar{q}_n; \varepsilon_N) = R_3(\bar{q}_n) + \varepsilon_N R_4(\bar{q}_n) \qquad (26)$$
$$g_2(\bar{q}_n; \varepsilon_N) = R_5(\bar{q}_n).$$

Putting this into the Eq. (25) we determine the functions $R_1$ to $R_5$ in terms of $V$. The $p = 3$ level solution equals

$$g_0 = V + \varepsilon_N \frac{V''}{12} + \varepsilon_N^2 \left[ -\frac{V'^2}{24} + \frac{V^{(4)}}{240} \right]$$

$$g_1 = \frac{V''}{24} + \varepsilon_N \frac{V^{(4)}}{480} \qquad (27)$$

$$g_2 = \frac{V^{(4)}}{1920}.$$

Note that $W^*$ depends only on the initial potential $V$ and its derivatives (as well as on $\varepsilon_N$). One can similarly calculate the effective action $S^*$ to any desired level $p$. We denote the $p$ level truncation of the effective action as $S^{(p)}$. $S^{(p)}$ has the property that its $N$-fold amplitudes deviate from the continuum expressions as $O(\varepsilon_N^p)$

$$A(a,b;T) = A_N^{(p)}(a,b;T) + O(\varepsilon_N^p). \qquad (28)$$

Comparing this to eq. (6) we see that we have just derived the generalization of the Euler summation formula to path integrals. Just as with the ordinary Euler formula it gives the relation between path integrals and their discretizations to any given precision.

It is important to note that one solves for the effective action at level $p$ but once for all theories, i.e. the solution that is found holds for all initial potentials. The only requirement for the level $p$ solution is that the starting potential is differentiable $2p-2$ times. Solutions for larger values of $p$ are a bit more cumbersome, however, they are just as easy to use in simulations. We have found that the growth in complexity of the effective actions with increasing $p$ has little effect on computation time for $p \leq 4$, while simulations with $p = 9$ are roughly ten times slower due to this. However, this is an extremely small price to pay for a gain of eight orders of magnitude in the speed of convergence. Expressions for effective actions up to $p = 9$ can be found on our web site [16]. The analytical derivations presented work equally well in both the Euclidean

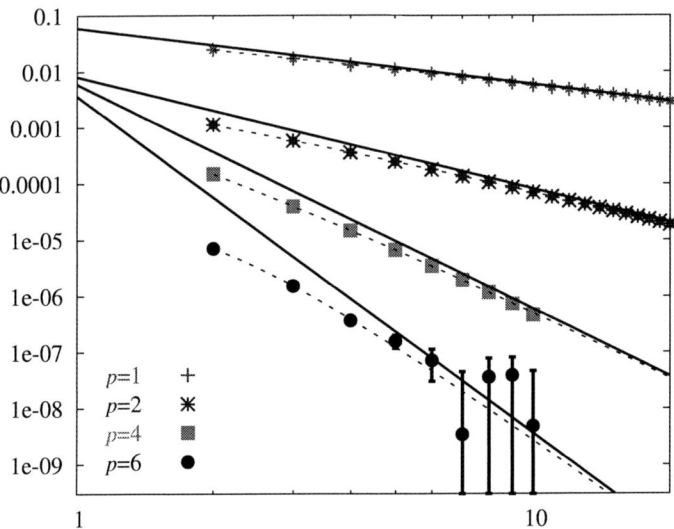

**FIGURE 1.** Deviations from the continuum limit $|A_N^{(p)} - A|$ as functions of $N$ for $p = 1,2,4$ and 6 for an anharmonic oscillator with quartic coupling $\lambda = 10$, time of propagation $T = 1$ from $a = 0$ to $b = 1$. $N_{MC}$ was $9.2 \cdot 10^9$ for $p = 1, 2$, $9.2 \cdot 10^{10}$ for $p = 4$, and $3.68 \cdot 10^{11}$ for $p = 6$. Dashed lines correspond to appropriate $1/N$ polynomial fits to the data. Solid lines give the leading $1/N$ behavior. The level $p$ curve has a $1/N^p$ leading behavior.

and Minkowski formalism (with appropriate $i\varepsilon_N$ regularization), i.e. they are directly applicable to quantum systems as well as to statistical ones. However, the Monte Carlo simulations used to numerically document our analytical results necessarily needed to be done in the Euclidean formalism. We analyzed in detail several models: the anharmonic oscillator with quartic coupling $V = \frac{1}{2}q^2 + \frac{\lambda}{4!}q^4$ and a particle moving in a modified Pöschl-Teller potential over a wide range of parameters. In all cases we found agreement with eq. (28). Fig. 1 illustrates this behavior in the case of an anharmonic oscillator. We see that the $p$ level data indeed differs from the continuum amplitudes as a polynomial starting with $1/N^p$. The deviations from the continuum limit $|A_N^{(p)} - A|$ become exceedingly small for larger values of $p$ making it necessary to use ever larger values of $N_{MC}$ so that the MC statistical error does not mask these extremely small deviations. For $p = 6$ we see that although we used an extremely large number of MC samples ($N_{MC} = 3.68 \cdot 10^{11}$) the statistical errors become of the same order as the deviations already at $N \gtrsim 8$. For $p = 9$ this is the case even for $N = 2$, i.e. we already get the continuum limit within a MC error of around $10^{-8}$.

## 5. CONCLUSION

A general quantum theory may be written in terms of a quantity which we designate the path integral ideal. We have determined the integral equation satisfied by ideals and have solved it in terms of an asymptotic series in discretized time step. This solution represents a generalization of the well known Euler summation formula to path integrals and leads to the speedup of numerical simulations of path integrals of a general theory by many orders of magnitude. We have also briefly commented on the use of ideals as a natural starting point for extensions of quantization to more complex settings such as p-adic theories.

## ACKNOWLEDGMENTS

We acknowledge the financial support from the Ministry of Science and Environmental Protection of the Republic of Serbia through projects. No. 1486 and No. 1899.

## REFERENCES

1. R.P. Feynman and A.R. Hibbs, *Quantum Mechanics and Path Integrals*, McGraw-Hill, New York, 1965.
2. R.P. Feynman, *Statistical Mechanics*, W.A. Benjamin, New York, 1972.
3. C. Itzykson and J.B. Zuber, *Quantum Field Theory*, McGraw-Hill, New York, 1980.
4. S. Coleman, *Aspects of Symmetry*, Cambridge University Press, 1985.
5. C. Itzykson and J.M. Drouffe, *Statistical Field Theory* Cambridge University Press, 1991.
6. G. Parisi, *Statistical Field Theory* Addison Wesley, New York, 1988.
7. J.A. Barker and D. Henderson, *Rev. Mod. Phys.* **48**, 587 (1976).
8. J.A. Barker, *J. Chem. Phys.* **70**, 2914 (1979).
9. E.L. Pollock and D.M. Ceperley, *Phys. Rev. B* **30**, 2555 (1984).
10. D.M. Ceperley, *Rev. Mod. Phys.* **67**, 279 (1995).
11. H. Kleinert, *Path Integrals in Quantum Mechanics, Statistics, Polymer Physics, and Financial Markets*, World Scientific, 2004.
12. H. Kröger, *Phys. Rep.* **323**, 81 (2000).
13. A. Bogojević, A. Balaž and A. Belić, *Phys. Lett. A*, **345**, 258 (2005).
14. A. Bogojević, A. Balaž and A. Belić, *Phys. Rev. Lett.* **94**, 180403 (2005).
15. A. Bogojević, A. Balaž and A. Belić, *Phys. Rev. B* **72**, 064302 (2005).
16. URL http://scl.phy.bg.ac.yu/speedup/.
17. A. Bogojević, A. Balaž and A. Belić, *Phys. Rev. E* **72**, 036128 (2005).
18. A. Bogojević, A. Balaž and A. Belić, *Phys. Lett. A* **344**, 84 (2005).
19. A. Erdelyi, *Asyptotic Expansions*, Dover Publications, New York, 1956.

# Analysis of Business Connections Utilizing Theory of Topology of Random Graphs

Jennifer Q. Trelewicz* and Igor V. Volovich[†]

*IBM Almaden Research Center, 650 Harry Road, San Jose, CA 95120, USA
[†]Steklov Mathematical Institute, Russian Academy of Sciences,
Gubkin St. 8, GSP-1, 119991, Moscow, RUSSIA
email: volovich@mi.ras.ru

**Abstract.** A *business ecosystem* is a system that describes interactions between organizations. In this paper, we build a theoretical framework that defines a model which can be used to analyze the business ecosystem. The basic concepts within the framework are *organizations*, *business connections*, and *market*, that are all defined in the paper. Many researchers analyze the performance and structure of business using the workflow of the business. Our work in business connections answers a different set of questions, concerning the monetary value in the business ecosystem, rather than the task-interaction view that is provided by workflow analysis.

We apply methods for analysis of the topology of complex networks, characterized by the concepts of small path length, clustering, and scale-free degree distributions. To model the dynamics of the business ecosystem we analyze the notion of the state of an organization at a given instant of time. We point out that the notion of state in this case is fundamentally different from the concept of state of the system which is used in classical or quantum physics. To describe the state of the organization at a given time one has to know the probability of payments to contracts which in fact depend on the future behavior of the agents on the market. Therefore methods of p-adic analysis are appropriate to explore such a behavior. Microeconomic and macroeconomic factors are indivisible and moreover the actual state of the organization depends on the future. In this framework some simple models are analyzed in detail. Company strategy can be influenced by analysis of models, which can provide a probabilistic understanding of the market, giving degrees of predictability.

**Keywords:** Social organizations, Economics, business and management.
**PACS:** 89.65.Ef, 89.65.Gh, 89.75.-k.

## 1. INTRODUCTION

To analyze business ecosystems, models such as workflow diagrams, Petri nets, and others are generally employed. A general property of these approaches is that they represent complex networks. Recent empirical and theoretical studies of some real complex networks such as the Internet, ecological networks, cellular networks, phone-call networks etc., uncover some remarkable general properties of the networks [1]. In this paper, we develop a model of the business ecosystem as a complex network that is represented as a generalized random graph. This allows the application of methods for analysis of the topology of complex networks, characterized by the concepts of small path length ("Small Worlds" [2]), clustering, and scale-free degree distributions.

A *business ecosystem* is a system that describes interactions between organizations.

The concept of business ecosystems has been mentioned in the literature, and is often credited to Iansiti [3]. The basic concepts of *organizations*, *business connections*, and *market* are defined in what follows. For now, we propose some analytical questions of the business ecosystem in a more natural context, to motivate the theoretical development that follows. We consider only those interactions which are concerned with the flow of monetary value, where monetary value is defined in Section 4.2.

It should be noted that trying to predict the market is foolhardy; e.g., will the value of a particular traded good in the market go up or down at a given time? Analysis of models can, however, provide a probabilistic understanding of the market, giving degrees of predictability (e.g., confidence intervals) of the models. This can be used to make business decisions for resiliency or adaptability, etc. (See Section 3.2.)

Let the business ecosystem be given. For a given time interval, the following questions concern the dynamical behavior of the business ecosystem. If the interval consists of a single point in time, the following questions are static considerations. All of these questions can be answered for one or more given markets to learn how to describe the business ecosystem at a policy level, e.g., do the dynamics display cyclic behavior that can be used to influence macro-economic policy?

- How many organizations are in the business ecosystem during the interval? This gives one measure of the approximate size of the market(s), and its change during the interval.
- How many business connections are in the business ecosystem? This gives one measure of the approximate interactivity of the market(s), and the change during the interval.
- What is the graphical structure of a typical organization, e.g., the typical number of business connections?
- How should this dynamical behavior be normalized for increases in world population, globalization, or other factors of changing environment external to the business ecosystem?

Of course, many more questions of the business ecosystem may be formulated, but these are intended to give a flavor to the analysis and model that we propose. Some of these questions are examined in more detail in Section 3.2. A recent paper [4] discusses the insufficiency of business process analysis for answering certain types of questions, and these questions demonstrate further scenarios where this insufficiency holds.

All of these questions may concern either the extraction of a generalized model from empirical data, or the construction of a general mathematical theory of business ecosystems. We are addressing both in our research. The goal of our research is to provide models and analysis that can augment existing methods for formulating micro- and macro-economic policies.

Analyzing the business ecosystem for these questions can be facilitated by a model with a structure that allows us to map finding the answers these questions to the application of specific algorithms to the model.

# 2. TERMINOLOGY

In this section we will define the terminology and structures that will define the model that we use to answer certain questions about businesses.

## 2.1. Business ecosystem graph

The business ecosystem is represented as a graph. Organizations are represented by the components of the graph, that possesses additional structure, described in Section 2.2.

Define an organization to be a collection of people that can make legal agreements (contracts) with other, disjoint collections of people. A contract is defined as a legal document concerning the possibility of transaction of monetary value. At one level of hierarchy, organizations can include for-profit businesses, non-profit businesses, individual people as consumers (i.e., separate from a single-person business that a person may create), and the State. The State is a special organization, in that the State defines the legal framework under which contracts can be made, businesses registered, and taxes paid. More is said about the State in Section 4.1, where we explain that the State is represented by a vertex that is connected by business connection edge to all other vertices. This means that if orientation of these edges is not considered, then every business ecosystem within such a State-specified legal framework is represented by a connected graph.

The set of all organizations under consideration is denoted $V$. We only consider in this version of our model legal organizations (i.e., those registered within the legal registration framework of the State) and legal business transactions. The State maintains a list of registered organizations, and these registration indices are required in our model to be unique. Arbitrarily, we denote the State by the index 1, so that the State is denoted $v_1$, and any other organization is denoted $v_n$, where $n \in \mathbb{N} \setminus \{1\}$ is its registration number. Put $N \subset \mathbb{N}$ the set of all registration numbers.

Define a *business connection* to be a relationship between organizations, of the two general classes that follow:

- *Collaborative relationships* may be defined by the flow of monetary value between organizations, and may be measured by the amount and direction of monetary value flow. Note that business connections can occur between organizations that are not for-profit businesses; e.g., the payment of taxes by an individual person to the State is considered to be a business connection in this model. There are many methods to measure flow of monetary value. For the purposes of our model, we use gross revenue associated with the transaction. Section 4.3 contains a longer discussion of sales and business transactions.
  Business connections are oriented edges, reflecting the direction of primary flow of monetary value, which can be paid for goods or services. The collection of all such edges is denoted $C_{M_0}$.
  We define this context implicitly as the *collaborative market*, which is important in the context of competitive markets in what follows.
- *Competitive relationships* are defined by competition on the market or in a tender.

Competition between two sellers involves monetary value flow to one seller that displaces some part of the monetary value flow to the other seller. The competitive relationship does not, in our definition, involve monetary value flow directly between the two sellers. For this reason, competitive relations between organizations are, in practice, more difficult to measure than collaborative relationships, but very important in obtaining meaningful analysis from the model.

The context in which these monetary value flows interact between sellers and buyers, is called the *market* for the competition. The concept of market is not needed rigorously for the theory, but is necessary for interpretation of the model.

A competitive relationship between two organizations is represented by an edge between the two vertices representing the organizations. This edge has neither orientation nor weight, since the edge indicates that both organizations are both selling in one market for a particular good. The set of all such edges for a particular market $M$ under consideration will be denoted $C_M$. If the model is used to analyze the business ecosystem in the context of several competitive markets at once, then there may be several such $C_M$, one for each market.

Competition is more difficult to measure accurately than contracts, but can still be measured to an order meaningful for the model. Competition in a tender is recorded formally, making such competition straightforward to measure. Competition on the market is not necessarily as formally registered; however, organizations usually announce their intention to participate in a market through advertising, and this intention is even now measured by market analysts.

Note that a business connection is a member of $N \times N$. In the competitive markets, we maintain uniqueness of the specification of a business connection by requiring that for a pair $(n_1, n_2) \in N \times N$ that $n_1 \leq n_2$.

The collection $G = \{V, C_{M_0}, \ldots, C_{M_n}\}$ (for $n \geq 0$ competitive markets $M_1 \ldots M_n$) comprises the graph of the business ecosystem.

The application of the model is not restricted to markets for the trade of tangible goods between separate enterprises. In Section 4.4, we describe several other examples of collaborative and competitive relations that can be represented in the model.

## 2.2. Graph dynamics

In the prequel, we described the static structure of the model. In this section we introduce the dynamics of the model, which allows time dependence on the set of vertices and the sets of edges. Consider $T \subset \mathbb{R}$, the time interval of the model. We assume that the State maintains not only a list but a history of registered organizations through $T$, and that a registration is unique through all of $T$. Note that intuitively, a new or defunct organization can have no formal relationship with the State before its inception or after its demise, allowing for organizations to exist in the model only for some interval subset of $T$.

In terms of time scale, one may explore the question of what is the meaning of a very small time interval versus a very long time interval, since a very short time interval will show only a snapshot of business connections. This may be understood as follows:

- A short time scale showing a kind of kinetic state of the ecosystem, with some irreversibility in the interactions between organizations. In physics, this would be governed by the Boltzmann equation.
- A long time scale can show a type of heterodynamics of the ecosystem, including equilibrium states of the ecosystem.

Put $\mathscr{P}(N)$ the power set (the family of all subsets) of $N$. Introduce $\mathscr{F} = \{F \in \mathscr{P}(N) : v_1 \in F\}$, where $v_1$ is the State. The set of vertices in the market at any given time is a member of $\mathscr{F}$ given by $V : T \to \mathscr{F}$. Then $V(t)$ is the natural dynamic extension of $V$. Note that the market is connected with these collections of organizations $V(t)$, since $V(t)$ describes all organizations that have the potential for collaborative or competitive position at given time $t$.

Let $\mathscr{C} = \mathscr{P}(N \times N)$. Then the set of all business connections (collaborative or competitive) at a given time $t$ is a member of $\mathscr{C}$. The set of business connections in a given market $M_j$ at any given time is a member of $\mathscr{C}$ given by $C_{M_j} : T \to \mathscr{C}$ for $j \in 0 \ldots n$ and $n \geq 0$ competitive markets $M_1 \ldots M_n$. Note that $C_{M_0}$ has the additional property that $\forall t \in T$ and $\forall v \in V(t)$ with $v \neq v_1$, $(v_1, v) \in C_{M_0}$; i.e., the State has a collaborative relationship with every organization.

Now we consider how to define the business connections between two organizations in $V(t)$. Let $t_1, t_2 \in T$ be given such that $\exists v_1, v_2 \in V(t_1) \cap V(t_2)$. If for $T' \subset [t_1, t_2]$ there exists continuously a business connection between $v_1, v_2$ that is defined by some edge $\gamma$, then we identify $\gamma(t')$ for all $t' \in T'$. Note that we allow that for some $\gamma$ at time $t \in T'$, the monetary flow on $\gamma$ may be zero. For example, a warranty on a good or service carries with it the potential for flow of negative monetary value, from the seller to the buyer, and thus extends the lifetime of the business connection until the warranty contract is expired, even if the flow of monetary value after the sale is zero.

This naturally introduces for $t \in T$

$$G(t) = \{V(t), C_{M_0}(t), \ldots, C_{M_n}(t)\}$$

for $n \geq 0$ competitive markets $M_1 \ldots M_n$. The static model describes only the business ecosystem at one infinitesimal point in time, but does not capture the important relationships in the ecosystem that change with time, such as multiple monetary flows between organizations.

Consider the set $C_{M_0}(t)$ of collaborative business connections. There is additional structure that can be introduced into this time-dependent set of collaborative edges. Consider $\rho \in \cup_{t \in T} C_{M_0}(t)$, where the union is the set of all edges that exist at one or more points in time. Note that this union is still countable, which can facilitate the application of certain algorithms to the model. Countability follows from the countability of $N \times N$, so that any members of $\mathscr{C}$, which includes the union of any subset of $\mathscr{C}$, is thus countable. Each such edge $\rho$ has associated with it an indicator function $\varphi_\rho : T \to \{0, 1\}$ defined by

$$\varphi_\rho(t) = \begin{cases} 0, & \rho \notin C_{M_0}(t) \\ 1, & \rho \in C_{M_0}(t) \end{cases}$$

Define $T_\rho = \{t \in T : \rho \in C_{M_0}(t)\}$. There is also associated with $\rho$ a revenue function $r_\rho : T \to \mathbb{R}$, defined by the monetary value flowing on the edge $\rho$ $\forall t \in T_\rho$, and with $r_\rho$ defined to be zero $\forall t \in T \setminus T_\rho$. Note that if $r_\rho(t) < 0$, this indicates a flow of monetary value in the direction opposite of the edge orientation at time $t$. The *lifetime* of the edge is described by $\varphi_\rho$, where each interval on which $\varphi_\rho(t) = 1$ indicates the existence of a contract that can facilitate the flow of monetary value between organizations.

This $\varphi_\rho(t)$ is not simply the indicator function of the smallest single interval of $T$ that contains supp $r_\rho$. Consider the warranty example above. Another example of a collaborative business connection is long-term purchasing agreements, or futures, in which a series of discrete payments is made during the life of a contract. In this case, $\varphi(t) = 1$ during the length of the contract, even when no actual payment is being made.

In principle, it is possible to define $\rho(t)$ as an edge that changes, for example, from $(v_1, v_2)$ to $(v_1, v_3)$ with time, indicating an impact of a competitive relationship between $v_2$ and $v_3$. We do not define such $\rho(t)$ in our model, because of the difficulty of reliably identifying such competitive relationship impacts with real data. This connects to the discussion above about the difficulty of measurement of competitive relationships.

## 2.3. Stochastic graph dynamics

Assume that the State allows random behavior in the market; specifically, that the State allows some degree of independence in organizations, so that the creation and demise of organizations and/or their business connections is not completely under the control of the State. Then it is not possible to deterministically predict $G(t')$ from $\{G(t) : t < t'\}$.

This random behavior suggests introducing $\{\Omega, \Gamma, P\}$, where $\Omega$ is probability space, $\Gamma$ is the set of subsets of $\Omega$, $P$ is a probability measure on $\Gamma$. This is a classical Kolmogorov probability space. Then the collection of organizations is more correctly defined by $V : T \times \Omega \to \mathscr{F}$, thus $V(t, \omega)$. By natural extension, we thus have

$$G(t, \omega) = \{V(t, \omega), C_{M_0}(t, \omega), \ldots, C_{M_n}(t, \omega)\}.$$

Thus, we have represented the business ecosystem by a stochastical, dynamical system on a star-like graph.

From this extension, we may use stochastic approximation in non-commutative probability [5].

## 2.4. Quantization

Both time and money values in the model will exhibit some degree of quantization, especially where empirical measurements in the market are concerned. We describe this possibility in this section. However, it should be noted that for certain types of analysis, continuous models are more useful.

Time is quantized by the nature of measurement in the ecosystem. The smallest time unit is the smallest unit for which the information for the data can be collected. This specific unit will depend on the organizations and business connections in the model, since the data required for the model will depend on these.

One may think that such measurement can be made simple by taking quarterly measurements of the performances of publicly-traded companies, for example, if the rules of the businesses' bourses require quarterly reports. However, it may be that these bourses allow businesses to define the boundaries of the fiscal reporting year, and thus the boundaries of the fiscal reporting quarter. This requires some renormalization of time between organizations in the ecosystem. However, given the possibility that enterprises may be able to exploit peculiarities of reporting rules to manipulate the numbers in their reports, one should realize that any model of a business connection ecosystem will be, by its very nature, both incomplete and inaccurate. The essence of the model is to obtain understanding of the interactions in the model without requiring perfect measurement. Physical models are no different, for example, because of sources of noise in measurement.

Money is also quantized in the model. The quantization unit may be the smallest unit of currency in the economy; e.g., the Euro cent. However, business results reporting rules, and tax laws, may impose different minimal units, such as whole units of currency, such as one Euro. This unit will differ by the economic and regulatory context of the ecosystem, and should be considered by anyone building such a model.

Discretization of time allows for analyzing the behavior of $G(t)$ to find if there is a Markovian property; i.e., if

$$P(G(t)|G(t-1)) = P(G(t)|G(t-1), G(t-2), \ldots),$$

where $P(A|B)$ denotes probability of $A$ given prior information $B$.

Such analysis also lead naturally to attempts to predict some degree of market behavior, at least to the degree that can provide guidance to the managers of the enterprises in making market-related decisions.

## 3. ILLUSTRATIVE EXAMPLE AND CONTEXT

In this section, we give an example of stochastic market models, to help to illustrate the concepts. We also discuss briefly how the application of mathematical models for the stochastic behavior imply certain characteristics on the context of the business ecosystem. This consideration is important for the analysis of real systems.

### 3.1. Non-competitive market

Consider a market where the creation and demise of organizations is permitted by the State to be random, but where no competition or collaboration exists between the organizations. An example of such a system might be a closed, agricultural economy, where all agricultural products are sold only to the State under fixed prices, and subsequently distributed or consumed by the State. Furthermore, new farms are permitted to be created, and farms are permitted to fail. Each $G(t)$ is a "star" graph, with connections only to the State.

We then have
$$G(t,\omega) = \{V(t,\omega), C_{M_0}(t,\omega)\},$$
where all edges in $C_{M_0}(t,\omega)$ include the State $v_1$. Then we can analyze, for example, $\xi(t) = |V(t)|$, defined to be the number of producers in the market at time $t$.

Assume that $\xi(t)$ satisfies the following properties:

- The probability of creation of one more producer in interval $\Delta t$ is $\lambda \Delta t + q(\Delta t)$, where the second term is negligibly small, so that the probability is proportional to $\Delta t$.
- The probability of more than one creation event is $O(\Delta t)$.
- Suppose that $\xi(\Delta_1)$ and $\xi(\Delta_2)$ are independent if $\Delta_1 \cap \Delta_2 = \emptyset$.

Let $\xi(t) = \sum_{k=1}^{n} \xi_k(\Delta_t)$ with interval $(0,t) = \cup_{k=1}^{n} \Delta_k$. Then $\xi_t$ exhibits a Poisson process; i.e.,
$$P(\xi_t = n) = \frac{(\lambda t)^n}{n!} e^{-\lambda t}.$$

This is a Markovian process, so that the market exhibits short memory of past markets. However, in our agricultural example, this implies that the failure of past farms does not have a long-term impact on the creation and failure of new farms, which may not reflect actual agricultural economies – for example, if a farm fails because of unsuitability of the land, the probability of creation or failure of a new farm on the land should exhibit a longer memory.

## 3.2. Practical application

In this section, we discuss some motivating scenarios for why answering these types of questions on such models are interesting in business, management, and economics. We consider some of the questions that we asked in the prequel.

*How many organizations are in the business ecosystem during the interval?*

This is an important question for a company that is receiving consulting for how to structure in a strongly cyclic market. A company in a strongly cyclic market should structure their financing to carry through "bearish" times, their employment for cyclic needs for workers, and their approach to new product development with the cyclic behavior to ensure that new products are ready when the market becomes bullish again, for best capture of market share.

Such structuring is semi-optimized for the behavior of the market. A company that is structured for a strongly cyclic market might not be well structured for a steady market and vice-versa. Consider a company in a steady market that is structured for a strongly cyclic market, by keeping the bulk of their employees as temporary contractors. Such employees may be more expensive in the short term, because of the contracting fees.

Also, these employees are likely to have reduced loyalty, and may be quick to move to competing companies for permanent positions. This company is in a situation that may be viewed as being analogous to purchasing the wrong type or too much insurance, which is a waste of money that could be better invested.

A company in a strongly cyclic market that is structured for a very steady market is in a situation analogous to not having enough insurance against the possibility of problems, which leaves them vulnerable to downturns in the market. For example, if the company invests its excess cash in non-liquid assets, rather than keeping enough liquid assets for a bearish market, the bearish market may force the company to take loans or sell critical capital assets, damaging the company's long-term financial health.

*How many business connections are in the business ecosystem?*

The answer to this question could give some measure of the "state" of the market, which can be studied to understand how the state and the maturity of the market are related with each other, which is an important economic question. In the case where we are evaluating one large company as the market and the departments inside of it as the organizations, this can be an important organizational question. The answer may predict, for example, if the company should be expected to undergo a significant (or disruptive) organizational or strategic change, based on the degree of dysfunctional interactions occurring inside the company. This analysis may be based on transitions in random graphs, which was started in mathematics; e.g., [6]. We are exploring this question in more detail in our subsequent research.

# 4. APPENDIX

Certain assumptions are required to make the model rigorous, some of which are described in this section.

## 4.1. Role of the State

The State plays a special role of all of the organizations in the ecosystem, beyond the definition of the legal framework. Specifically, the State is represented by a vertex that is connected by business connection edge to all other vertices. This means that every business ecosystem within such a State-specified legal framework is represented by a very specific topology of connected graph, where there exists one vertex that is connected to every other vertex. These are sometimes called graphs with a "star-like" topology. This restricted domain allows some interesting results that are not possible for random or general-connected graphs.

Even if non-profit businesses do not pay taxes or receive funds from the State, there is still a business connection with the State. Specifically, if the non-profit should violate the conditions of its tax-free status, there will be payment of a fine or some other funds.

This possibility is governed by a business connection to the State during the lifetime of the non-profit business.

Because of this connectedness, the State is the only organization in the ecosystem that has both "visibility" into all organizations in the ecosystem (and, by connection, their business transactions within the legal framework), and the ability to adjust the legal framework to influence those transactions. Specifically, the State can impose anti-trust (i.e., anti-monopoly) laws, tariffs, and subsidies to certain industries. Tariffs and subsidies may be represented by flows of monetary value to or from the state, respectively. The impact of anti-trust law is not explicitly reflected in this version of our model.

It should also be noted that on an edge between any organization and the State, the lifetime $\varphi(t)$ will be the support of the duration of the organization – the organization has a business connection with the State from the organization's inception until its demise. This imposes a very specific topology on the graph.

## 4.2. Definition of money

The reader should note that there is an extensive literature concerned with the definition of money and monetary value in an economy (e.g., [7]). In our research connected with this model, we have drawn on this literature and our own analysis to provide a rigorous definition of money that feeds our model. In this section, we cover an abridged version of that definition for the purposes of providing rigour to the model described in this paper.

The USA Federal Reserve Board measures the money supply using the following measures [8]:

- *M1* is money that can be spent immediately, including cash and checking accounts.
- *M2* is M1 plus assets that are invested for the short term. These assets include money- market accounts and money-market mutual funds.
- *M3* is M2 plus big deposits, including institutional money-market funds and agreements among banks.

In this paper, we use the terminology *M1'* for cash, and *M3'* for all other forms of money. This makes the distinction between that money which has immediate "street value", and that which is subject to contracts between deposit, investment, or credit institutions, and the depositor or creditor. M3' is effectively contract for payment, which includes checks, letters of credit, etc.

Microeconomic analysis is often used to drive future adjustments to strategy, to achieve desired business outcomes. Microeconomics provides a framework for parametric modeling of the "state" of the company, which is affected by changes in strategy. Analysis of this microeconomic model may show the leadership of the company that better market stability may be achieved by, for example, changing the competitive approach to the market. Effective use of microeconomic analysis, for the purpose of driving strategy, requires the use of a static model of the market (which can include information about expected market trajectory) with sufficient accuracy, that strategic decisions will track the market. This static model must be updated periodically, as the trajectory of the market changes.

More accurate strategic decisions can be made with the use of a macroeconomic model of the market, which accounts for critical parameters such as the value of money, and the broader market impact of movements of monetary value. In some sense, a traditional market model based on macroeconomics is a higher-order, broader model, covering a market, rather than only one company or organization.

Some economists have attempted to map these longer-term market trajectories by creating models of the market and its participating organizations, using longer time periods and more complicated parametric frameworks. If these attempts were successful, prediction of the market, and thus deterministic tracking of stock performance, would be possible. Clearly, no model has allowed any researcher to achieve this goal.

### 4.2.1. Microeconomic factors

This section introduces a non-Archimedean topology for money as used in transactions in our model. Our models are based on the assumption that money, or "monetary value" is conserved in each transaction. There are two types of money, legal and illegal (counterfeit) money. Note that the State creates legal money, which is eventually destroyed either by an organization or by the State (e.g., when it is worn). Counterfeit money is created by an organization (not the State, by definition), and destroyed by either another organization, or by the State (if discovered). Thus, these two types of money are topologically different:

- Legal money passes in a loop from State to State, or from State to organization.
- Illegal money passes from first organization either to second organization (which may be the same as the first), or to the State.

Define a Bank as an organization that converts to legal M1' from M3'. The Bank is the only such organization, which includes the Central Bank of the State. The Central Bank is the only organization that can create legal M1' from nothing.

Fraudulent transactions of M3' are considered the same as legal transactions with M3', since both represent letters of credit. There is not necessarily conservation of money within an organization, which allows for destruction of currency or purchase of consumable goods.

### 4.2.2. Macroeconomic factors

For certain questions it is interesting to discuss how much monetary value a given organization will have at one point in time. In the case of M1' (legal or illegal), this is a simple matter of counting, assuming that the face value of the currency expresses the true value of the note. However, in the case of M3', this is a question of credit rating; e.g., how many goods the organization can purchase on letters of credit at time

$t$ is dependent on ratings of past credit behavior, and may differ based on the selling organization. This time-dependent function also has a stochastic aspect. Furthermore, the organization may hold a number of letters of credit from other organizations, which have an implied, but illiquid, monetary value.

Thus, the value of money is a macroeconomic question. Because of this complexity, we avoid the discussion of the amount of money in each organization as a function of time, and discuss instead only the monetary value concerned with monetary transactions. We define monetary value such that monetary value is conserved in each and every transaction, making the assumption of the conservation of money and the time-stasis of money. Other scenarios are addressed in our subsequent research. Noting that as a result of the M1' topology, there is a probabilistic issue even with M1', not just with the value of the currency.

It thus can be concluded that *the classical mechanical model cannot be applied to business ecosystems, because of the influence of macroeconomic factors and the dependence of the model on these factors.* This problem was identified early by Marshall [9], who found what could be considered a (classical) mechanical model of the "state of a company", which Marshall found to be too static to capture the dynamics of real business. Note that this does not necessarily indicate that we need a quantum mechanical model. But it does imply that we need information from the macroeconomic level to quantify money at the microeconomic level.

This discussion and implicit "dependence on the future" leads to the conjecture that it might be appropriate to use a p-adic parametrization of time [10] to define the notion of the state of an organization at a given instant of time. This is because there is not a prescribed ordering (past or future) in p-adic number field. From the other side one could write p-adic differential equations to describe evolution of the organization. The one can use a kind of the coarse graining to go from p-adic parametrization of time to the ordinary parametrization in terms of real numbers.

As one example consider the Black - Scholes model [11] of the varying price over time of financial instruments, and in particular stocks. The law of evolution of the value of the option $V$ satisfies the following Black - Scholes differential equation:

$$\frac{\partial V}{\partial t} + \frac{1}{2}\sigma^2 S^2 \frac{\partial^2 V}{\partial S^2} + rS\frac{\partial V}{\partial S} - rV = 0.$$

Here $\sigma$ is the constant stock volatility, $r$ is the constant interest rate and $V$ is a function of the price $S$ and time $t$. The underlying price $S_t$ follows a geometric Brownian motion which satisfies the following stochastic differential equation:

$$dS_t = \mu S_t dt + \sigma S_t dW_t$$

where $W_t$ is the Brownian motion and $\mu$ is a constant. There is a theory of real valued stochastic processes with p-adic time [10] which can be used to develop a p-adic Black - Scholes model and which could help to formulate mathematically the above discussed idea on "depending on the future" state of organization.

It should be noted there are many cases where the classical model is sufficient. A simple example is the classical market which works only with M1' (e.g., farm market trade), which is generally reasonably immune to macroeconomic effects. This market is also relatively resistant to counterfeiting, because of fast and local enforcement by shop owners.

Furthermore, unlike in classical mechanics, a part of the system cannot indicate the state of the whole. In particular, the state of a part of the market will not necessarily indicate the state of the entire market, since the long-term state of any one company depends on the state of the entire economy. These entangled states are one of the most crucial points where classical and quantum mechanics differ from each other, providing one large source of uncertainty in the state of any one company.

Our subsequent research addresses specific analogies to and differences from quantum mechanics in this market model.

## 4.3. Business transactions

For the model to be useful for more than theory, it is important that the business transactions discussed in this paper can be measured. We consider those business transactions which have some registration that can be measured.

- Business transactions governed by formal contracts are registered with the courts upon the registration of the formal contract.
- Business transactions performed using traceable instruments such as checks, credit cards, or other cash-replacement instruments, are recorded by the banks that process the transactions.
- Business transactions recorded in a ledger or on a printed receipt are recorded by the organizations involved in the transaction. Although these may be more difficult to measure centrally in an economy, this information often must be reported to tax authorities, which constitutes a central registration of the aggregate transactions, if not the specific organizations involved in each particular transaction.

In some countries, a sale is not considered to be "legal" without a printed receipt. In such systems, an unrecorded cash transaction with no receipt is untraceable, thus unmeasurable, and thus not considered in our model.

## 4.4. Additional Contexts

The application of the model is not restricted to markets of the trade of tangible goods between enterprises. In this section, we give several examples of other types of collaborative and competitive relations, to help to illustrate the concepts of these types of business connections.

- At a different level of hierarchy, the business ecosystem may consist of collections of people inside of *one organization*; e.g., departments t hat comprise one company. In this case, the headquarters office plays the role of the State, in that the State determines legal framework for all constituents of the ecosystem. We restrict our model to considering levels of hierarchy where each such collection of people has an authority to make a legal agreement with other organizations, where "legal" is defined with respect to the framework set by the headquarters. For example, an individual worker in a company normally does not have the authority to make company-binding agreements with other workers or organizations in the company, but a certain level of management generally will have such an authority.

- Consider *stock ownership* of public companies. The organizations in this case include the company offering the stock, zero or more intermediate brokers trading the stock, and end buyers and sellers of the stock, which can be fund companies or individual people, for example. In this case, the buyer of the stock has a collaborative relationship with the broker who sells the stock, as well a collaborative relationship with the company offering the stock, for example, if the company ever pays dividends for the stock. The buyer has a competitive relationship with other buyers, since other buyers can affect the price.

## 5. FURTHER CONSIDERATIONS

Our current research work concerns boundary conditions of resources, inputs, and outputs of the model. We are also extending our analysis of the model to explore the model's time-dependent behavior. In parallel, we are developing applications of this model for concrete economic data, to demonstrate use of the model in the context of the development of business strategy.

## ACKNOWLEDGMENTS

This work was funded by an IBM Research Division Innovation Grant, with the assistance of the International Foundation of Technology and Investment.

## REFERENCES

1. R. Albert, and A.L. Barabasi, *Statistical mechanics of Complex Networks*, 2002.
2. D.J. Watts, *Six Degrees: the Science of a Connected Age*, W.W. Norton and Company, New York, 2003.
3. M. Iansiti, and R. Levien, *The Keystone Advantage: What the New Dynamics of Business Ecosystems Mean for Strategy, Innovation, and Sustainability*, Harvard Business School Press, 2004.
4. J.Q. Trelewicz, J.L.C. Sanz, D.W. McDavid, A. Chandra, and S.C. Bell, Informatics for business is more than process automation: i-business > e-process, *Information Technologies and Control* **2**, 2–7 (2004).

5. L. Accardi, Y.G. Lu, and I.V. Volovich, *Quantum Theory and its Stochastic Limit*, Springer, 2002.
6. P. Erdös, and A. Renyi, On the evolution of random graphs, *Publ. Math. Inst. Hung. Acad. Sci.*, 17–61 (1960).
7. B. Lietaer, *The Future of Money : Creating New Wealth, Work and a Wiser World*, Arrow Books Ltd., 2001.
8. D.M. Nichols, *Modern Money Mechanics*, Federal Reserve Bank of Chicago, 1992.
9. A. Marshall, *Principles of Economics, 8th ed.*, Macmillan, New York, 1948.
10. V.S. Vladimirov, I.V. Volovich and E.I. Zelenov, *p-Adic Analysis and Mathematical Physics*, World Scientific, 2002...
11. F. Black and M.Scholes, *The Pricing of Options and Corporate Liabilities*, Journal of Political Economy, 81, pp. 637 - 654, 1973.

# On Quantum Cryptography and Number Theory

Anton S. Trushechkin* and Igor V. Volovich[†]

*Moscow Engineering Physics Institute, Kashirskoe sh. 31, 115409 Moscow, RUSSIA
email: trushechkin@mail.ru

[†]Steklov Mathematical Institute of the Russian Academy of Sciences,
Gubkin st. 8, 119991 Moscow, RUSSIA
email: volovich@mi.ras.ru

**Abstract.** Relations between number theory and cryptography are discussed. A general mathematical framework for quantum key distribution based on the concepts of quantum channel and Turing machine is suggested. The security for its special case is proved. The assumption is that the adversary can perform only individual (in essence, classical) attacks. For this case an advantage of quantum key distribution over classical one is shown.

**Keywords:** Quantum cryptography, Quantum key distribution, Quantum channel.
**PACS:** 01.30.Cc.

## 1. INTRODUCTION

Cryptography is considered to appear as a science in 1948, when the paper by Shannon [1] was published. In [1] the notion of *cipher system* is introduced. From the practical point of view a cipher system consists of a *message (plaintext) source*, a *key source*, an *encipherer*, which transforms a *plaintext* to a *ciphertext* using a *key*, and a *decipherer*, which transforms the received ciphertext to the plaintext using the same key. The adversary analysis the ciphertext in order to get as much information as possible. The cipher system is called *perfectly secure* if the ciphertext provides no information about the plaintext or, equivalently, ciphertext and plaintext are statistically independent. A simple example of the perfectly secure cipher system is the so-called one-time pad proposed by Vernam where the encipherer transforms the plaintext to its binary representation and then adds modulo 2 (binary XOR) a secret random key to it. In the same way decipherer gets the plaintext using the key.

But there is also a theorem in [1] which asserts that perfect security demands the key to be at least as long as the message. This makes perfectly secure cipher systems impractical, since the two legal parties have not any way to get distantly a common secret key.

That's why the so called computational security became the main paradigm in cryptography. The system is called computationally secure if the adversary using the modern computational devices cannot get the plaintext from the ciphertext in a reasonable time.

Cryptography and number theory are closely related. Applications of number theory in cryptography are important in constructions of public key cryptosystems. The most popular public key cryptosystems (RSA, ElGamal) are based on the problem of factorization of large integers and discrete logarithm problem in finite groups, in particular in the multiplicative group of finite field and the group of points on elliptic curve over finite field [2, 3]. Important applications of $p$-adic analysis in cryptography are given in [4].

The best classical factoring algorithm which is currently known is the number field sieve which is exponential in $n^{\frac{1}{3}}$, where $n$ is the number of digits. While there is no known efficient classical algorithm for factoring, the quantum polynomial algorithm exists [5, 6].

The idea of quantum cryptography was proposed in the 1970's by Wiesner [7] and Bennett and Brassard [8] (their method is called the BB84 protocol). They used the sending of single quantum particles to solve the key distribution problem. Moreover, it can detect the presence of an eavesdropper. The phenomenon of entanglement and Bell's inequality are also used in quantum cryptography [9]. Recently the first commercial quantum key distribution systems became available on the market [10, 11].

In a number of papers, several particular quantum key distribution protocols were suggested, most of which are based on the BB84 protocol [8]. Its security was considered, e.g., in [12, 14, 13, 15, 16, 3]. Practical realizations of quantum key distribution protocols were described, for example, in [16, 17]. A sufficiently general mathematically rigorous approach to quantum key distribution is lacking at present. For recent discussions of these problems see [18, 19, 20].

In this paper we suggest a rather general mathematical model of quantum key distribution. We also prove its security in a certain special case. The security is proved on the assumption that the adversary can perform only individual (in essence, classical) attacks. For this case, an advantage of the quantum key distribution over the classical one is shown.

$p$-Adic quantum mechanics is developed in [21, 22]. Adelic quantum mechanics is constructed by Branko Dragovich [23]. It would be interesting to construct $p$-adic quantum cryptography, in particular, to investigate the question how the security depends on the choice of field of real or $p$-adic numbers.

We are very happy to dedicate this paper to the 60th birthday of Professor Branko Dragovich.

## 2. NOTATIONS

$\mathcal{H}$ is a Hilbert space, and $\mathcal{S}(\mathcal{H})$ is a convex set of quantum states (density operators) on $\mathcal{H}$. Let $\mathcal{H}_A$ and $\mathcal{H}_B$ be a pair of Hilbert spaces. A channel $\Theta$ from $\mathcal{H}_A$ to $\mathcal{H}_B$ is an affine map from $\mathcal{S}(\mathcal{H}_A)$ to $\mathcal{S}(\mathcal{H}_B)$ such that its linear extension has a completely positive conjugate map [24, 25]. The sequence of channels

$$\Theta^n : \mathcal{S}(\mathcal{H}_A^{\otimes n}) \to \mathcal{S}(\mathcal{H}_B^{\otimes n}), \quad n \in \mathbb{N}$$

is associated with the channel $\Theta$ by the formula

$$\Theta^n(\rho_1 \otimes \ldots \otimes \rho_n) = \Theta(\rho_1) \otimes \ldots \otimes \Theta(\rho_n), \quad \rho_i \in \mathscr{S}(\mathscr{H}_A), \quad i = 1, \ldots, n.$$

If $\mathscr{A}$ and $\mathscr{B}$ are finite sets and $\mathscr{P}(\mathscr{A})$ and $\mathscr{P}(\mathscr{B})$ are probability distributions on $\mathscr{A}$ and $\mathscr{B}$, then an affine map $V : \mathscr{P}(\mathscr{A}) \to \mathscr{P}(\mathscr{B})$ specifies a classical channel. If $P \in \mathscr{P}(\mathscr{A})$, then $I(P,V)$ denotes the Shannon mutual information between the input and output of the channel $V$ if $P$ is a distribution at the input of $V$. A channel $V$ can be also specified by a mapping of the corresponding random variables. If $X$ and $Y$ are random variables, then $I(X;Y)$ denotes the mutual information between them.

By $V^n : \mathscr{P}(\mathscr{A}^n) \to \mathscr{P}(\mathscr{B}^n)$ denote the discrete memoryless channel corresponding to $V$. An affine map $\Xi : \mathscr{P}(\mathscr{A}) \to \mathscr{S}(\mathscr{H})$ can be specified by a function $\xi : \mathscr{A} \to \mathscr{S}(\mathscr{H})$. Let $\rho \in \mathscr{S}(\mathscr{H})$ be a quantum state. Then, the von Neumann entropy is defined by the formula

$$H(\rho) = -\operatorname{Tr}\rho\log\rho.$$

Let $\mathscr{A}$ be a finite set, $P \in \mathscr{P}(\mathscr{A})$ be a distribution, $\mathscr{H}$ be a Hilbert space, $\xi : \mathscr{A} \to \mathscr{S}(\mathscr{H})$ be a function. We define

$$C(\xi) = \max_P \left[ H\left( \sum_{a \in \mathscr{A}} P(a)\xi(a) \right) - \sum_{a \in \mathscr{A}} P(a) H(\xi(a)) \right].$$

If $\{M(b)\}$ is a positive operator-valued measure (POVM) (i.e., an observable) on $\mathscr{H}$, then the formula

$$P(b) = \operatorname{Tr} M(b)\rho, \quad \rho \in \mathscr{S}(\mathscr{H}),$$

specifies an affine map (channel) from $\mathscr{S}(\mathscr{H})$ to $\mathscr{P}(\mathscr{B})$. The space of POVM on $\mathscr{H}$ taking values on $\mathscr{B}$ is denoted by $\mathscr{M}(\mathscr{H};\mathscr{B})$. An observable from

$$\mathscr{M}(\mathscr{H};\mathscr{B})^{\otimes n} \subset \mathscr{M}(\mathscr{H}^{\otimes n};\mathscr{B}^n)$$

is called an individual observable on the space $\mathscr{H}^{\otimes n}$ taking values on $\mathscr{B}^n$. An observable that does not belong to this class is called a joint observable. By

$$\mathscr{B}^\mathscr{A} \circ \mathscr{M}(\mathscr{H};\mathscr{A}) \subset \mathscr{M}(\mathscr{H};\mathscr{B}),$$

denote the class of observables of the form

$$\{F(b) = \sum_{a \in f^{-1}(b)} M(a)\}_{b \in \mathscr{B}}, \quad \{M(a)\}_{a \in \mathscr{A}} \in \mathscr{M}(\mathscr{H};\mathscr{A}),$$

where $f$ is an element of the set $\mathscr{B}^\mathscr{A}$ of functions from $\mathscr{B}$ to $\mathscr{A}$.

# 3. GENERAL MODEL OF QUANTUM KEY DISTRIBUTION

In this section a definition of the general mathematical model of quantum key distribution is outlined. We consider the following problem of key distribution. Two parties, Alice and Bob, want to get a pair of keys (one key for Alice and another one for Bob) using communication channels. A key is regarded as a realization of a certain random variable on a finite set $\mathcal{K}$, or this random variable itself. If Alice's and Bob's keys are identical with high probability and Eve's information about the keys is negligibly small, then the problem of key distribution is considered to be solved with some *security degree*.

We will model the parties using extended Turing machines, which form the model of classical computers interacting with external devices and communicating with each other through the classical and quantum channels.

Let $(\mathcal{A}, \mathcal{Q}_A, \tau_A)$ be a Turing machine describing Alice, where $\mathcal{A}$ is an alphabet, $\mathcal{Q}_A$ is a set of states with additional chosen elements, and $\tau_A$ is a transition function. Furthermore, let $(\mathcal{A}, \mathcal{Q}_B, \tau_B)$ and $(\mathcal{A}, \mathcal{Q}_E, \tau_E)$ be Turing machines describing Bob and Eve, respectively.

**Definition 1.** A *system of quantum key distribution* is a triple of objects (*extended Turing machines*)

$$(\mathrm{ETM}_A, \mathrm{ETM}_B, \mathrm{ETM}_E),$$

where

$$\mathrm{ETM}_A = (\mathcal{A}, \mathcal{Q}_A, \tau_A, P_A, \xi_A),$$
$$\mathrm{ETM}_B = (\mathcal{A}, \mathcal{Q}_B, \tau_B, P_B, \{\xi_B^{(n)}\}_{n \in \mathbb{N}}),$$
$$\mathrm{ETM}_E = (\mathcal{A}, \mathcal{Q}_E, \tau_E, P_E, \{\xi_E^{(n)}\}_{n \in \mathbb{N}}, \mu_E).$$

Here, $P_A, P_B, P_E$ are probability distributions on $\mathcal{A}^+ = \bigcup_{i=1}^{\infty} \mathcal{A}^i$,

$$\xi_A : \mathcal{A}^+ \to \mathcal{S}_A \subset \bigcup_{i=1}^{\infty} \mathcal{S}(\mathcal{H}_A^{\otimes i}),$$
$$\xi_B^{(n)} : \mathcal{A}^+ \to \mathcal{M}_B^{(n)} \subset \mathcal{M}(\mathcal{H}_B^{\otimes n}; \mathcal{A}^+),$$
$$\xi_E^{(n)} : \mathcal{A}^+ \to \Omega^{(n)},$$
$$\mu_E : \mathcal{A}^+ \to \mathcal{M}_E \subset \mathcal{M}(\mathcal{H}_E; \mathcal{A}^+),$$

where $\mathcal{H}_A, \mathcal{H}_B$, and $\mathcal{H}_E$ are Hilbert spaces, $\Omega^{(n)} = \{\Theta_i^n\}_{i \in I}$,

$$\Theta_i^n : \mathcal{S}(\mathcal{H}_A^{\otimes n} \otimes \mathcal{H}_E) \to \mathcal{S}(\mathcal{H}_B^{\otimes n} \otimes \mathcal{H}_E).$$

Some restrictions are imposed on the functions $\xi_A, \xi_B^{(n)}$, and $\xi_E^{(n)}, \mu_E$. They will be presented in the further paper.

The parties interact through quantum and classical channels by using the additional chosen elements in $\mathcal{Q}_A, \mathcal{Q}_B$, and $\mathcal{Q}_E$ and the functions introduced above.

## 4. SPECIAL MODEL OF QUANTUM KEY DISTRIBUTION

In this paper we will consider only a special case of the described model.

**Definition 2.** A *system G of quantum key distribution* is a family of the objects

$$G = \left( \mathcal{K}, \mathcal{H}_A, \mathcal{H}_B, \mathcal{H}_E, \Theta, \{q^{(n)}\}_{n \in \mathbb{N}}, \{M_B^{(n)}\}_{n \in \mathbb{N}}, \{\mathcal{M}_E^{(n)}\}_{n \in \mathbb{N}} \right). \quad (1)$$

Here, $\mathcal{K}$ is a finite set (set of keys); $\mathcal{H}_A, \mathcal{H}_B$, and $\mathcal{H}_E$ are Hilbert spaces;

$$\Theta : \mathcal{S}(\mathcal{H}_A) \to \mathcal{S}(\mathcal{H}_B \otimes \mathcal{H}_E)$$

is a channel; the functions

$$q^{(n)} : \mathcal{K} \to \mathcal{S}(\mathcal{H}_A^{\otimes n})$$

specify channels $Q^{(n)}$;

$$M_B^{(n)} \in \mathcal{M}(\mathcal{H}_B^{\otimes n}; \mathcal{K}), \quad \mathcal{M}_E^{(n)} \subset \mathcal{M}(\mathcal{H}_E^{\otimes n}; \mathcal{K}).$$

For any $n \in \mathbb{N}$ and $M_E^{(n)} \in \mathcal{M}_E^{(n)}$, we define the channel

$$\Lambda_n = (M_B^{(n)} \otimes M_E^{(n)}) \circ \Theta^n \circ Q^{(n)}$$

with the input alphabet $\mathcal{K}$ and the output alphabet $\mathcal{K}^2$.

Let $K_A$ denote a random value uniformly distributed on $\mathcal{K}$ (a key). By $K_B$ and $K_E$, we denote random variables taking values on $\mathcal{K}$ and related to $K_A$ by the channel $\Lambda_n$ for some $n$; i.e.,

$$(K_B, K_E) = \Lambda_n(K_A).$$

A schematic view of the quantum key distribution process is presented in Fig. 1.

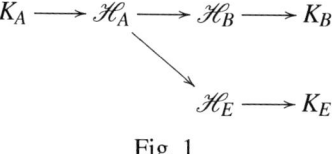

Fig. 1

The proposed model is a quantum analogue of the classical key distribution model considered in [26], where the following theorem is proved.

**Theorem 1.** *Let $\mathcal{K}, \mathcal{A}, \mathcal{B}, \mathcal{E}$ be finite sets and a pair of channels*

$$V : \mathcal{P}(\mathcal{A}) \to \mathcal{P}(\mathcal{B}), \quad W : \mathcal{P}(\mathcal{A}) \to \mathcal{P}(\mathcal{E})$$

*be given. Furthermore, let*

$$\max_{P \in \mathcal{P}(\mathcal{A})} [I(P,V) - I(P,W)] > 0. \quad (2)$$

Then, for any $\alpha, \beta \in (0,1)$ and any sufficiently large $n \in \mathbb{N}$, there exists a channel (random coder)
$$F_A : \mathscr{P}(\mathscr{K}) \to \mathscr{P}(\mathscr{A}^n)$$
and a function (decoder)
$$f_B : \mathscr{B}^n \to \mathscr{K}$$
such that, for any function (decoder)
$$f_E : \mathscr{E}^n \to \mathscr{K},$$
we have:

(i) $\Pr[K_A = K_B] \geq \alpha$; (ii) $I(K_A; K_E) \leq 1 - \beta$,

where
$$K_B = f_B \circ V^n \circ F_A(K_A)$$
and
$$K_E = f_E \circ W^n \circ F_A(K_A).$$

If, for any P, the condition
$$I(P,V) \geq I(P,W)$$
is true, then condition (2) is not only sufficient but also necessary for existence of $F_A$ and $f_B$ with the specified properties for sufficiently large n.

Thus, condition (2) can be considered as a condition for the possibility of a classical key distribution with an arbitrary security degree. The security degree is a pair of numbers $(\alpha, \beta)$.

Now, we formulate our main theorem.

**Theorem 2.** Let $\mathscr{H}_A, \mathscr{H}_B, \mathscr{H}_E$, and $\Theta$ be fixed in the model G of quantum key distribution (1). For any $n \in \mathbb{N}$, we define
$$\mathscr{M}_E^{(n)} = \mathscr{K}^{\mathscr{E}} \circ \mathscr{M}(\mathscr{H}_E; \mathscr{E})^{\otimes n},$$
where $\mathscr{E}$ is a finite set.

Suppose that there exists a finite set $\mathscr{A}$ and a channel $\Xi$ specified by a function $\xi : \mathscr{A} \to \mathscr{S}(\mathscr{H}_A)$ that obey the property
$$C(\Theta_B \circ \xi) > C_1(\Theta_E \circ \xi), \tag{3}$$
where $\Theta_B = \operatorname{Tr}_{\mathscr{H}_E} \Theta$, $\Theta_E = \operatorname{Tr}_{\mathscr{H}_B} \Theta$, and
$$C_1(\Theta_E \circ \xi) = \max_{P \in \mathscr{P}(\mathscr{A}), M \in \mathscr{M}(\mathscr{H}_B; \mathscr{K})} I(P, M \circ \Theta_E \circ \Xi).$$

Then, for any $\alpha, \beta \in (0,1)$ and any sufficiently large $n \in \mathbb{N}$, there exist a channel (random coder)
$$F_A : \mathscr{P}(\mathscr{K}) \to \mathscr{P}(\mathscr{A}^n)$$

and an observable
$$M_B \in \mathcal{M}(\mathcal{H}_B^{\otimes n}; \mathcal{K})$$
such that $\forall M_E^{(n)} \in \mathcal{M}_E^{(n)}$, the random variables $K_A$, $K_B$, and $K_E$, where $(K_B, K_E) = \Lambda_n(K_A)$, obey the following properties:

(i) $\Pr[K_A = K_B] \geq \alpha$; (ii) $I(K_A; K_E) \leq 1 - \beta$.

Here,
$$\Lambda_n(K_A) = (M_B^{(n)} \otimes M_E^{(n)}) \circ \Theta^n \circ Q^{(n)},$$
and
$$Q^{(n)} = \Xi^n \circ F_A.$$

*Outline of proof.* Let us define
$$C_k(\Theta_B^k \circ \xi^{(k)}) = \max_{P \in \mathcal{P}(\mathcal{A}), M \in \mathcal{M}(\mathcal{H}_B^{\otimes k}; \mathcal{K})} I(P, M \circ \Theta_B^k \circ \Xi^k),$$
where
$$\xi^{(k)} : \mathcal{A}^k \to \mathcal{S}(\mathcal{H}_A)$$
is the function that specifies the memoryless channel $\Xi^k$ corresponding to the channel $\Xi$; i.e.,
$$\xi^{(k)}(a_1, \ldots, a_k) = \xi(a_1) \otimes \ldots \otimes \xi(a_k).$$
Since [25]
$$C(\Theta_B \circ \xi) = \lim_{k \to \infty} \frac{1}{k} C_k(\Theta_B^k \circ \xi^{(k)}),$$
in view of (3),
$$\exists n \; C_n(\Theta_B^n \circ \xi^{(n)}) > n C_1(\Theta_E \circ \xi).$$
Since, for any $k$, in view of the conditions imposed on the adversary's measurements class, we have
$$\max_{\substack{P \in \mathcal{P}(\mathcal{A}^k) \\ M \in \mathcal{M}(\mathcal{H}_E; \mathcal{E})^{\otimes k}}} I(P, M \circ \Theta_E^k \circ \xi^{(k)}) = k C_1(\Theta_B \circ \xi),$$
this implies that
$$\max_{\substack{P \in \mathcal{P}(\mathcal{A}^n) \\ M \in \mathcal{M}(\mathcal{H}_B^{\otimes n}; \mathcal{K})}} I(P, M \circ \Theta_B^n \circ \xi^{(n)}) > \max_{\substack{P \in \mathcal{P}(\mathcal{A}^n) \\ M \in \mathcal{M}(\mathcal{H}_E; \mathcal{E})^{\otimes n}}} I(P, M \circ \Theta_E^n \circ \xi^{(n)}).$$

Then, since the adversary can perform only individual measurements with subsequent classical processing, the conditions of Theorem 1 are satisfied. □

# 5. ON ADVANTAGES OF THE QUANTUM KEY DISTRIBUTION OVER THE CLASSICAL ONE

Thus, it is possible that, for a single use of a quantum channel, classical condition (2) of key distribution is not satisfied, but, with multiple repetitions, Bob uses an essentially quantum property – joint measurements – in contrast to the adversary Eve, who performs, in essence, classical attacks, i.e., individual measurements. As a result, Bob achieves advantage over Eve and the subsequent key distribution according to the classical scheme becomes possible.

To demonstrate this possibility, we give an example of the quantum key distribution system, in which, at every run of the quantum channel, Eve receive the same state as Bob; i.e., there is an ideal eavesdropping. However, as a consequence of the effect of joint measurements applied by Bob with the multiple use of the channel, he receives more information than Eve, and the key distribution becomes possible. The effect of joint measurements is caused by the use of nonorthogonal states.

**Example.** Let $\mathcal{H}_A = \mathbb{C}^2 \otimes \mathbb{C}^2$, $\mathcal{H}_B = \mathcal{H}_E = \mathbb{C}^2$, and $\Theta$ be the identity map. $\mathcal{A} = \{0,1\}$,

$$\xi(0) = (|\varphi\rangle\langle\varphi|)^{\otimes 2},$$

and

$$\xi(1) = (|\psi\rangle\langle\psi|)^{\otimes 2},$$

where

$$|\varphi\rangle, |\psi\rangle \in \mathbb{C}^2,$$

and

$$\langle\varphi|\psi\rangle \neq 0.$$

If, for $P \in \mathscr{P}(\mathcal{A})$, the condition $P(0) = 0$ or $P(1) = 0$ is satisfied, then

$$\forall M_E \in \mathscr{M}(\mathcal{H}_E; \mathscr{E})$$

we have

$$I(P, M_E \circ \Theta_E \circ \Xi) = 0,$$

so

$$\max_P I(P, M_E \circ \Theta_E \circ \Xi)$$

is achieved when $P(0), P(1) > 0$. Then, since the operators $P(0)\xi(0)$ and $P(1)\xi(1)$ do not commute, condition (3) is satisfied. So, by Theorem 2, the key distribution is possible.

In classical cryptography, there are no analogues to this effect, in which case the key distribution is possible even under ideal eavesdropping. In this sense, we can talk about advantages of quantum cryptography over classical one.

# 6. ON SPECIFICATION OF QKD SYSTEMS

The quantum channel $\Theta$ is fixed in our model $G$ and Eve cannot choose a channel, whereas in the realistic case she can (and this is reflected in the general model). So, the model $G$ is a some kind of a "toy model", which can be useful in the security analysis of more complicated and realistic protocols.

If Eve can choose different channels (Alice and Bob do not know about her choice), Alice and Bob have to use the methods of mathematical statistics in order to estimate the amount of information that Eve has received (the amount of this information depends on a channel chosen by Eve).

In the protocols based on the BB84 protocol (or on modifications of the BB84 protocol, for example, the six-state protocol) Alice and Bob estimate the so called *quantum bit error rate* (*QBER*) between the data sent by Alice and received by Bob. Then they use quantum mechanics in order to estimate the Eve's information about these data. This analysis is based on the property of quantum world that the measurement of a quantum system changes the state of this system and the noise is introduced. There exists the maximal value $\text{QBER}_m$ of the QBER with the following property. If $\text{QBER} < \text{QBER}_m$, then the conditions of Theorem 1 are satisfied (the classical channels $V$ and $W$ resulting from the measurements of the states transmitted through the quantum channel $\Theta$ satisfy the condition (2)) and the key distribution is possible. If $\text{QBER} \geq \text{QBER}_m$, then the conditions of Theorem 1 are not satisfied and the key distribution is impossible.

And a protocol is characterized by its $\text{QBER}_m$, which is also used to compare the different protocols. In our opinion, $\text{QBER}_m$ is, of course, an important characteristic of a QKD protocol, but not an exhaustive one. Theorem 1 tells that if its conditions are satisfied (in our case this means that $\text{QBER} < \text{QBER}_m$), then the key distribution with the desired security degree $(\alpha, \beta)$ is possible *only for sufficiently large n* (in our case $n$ is a number of quantum particles, e.g., photons, transmitted through the quantum channel). But in the practical case it is important to know, how many photons we have to transmit in order to get a pair of keys with the desired length $k$ and security degree $(\alpha, \beta)$. So, there is a function $n(\text{QBER}, k, \alpha, \beta)$ and $n(\text{QBER}_m, k, \alpha, \beta) = 0$ for any $k$ and any sufficiently close to one $\alpha$ and $\beta$. But in addition to $\text{QBER}_m$ we also must know some other information about the function $n(\text{QBER}, k, \alpha, \beta)$. For example, how fast $n(\text{QBER}, k, \alpha, \beta)$ with fixed $k$ and $(\alpha, \beta)$ decreases as $\text{QBER} < \text{QBER}_m$ moves away from $\text{QBER}_m$.

These problems are discussed in our paper [27].

# ACKNOWLEDGMENTS

This work was supported in part by the Russian Foundation for Basic Research (project No. 05-01-00884), a grant from the President of the Russian Federation (project No. NSh-1542.2003.1), and the program "Modern Problems in Theoretical Mathematics" of the Department of Mathematical Sciences of the Russian Academy of Sciences.

# REFERENCES

1. C.E. Shannon, *Bell System Technical Journal* **28**, 656–715 (1948).
2. N. Koblitz, *A Course in Number Thoery and Cryptography*, Springer-Verlag, New York, 1994.
3. I.V. Volovich, and Ya.I. Volovich, On classical and quantum cryptography, `quant-ph/0108133`.
4. V.S. Anashin, *Discret. Mat.* **14**, no. 4, 3–64 (2002) [in Russian]; English translation in *Discrete MAth. Appl.* **12**, no. 6, 527–590 (2002); `math.NT/0209407`.
5. P.W. Shor, in *Proc. of the 35th Annual Symposium on Foundations of Computer Science*, Santa Fe, pp. 124–134 (1994).
6. I.V. Volovich, Quantum Computing and Shor's Factoring Algorithm, `quant-ph/0109004`.
7. S. Wiesner, *SIGACT News* **15**, no. 1, 77–88 (1983); originally written c. 1970 but unpublished.
8. C.H. Bennett, and G. Brassard, in *Proceedings of IEEE International Conference on Computers, Systems and Signal Processing*, Bangalore, India, pp. 175–179 (1984).
9. A. K. Eckert, *Phys. Rev. Lett.* **67**, no. 6, 661–663 (1994).
10. id Quantique SA, Geneva, Switzerland, `http://www.idquantique.com/`.
11. MagiQ Technologies, New York, USA, `http://www.magiqtech.com/`.
12. D. Mayers, Unconditional security in quantum cryptography, `quant-ph/9802025`.
13. E. Biham, M. Boyer, P.O. Boykin, T. Mor, and V. Roychowdhurry, A proof of the security of quantum key distribution, `quant-ph/9912053`.
14. H.K. Lo, and H.F. Chau, *Science* **283**, 2050–2056 (1999); `quant-ph/9803006`.
15. P.W. Shor, and J. Preskill, *Phys. Rev. Lett.* **85**, 441–444 (2000); `quant-ph/0003004`.
16. N. Gisin et al., Quantum cryptography, `quant-ph/0101098`.
17. A. Poppe et al., Practical quantum key distribution with polarization Entangled Photons, `quant-ph/0404115`.
18. M. Ben-Or, and D. Mayers, General security definition and composability for quantum and classical protocols, `quant-ph/0409062`.
19. M. Christandl, R. Renner, and A. Ekert, A generic security proof for quantum key distribution, `quant-ph/0402131`.
20. R. Renner, N. Gisin, and B. Kraus, An information-theoretic security proof for QKD protocols, `quant-ph/0502064`.
21. V.S. Vladimirov, I.V. Volovich, and E.I. Zelenov, *p-Adic Analysis and Mathematical Physics*, World Scientific, Singapoure, 1994.
22. A.Yu. Khrennikov, *Journal of Math. Phys.* **32**, 932–937 (1991).
23. B. Dragovich, *International Journal of Modern Physics* **A**10, 2349–2365 (1995).
24. M. Ohya, and D. Petz, *Quantum entropy and its use*, Berlin, Springer-Verlag, 1993.
25. A.S. Kholevo, Introduction to quantum information theory, Moscow, MTsNMO, 2002 [in Russian].
26. I. Csiszár, and J. Körner, *IEEE Transactions on Information Theory* **24**, no. 3, 339–348 (1978).
27. A.S. Trushechkin, and I.V. Volovich, On Standards and Specifications in Quantum Cryptography, `quant-ph/0508156`.

# APPENDIX

# Appendix: Open Problems

## Open problem proposed by *Vladimir Anashin*

**Problem.** *For a p-adic integer $a = a_0 + a_1 \cdot p + a_2 \cdot p^2 + \cdots$ ($a_i \in \{0, 1, \ldots, p-1\}$) denote $a \bmod p^n = a_0 + a_1 \cdot p + a_2 \cdot p^2 + \cdots + a_{n-1} \cdot p^{n-1}$, the smallest non-negative residue modulo $p^n$.*

*Does there exist a mapping $g \colon \mathbb{Z}_2 \longrightarrow \mathbb{Z}_2$ that satisfy the 2-adic Lipschitz condition with a constant 1 such that every mapping*

$$z \mapsto g\left(\frac{z^2 + z}{2}\right) \bmod 2^n \qquad (z \in \{0, 1, 2, \ldots, 2^n - 1\})$$

*is a single cycle permutation on $\{0, 1, 2, \ldots, 2^n - 1\}$, for all $n = 1, 2, \ldots$? If yes, could (some of) these mappings be represented in an explicit form?*

**Comments:** Mappings $f \colon \mathbb{Z}_p \longrightarrow \mathbb{Z}_p$ that satisfy $p$-adic Lipschitz condition with a constant 1, and such that every induced mapping $\bar{f}_n \colon z \mapsto f(z) \bmod p^n$, ($z \in \{0, 1, \ldots, p^n - 1\}$) is a single cycle permutation on $\{0, 1, 2, \ldots, p^n - 1\}$ are described: They are ergodic isometries of $\mathbb{Z}_p$. These mappings are used in cryptography to construct fast pseudorandom number generators (and stream ciphers), especially for $p = 2$, see [1] and [2] for details.

In [4] Yurov proved that continuous mappings that induce permutations (which are not necessarily single cycles) on $\mathbb{Z}/p^n\mathbb{Z}$ for all $n$, do exist only if $p = 2$; in this case they are of the form

$$z \mapsto f\left(\frac{z^2 + z}{2}\right),$$

where $f$ preserves the normalized Haar measure on $\mathbb{Z}_2$ and satisfies the 2-adic Lipschitz condition with a constant 1. All the latter mappings $f$ are completely characterized in [2].

In [3] Woodcock and Smart consider a chaotic mapping $z \mapsto \frac{z^2 - z}{2}$ on $\mathbb{Z}_2$ in order to produce pseudorandom sequences modulo $2^n$. However, the corresponding mappings modulo $2^n$ contain short cycles, which are highly undesirable in cryptography.

If 'good' (explicit, and not too complicated) mappings $g$ under the question exist, they could be used in various constructions of stream ciphers and pseudorandom generators.

# REFERENCES

1. V.S. Anashin, Uniformly distributed sequences over $p$-adic integers, *Number theoretic and algebraic methods in computer science. Proceedings of the Int'l Conference (Moscow, June–July, 1993)* (A.J. van der Poorten, I. Shparlinsky and H.G. Zimmer, eds.), World Scientific, pp. 1–18 (1995).
2. V. S. Anashin, Uniformly distributed sequences of $p$-adic integers II, (Russian), *Diskret. Mat.* **14** (2002), no. 4, 3–64; English translation in *Discrete Math. Appl.* **12**, no. 6, 527–590 (2002), `math.NT/0209407`.
3. C.F. Woodcock and N.P. Smart, $p$-adic chaos and random number generation, *Experimental Mathematics* **7**, no. 4, (1998).
4. I.A. Yurov, On $p$-adic functions which preserve Haar measure, *Mathematical Notes* **63**, no. 6, (1998).

# Open problems proposed by *Andrei Khrennikov*

**1. Cauchy problem.** We recall that the Cauchy problem for a partial differential equation is well posed in a functional space if there exists the unique solution in this space which is continuously dependent on initial conditions. Partial differential equations with respect to functions $u(t,x)$, where the variables $t,x$ as well as values of $u$ belong to a complete non-Archimedean field $K$ were studied in [1]. There was shown that the Cauchy problem for a large class of linear partial differential equations is well posed in the space of entire analytic functions. On one hand, analytic solutions play an important role in applications. Therefore it was important to show that the Cauchy problem is well posed in the class of analytic functions. On the other hand, consideration of only analytic functions is not so much interesting from purely mathematical viewpoint: this is more or less manipulation with power series and the non-Archimedean specific is reduced to using of the strong triangle inequality. Therefore it would be interesting to find natural functional spaces containing nonanalytic functions in that the Cauchy problem for some classes of partial differential equations (e.g., linear and with constant coefficients) would be well posed. This problem is really nontrivial due to existence in the non-Archimedean case nontrivial smooth functions having the derivative, which is identically equal to zero.

**2. Stable distributions.** In a series of works of the author (see, e.g., the paper in this volume) there was developed a kind of probability theory in that "probabilities" take values in the field of $p$-adic numbers. In particular, there were obtained $p$-adic analogues of the law of large numbers and the central limit theorem. An interesting open problem is to find the general form of stable $p$-adic valued probability distributions (an analogue of the Khintchine-Levy classification in the real case).

**3. Analysis over non-Archimedean rings.** In applications to cognitive sciences and psychology there can be used not only $p$-adic trees for a prime number $p > 1$, but also $m$-adic trees for an arbitrary natural number $m > 1$, see [2]. There is well developed analysis of maps from $Q_p$ into $Q_p$. However, analysis for map from $Q_m$ to $Q_m$ was never presented in a regular way. In spite common declarations that such an analysis could be developed in a similar way to the analysis over non-Archimedean rings, it is not clear which theorems for differential and analytic functions are still valid in the $m$-adic case or how their should be modified. It is an interesting open problem to develop analysis over non-Archimedean rings (in particular, $Q_m$ and, moreover, maps from $Q_{m_1}$ into $Q_{m_2}$, where in general $m_1 \neq m_2$). In applications to cognitive sciences and psychology there can be even used nonhomogeneous trees which have only the structure of an additive group. It seems that analysis for maps between such trees was not developed anywhere. In [2] human subconscious was modeled with the aid of $p$-adic dynamical systems. However, in [2] it was emphasized that the $p$-adic case was considered only by mathematical reasons. By neurophysiological reasons it would be more natural to model subconsciousness by using general $m$-adic or even nonhomogeneous trees. Therefore dynamical systems on such trees (in particular, dynamical systems in non-Archimedean rings) are of the great interest for applications.

**4. Particles with the "$p$-adic charges."** It is well known that in quantum field theory over the field of complex numbers the operation of the field conjugation, $\phi \to \bar{\phi}$, provides the symmetry between particles and antiparticles, e.g., electron and positron. In a theoretical quantum model [1] over the field of p-adic numbers (with p-adic valued wave functions and fields) there was noticed that there exist Galois extensions of $Q_p$ of an arbitrary degree n. Therefore it is possible to construct theoretical models in that the Galois group G plays the role of the group of involutions. In such models fields take values in algebraic extensions of $Q_p$. The construction in [1] was of purely mathematical interest. It is an interesting problem to find real physics behind this mathematical construction. And it would be great to find some possibilities of experimental verification of the existence of "multi-charges" (models with $n > 1$ antiparticles).

## REFERENCES

1. A. Yu. Khrennikov, *p-adic valued distributions in mathematical physics*, Kluwer, Dordrecht, 1994.
2. A. Yu. Khrennikov, *Information dynamics in cognitive, psychological, social, and anomalous phenomena*, Kluwer, Dordrecht, 1999.

# Open problems proposed by
# *Moukadas Missarov and Roman Stepanov*

**Problem 1.** *We see a nice similarity in the $(\alpha - 3/2d)$-expansions for the critical exponent v in p-adic and Euclidean model up to the second order of perturbation theory. Here $\alpha$ is the renormalization group parameter and d is the dimension of the space. Is this analogy true in higher orders of perturbation theory?*

**Problem 2.** *Will be very interesting to give rigorous and non-perturbative explication of the fact that anomalous dimension in $(4-d)$-expansion is non-zero in the Euclidean case and is zero in the p-version.*

# Open problems proposed by *F. M. Mukhamedov*

## ON POINCARE RECURRENCE THEOREM FOR *p*-ADIC DYNAMICAL SYSTEMS

It is known [1] that the study of *p*-adic dynamical systems arises in Diophantine geometry in the constructions of canonical heights, used for counting rational points on algebraic vertices over a number field. In [4] some applications of *p*-adic dynamical systems was given. In [2] an ergodic behavior of a *p*-adic dynamical system $f(x) = x^n$ in the fields of *p*-adic numbers $Q_p$ with respect to Haar measure was investigated. One of the more important theorem in the theory of dynamical systems is so called Poincare recurrence theorem [5]. It would be worth to prove such kind of theorem in a *p*-adic setting.

Before formulating a problem exactly let us recall some definitions [6].

Let $(X, \mathscr{B})$ be a space, where $\mathscr{B}$ is an algebra of subsets $X$. Consider a complete filed $K$ with a non-trivial non-Archimedean valuation. Suppose that $\mathscr{S}$ is a subfamily of $\mathscr{B}$ such that for $A, B \in \mathscr{S}$ there exists $C \in \mathscr{S}$ with $C \subset A \cap B$, then $\mathscr{S}$ is called *shrinking*. For a function $f : \mathscr{B} \to K$ the notation $\lim_{A \in \mathscr{S}} f(A) = 0$ means that for each $\varepsilon > 0$ there exists $B \in \mathscr{S}$ such that $|f(B)| \leq \varepsilon$ for each $A \in \mathscr{S}$ with $A \subset B$.

A function $\mu : \mathscr{B} \to Q_p$ is said to be a *p-adic measure* if it satisfies the following conditions:

1. for any $A_1, ..., A_n \subset \mathscr{B}$ such that $A_i \cap A_j = \emptyset$ ($i \neq j$)

$$\mu\left(\bigcup_{j=1}^n A_j\right) = \sum_{j=1}^n \mu(A_j)$$

2. for each $A \in \mathscr{B}$ the value $\sup\{|\mu(B)| : A \in \mathscr{B}, B \subset A\} < \infty$;
3. if $\mathscr{S} \subset \mathscr{B}$ is shrinking and $\bigcap_{A \in \mathscr{S}} A = \emptyset$ then $\lim_{A \in \mathscr{S}} \mu(A) = 0$.

A *p*-adic measure is called *a probability measure* if $\mu(X) = 1$.

Let us consider $(X, \mathscr{B})$ with a *p*-adic probability measure $\mu$. Let $T : X \to X$ be a measurable transformation (i.e. $T^{-1}(A) \in \mathscr{B}$ for any $A \in \mathscr{B}$), which preserves the measure $\mu$, i.e. $\mu(T^{-1}(A)) = \mu(A)$ for every $A \in \mathscr{B}$.

**Problem 1.** *Let $(X, \mathscr{B}, \mu)$ and $T$ be as above. Denote*

$$A_r = \{x \in A : T^n x \in A \text{ for some } n \in \mathbb{N}\}$$

*for every $A \in \mathscr{B}$. Then does the following equality hold $\mu(A_r) = \mu(A)$ for every $A \in \mathscr{B}$ with $|\mu(A)| > 0$.*

**Problem 2.** *If, in general, problem 1 is not valid, then find some conditions on $T$ or $\mu$ for which that problem is true.*

For example, Problem 1 is valid for the identical transformation of $X$ with respect to any non-trivial measure given on $(X, \mathscr{B})$.

# REFERENCES

1. G. Call and J. Silverman, *Composito Math.* **89**, 163–205 (1993).
2. V.M. Gundlach, A. Khrennikov and K.O. Lindahl, *Infin. Dimen. Anal. Quantum Probab. Relat. Top.* **4**, 569–577 (2001).
3. A.Yu. Khrennikov, *p-adic Valued Distributions in Mathematical Physics*, Kluwer Academic Publisher, Dordrecht, 1994.
4. A.Yu. Khrennikov, M. Nilsson, *p-adic deterministic and random dynamical systems*, Kluwer, Dordrecht, 2004.
5. I.P. Kornfeld, Ya.G. Sinai, and S.V. Fomin, *Ergodic Theory*, Springer, Berlin–Heidelberg–New York, 1982.
6. van A.C.M. Rooij, *Non-Archimedean functional analysis*, Ser. Pure and Appl. Math. V.**51**, New-York, Marcel Dekker, 1978.

## ON CORRESPONDENCE OF MEASURES

Throughout $p$ will be a fixed prime number. Consider a set $\Phi = \{1, 2, \cdots, p\}$. Denote $\Omega = \Phi^{\mathbf{N}}$. We consider a standard $\sigma$-algebra $\mathscr{F}$ of subsets of $\Omega$ generated by cylinder subsets, all probability measures are considered on $(\Omega, \mathscr{F})$. Let $\mu_p$ be the uniform distribution on $\Phi$, i.e. $\mu_p(k) = 1/p$ for every $k \in \Phi$. Now on $\Omega$ consider a product measure

$$\mu^{(p)} = \bigotimes_{\mathbf{N}} \mu_p.$$

Note $\mu^{(p)}$ is called *Bernoulli measure*. For given stochastic matrix $Q = (q_{ij})_{i,j=1}^{p}$ on $\Omega$ one can define a *Markov measure* as follows

$$\mu_Q(A(\varepsilon_1, \varepsilon_2 \cdots, \varepsilon_k)) = q_{\varepsilon_1} q_{\varepsilon_1, \varepsilon_2} \cdots q_{\varepsilon_{k-1}, \varepsilon_k},$$

here $A(\varepsilon_1, \varepsilon_2 \cdots, \varepsilon_k) = \{\omega = (\omega_k) \in \Omega : \omega_i = \varepsilon_i, \ i = 1, \cdots, k\}$.
It is clear if $q_{ij} = 1/p$ for every $i,j$ then $\mu_Q$ coincides with $\mu^{(p)}$. These two $\mu_Q$ and $\mu^{(p)}$ measures play important role in the theory of probability and ergodic theory [5],[4].
Let $\mathbf{Q}_p$ be the filed of $p$-adic numbers. Let $\mathbf{Z}_p$ be the set of integers of $\mathbf{Q}_p$, i.e.

$$\mathbf{Z}_p = \{x \in \mathbf{Q}_p : |x|_p \leq 1\}.$$

Consider $(\mathbf{Z}_p, \mathscr{B})$ a measurable space, here $\mathscr{B}$ is a $\sigma$-algebra generated by clopen subsets of $\mathbf{Z}_p$. On $(\mathbf{Z}_p, \mathscr{B})$ one can define $\mu_H : \mathscr{S} \to \mathbf{R}$ - *Haar measure* as follows (see for more details [3]): for every $B(a,r) = \{x \in \mathbf{Q}_p : |x - a|_p \leq r\}$ put

$$\mu_H(B_{p^{-l}}(a)) = \frac{1}{p^l}.$$

One can naturally naturally one-to-one transformation $\varphi : \Omega \to \mathbf{Z}_p$. For this $\varphi$ we have the equality

$$\mu^{(p)}(\varphi^{-1}(A)) = \mu_H(A) \ \text{ for every } A \in \mathscr{B}. \tag{1}$$

**Problem 1.** *What kind of measures will appear on $(\mathbf{Z}_p, \mathscr{B})$ if we replace in (1) the measure $\mu^{(p)}$ with $\mu_Q$?*

**Problem 2.** *Describe the following set:*

$$\{\mu_Q \circ \varphi^{-1} : Q \text{ is a stochastic matrix}\}.$$

In [1] a law of large numbers was proved for the measure $\mu^{(p)}$. We think that solution of the problems will allow to prove certain limit theorem on $(\mathbf{Z}_p, \mathscr{B})$ with respect to the Markov measures. Therefore we can formulate the following

**Problem 3.** *Prove a a law of large numbers on $(\mathbf{Z}_p, \mathscr{B})$ with respect to a measure $\mu_Q \circ \varphi^{-1}$ for some stochastic matrix Q.*

## REFERENCES

1. A.Yu. Khrennikov, *Izv. Math.* **64**, 207–219 (2000).
2. I.P. Kornfeld, Ya.G. Sinai and S.V. Fomin, *Ergodic Theory*, Springer, Berlin–Heidelberg–New York, 1982.
3. W. Schikhof, *Ultrametric Calculus*, Cambridge Studies in Advanced Math. **4**, Cambridge: Cambridge Univ. Press, 1984.
4. A.N. Shiryaev, *Probability*, Nauka, Moscow 1980.

# Open problems proposed by Fionn Murtagh

A good deal of our contribution in these proceedings has been in the application area of search and matching. In different ways, clusters or equivalence classes can be of help for such purposes. So we are exploiting symmetries in the data for our ends.

On many occasions I have come back to the question as to whether or not this methodology can be easily characterized as statistics or machine learning. I would say no for the following reason.

Traditionally, in pattern recognition and other fields, there has been the distinction between supervised and unsupervised classification. Supervised classification, with its a priori knowledge of class memberships, has typified machine learning. In statistical data analysis, the point of departure too is an a priori model.

But the other strong tradition of unsupervised classification and learning is that of "letting the data itself speak", it is the domain of data mining, and it is the area of finding symmetries in very large data sets.

Hence the immediate problem to be addressed: Pursue some or all of the areas described in my contribution to these proceedings so that the algebraic (p-adic) and/or geometric (ultrametric) perspective becomes central to data interpretation and computation.

The links with physics need hardly be stressed: (i) computation and physics have many points of overlap; (ii) some prime application areas, such as cosmology, are central to physics; and (iii) interdisciplinarity has its own rewards and benefits.

And a further problem to be addressed in the medium term future is that of the incorporation of resolution scale. In signal and image processing the wavelet and other multiresolution transforms have been used widely for this purpose. The terms registration or fusion are used in fields such as image and signal processing to indicate merging of very different data sets. Consider, for example, fMRI imagery and results of subject thinking or reasoning. It would appear that an ultrametric approach has much to offer for resolution scale based information fusion.

# List of Participants

1. Mikhail Altaisky, *Inst. for Nuclear Research, Dubna*, `altaisky@mx.iki.rssi.ru`
2. Vladimir Anashin, *Russian State Univ., Moscow*, `vs-anashin@yandex.ru`
3. Vladica Andrejić, *Faculty of Mathematics, Belgrade*, `andrew@matf.bg.ac.yu`
4. Irina Ya. Arefeva, *Steklov Mathematical Institute, Moscow*, `arefeva@mi.ras.ru`
5. Muharem Avdispahić, *University of Sarajevo, Sarajevo*, `mavdispa@pmf.unsa.ba`
6. Antun Balaž, *Institute of Physics, Belgrade*, `antun@phy.bg.ac.yu`
7. Aleksandar Belić, *Institute of Physics, Belgrade*, `abelic@phy.bg.ac.yu`
8. Aleksandar Bogojević, *Institute of Physics, Belgrade*, `alex@phy.bg.ac.yu`
9. Miroljub Dugić, *Faculty of Sci. & Math., Kragujevac*, `dugic@knez.uis.kg.ac.yu`
10. Carlos Castro, *Clark Atlanta University, Atlanta*, `perelmanc@hotmail.com`
11. Nicusor Dan, *Institute of Mathematics, Bucuresti*, `Nicusor.Dan@imar.ro`
12. Nicole De Grande-De Kimpe, *Vrije Universiteit Brussel, Brussel*
13. Alexandra Dragovich, *Vavilov Institute of Genetics, Moscow*, `dragovich@vigg.ru`
14. Branko Dragovich, *Institute of Physics, Belgrade*, `dragovich@phy.bg.ac.yu`
15. Peter G.O. Freund, *Enrico Fermi Institute, Chicago*, `freund@theory.uchicago.edu`
16. Hidekazu Furusho, *Nagoya University, Nagoya*, `furusho@math.nagoya-u.ac.jp`
17. H. Glockner, *Darmstadt Univ. of Techn.*, `gloeckner@mathematik.tu-darmstadt.de`
18. Neven Grbac, *Faculty of EEC, Zagreb*, `ngrbac@msun.zpm.fer.hr`
19. Shamgar Gurevich, *Israel Institute of Technology, Tel Aviv*, `shamgar@math.tau.ac.il`
20. Ronny Hadani, *Israel Institute of Technology, Tel Aviv*, `nogaporat@hotmail.com`
21. Marcela Hanzer, *University of Zagreb, Zagreb*, `hanmar@math.hr`
22. Shay Haran, *Israel Institute of Technology, Haifa*, `haran@techunix.technion.ac.il`
23. Dejan Joković, *Institute of Physics, Belgrade*, `yokovic@phy.bg.ac.yu`
24. Hiroshi Kaneko, *Tokyo University of Science, Tokyo*, `kaneko@home.email.ne.jp`
25. Andrei Khrennikov, *University of Växjö, Växjö*, `andrei.khrennikov@msi.vxu.se`
26. Anatoly Kochubei, *Institute of Mathematics, NASU, Kiev*, `kochubei@i.com.ua`
27. Sergey Kozyrev, *Steklov Mathematical Institute, Moscow*, `kozyrev@mi.ras.ru`
28. K. Lukierska-Walasek, *Inst. of Phys., Zielona Góra*, `klukie@proton.if.uz.zgora.pl`
29. N. Mainetti, *LLAIC IUT Univ. d t' Auvergne, Aurillac*, `mainetti@iut.u-clermont1.fr`
30. Hamza Menken, *Mersin University, Mersin*, `hmenken@mersin.edu.tr`

31. Žarko Mijajlović, *Faculty of Mathematics, Belgrade*, `zarkom@matf.bg.ac.yu`
32. Darko Milinković, *Faculty of Mathematics, Belgrade*, `milinko@matf.bg.ac.yu`
33. Miloš Milošević, *Mathematical Institute of SANU, Belgrade*, `mionamil@eunet.yu`
34. Moukadas Missarov, *Kazan State University, Kazan*, `Moukadas.Missarov@ksu.ru`
35. Fionn Murtagh, *University of London, London*, `fmurtagh@acm.org`
36. Aleksandar Perović, *Math. Inst. of SANU, Belgrade*, `peramail314@yahoo.com`
37. Karmelita Pjanić, *University of Sarajevo, Sarajevo*
38. Vitaly Pukhalsky, *Vavilov Institute of Genetics, Moscow*
39. Aliaksandr Radyna, *Belarusian State University, Minsk*, `Radyna@bsu.by`
40. Yauhen Radyna, *Belarusian State University, Minsk*, `yauhenradyna@tut.by`
41. Yakov Radyno, *Belarusian State University, Minsk*, `yakovradyno@tut.by`
42. Zoran Rakić, *Faculty of Mathematics, Belgrade*, `zrakic@matf.bg.ac.yu`
43. Wim Schikhof, *University of Nijmegen, Nijmegen*, `schikhof@math.kun.nl`
44. George Shabat, *Russian State University, Moscow*, `george@shabat.mccme.ru`
45. Dino Sejdinović, *University of Sarajevo, Sarajevo*, `dinosejdo@yahoo.com`
46. Paul Sorba, *LAPTH-CNRS, Annecy le Vieux Cedex*, `sorba@lapp.in2p3.fr`
47. Roman Stepanov, *Kazan State University, Kazan*, `rstepanov@mail.ru`
48. Lenny Taelman, *Rijksuniversiteit Groningen, Groningen* `lenny@math.rug.nl`
49. Anton Trushechkin, *Steklov Mathematical Institute, Moscow*, `trushechkin@mail.ru`
50. Branko Urošević, *Univ. Pompeu Fabra, Barcelona*, `branko.urosevic@econ.upf.es`
51. Franco Vivaldi, *University of London, London*, `f.vivaldi@qmul.ac.uk`
52. Igor V. Volovich, *Steklov Mathematical Institute, Moscow*, `volovich@mi.ras.ru`
53. Mirjana Vuković, *University of Sarajevo, Sarajevo*, `mirjana@yahoo.com`
54. Polina Vytnova, *Moscow State University, Moscow*, `vytnova@mccme.ru`
55. Rade Živaljević, *Mathematical Institute of SANU, Belgrade*, `rade@mi.sanu.ac.yu`

## Author Index

### A

Albeverio, S., 195
Anashin, V., 3
Aref'eva, I. Ya., 301
Avdispahić, M., 312

### B

Balaž, A., 320
Belić, A., 320
Bogojević, A., 320

### D

De Grande-De Kimpe, N., 206
Dragovich, B., 25
Dremov, V., 43

### E

Escassut, A., 214
Ezhov, A. A., 55

### F

Fischenko, S., 174
Freund, P. G. O., 65
Furusho, H., 222

### G

Glöckner, H., 237
Gurevich, S., 74

### H

Hadani, R., 74

### K

Kaneko, H., 91
Khrennikov, A. Yu., 55, 105, 195
Kochubei, A. N., 254
Kozyrev, S. V., 121

### L

Lukierska-Walasek, K., 81

### M

Maïnetti, N., 214
Mamedov, K. R., 267
Mendes, J. F. F., 140
Menken, H., 267
Mijajlović, Ž., 274
Milošević, M., 274
Missarov, M. D., 129
Mukhamedov, F. M., 140
Murtagh, F., 151

### P

Perović, A., 274

### R

Rozikov, U. A., 140

### S

Schikhof, W. H., 280
Shabat, G., 43
Shelkovich, V. M., 195
Smajlović, L., 312
Stepanov, R. G., 129

## T

Taelman, L., 291
Topolski, K., 81
Trelewicz, J. Q., 330
Trushechkin, A. S., 345

## V

van der Put, M., 291
Vivaldi, F., 162
Volovich, I. V., 330, 345

Vytnova, P., 43

## Z

Zelenov, E., 174